Two Manuals Combined into One Book

Mine Countermine Operations

And

Electronic Warfare in Operations

Department of Defense

Pentagon
Publishing

*FM 20-32

Field Manual
No. 20-32

Headquarters
Department of the Army
Washington, DC, 29 May 1998

MINE/COUNTERMINE OPERATIONS

Table of Contents

Page

Part One. Mine Operations

DISTRIBUTION RESTRICTION: Approved for public release; distribution is unlimited.

*This manual supersedes FM 20-32, 30 September 1992.

Page

Page

Page

Part Two. Counteroperations

Page

Page

Part Three. Special Mining Operations

Page

LIST OF ILLUSTRATIONS

Figures

Page

Page

Page

Page

Page

Tables

Preface

Field Manual (FM) 20-32 provides United States (US) armed forces with tactical, technical, and procedural guidance for conducting mine and countermine operations. It applies to all elements of the combined arms team for maneuver and engineer staff planning and coordination. The manual is presented in three parts—mine operations, counteroperations, and special-mining operations.

The guidance provided focuses on individual skills of emplacing and removing mines, team and squad tasks, platoon and company organization and planning, and battalion/task force (TF) organization and coordination for successful obstacle reduction and breaching operations.

The provisions of this publication support existing doctrine established by FMs 5-34, 5-100, 90-7, and 90-13-1. It also contains new and improved techniques for emplacing row mines; marking, reporting, and recording minefields; reducing simple and complex obstacles; and emplacing a standard-pattern minefield. This manual reflects new doctrine from FMs 5-10, 5-71-2, and 5-71-3.

This publication implements the following International Standardization Agreements (STANAGs) between North Atlantic Treaty Organization (NATO) forces:

- STANAG 2036. *Land Minefield Laying, Marking, Recording, and Reporting Procedures.* Edition 5.
- STANAG 2889. *Marking of Hazardous Areas and Routes Through Them.* Edition 3.
- STANAG 2990. *Principles and Procedures for the Employment in Land Warfare of Scatterable Mines with a Limited Laid Life.* Edition 1.

NOTE: US policy regarding the use and employment of antipersonnel land mines (APLs) outlined in this FM is subject to the Convention on Certain Conventional Weapons and Executive Orders. Current US policy limits the use of non-self-destructing APLs to (1) defending the US and its allies from armed aggression across the Korean demilitarized zone and (2) training personnel engaged in demining and countermine operations. The use of the M18A1 claymore in the command-detonation mode is not restricted under international law or Executive Order.

All references to US employment of non-self-destructing APLs (such as row mining) in this manual are intended to provide doctrine for use in Korea only. This information is provided in bold lettering throughout the manual. Detailed doctrine on APLs is also provided to ensure that US forces recognize how the enemy can employ these weapons.

As the US military seeks to end its reliance on APLs, commanders must consider the increased use of other systems such as the M18A1 claymore, nonlethal barriers (such as wire obstacles), sensors and surveillance platforms, and direct and indirect fires.

This publication includes the following appendixes:

- Appendix A. Installation and Removal of US Mines and Firing Devices.
- Appendix B. Controls and Components of Special-Purpose Munitions.
- Appendix C. Threat Mine/Countermine Operations.
- Appendix D. Air Volcano.
- Appendix E. Safety and Training.
- Appendix F. Mine Awareness.
- Appendix G. Countermine Data.
- Appendix H. Metric Conversion Chart.

The proponent for this publication is Headquarters, US Army Training and Doctrine Command (TRADOC). Forward comments and recommendations on Department of the Army (DA) Form 2028 to Commandant, US Army Engineer School, ATTN: ATSE-DME-MWF, Fort Leonard Wood, Missouri 65473-5000.

Unless this publication states otherwise, nouns and pronouns do not refer exclusively to men.

Chapter 1
Introduction

This chapter provides the mechanics and characteristics of antitank (AT) mines and munitions, antipersonnel (AP) mines and munitions, and antihandling devices (AHDs). The information contained in this chapter also provides a foundation for the rest of the manual.

Land-based mines and munitions are hand-emplaced, remote-delivered, ground-delivered, or air-delivered:

- Hand-emplaced mines and munitions require manual arming and are labor-, resource-, and transport-intensive.
- Remote- and air-delivered mines and munitions require less time and labor; however, they are not as precisely placed as hand-emplaced mines and munitions.
- Ground-delivered mines are less resource-intensive than hand-emplaced mines. They are not precisely placed; however, the minefield boundaries are.

Soldiers can surface lay or bury mines and munitions and can place AHDs on hand-emplaced AT mines.

NOTE: Some countries employ AHDs on AP mines, but US forces are not authorized to employ AHDs on any type of AP mine.

MECHANICS OF MINES

CHARACTERISTICS AND FUNCTIONING

A land mine is an explosive device that is designed to destroy or damage equipment or personnel. Equipment targets include ground vehicles, boats, and aircraft. A mine is detonated by the action of its target, the passage of time, or controlled means. There are two types of land-based mines—AT and AP. Mines generally consist of the following parts (Figure 1-1, page 1-2):

- Firing mechanism or other device (sets off the detonator or igniter charge).
- Detonator or igniter (sets off the booster charge).
- Booster charge (may be attached to the fuse or the igniter or be part of the main charge).
- Main charge (in a container; usually forms the body of the mine).
- Casing (contains all the above parts).

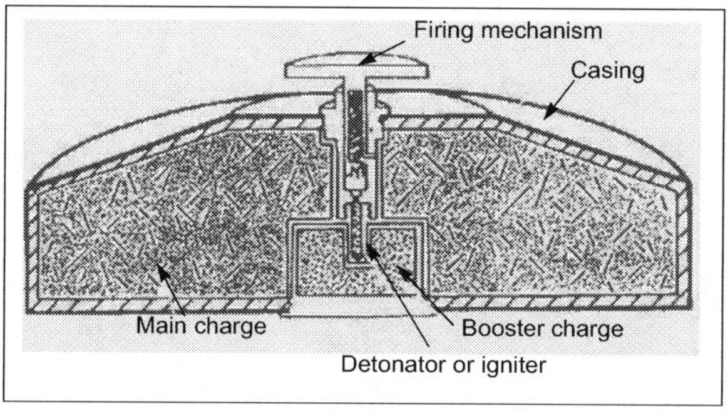

Figure 1-1. Mine components

COMPONENTS AND INITIATING ACTIONS

A firing mechanism prevents the mine from exploding until it makes contact with, or is influenced by, its target. Once a mine has been armed, the firing mechanism may be actuated by the following methods (Figure 1-2):

- Applying pressure (including tilt rod).
- Pulling a trip wire.
- Releasing tension or breaking a trip wire.
- Releasing pressure.
- Passage of time (time-delay mechanism).
- Impulses.
  - Electrical.
  - Vibration.
  - Magnetic-influence.
  - Electromagnetic-frequency.
  - Infrared-sensored.
  - Acoustic.

To arm some mines, you must position the igniter, set the mechanism properly, and disengage the safety device (usually by removing a safety pin). The fuse is the initial component in the firing chain; it has a low-explosive (LE) powder but is highly sensitive. The fuse is actuated by an initiating action. Although mines are issued with a standard fuse, alternate fuses are issued separately for some mines.

The four main fuse types are shown in Figure 1-3, page 1-4.

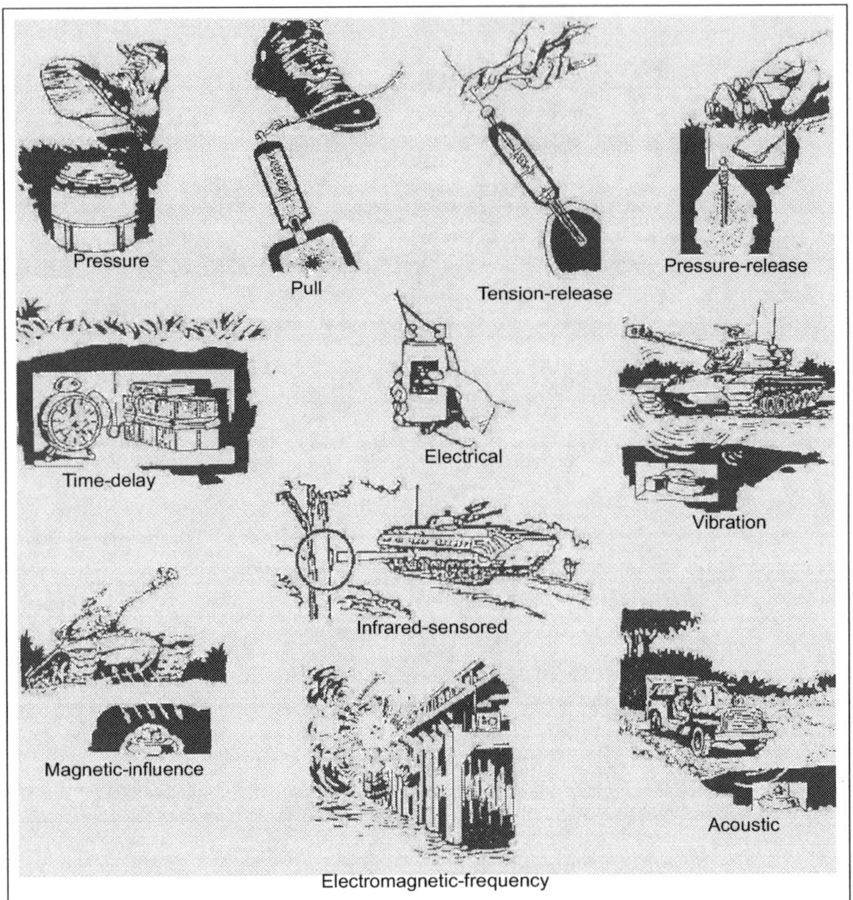

Figure 1-2. Methods of actuating mines

Mechanical. A spring drives a striker against a percussion cap, which fires the detonator.	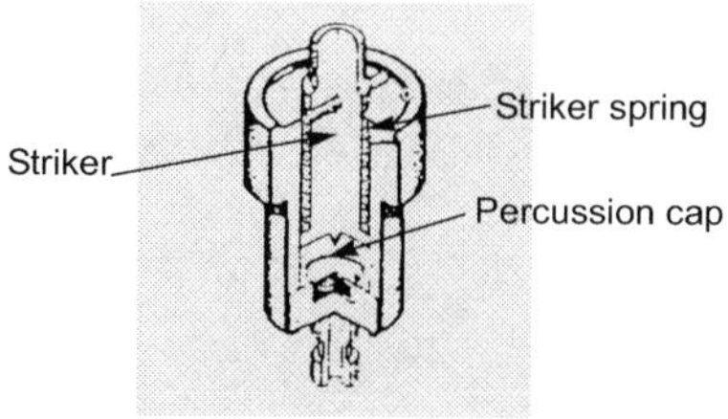
Chemical. A small container of a chemical compound is broken by the initiating action. The chemical compound reacts with another substance to generate heat, which ignites the detonator.	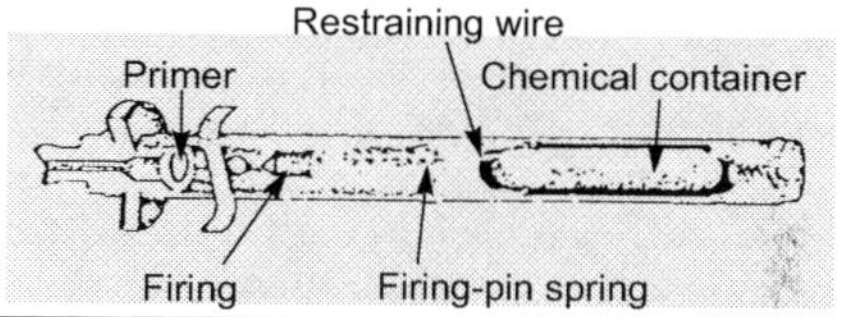
Friction. The initiating action ignites substances inside the fuse by friction. The flame fires the detonator.	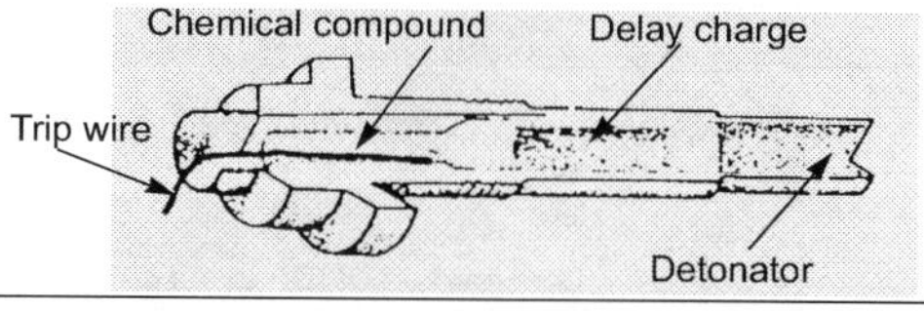
Electrical. The initiating action closes an electrical circuit, which functions an electrical detonator.	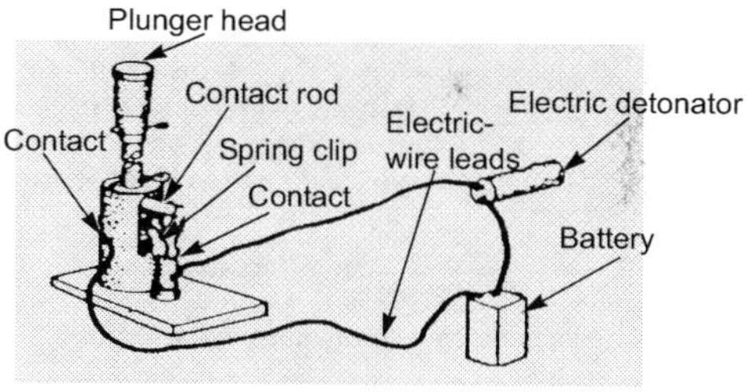

Figure 1-3. Types of fuses

ANTITANK MINES

AT mines are designed to immobilize or destroy vehicles and their occupants.

TYPES OF KILLS

An AT mine produces a mobility kill (M-Kill) or a catastrophic kill (K-Kill). An M-Kill destroys one or more of the vehicle's vital drive components (for example, breaks a track on a tank) and immobilizes the target. An M-Kill does not always destroy the weapon system and the crew; they may continue to function. In a K-Kill, the weapon system and/or the crew is destroyed.

Types of Sensing

AT fuses fall into three design categories:

- Track-width. Usually pressure-actuated, requiring contact with the wheels or tracks of a vehicle.
- Full-width. Activated by several methods—acoustics, magnetic-influence, tilt-rod, radio-frequency, infrared-sensored, command, or vibration. Tilt-rod or magnetic-influence fuses are the most common. Full-width fuses are designed to be effective over the entire target width and can cause a K-Kill from penetration and spalling metal or from secondary explosions. When a full-width fuse is activated solely by contact with the wheels or tracks of the target vehicle, it usually causes an M-Kill because most of the energy is absorbed by the wheels or tracks.
- Off-route. Designed to be placed along the side of a route likely to be taken by armored vehicles. It has numerous fuzing possibilities, including infrared, seismic, break wire, and magnetic. It produces an M-Kill or a K-Kill, depending on the location of the target at the time of mine detonation.

Types of Warheads

AT mines can be identified by their warheads:

- Blast AT mines derive their effectiveness from the force generated by high-explosive (HE) detonation. They usually produce an M-Kill when the blast damages the track or the vehicle, but a K-Kill is also possible.
- Shaped-charge mines use a directed-energy warhead. A shaped charge is formed by detonating an explosive charge behind a cone of dense metal or other material. Upon detonation, the cone collapses and forms a metal slug and a gaseous metal jet that penetrate the target. A K-Kill is probable if the crew or ammunition compartment is hit.
- Explosive-formed penetrating (EFP) mines have an explosive charge with a metal plate in front. Upon detonation, the plate forms into an inverted disk, a slug, or a long rod. A K-Kill is probable if the crew or ammunition compartment is hit.

ANTIPERSONNEL MINES

Types of Kills

AP mines can kill or incapacitate their victims. The injuries and deaths they cause commit medical resources, degrade unit morale, and damage nonarmored vehicles. Some types of AP mines may break or damage the track on armored vehicles.

TYPES OF SENSING

AP mines can be fused in many ways, to include pressure, seismic, wire, or command detonation:

- Pressure fuses usually activate an AP mine when a load is placed on the fuse.
- Seismic fuses activate an AP mine when the sensor detects vibrations.
- Trip wires or break wires activate an AP mine when something disturbs barely visible wires.
- Command-detonated mines are activated by a soldier when he detects the enemy in the mines' blast area.

TYPES OF EFFECTS

AP mines contain five types of effects:

- Blast. Cripples the foot or leg of a soldier who steps on it; can also burst the tires of a wheeled vehicle that passes over it.
- Bounding-fragmentation. Throws a canister into the air; the canister bursts and scatters shrapnel throughout the immediate area.
- Direct-fragmentation. Propels fragments in the general direction of enemy soldiers.
- Stake-fragmentation. Bursts and scatters shrapnel in all general directions.
- Chemical. Disperses a chemical agent to whoever activates it; contaminates the surrounding area.

ANTIHANDLING DEVICES

AHDs perform the function of a mine fuse if someone attempts to tamper with the mine. They are intended to prevent moving or removing the mine, not to prevent reduction of the minefield by enemy dismounts. An AHD usually consists of an explosive charge that is connected to, placed next to, or manufactured in the mine. The device can be attached to the mine body and activated by a wire that is attached to a firing mechanism. US forces can employ AHDs on conventional AT mines only. Other countries employ AHDs on AT and AP mines.

Some mines have extra fuse wells that make it easier to install AHDs (Figure 1-4). An AHD does not have to be attached to the mine; it can be placed underneath the mine (Figure 1-5). Mines with AHDs are sometimes incorrectly called *booby-trapped* mines.

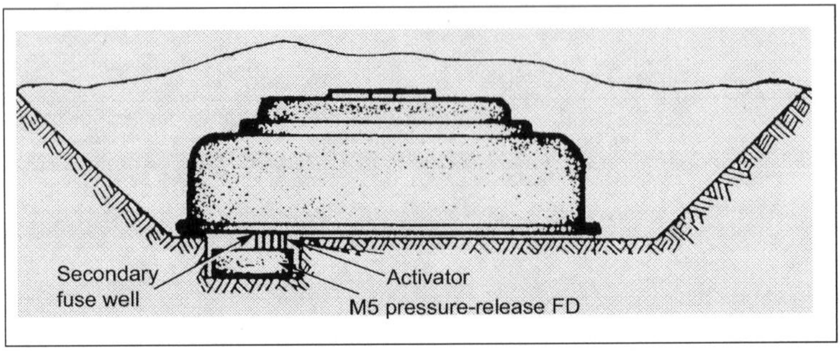

Figure 1-4. AHD incorporating a release mechanism

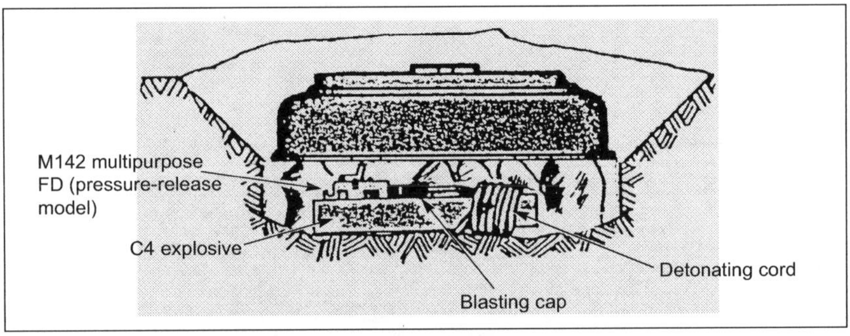

Figure 1-5. AHD not attached to the mine

The following hand-emplaced AHDs are used by US forces (Figure 1-6, page 1-8):

- M5 pressure-release firing device (FD).
- M142 multipurpose FD.

These devices use a spring-loaded striker with a standard base, and they function in one or more modes—pressure, pressure-release, tension, and/or tension-release. When an FD is employed as an AHD on certain AT mines, it requires the use of an M1 or M2 activator. FDs and activators are described in Appendix A.

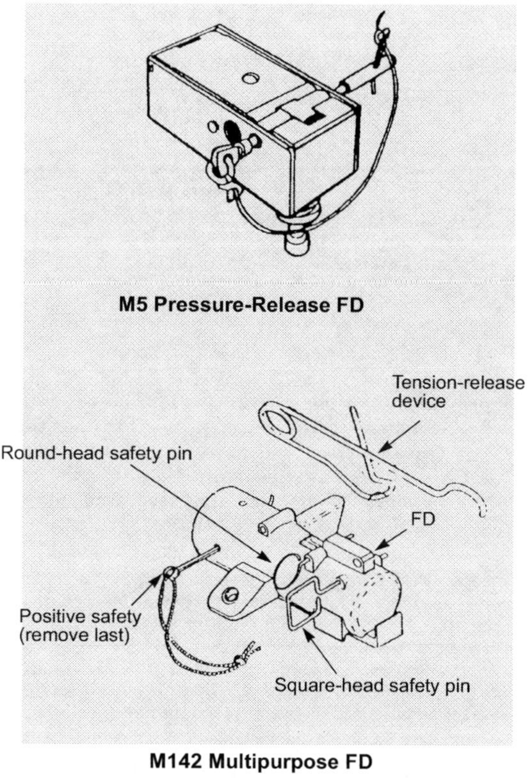

Figure 1-6. Hand-emplaced US AHDs

PART ONE

Mine Operations

This part of the manual provides tactical and technical information on mines, minefields, and mine-delivery systems; emplacement and employment methods and responsibilities; and reporting and recording procedures for US mine operations. Current US policy for restrictions on AP mines and other devices is also discussed.

Chapter 2

Mine-Warfare Principles

This chapter provides guidance to staff personnel who must plan the employment of minefields for tactical operations. It defines the four types of minefields—protective, tactical, nuisance, and phony. The remainder of the chapter provides guidance on the employment of tactical minefields—specifically their functions, designs, and integration principles.

MINE-WARFARE CONCEPTS

Mines are explosive devices that are emplaced to kill, destroy, or incapacitate enemy personnel and/or equipment. They can be employed in quantity within a specified area to form a minefield, or they can be used individually to reinforce nonexplosive obstacles. They can also be emplaced individually or in groups to demoralize an enemy force. A minefield is an area of ground that contains mines or an area of ground that is **perceived** to contain mines (a phony minefield). Minefields may contain any type, mix, or number of AT and/ or AP mines. Minefields are used to—

- Produce a vulnerability on enemy maneuver that can be exploited by friendly forces.
- Cause the enemy to piecemeal his forces.
- Interfere with enemy command and control (C^2).
- Inflict damage to enemy personnel and equipment.
- Exploit the capabilities of other weapon systems by delaying enemy forces in an engagement area (EA).
- Protect friendly forces from enemy maneuver and infiltration.

TYPES OF MINEFIELDS

There are four general types of minefields—protective, tactical, nuisance, and phony. Each type is determined by its distinct battlefield purpose. Therefore, minefields are employed differently, and they target the enemy in unique ways that support the overall concept of the operation.

- Protective minefields are employed to protect soldiers, equipment, supplies, and facilities from enemy attacks or other threats.
- Tactical minefields directly effect the enemy's maneuver in a way that gives the defending force a positional advantage.
- Nuisance minefields impose caution on enemy forces and disrupt, delay, and sometimes weaken or destroy follow-on echelons.
- Phony minefields deceive the enemy about the exact location of real minefields. They cause the attacker to question his decision to breach and may cause him to expend his reduction assets wastefully. Phony minefields may be employed in conjunction with other minefields, but should be used only after the enemy has become mine-sensitive.

It is important to distinguish the difference between the types of minefield and the means of emplacement. Volcano, Modular Pack Mine System (MOPMS), standard-pattern, and row mining are not types of minefields; they are just some of the means used to emplace tactical, nuisance, and protective minefields. They may also be the method of emplacement that is replicated by a phony minefield.

Protective Minefields

Protective minefields are employed to protect soldiers, equipment, supplies, and facilities from enemy attacks or other threats. Other threats range from enemy surveillance to theft of supplies and equipment. In tactical operations, protective minefields provide friendly forces with close-in protection and defeat the enemy's ability to maneuver or utilize the tenants of offense. They deny mechanized penetration and dismounted infiltration. In military operations other than war (MOOTW), protective minefields may focus on preventing unauthorized access to facilities and installations, rather than assisting in the destruction of an enemy force.

Protective minefields are usually employed and emplaced at the small-unit level (platoon or company/team). The authority to emplace protective minefields is normally delegated to the company/team commander. In some cases, such as a hasty defense, protective minefields are laid on short notice by units that use mines from their basic load or local stock. More commonly, protective minefields are used as part of a unit's deliberate defense. The mines are laid so that they are easy to detect and recover by the laying unit.

An important aspect of protective minefields is the requirement to recover them before leaving the area. This is often overlooked and is difficult to control because they are emplaced at the small-unit level. When a unit is being relieved in place by an adjacent unit, protective minefields are turned over to the relieving unit (minefield turnover is further defined later in this chapter). The decentralized emplacement of protective obstacles makes consolidating reports and records difficult and requires command involvement.

Much like final protective fires (FPF), protective minefields provide the defender with close-in protection during the enemy's final assault. Protective minefields serve two purposes. First, they impose a delay on an attacker that allows the defender time to break contact as the unit displaces to another battle position. Secondly, they break up the enemy's assault to complete its destruction. The composition of a protective minefield is driven by the vulnerability of the defender:

- Dismounted infantry is the greatest close-combat threat to a defending tank company/team. Protective minefields encountered in this case consist predominantly of AP mines that limit enemy dismounts from closing with the armor defender.
- A tank force is the greatest threat to an infantry defense. Protective minefields in this case consist predominantly of AT mines that reduce the enemy's ability to close quickly onto the infantry's position.

Neither AP nor AT mines are used in isolation. The preponderance of mine composition is designed against the most severe close-combat threat and the likelihood of that threat.

A protective minefield may take many forms. It may be only a few mines in front of a platoon, or it may be a standard-pattern minefield around an airfield. Protective minefields are used in both close and rear operations, and they are classified as either hasty or deliberate:

- Hasty protective minefields are temporary in nature and are used as part of a unit's defense perimeter. They are usually laid by units using mines from their basic load. If time permits, mines should be buried to increase their effectiveness; but they can be laid on top of the ground. AHDs and low-metallic mines are not used so that the minefield can be easily recovered. Mines are employed outside the hand-grenade range but within the range of small-caliber weapons. All mines are picked up by the emplacing unit upon leaving the area, unless enemy pressure prevents mine retrieval or the minefield is being transferred to a relieving commander. The brigade commander has the initial authority to employ hasty protective minefields. This authority may be delegated to a battalion or company commander on a mission basis. Procedures for emplacing a hasty protective row minefield are contained in Chapter 6.
- Deliberate protective minefields are more permanent, require more detailed planning, and usually require more resources. They are commonly used to protect static assets (vital sites)—logistical sites, communication nodes, depots, airfields, missile sites, air-defense artillery (ADA) sites, and permanent-unit locations. A typical deliberate protective minefield is the standard-pattern minefield; however, a row minefield can also be used. Deliberate protective minefields are usually emplaced for extended periods of time and can be transferred to another unit. Techniques for emplacing deliberate protective minefields are discussed in Chapters 6 and 7.

Tactical Minefields

Tactical minefields are employed to directly attack enemy maneuver and to give the defender a positional advantage over the attacker. Tactical minefields may be employed by themselves or in conjunction with other types of tactical obstacles. They attack the enemy's maneuver by disrupting its combat formations, interfering with its C^2, reducing its ability to mass fires, causing him to prematurely commit limited breaching resources, and reducing his ability to reinforce. The defender masses fires and maneuver to exploit the positional advantage created in part by tactical obstacles.

Tactical minefields add an offensive dimension to the defense. They are a commander's tool for recapturing and maintaining the initiative that is normally afforded to an attacker. Combined with fires, tactical obstacles force the attacker to conform to the defender's plan.

Tactical minefields may be emplaced during offensive operations to protect exposed flanks, isolate the objective area, deny enemy counterattack routes, and disrupt enemy retrograde. This chapter further discusses the principles behind designing, integrating, siting, and emplacing tactical minefields.

Nuisance Minefields

Nuisance minefields are a form of tactical minefields. They are mainly used to impose caution on enemy forces and to disrupt, delay, and sometimes destroy follow-on echelons. Once nuisance minefields are emplaced, they do not require cover by observation or direct fire. Nuisance minefields are usually irregular in size and shape; they can be a single group of mines or a series of mined areas. They can be used to reinforce existing obstacles and can also be rapidly emplaced on main avenues of approach (AAs). Conventional mines and scatterable mines (SCATMINEs) may be used in nuisance minefields.

Phony Minefields

Phony minefields are areas of ground that are altered to give the same signature as a real minefield and thereby deceive the enemy. Phony minefields serve two primary functions. First, they confuse an attacker's breach cycle and cause him to question his breach decision. Secondly, they may cause an attacker to wastefully expend reduction assets to reduce mines that are not really there.

The success of phony minefields depends on the enemy's state of mind. The bluff succeeds best when the enemy is mine-conscious and has already suffered the consequences of a mine encounter. A fear of mines can quickly evolve into paranoia and break the momentum of the enemy's attack. Therefore, phony minefields are normally employed in conjunction with real minefields and are seldom employed alone. Once the enemy has become mine-conscious, phony minefields may produce considerable tactical effects with very little investment in time, labor, and material. Phony minefields may also be used to extend the front and depth of live minefields when mines or labor are in short supply or when time is restricted. They may be used to conceal minefield gaps through live minefields. There is **no** guarantee that phony minefields will achieve their purpose.

There are two mission-essential tasks inherent in the employment of a phony minefield:

- The phony minefield must completely replicate a live minefield in every detail, using a specific method of emplacement as a model. This becomes the deception story, and every aspect of the phony minefield must support the deception story. For example, if the deception story is a buried row minefield, the depth, front, and marking must be similar to that of a live buried row minefield. The ground should be disturbed, and tracks should be made on the ground in the same pattern as other minefields to give the ground the same signature. Occasional empty mine crates, discarded fuses, or other mine-laying supplies add to the deception.

- The deception story must never be compromised. Once emplaced, the phony minefield must be regarded by friendly forces as live until the tactical situation no longer warrants maintaining the deception. This can be extremely painful for the friendly unit. There is great temptation to drive through, rather than around, a known phony minefield—particularly if it is intended to be a gap between live minefields. However, one vehicle driving through a phony minefield and observed by enemy reconnaissance compromises the minefield's effectiveness.

Live mines are never laid in a phony minefield. A minefield designated as phony implies that the area contains no live mines. Emplacing even a single live mine within a phony minefield makes it a live minefield. Empty tins and such may be laid in a phony minefield but is seldom worthwhile. Minefield marking and covering fire should be the same as for a live minefield. Employment authority and reporting requirements are the same as for the minefield being simulated.

PROTECTIVE VERSUS TACTICAL MINEFIELDS

As discussed, minefields can be tactical or protective obstacles (Figure 2-1, page 2-6). Tactical and protective obstacles have different purposes with regard to the enemy's maneuver. This difference causes them to have a particular relative place on the battlefield. Tactical obstacles attack enemy maneuver and are placed on the battlefield where the enemy maneuvers from march, prebattle, and attack formations. Protective obstacles are used to protect the force from the enemy's final assault onto the force's position. Protective obstacles are close to defensive positions and are tied in with the FPF of the defending unit. Additional information on obstacles can be found in FM 90-7.

TACTICAL MINEFIELDS

Tactical minefields are designed, sited, emplaced, and integrated with fires to produce four specific tactical-obstacle effects—disrupt, turn, fix, and block (Figure 2-2, page 2-6). Each obstacle effect has a specific impact on an enemy's ability to maneuver, mass, and reinforce. Obstacle effects also increase the enemy's vulnerability to friendly fires. They support the friendly scheme of maneuver by manipulating the enemy in a way that is critical to the commander's intent. Minefield design is the means by which an emplacing unit varies minefield width, minefield depth, mine density, mine composition, the use of AHDs, and the irregular outer edge (IOE) to best achieve one of the four tactical-obstacle effects. Modifying these variables is at the heart of tactical minefield employment principles.

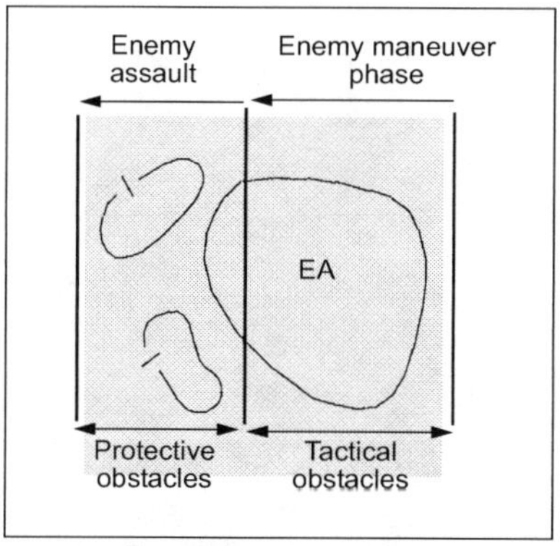

Figure 2-1. Tactical versus protective obstacles

Obstacle Effect	Application	Examples Conveying Intent	Description
Disrupt 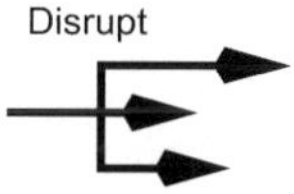	The short arrow indicates where an enemy is attacked by obstacles. The longer arrows indicate where the bypass is allowed and attacked by fires.	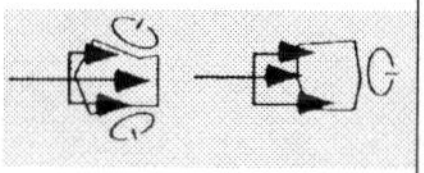	Causes an enemy to break up its formation and tempo, interrupt its timetable, commit its reduction assets prematurely, and piecemeal the attack.
Turn 	The heel of the arrow is the anchor point. The direction of the arrow indicates the desired direction of turn.		Manipulates an enemy's maneuver in a desired direction.
Fix 	The irregular part of the arrow indicates where an enemy advance is slowed by obstacles.	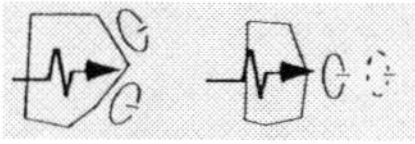	Slows an enemy within a specific area, normally an EA. Gives the defender time to acquire, target, and destroy the attacking enemy throughout the depth of an EA.
Block 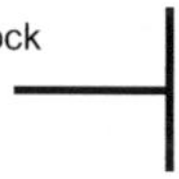	The ends of the vertical line indicate the limit of an enemy's advance and where obstacles tie in to no-go terrain.	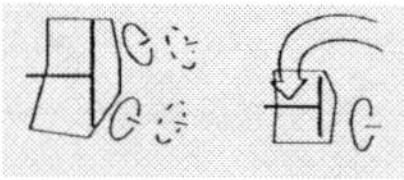	Stops an enemy along a specific AA or prevents it from passing through an EA.

NOTES:
1. Arrows indicate the direction of enemy attack.
2. Minefields must be integrated with fires to achieve the desired effect.

Figure 2-2. Tactical-obstacle effects

Minefield Variables

First, it is important to understand how the variables relate to minefield effects. Figure 2-3 clearly defines some of the terms used to discuss minefield variables.

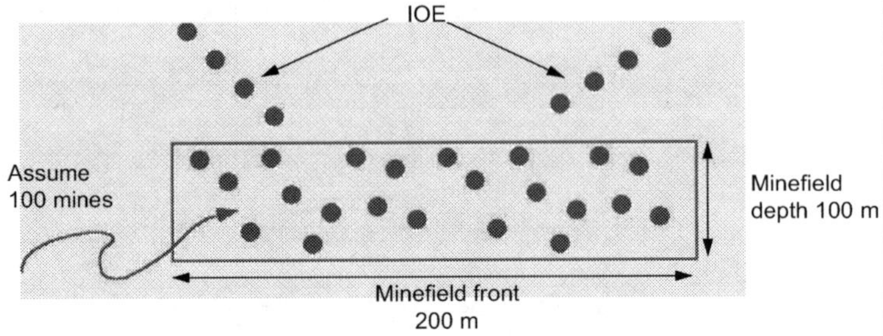

Linear density: 100 mines ÷ 200 meters = 0.5 mine per meter of front
Area density: 100 mines ÷ (100 x 200) = 0.005 mine per square meter

Figure 2-3. Minefield variables

Front

Minefield front is the dimension of a minefield that defines how much of the attacking enemy formation is affected by the minefield. The front of a minefield is based on the desired obstacle effect (disrupt, turn, fix, or block) and the attack front of a company-size enemy force. The front of an attacking enemy depends largely on the type of enemy force (armored, motorized, or dismounted infantry) and norms by which the enemy army fights. For armored warfare, the minefield front is based on effecting a doctrinal company attack front of 500 meters (13 to 18 combat vehicles). For dismounted warfare, the minefield front is based on effecting a company attack front of 150 meters. The front may vary and require a study of enemy force and terrain. Groups of individual minefields are employed to achieve a larger front for battalion and larger enemy formations. For example, a battalion consisting of 52 to 72 combat vehicles has a front of 1,500 meters and requires more minefields.

Depth

Minefield depth is based on the amount of reduction assets that will be required by an enemy to reduce a lane. The standard should start with 100 meters and increase in depth if denying the enemy the use of a mobility corridor (MC) is the intent (turn or block).

Density

Minefield density depicts the number of mines in a minefield. It is expressed in linear or area density:

- Linear density is the average number of mines within a 1-meter-wide path through the minefield's depth, anywhere along the front. In Figure 2-3, page 2-7, the minefield contains 100 mines, with a minefield front of 200 meters. The linear density is 0.5 mine per meter of front (100 mines/200 meters of front).
- Area density is the average number of mines within a square meter, anywhere in the minefield. In Figure 2-3, the minefield contains 100 mines within a 20,000-square-meter area. The area density is 0.005 mine per square meter (100 mines/[200 meters x 100 meters]). Area density is normally used to express the density of scatterable minefields.

Mine Composition

This variable includes the effective use of different types of mines. By using full-width mines, the probability of kill increases for the minefield. AT mines with AHDs **(Korea Only: as well as AP mines)** are used where the enemy is expected to use dismounted reduction techniques.

Antitank Mines

If the enemy is an armored force, tactical obstacles are predominantly AT mines. Track-width mines (M15s with the M603 fuse) have a lower probability of kill (M-Kill or K-Kill) than full-width mines (M21s and M15s with the M624 fuse). The ratio of full-width versus track-width mines in a minefield depends on the kill required. In general, a track-width minefield does not adequately affect the enemy's maneuver.

Antipersonnel Mines

AP mines target dismounted soldiers. Their composition in tactical minefields depends on the threat and the enemy's reduction assets. Based on current technology, most breaching operations are accomplished by mechanical or explosive means. **(Korea Only: If the minefield group's intent is to exhaust the enemy's breaching assets, AP mines should be integrated to attack its dismounted reduction ability.)**

Probabilities of Encounter and Kill

The probabilities of encounter and kill measure a minefield's lethality.

Probability of Encounter

The probability of encounter is measured by the chance (in percent) that a vehicle, blindly moving through a minefield, will detonate a mine. The probability of encounter is based on mine density, the type of mine, and the type of enemy vehicle. In short, the more dense a minefield, the higher the probability is of encountering a mine. Probability of encounter also depends on the fuse capability of the mines. Tilt-rod and magnetic-influence mines will detonate if they are encountered anywhere along the width of the enemy vehicle. Pressure-fused mines detonate if a vehicle's track or wheel actually runs over them. The probability of encounter is also affected by the type of enemy vehicle. The smaller the width or track signature of the vehicle, the less likely it will encounter and detonate a mine.

Figure 2-4 illustrates the relationship between mine density and the probability of encounter for light versus heavy tracked vehicles and for track-width versus full-width mines. Figure 2-4 also provides general guidance for varying the mine density to yield the necessary probability of encounter when developing disrupt, fix, turn, and block minefields. Varying mine density is further discussed later in this chapter.

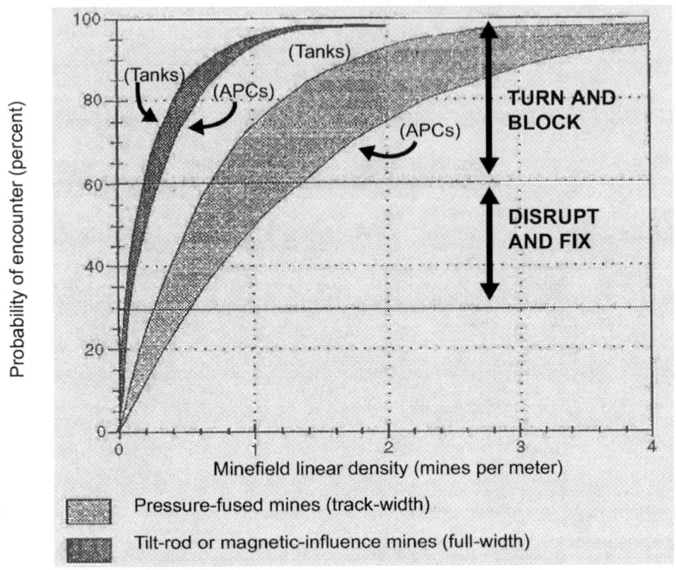

Figure 2-4. Vehicle mine encounter probability versus minefield density

Probability of Kill

The probability of kill is measured by the chance (in percent) that a vehicle will no longer be mission-capable (M-Kill or K-Kill) because of mine effects. It is a function of the combined probability that a vehicle will encounter a mine and the probability that the mine effect will produce an M-Kill or a K-Kill.

Antihandling Devices

Emplacing AHDs on mines is time-intensive. AHDs are added to a minefield to discourage manual removal and reuse of mines by the enemy and to demoralize the enemy who is attempting to reduce the minefield. AHDs do not prevent an enemy from reducing the minefield; they only discourage manual reduction methods.

Irregular Outer Edge

An IOE is a strip/row or multiple strips/rows of mines that normally extend toward the enemy from the first (enemy side) row of mines. An IOE is employed to break up the otherwise regular pattern of a minefield. It is used to confuse the enemy about the exact limits of the minefield, particularly its leading edge. An IOE adds an unknown quality to a minefield that makes the enemy's decision of whether to breach or bypass more difficult. The effect an IOE has on enemy actions may increase the overall lethality of a minefield.

DESIGN

Modifying minefield variables to achieve the desired obstacle effect is a challenge for the engineer, both technically (resourcing and designing) and tactically (supporting the maneuver scheme). Experience will provide the best basis for designing minefields. Figures 2-5 through 2-8, pages 2-10 through 2-13, provide guidelines for varying minefield depth, front, density, and composition to best achieve disrupt, fix, turn, and block effects.

These are guidelines, not fixed rules. Minefield designs must be based on a threat analysis. The designs are simply considerations or parameters to use when designing tactical minefields, regardless of the emplacement method. They apply to conventional mine-laying techniques as well as the employment of SCATMINE dispensers. These parameters give the engineer the flexibility to design and emplace tactical minefields based on mission, enemy, terrain, troops, time available, and civilian considerations (METT-TC) (particularly resources and terrain) and still achieve the required effect. These norms are also the basis for developing minefield packages and emplacement procedures outlined throughout this manual. Chapter 3 discusses the characteristics and emplacement procedures for each of the SCATMINE systems, Chapter 6 outlines procedures for row mining using conventional mines, and Chapter 7 is dedicated to the standard-pattern minefield. Each chapter describes standard disrupt, fix, turn, and block minefield packages particular to that method of emplacement or dispensing system.

Each tactical-obstacle effect has a specific resourcing factor. In short, this numeric value helps determine the amount of linear obstacle effort that is needed to achieve the desired effect. The resource factor is multiplied by the width of the AA or MC to get the total amount of linear obstacle effort required. The linear obstacle effort is then divided by the minefield front norm for the specific effect (rounded up) to yield the number of individual minefields required in the obstacle group.

Disrupt

A disrupt effect (Figure 2-5) focuses fire planning and obstacle effort to cause the enemy to break up its formation and tempo, interrupt its timetable, commit reduction assets prematurely, and piecemeal the attack. It also deceives the enemy about the location of friendly defensive positions, separates combat echelons, or separates combat forces from their logistical support. A disrupt effect should not be time-, manpower-, or resource-intensive. It should not be visible at long range but easily detected as the enemy nears it. Commanders normally use the disrupt effect forward of EAs.

Resource factor	0.5 (3 point obstacles) x AA
Group dimensions	W = 0.5 x AA
Probability of kill	50%
Minefield front	250 m
Minefield depth	100 m
AT mines	Yes (pressure/tilt)
AP mines	No **(Korea Only: optional, based on threat analysis)**
AHD	Optional, based on threat analysis
IOE	No

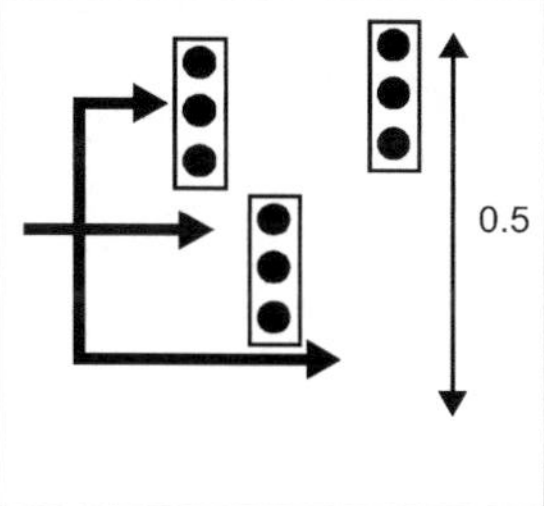

Figure 2-5. Disrupt-effect group

Normally, only half of the enemy's AA is attacked with minefields or other tactical obstacles to achieve a disrupt effect. For a minefield to disrupt an enemy company, half of the formation must react to the minefield. The typical width of a disrupt-effect minefield is 250 meters with a standard minefield depth of 100 meters. When designing a disrupt effect to attack an enemy battalion, three disrupt-effect minefields are arrayed in a group to achieve a width that is about half the size of the battalion's attack front (750 meters of minefield). Multiply the resource factor of 0.5 by the width of the AA to provide the amount of linear obstacle effort required for the disrupt effect. When the AA is narrow, an alternative disrupt group is three point obstacles along the AA.

Disrupt-effect minefields should be designed with approximately 50 percent probability of mine encounter to achieve the desired disrupt effect (see Figure 2-4, page 2-9). They should contain predominantly track-width AT mines and include full-width AT mines at the leading edge of the minefield to increase the probability of mine encounter. This should cause the enemy to commit its reduction assets.

AHDs can be added to disrupt-effect minefields to frustrate the enemy's breaching and clearing operations. However, adding AHDs may be too resource-intensive for the return in effect. An IOE is not required.

Fix

A fix effect (Figure 2-6) focuses fire planning and obstacle effort to slow an attacker within a specific area, normally an EA. The fix effect is primarily used to give the defender time to acquire, target, and destroy the attacking enemy throughout the depth of an EA or AA. A fix effect may be used to generate the time necessary for the force to break contact and disengage as the enemy maneuvers into the area (typically used for delays). Fix-effect minefields in the obstacle group must be employed in depth, causing the enemy formation to react and breach repeatedly. Fix-obstacle groups must span the entire width of the AA. Commanders normally use the fix effect inside the EA.

Resource factor	1.0 x AA
Group dimensions	W = 1.0 x AA
Probability of kill	50%
Minefield front	250 m
Minefield depth	120 m
AT mines	Yes (pressure/tilt)
AP mines	No **(Korea Only: optional, based on threat analysis)**
AHD	Optional, based on threat analysis
IOE	Yes

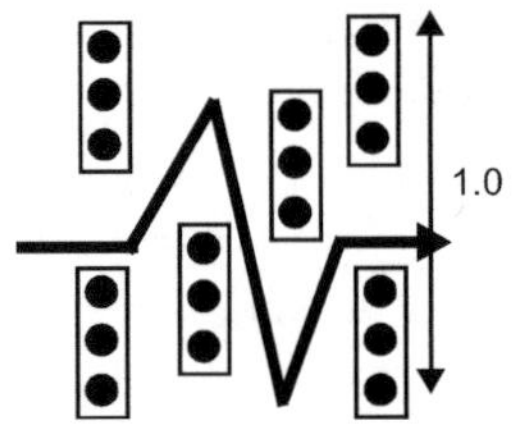

Figure 2-6. Fix-effect group

Individual fix-effect minefields must not appear too difficult to reduce. The enemy should be enticed into the area. The concept is to employ multiple minefields that individually attack a portion of a deploying company formation. Therefore, the fix-effect minefield front is 250 meters. It takes on the characteristics of a disrupt-effect minefield with a similar density, composition, and probability of encounter (Figure 2-4, page 2-9), with two exceptions. First, AHDs are not used because the application of massed direct and indirect fires complicate the enemy's breaching effort. Secondly, an IOE is added to further delay the enemy and confuse the attacker on the exact orientation of individual minefields. This also serves to increase the effective lethality of the minefield. The majority of mines are track-width AT, but full-width AT mines are used in the IOE and the leading edge of the minefield because they are the most lethal.

While individual minefields are designed to attack only portions of an enemy company formation, the fix-obstacle group is resourced, arranged, and sited to attack the entire front of an enemy battalion. Figure 2-6, page 2-11, depicts a fix group effect on an attacking enemy battalion. In this case, six fix-effect minefields are arrayed in an area the full width of the battalion AA (1,500 meters) by 1,500 meters deep. Accordingly, the resource factor for a fix-effect minefield group is one; the amount of linear minefield that should be resourced equals the width of the AA.

Turn

A turn effect (Figure 2-7) manipulates the enemy's maneuver in a desired direction. One technique or a combination of techniques aids in achieving the turn effect. First, in order to entice the enemy to maneuver in the desired direction rather than reduce the obstacle, the obstacle must have a subtle orientation relative to the enemy's approach. Secondly, the obstacle and fires must allow bypass in the direction desired by the friendly scheme of maneuver. Obstacles in the start of the turn are visible and look more complex than those in the direction of the turn. Finally, the obstacle is tied into severely restricted terrain (restricted terrain as a minimum) at the initial point of the turn. The point where the severely restricted terrain feature and the minefield meet is known as the *anchor point*. Commanders normally use the turn effect on the flanks of an EA.

Resource factor	1.2 x AA
Group dimensions	W = 1.0 x AA
Probability of kill	75%
Minefield front	500 m
Minefield depth	300 m
AT mines	Yes (pressure/tilt)
AP mines	No **(Korea Only: optional, based on threat analysis)**
AHD	Optional, based on threat analysis
IOE	No

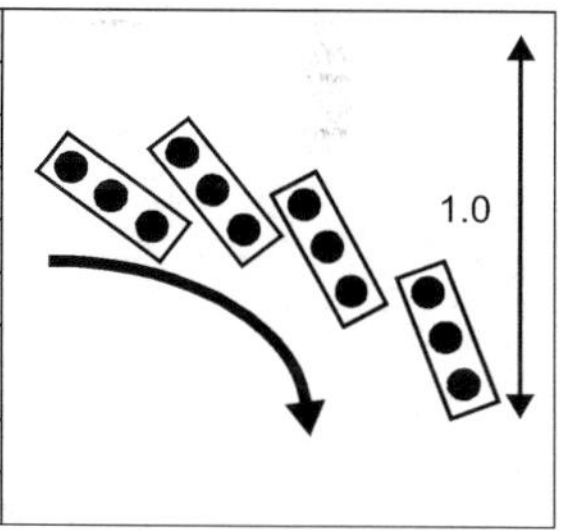

Figure 2-7. Turn-effect group

The standard turn-effect minefield has a width of 500 meters and a depth of 300 meters. One turn-effect minefield affects the entire width of an enemy company's front. It must be deep enough to cause multiple applications of line-charge reduction assets. In short, the minefield must discourage any attempts to breach and must entice the enemy to bypass rather than reduce. Figure 2-7 depicts a turn effect on an attacking battalion, with turn-effect minefields arrayed in a group across a 1,500-meter AA. The angle of the minefields should be subtle, encouraging the enemy to bypass individual minefields. Each minefield overlaps another one to tie the minefields together and prevent gaps. This is considered in the resource factor (1.2) for a turn-obstacle group. This factor, multiplied by the width of the AA, equals the amount of linear obstacle effort required for this turn effect.

As shown in Figure 2-4, page 2-9, turn-effect minefields must be extremely lethal and achieve approximately 80 percent probability of encounter. In other words, an enemy vehicle attempting to reduce or pass through the minefield will likely encounter a mine. This forces the small-unit commander to make an immediate decision—breach or bypass. A lethal minefield that is covered by intense fires and has an easily detectable bypass reduces breach decisions to instinct and causes the enemy to choose the bypass (turn). To produce this lethality, the majority of mines should be full-width AT. Full-width mines in the first rows the enemy encounters and in the depth of the minefield either exhaust the enemy's breaching assets or convince him to bypass early. AHDs are not required because the enemy force will seldom commit to dismounted breach when faced with intense direct and indirect fires. An IOE should not be used because the enemy must be able to determine the orientation of the minefield and the bypass.

Block

A block effect (Figure 2-8) integrates fire planning and obstacle effort to stop an attacker along a specific AA or prevent him from passing through an EA. Block obstacles are complex and are integrated with intense fires; they do not stop an attacker by themselves. Individual block obstacles are employed successively in a relatively shallow area. When the enemy reduces one block obstacle, it is critical that he encounters another, thus denying him to project combat power and maintain momentum. Block obstacles must defeat the enemy's breaching effort (mounted and dismounted) as well as his maneuver. A block effect must span the entire width of an AA and prevent a bypass.

Resource factor	2.4 x AA
Group dimensions	W = 1.0 x AA
Probability of kill	80%
Minefield front	500 m
Minefield depth	300+ m
AT mines	Yes (pressure/tilt)
AP mines	No **(Korea Only: Yes)**
AHD	Yes
IOE	Yes

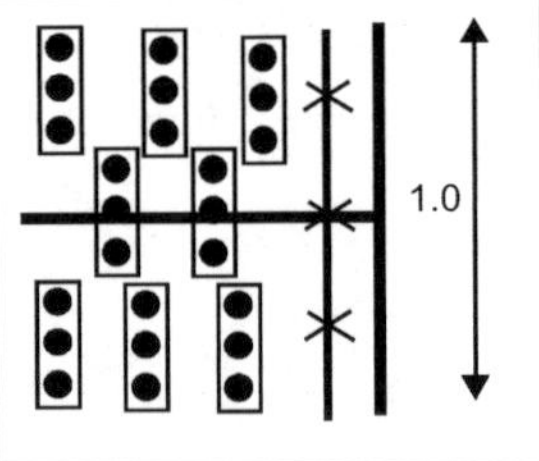

Figure 2-8. Block-effect group

The typical block-effect minefield is 500 meters wide and 300+ meters deep (includes an IOE). Figure 2-8 depicts a block effect on an attacking battalion. Note how individual minefields are arrayed to affect the entire width of the

AA but in a relatively shallow depth. Eight block-effect minefields are required in this example to achieve the necessary depth and width. The block group is the most resource-intensive. A resource factor of 2.4 is multiplied by the width of the AA to determine the linear obstacle effort required.

The lethality of a block-effect minefield (80 percent or higher) is similar to that of a turn-effect minefield (see Figure 2-4, page 2-9). The lethality of the group is considerably higher, since there are enough minefields in the group to cover more than twice the width of the AA. This lethality is produced by a density slightly greater than one mine per meter of front and the use of predominantly full-width AT mines.

A block-effect minefield must be capable of defeating mechanical and dismounted breach efforts. Therefore, AHDs **(Korea Only: and AP mines)** are used to target dismounted breaching. An IOE confuses the attacker about the exact minefield limits and complicates his employment of mechanical reduction assets. The depth of the block-effect minefield requires employing multiple line charges.

The above minefields are not standard solutions to every situation. The terrain could dictate a decrease or an increase in the effort required. Incorporating other reinforcing obstacles (AT ditches, road craters, wire, SCATMINEs) aid in attacking the different reduction assets.

TACTICAL-OBSTACLE INTEGRATION PRINCIPLES

Tactical minefields are considered tactical obstacles and follow the same basic integration principles. Tactical obstacle C^2 focuses on obstacle emplacement authority and obstacle control.

OBSTACLE EMPLACEMENT AUTHORITY

Obstacle emplacement authority is the jurisdiction that a unit commander has to emplace tactical obstacles. In a theater of operations (TO), theater commanders have the authority to emplace obstacles. In most cases, they delegate that authority to corps commanders who further delegate it to division commanders. Division commanders then have obstacle emplacement authority in their area of operations (AO), unless the authority is subsequently withheld or restricted by a higher commander. Commanders subordinate to corp and division do not have the authority to emplace obstacles unless the higher commander delegates authority for a current operation. Commanders use control measures and other specific guidance or orders to grant obstacle emplacement authority to subordinate commanders. Emplacement authority for SCATMINEs is covered in Chapter 3.

OBSTACLE CONTROL

Commanders exercise obstacle control to ensure that obstacles support current and future operations. Obstacle control ensures that subordinate commanders emplace obstacles to best support the higher commander's scheme of maneuver. It also ensures that subordinate commanders do not interfere with future operations. Commanders maintain obstacle control by focusing or withholding emplacement authority or by restricting the types or locations of obstacles. Commanders use control measures, specific guidance, and orders to maintain obstacle control.

Obstacle Control Measures

Obstacle control measures are specific control measures that simplify granting obstacle emplacement authority and providing obstacle control for all levels of command (Table 2-1). Obstacle control measures are classified as—

- Zone.
- Belt.
- Group.
- Restriction.

Table 2-1. Echelons of obstacle control and effect

Obstacle Control Measure	Echelon	Specific Obstacle Effects Assigned	Size of Enemy AA/MC		Planning Guidance
			Armored	Light vs Armored	
Zone	Corps or division	Optional	Division/ brigade	Brigade/ battalion	Requires anticipating belts and intents
Belt	Brigade	Optional but normal	Brigade/ battalion	Battalion/ company	Requires anticipating groups and intents
Group	Corps, brigade, division, or battalion/TF	Mandatory	Battalion/TF	Company/ platoon	Based on individual obstacle norms
Restriction	All	NA	NA	NA	Used only when necessary to support the scheme of maneuver

A protective obstacle is the only obstacle that can be employed outside a designated obstacle zone, belt, or group.

A specific obstacle effect (disrupt, turn, fix, or block) may be assigned to obstacle control measures. This enables the commander to direct the overall effect of obstacles within a designated zone, belt, or group to support his plan. At corps and division levels, assigning specific effects to obstacle zones is optional. At brigade level, the commander will normally assign a specific effect to the obstacle belts. At TF and battalion levels, obstacle effects for obstacle groups are required. This ensures that subordinate commanders emplace tactical obstacles that support the maneuver and fire plans. Assigning a specific obstacle effect to a control measure becomes obstacle intent, giving the obstacle effect, target, and location. Obstacle intent provides a direct link between the obstacle plan, the scheme of maneuver, the commander's intent, and the fire plan (direct and indirect). Obstacle intent is critical at brigade level and below, and it becomes the foundation for obstacle group development and design at the TF level.

Obstacle Zones

Obstacle zones (Figure 2-9, page 2-16) are graphic control measures that corps and division commanders use to grant obstacle emplacement authority to brigades (including armored cavalry regiments [ACRs] and other major subordinate units). Corps and division commanders use zones to ensure that subordinates emplace obstacles that support the higher commander's scheme of maneuver, and to ensure that the obstacles do not interfere with future operations.

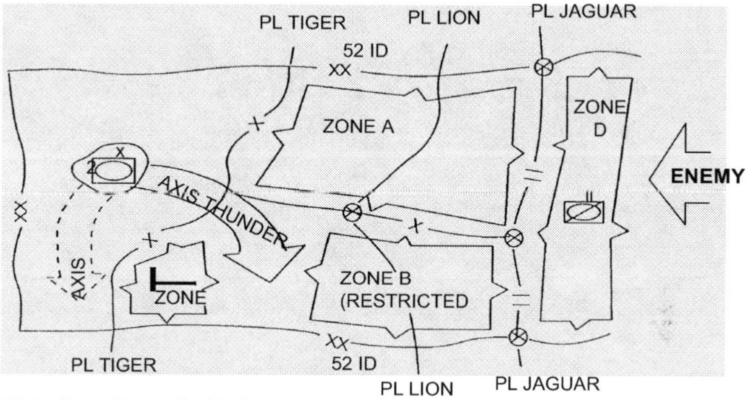

Obstacle number prefix: IO52

Unit	Zone	Priority	Remarks
1st Brigade	A	3	
3d Brigade	B	2	No SCATMINEs with SD after H+2
3d Brigade	C	1	Block intent
Cavalry Squadron	D	4	

Figure 2-9. Obstacle zones

Obstacle Belts

Obstacle belts (Figure 2-10) are graphic control measures that brigade commanders use to constrain tactical-obstacle employment, and the belts should not cross unit boundaries. Commanders plan obstacle belts within assigned obstacle zones to grant obstacle emplacement authority to their major subordinate units. This is normally the first level in which the commander assigns an intent to the obstacle plan. It gives TF commanders the necessary guidance on the overall effect of obstacles within a belt. It does not designate that all obstacle groups within the belt must be the same effect. It simply means that the sum effect of groups within the belt must achieve the assigned belt effect. This serves to synchronize the obstacle effort within the brigade, particularly between adjacent TFs. Obstacle belts also focus obstacles in support of the brigade scheme of maneuver and ensure that obstacles do not interfere with the maneuver of any higher headquarters.

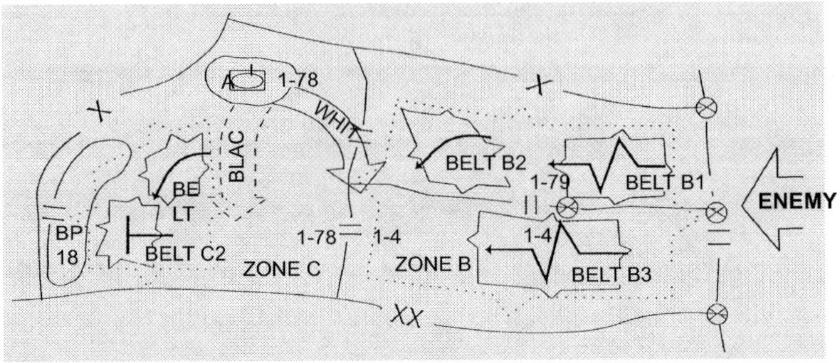

Unit	Belt	Intent	Priority	Remarks
1-78	C1	Turn	2	
1-78	C2	Block	1	
1-79	B1	Fix	5	No long-duration SCATMINEs; must SD by H+2
1-79	B2	Turn	3	No long-duration SCATMINEs; must SD by H+2
1-4	B3	Fix	4	No long-duration SCATMINEs; must SD by H+2

Figure 2-10. Obstacle belts

Belts are planned to attack enemy regiments based on an analysis of enemy battalion MCs. Brigades allocate maneuver companies based on the motorized rifle battalion (MRB) AAs and organize TFs to defeat the motorized rifle regiment (MRR). Obstacle belts and their intents are directed against MRR AAs. This provides the appropriate level of guidance while preserving the TF's need to refine the obstacle intent, based on how the TF will fight its allocated companies.

Obstacle Groups

Obstacle groups (Figure 2-11, page 2-18) contain one or more individual obstacles that are grouped together to provide a specific obstacle effect. TFs use obstacle groups to ensure that company teams emplace individual obstacles that support the TF scheme of maneuver. In rare cases, brigades, divisions, or even corps may use obstacle groups for specific tactical obstacles.

Also, units perform detailed integration of obstacle groups with direct- and indirect-fire plans.

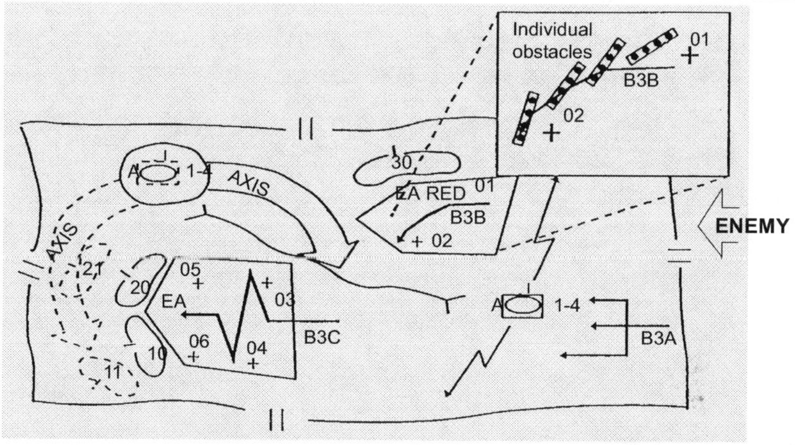

Unit/BP	Group	Intent	Priority	Remarks
Company A/Screen	B3A	Disrupt	3	All groups restricted: • Long-duration SCATMINEs are not allowed. • Short-duration SCATMINEs must be emplaced by H-3.
Team B/BP 30	B3B	Turn	1	
Team D/BP 10	B3C	Fix	2	

Figure 2-11. Obstacle groups

A TF allocates platoons against motorized rifle company (MRC) MCs and task-organize them into companies/teams to defeat MRBs. Likewise, direct-fire plans are designed based on the maneuver of MRBs and independent MRCs. Therefore, obstacle groups are used to attack the maneuver of MRB-size forces. Groups are designed specifically to support the direct-fire plan of the TF. The TF designates groups rather than obstacles because the location of individual obstacles hinges on siting at the company/team level. The group effect or obstacle intent drives obstacle siting and is therefore more important to convey to commanders. There can be more than one type of obstacle group to support the overall intent of an obstacle belt. This is because the belt design is based on the brigade's scheme of maneuver, without knowing the TF's direct-fire plan and scheme of maneuver. Groups are developed once the fire plan is established.

Obstacle Restrictions

Commanders at all levels may use obstacle restrictions to provide additional obstacle control. They may use obstacle restrictions to limit the specific types of obstacles used (for example, no buried mines and no SCATMINEs that do not self-destruct within 48 hours). These restrictions ensure that subordinates do not use obstacles with characteristics that impair future operations. It also allows commanders to focus the use of limited resources for the main effort by restricting their use elsewhere. Commanders may also use restrictions to prevent subordinates from emplacing obstacles in a certain area. This type of restriction may be shown graphically as an obstacle-restricted area.

FRATRICIDE PREVENTION

The modern tendency toward maneuver warfare and the disappearance of the linear battlefield places repositioning forces at an increased risk of fratricide by minefields. Obstacle control and the use of graphic obstacle control measures are vital in preventing minefield fratricide at every echelon.

Obstacle control is further facilitated by positive C^2 of all sustainment traffic, tactical repositioning, obstacle turnover, well-established and disseminated traffic plans with traffic control, and strict adherence to minefield marking procedures (discussed later in this chapter).

MANEUVER-PLAN SUPPORT

Commanders include obstacle planning in each level of the decision-making process. This ensures that obstacle integration is effective and that the obstacle plan is flexible enough to allow changes during the planning, preparation, and execution phases of an operation. The following method is used to integrate obstacle planning at the TF level; it uses the decision-making doctrine contained in FM 101-5. Obstacle planning for brigade and higher levels can be found in FM 90-7.

The focus of obstacle planning is to integrate obstacles into the maneuver direct- and indirect-fire plans. This planning is directive and detailed in nature and focuses on the determination of obstacle groups and the type and amount of prepared positions. Actual obstacle siting, emplacement, and position location are the purview of the company/team commander and are normally supported by an engineer platoon.

TF-level defensive planning is part of the military decision-making process. The engineer battlefield assessment (EBA) process provides the basis for integrating engineer issues into the decision-making process.

Mission Analysis

The key activities during mission analysis are to—

- Determine facts and assumptions.
- Analyze relative combat power.
- Analyze the engineer battalion's/brigade's mission and the commander's intent.
- Issue the commander's guidance.

Determine Facts and Assumptions

Defensive planning normally starts with the receipt of a warning order (WO) from the higher headquarters to defend. The company executive officer (XO) and the battlefield information control center (BICC) (or the TF engineer and the Intelligence Officer [US Army] [S2]) begin by developing a situation template (SITEMP) that includes a modified combined obstacle overlay (MCOO). The MCOO is a product developed during the intelligence preparation of the battlefield (IPB) process. The MCOO development is a joint effort of the engineer and the intelligence section of the TF's tactical operations center (TOC).

The MCOO should define the AAs and the MCs within the TF's AO. This information is vital to obstacle planning. Obstacles are placed on AAs to attack enemy maneuver. The AA analysis details potential EAs and indicates where forces can defend with limited survivability construction, because a reverse slope or undulating terrain provides natural concealment and cover.

The threat evaluation and the enemy course-of-action (COA) development detail how the enemy will potentially attack. They also provide an insight as to what and where the enemy's objective and routes might be. The SITEMP helps the engineer to understand how the enemy will traverse through the TF's sector and allows the engineer to gain an understanding of how and where he can best attack the enemy's maneuver.

The SITEMP also depicts how the enemy's reconnaissance forces will enter the sector. This is especially important when countering the enemy's ability to reconnoiter obstacle efforts.

The engineer must articulate the current capabilities of the engineer forces, its current combat power, and its ability to support the TF. Assumptions of future capability or potential reinforcement by other engineers should be analyzed. Specific characteristics of special engineer equipment and SCATMINE systems are detailed for the staff. An initial Class IV/V supply-point location and an operation plan should be developed with the TF staff (note that the TF has responsibility for Class IV/V supply-point operation). The XO/first sergeant (1SG) works with the TF and the engineer battalion Supply Officer (US Army) (S4) to ensure that delivery of Class IV/V barrier material supports the TF countermobility plan.

Analyze Relative Combat Power

The engineer compares friendly and enemy combat power and identifies possible obstacle requirements that offset potential enemy breaching and direct- and indirect-fire capabilities. The actual inclusion of the obstacles normally occurs after COA development. During this phase, the engineer finishes his EBA to gain an understanding for the engineer company's ability to support the TF.

Analyze the Engineer Battalion's/Brigade's Mission and the Commander's Intent

The staff analyzes and identifies information from the maneuver brigade order and the commander's intent that will potentially impact defensive planning. The engineer analyzes the maneuver brigade/engineer battalion commander's intent to determine potential obstacle placement, obstacle intent, and construction priority based on his concept of the operation.

The TF must identify tasks and limitations imposed from the brigade operation order (OPORD). These might include obstacle belts with or without specific intents, obstacle-restricted areas, or restrictions on the type of obstacles. Also, the brigade OPORD might specify reserve, situational, or directed obstacle groups.

The engineer must identify the TF's total obstacle capabilities. Available assets include engineer units, SCATMINE systems, and other units that can provide additional manpower for obstacle construction. Engineer-equipment status, work rates, and the time available must be evaluated in detail.

NOTE: Work rates should only be used in the absence of unit-developed planning factors.

Table 2-2 provides planning factors for the mine dump. Tables 2-3 through 2-6, pages 2-21 through 2-23, provide planning factors for obstacles. Special focus to limited visibility work rates and engineer squad strengths must be made when making total-capability estimates. Plan to accomplish obstacle siting and Class IV/V supply-point setup during daylight hours, and plan to emplace mines during limited visibility hours as much as possible.

Table 2-2. Planning factors for the mine dump

Number of Personnel	Quantity of Mines	Required Equipment
2-man team (2 minutes per mine)	25 mines per hour	Shears, metal cutting Grease, automotive and artillery Rags Work gloves Flashlight Night-vision goggles Pliers
Squad (7 soldiers and an NCO)	100 mines per hour	
Platoon (with leadership)	300 mines per hour; 3,600 mines per day	
Company	10,800 mines per day	
NOTE: Soldiers work 50 minutes per hour, 12 hours per day.		

Table 2-3. Planning factors for work rates

Survivability	Time Required to Construct		
	With D7F Dozer	**With ACE**	**With SEE**
Hull-defilade position	1 BTH	1.5 BTH	NA
Turret-defilade position	2.5 BTH	3.5 BTH	NA
HMMWV TOW position	1.5 BTH	2 BTH	NA
Vehicle-protective position	0.75 BTH	1 BTH	NA
Dismount-crew position	NA	NA	1 SEEH
Individual-fighting position	NA	NA	0.5 SEEH
Countermobility	**With D7F Dozer**	**With ACE**	**In Man-Hours**
Antitank ditch	1 BTH/70 m	1 BTH/50 m	NA
Standardized disrupt minefield	NA	NA	1.5 PH
Standardized fix minefield	NA	NA	1.5 PH
Standardized turn minefield	NA	NA	3.5 PH
Standardized block minefield	NA	NA	5 PH
Triple-standard concertina	NA	NA	1 PH/300 m
Road crater	NA	NA	1.5 SH
Point minefield	NA	NA	1 SH
Concertina roadblock	NA	NA	1 SH
Bridge demolition (massive)	NA	NA	2 SH
Bridge demolition (steel)	NA	NA	1 SH
Mine preparation at the TF Class IV/V supply point	NA	NA	1 SH/100 mines

LEGEND:

BTH (blade team hour). One blade team working for one hour. A blade team consists of two engineer blades (two dozers, two ACEs, or one ACE and one dozer). One vehicle digs (cutter) while the other spreads the spoil (striker). A dozer-ACE blade team uses the dozer BTH.

SEEH (SEE hour). One SEE working for one hour.

PH (platoon hour). One platoon (3 squads) working for one hour.

SH (squad hour). One squad working for one hour.

Table 2-4. Planning factors for standardized row minefields

Effect	Resource Factor	Front	Depth	Full-Width AT Mines	Track-Width AT Mines	Frag AP Mines
Disrupt	0.5	250 m	100 m	42	84	NA
Fix	1.0	250 m	120 m	63	84	NA
Turn	1.2	500 m	300 m	336	168	NA
Block	2.4	500 m	320 m	378	168	**84 (Korea Only)**

Table 2-5. Planning factors for scatterable minefields

System	Minefield Size	SD Time	Arming Time
ADAM	400 x 400 m 200 x 800 m	4 hr 48 hr	Within 1 min after ground impact
RAAM	400 x 400 m 200 x 800 m	4 hr 48 hr	2 min 45 sec
Volcano (one load = 160 canisters or 960 mines [800 AT and 160 AP])	Turn or block (1 per load): • Ground: 555 x 320 m • Air: 557 x 320 m Fix or disrupt (4 per load): • Ground: 277 x 120 m • Air: 278 x 120 m	4 hr 5 days 15 days	2 min
MOPMS	70 x 35 m	4 hr*	89 sec
*Can be recycled 3 times for a total of 13 hr			

Table 2-6. Ranges of common weapons

Weapon		Maximum Effective Range	Planning Range*
FRIENDLY WEAPON SYSTEMS			
M16A2		580 m	400 m
M249 SAW		1,000 m	800 m
M60		1,100 m	1,100 m
M203	Area	350 m	350 m
	Point	160 m	160 m
M2, .50 Cal	Area	1,830 m	1,830 m
	Point	1,200 m	1,200 m
MK19	Area	2,200 m	2,200 m
	Point	1,600 m	1,600 m
AT4		300 m	300 m
M47 Dragon		1,000 m	800 m
Javelin		2,000 m	2,000 m
M1 Abrams tank	105 mm	2,500 m	2,000 m
	120 mm	3,000 m	2,500 m
M2 Bradley ITV	25 mm (APDS)	3,000 m	1,700 m
	25 mm (HEI-T)	3,000 m	1,700 m
	TOW2	3,750 m	3,750 m
60-mm mortar	HE	3,400 m	50 m (min)
	WP	4,800 m	50 m (min)
	ILLUM	931 m	50 m (min)
81-mm mortar	HE	4,595 m	75 m (min)
	WP	4,595 m	75 m (min)
	ILLUM	3,150 m	75 m (min)
4.2-in mortar	HE	6,840 m	770 m (min)
	WP	5,650 m	920 m (min)
	ILLUM	5,490 m	400 m (min)
SOVIET-STYLE WEAPON SYSTEMS			
BMP, 73 mm		800 m	800 m
AT3 missile		3,000 m	3,000 m
AT5 missile		4,000 m	4,000 m
BMP-2		2,000 m	2,000 m
BTR, 14.5 mm		2,000 m	1,000 m
T-72 tank, 125 mm		2,100 m	2,000 m
T-80 tank, 125 mm		2,400 m	2,000 m
T-80 AT8		4,000 m	4,000 m

*The planning range is based on ideal weather conditions during daylight.

Issue the Commander's Guidance

The commander should be as specific as possible with his initial obstacle guidance. If the commander narrows the COA focus, he should also provide obstacle guidance. His guidance is a key factor to an early start and must be solicited if not offered.

Course-of-Action Development

Detailed planning begins following the COA development (Figure 2-12). The engineer focuses on four specifics of obstacle planning in the scheme of engineer operations (SOEO) for the defensive plan:

- Direct-/indirect-fire analysis.
- Obstacle-intent integration.
- Obstacle priority.
- Mobility requirements.

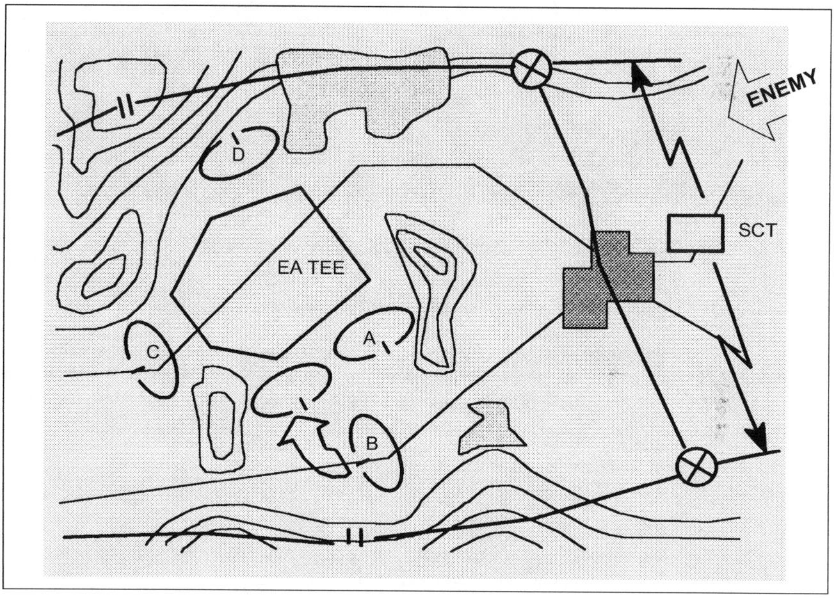

Figure 2-12. TF defense COA

Direct-/Indirect-Fire Analysis

The direct-/indirect-fire analysis examines how engineers can best use obstacles (within the commander's intent) to enhance the direct-/indirect-fire plan. Figure 2-13 shows a sample direct-fire analysis. This analysis can be used to formulate obstacle locations with the direct-fire plan. The engineer must have a fundamental understanding of the direct-/indirect-fire and maneuver plans and the TF's organization of the EA to effectively integrate obstacles with the direct-/indirect-fire plan.

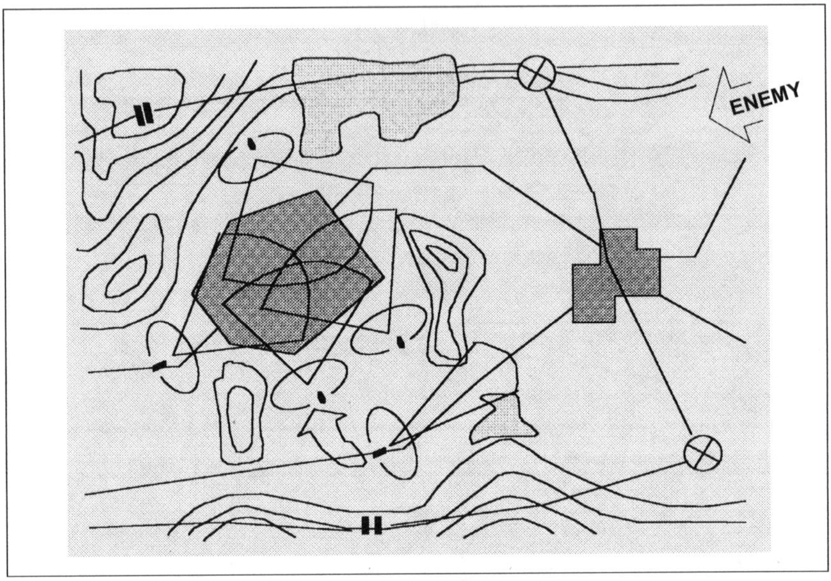

Figure 2-13. TF direct-fire analysis

Synchronization of direct and indirect fires with obstacles multiplies the relative effect on the enemy. An obstacle is an excellent location for preplanned artillery and mortar fires. These fires can eliminate dismounted breaching efforts. Indirect fires contribute to the threat's ability to breach, making the obstacle more effective and providing direct-fire systems a higher probability to kill.

Obstacle-Intent Integration

The engineer determines locations for the directed obstacle groups. Groups are placed on the COA overlay to support the maneuver plan. This location is for planning only and will normally be adjusted after the ground reconnaissance.

Obstacle groups target specific enemy elements based on the SITEMP. The engineer generally allocates an obstacle group against a battalion-size AA. This approach mirrors the staff's placement of a company/team against the same enemy force. The company's/team's fire responsibility drives the placement of the obstacle groups. The engineer advises the commander on which specific effect each directed obstacle group must achieve. He plans obstacle groups to—

- Disrupt the enemy.
- Turn the enemy into an area where friendly units can mass fires.
- Fix the enemy in the EA and enhance his direct-fire destruction.
- Block the enemy from using an AA.

The engineer integrates directed obstacle groups with the COA. The obstacle effects are shown on the COA overlay using obstacle-effect graphics (Figure 2-14). The engineer draws the obstacle-group graphic to reflect the location, the target, and the specific intent of the group as accurately as possible. The engineer should visualize how the terrain naturally effects maneuver. Terrain visualization is vital to proper obstacle-group design.

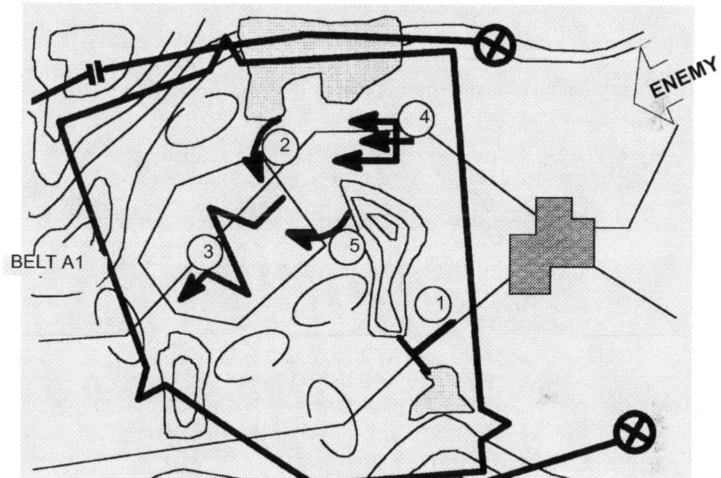

Figure 2-14. TF obstacle-intent integration and priorities

Note the placement and the effect of obstacle groups in Figure 2-14. First, the engineer must manipulate the MRB into the EA. The turn groups (2 and 5), combined with a heavy volume of AT fires from a company team at the turning point, achieves this. In the EA, particularly where the TF fires are massed, a fix group (3) slows the enemy and increases the effects of the fires. A block group (1) in the south, along with direct fires from a company team, will stop the advance of any element along the southern AA. A disrupt group (4), with indirect fires, will break up the C^2 and the tempo of the attacking force.

Protective obstacles in front of all team positions protect the teams from the enemy's final assault.

Obstacle Priority

The staff determines the priority of each group depicted on the overlay. Priority is established by the commander's intent and the most likely enemy COA. The obstacle priority should reflect the TF's greatest obstacle requirement. The primary obstacle effort can be with an economy of force where the commander needs more obstacles to overcome a shortage of direct-fire systems. The TF engineer should be cognizant of flank protection, weapon types and ranges, and the overall commander's intent for the entire force before placing obstacle priority on the main EA. Priorities assist the engineer in allocating resources and ensuring that the most critical obstacle groups are emplaced first.

In Figure 2-14, the first priority is to turn the enemy where the fires are massed. The second priority is to deny the enemy access (block effect) to the southern AA. The fix effect is the third priority because it enhances the TF fires in the EA but only slows the enemy). The block effect is a higher priority than the fix effect because it stops the enemy from flanking the TF. The disrupt effect is the last priority.

Mobility Requirements

The engineer identifies the TF's mobility requirements. Obstacle groups should not be arrayed along potential counterattack routes or where there is a potential to hamper unit repositioning. Mobility assets should be used to counter potential enemy situational obstacles and friendly obstacles that might hinder friendly maneuver. The TF engineer must consider the commander's mobility requirements and plan for mobility assets so he can be ready when and where he is needed.

Course-of-Action Analysis

The staff war-games the COAs to determine their viability and recommends the best COA to the commander. The engineer refines the SOEO during this process as well. Obstacles should be considered within the context of the maneuver COA (Figure 2-15, page 2-28).

The engineer staff officer should consider the following:

- Enemy reactions at the obstacle groups (breaching or bypassing capability) versus the desired obstacle effect.
- Enemy breaching capabilities that make one obstacle type preferable to another (such as an AT ditch versus a minefield).
- Obstacle locations that hinder friendly maneuver.
- The compatibility of obstacle effects and weapon-system capabilities.
- Adequate direct-/indirect-fire control measures and targeting that support the obstacle effect. The effects of artillery and obstacles must be synchronized to gain the desired effect on the enemy's maneuver.
- Locations and types of enemy situational obstacles that make one type of breaching asset preferable to another.

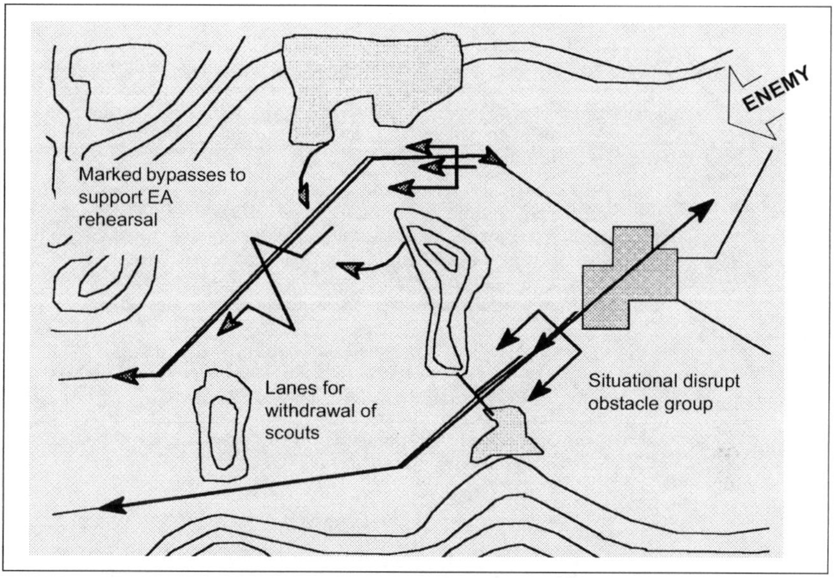

Figure 2-15. Obstacle-plan refinement

After war gaming, the staff adjusts the COA (including the obstacle plan) by—

- Changing the location of obstacle groups.
- Changing the obstacle effects.
- Adding more situational obstacles.
- Adding more reserve obstacles.
- Identifying other mobility requirements.
- Refining artillery targets based on obstacle-group changes.

Mobility Requirements

The staff determines which obstacles require lanes and determines the closure criteria for the lanes. It also determines obstacle-restricted areas that support the TF's maneuver. Lanes and bypasses are determined using tactical repositioning requirements developed during the COA analysis. Requirements for rehearsal movement, placement of the target reference point (TRP), and logistical support of forward TF elements are also considered in lane development. Mobility requirements identified during COA development are synchronized and refined during COA analysis. Additional mobility requirements identified during war gaming are resourced and planned. (Lane marking is discussed later in this chapter.)

Obstacle Design/Resourcing

After the COA analysis, the engineer conducts a detailed study of the obstacle plan to determine resource requirements. Groups are resourced using the methods previously discussed. The TF engineer resources the obstacle groups based on their assigned priorities. Once the engineer has developed the resource requirements for the obstacle groups, he plans the individual obstacles within the group.

If time permits, a detailed ground reconnaissance of the obstacle-group location can be conducted. This will allow a more detailed analysis of the obstacle requirement for that AA, and then the engineer can plan individual obstacles. The TF engineer usually designates the intent to guide the companies/teams; the company/team commanders and their supporting engineers complete the actual design of the obstacle groups.

Decision and Execution

The engineer makes adjustments to the SOEO based on the COA that the commander approves. The engineer then provides oral, written, and/or graphical orders with sufficient detail to allow the subordinate units to conduct the operation. The engineer provides critical information using the scheme-of-obstacle overlay and the obstacle-execution matrix.

Scheme-of-Obstacle Overlay

The scheme-of-obstacle overlay (Figure 2-16, page 2-30) depicts the location of the TF's obstacle groups, brigade-directed obstacle groups (if any), and obstacle belts within the TF's sector. The overlay also includes any obstacle restrictions dictated from a higher headquarters. The overlay depicts the obstacle groups using the standardized obstacle-effect symbols. The overlay does not generally show individual obstacles unless the engineer has had sufficient time to conduct a thorough ground reconnaissance where exact obstacle locations have been identified. The engineer must exercise extreme caution if he uses individual obstacles on the overlay. He must ensure that inexperienced leaders do not attempt to emplace obstacles exactly as shown on the overlay, but instead, properly site the obstacle with the company/team commander. The scheme-of-obstacle overlay graphically depicts how the commander seeks to influence enemy maneuver through obstacles.

Obstacle-Execution Matrix

The obstacle-execution matrix includes specific instructions and detailed information concerning the obstacle groups shown on the scheme-of-obstacle overlay. Develop an obstacle-execution matrix for all situational, reserve, and direct obstacles within the TF area. As a minimum, the matrix should include the information shown in Figure 2-17, page 2-31.

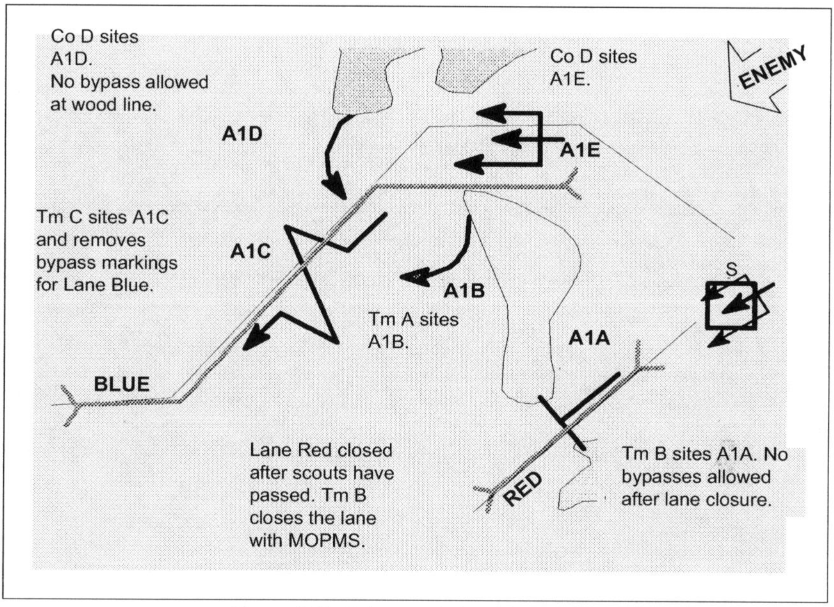

Figure 2-16. Scheme-of-obstacle overlay

Percent Complete	Priority	Standard Obstacle Number				Obstacle Location	Laying Unit	Overwatch Unit	Class IV/V Planned and Pre-position Grid	Start Time	Finish Time	Lane Grid (Open or Closed)	Remarks
		Obstacle Authority	Zone, Belt, or Group	Obstacle Type and Number	Status								

Figure 2-17. Sample obstacle-execution matrix

SITING AND EMPLACING TACTICAL MINEFIELDS

This section outlines the principles for siting tactical minefields to support the company/team. These principles apply to all methods of emplacement—standard pattern, row mining, and SCATMINE systems. The focal point of the discussion on siting is the coordination that must occur between the emplacing engineer (normally the engineer platoon leader) and the maneuver company commander. Coordination between the engineer platoon and the maneuver company is perhaps the most vital component of effective obstacle integration, and it is also a vital component in EA development. Obstacles are directly integrated with weapon effects, capability, and the fire plan at this level. The two subcomponents of tactical-obstacle siting are coordinating with the maneuver commander and siting the minefield.

Coordinating with the Maneuver Commander

Effective coordination with the maneuver company/team commander who will fight the obstacle(s) is essential to realizing the full potential of minefields as a combat multiplier. In short, the emplacing engineer becomes the maneuver company/team commander's team engineer for the mission. The engineer, the fire-support team (FIST), and the maneuver commander must work closely to ensure complete integration of the minefield into all aspects of the company plan. The engineer must be integrated into the maneuver company/team EA development process. Throughout each step of the process, the engineer must provide the maneuver commander and the FIST with the engineer expertise necessary to ensure complete and effective obstacle integration.

Before the emplacing engineer can conduct effective coordination, he must have tools and information from the TF order that serves as common ground between the emplacing engineer, the FIST, and the maneuver commander. The order drives the integration of tactical obstacles into the fire plan and ensures that the obstacles affect their intended enemy target in a way that supports the scheme of maneuver.

Modified Combined Obstacle Overlay

The MCOO is a product from the IPB process that graphically depicts the maneuverability of the terrain. It depicts slow-go and no-go terrain relative to the type of enemy force. It also defines AAs and MCs that the enemy may use for its attack. Since tactical obstacles attack the enemy's maneuver and must complement the existing terrain, the MCOO is vital to obstacle siting. It helps ensure that obstacles correctly address the enemy AAs and MCs. It also helps select how and what part of the enemy formation will be directly attacked by obstacles, and it shows the effect the obstacles will have on the enemy's maneuver.

Situation Template

The SITEMP is developed by the maneuver battalion S2 and the TF engineer during the IPB. It estimates how the enemy will attack, in terms of size and type of units, and the formations it will use. Tactical obstacles are employed to produce specific effects on specific enemy targets. Therefore, the SITEMP helps the engineer and the maneuver commander site and emplace obstacles in a way that attacks the intended target. The SITEMP may also depict the likely routes for enemy reconnaissance elements. This helps the engineer and the maneuver commander analyze requirements for reconnaissance and surveillance (R&S) patrols that defeat enemy attempts to reconnoiter the obstacles and reduce enemy effectiveness before the attack. The type of formations the S2 expects the enemy to use during the entire course of the attack is also vital information. The SITEMP should identify when the enemy is in march, prebattle, and attack formations. The enemy formation may impact on the necessary front of obstacle groups and the obstacle groups' effectiveness in achieving the intended effect on enemy maneuver.

Commander's Intent

The emplacing engineer, the FIST, and the maneuver commander must have a common understanding of the battalion commander's intent. The battalion commander's intent is his vision of the battle and normally outlines what actions the unit must do to accomplish the mission. The commander's intent may include key aspects of the plan that he wants to emphasize to subordinates to synchronize the actions of subordinates toward a single purpose. The engineer must understand the commander's intent and how it relates to integrating obstacles. The engineer should always ensure that the obstacles he is emplacing support the commander's overall intent.

Maneuver Graphics and the Fire Plan

In order to fully support the scheme of maneuver, the engineer must have and understand the maneuver graphics on the battalion's operational overlay. The maneuver graphics use symbols to depict the missions of each subunit within the battalion. Maneuver control measures such as battle positions, sectors, phase lines, passage lanes/points, and counterattack axis are vitally important to understanding the plan and integrating tactical obstacles. The maneuver graphics may include direct-fire control measures that direct how and where combat forces will mass, shift, and lift fires to destroy the enemy. Direct-fire control measures include EAs, trigger lines, and TRP and unit boundaries. In short, they dictate the direct-fire responsibilities of each subordinate. Understanding the direct-fire plan and the organization of the engagement is fundamental to integrating obstacles with fires. The maneuver graphics also give the engineer an appreciation of how tactical obstacles supporting one unit must complement the adjacent units. This is particularly true of adjacent EAs or plans requiring any tactical repositioning of forces.

Obstacle-Execution Matrix

The obstacle-execution matrix includes specific instructions and detailed information concerning the obstacle groups shown on the scheme-of-obstacle overlay. This matrix gives the engineer critical information on minefield groups that will be emplaced within the company's/team's AO. As a minimum, the obstacle-execution matrix should include the information shown in Figure 2-17, page 2-31.

Scheme-of-Obstacle Overlay

At the maneuver battalion level, the scheme-of-obstacle overlay depicts the location of brigade-directed belts, TF obstacle groups, and any directed obstacles within the battalion sector. Any obstacle restrictions attached to an obstacle control measure (belt or group) that preclude the employment of certain types of obstacles are annotated on the overlay. A scheme-of-obstacle overlay is a graphic control measure that defines the general location of the obstacle groups and the effect to be achieved by them. The scheme-of-obstacle overlay does not normally depict individual obstacle locations. The location of individual obstacles within a group is determined during the siting process between the emplacing engineer and the maneuver company commander. When overlaid on the maneuver graphics and the SITEMP, the scheme-of-obstacle overlay should depict the essential elements of obstacle integration:

- Enemy targeted by the obstacle.
- Location of the obstacle on the battlefield.
- Unit covering the obstacle.
- Directed link between the obstacle effects and the fire plan.

Fire-Support Plan

The emplacing engineer should be familiar with the key elements of the fire-support plan. He must understand the general scheme of fires and how the fires support the scheme of maneuver, the commander's intent, and the obstacle plan. Normally, the emplacing engineer does not need the entire fire-support overlay depicting the location of all targets. However, he should know the location of fire-support targets directed by the battalion to cover obstacles. The emplacing engineer should know who has the priority of fires for each phase of the battle. The emplacing engineer should also know the location and the type of priority targets or FPF allocated to the maneuver company he is supporting. During coordination with the maneuver company, the emplacing engineer should discuss the fire-support plan with the company FIST and get updates on changes to the plan as well as any company-level fire-support plans that may impact on the integration of obstacles.

Combat Service Support

The engineer must be familiar with the plan for combat service support (CSS). In particular, the engineer must know the location of major supply routes into and through the battalion area, the location of the battalion logistics release point (LRP), what routes the maneuver company will take from its position to the LRP, and the location of key battalion logistics nodes. The emplacing engineer must always be cognizant of the sustainment traffic flow and the impact obstacle emplacement has on sustainment operations. Of particular concern to the engineer is the location of the Class IV/V supply point and the routes to it.

Battlefield Operating System

During coordination with the maneuver company commander, a checklist or a framework is a useful tool for organizing thoughts and formulating questions. Below is a list of considerations or points of coordination that should drive the integration of the emplacing engineer and the maneuver company commander. The list is organized using the Battlefield Operating System (BOS) because it provides a logical sequence and a framework that is easily remembered.

- Intelligence.

- Enemy AAs and MCs (mounted and dismounted).
- Likely enemy COA and possible reactions to the obstacles.
- Enemy breach capability.
- Enemy reconnaissance routes and friendly counterreconnaissance or R&S plans, particularly company-level patrols.
- Likely enemy formations and transitions between formations.

- Maneuver.
  - Specified, implied, and essential tasks of the maneuver company.
  - Higher commander's intent.
  - Organization of the defensive position, including—
    - Task organization (type of weapons).
    - Decisive point or defeat mechanism.
    - Organization of direct fires in the EA.
    - Location and marking of direct-fire control measures.
    - Position of weapons to cover assigned direct-fire responsibilities.
  - Tactical mobility requirements of the maneuver company and any adjacent units, including—
    - Counterattack axis.
    - Repositioning of forces and their routes to alternate, supplementary, or subsequent battle positions.
    - Employment of reserves.
    - Passage of lines.
  - Obstacle protection measures.
- Mobility/survivability.
  - Intent of tactical obstacles covered by the maneuver company (enemy target, obstacle location, and obstacle effect).
  - Obstacle control measures and restrictions imposed by higher headquarters.
  - Mobility requirements (lanes/gaps), as identified above in maneuver.
  - Mutual support between the obstacle location, the fire plan, obstacle effects, and survivability positions.
  - Security for engineers provided by the maneuver unit supported.
- Fire support.
  - Location of the company FIST and frequency of fire support.
  - Updates on the tentative fire-support plan.

- Allocation of fires to the company, including—
  - Artillery or mortar targets.
  - Priority targets, types of targets, and the FPF.
- Covering obstacles and their effects with indirect fires.
- Indirect-fire control measures to synchronize direct fires, indirect fires, and obstacles.
- Area-denial artillery munition (ADAM)/remote antiarmor mine (RAAM) use (lane closure and breached obstacle repair).
- Registering fires. (Deconflict with obstacle emplacement; registration should occur after obstacles are sited but before emplacement.)
- Company fire-support execution matrix.
- Means for obtaining fire support, if enemy contact is made during emplacement.

• Air defense.

- Enemy air AAs during emplacement.
- Update on changes to air-defense warning and weapons status.
- Location of air-defense systems that can cover engineers emplacing obstacles.
- Method of obtaining early air-defense warning.

• CSS.

- Tentative location of the mine dump, if used, within the company position and routes from the mine dump to obstacles.
- Routes the company plans on using to conduct logistics package (LOGPAC) operations that must remain open.
- Manpower assistance for operations at the Class IV/V supply point and the mine dump.
- Casualty evacuation routes for scouts, observation posts (OPs), and ADA systems.

• C^2.

- Location of the commander during defensive preparation.
- Frequency-modulated (FM) net of the supported company and the means of communication.
- Unit boundaries affecting obstacle emplacement.
- Time and place of the company/team order.
- Coordination that must occur with adjacent units.
- Obstacle reporting and recording requirements.

- Control and execution of situational and reserve obstacles.
- Lane-closure responsibilities and procedures.
- Time and method of obstacle turnover, including lanes.
- Company/team understanding of the obstacle intent.

SITING THE MINEFIELD

The emplacing engineer and the company/team commander site individual obstacles to achieve synchronization between the obstacle effect and fires. Siting is a key component to the EA development process, and it represents the final adjustments to the obstacle location and the fire control measures before emplacement.

Certain preconditions are necessary to site individual obstacles. First, the company/team commander decides where he plans to mass fires and marks the necessary fire control measures on the ground. The location of these control measures must be clear, since they are the basis for obstacle siting. The commander then identifies tentative locations for key weapons within the position or the sector. Finally, the commander and the engineer must both understand the intent of the obstacle group.

Obstacle siting concentrates on marking the obstacle group as a whole instead of marking each individual obstacle. In broken terrain, however, it may be easier to site individual obstacles. The company/team commander and the emplacing engineer use vehicles or soldiers from the company/team, the engineer platoon, or both to simulate the enemy force and do the physical marking. The simulated enemy force moves into the EA to the enemy side of the obstacle group. The engineer platoon leader and the company/team commander collocate near the weapons covering the obstacle. As a technique, one or all of the tanks, Bradleys, and other crew-served weapons may occupy their position and contribute to the siting process. All participants in the siting process use a common FM net to communicate during siting.

The simulated enemy force moves into the EA, simulating the enemy's attack. It deploys into a formation of front that is similar to the expected enemy formation. Once it is near the marked fire control measures, it places markers at intervals as it drives the trace of the obstacle-group effect (or individual obstacles in broken terrain). It remains oriented on key fire control measures to ensure that obstacle location and effect are synchronized with fires. During the process, each participant verifies that he can cover the obstacle, notes the location of fire control measures and obstacles, and records the appropriate data on range cards. As the simulated force drives the obstacle trace, siting participants also identify dead space and requirements to refine the location of the obstacle group and fire control measures. Figure 2-18, page 2-38, illustrates how the engineer and the company/team commander work together to site turn- and fix-obstacle groups.

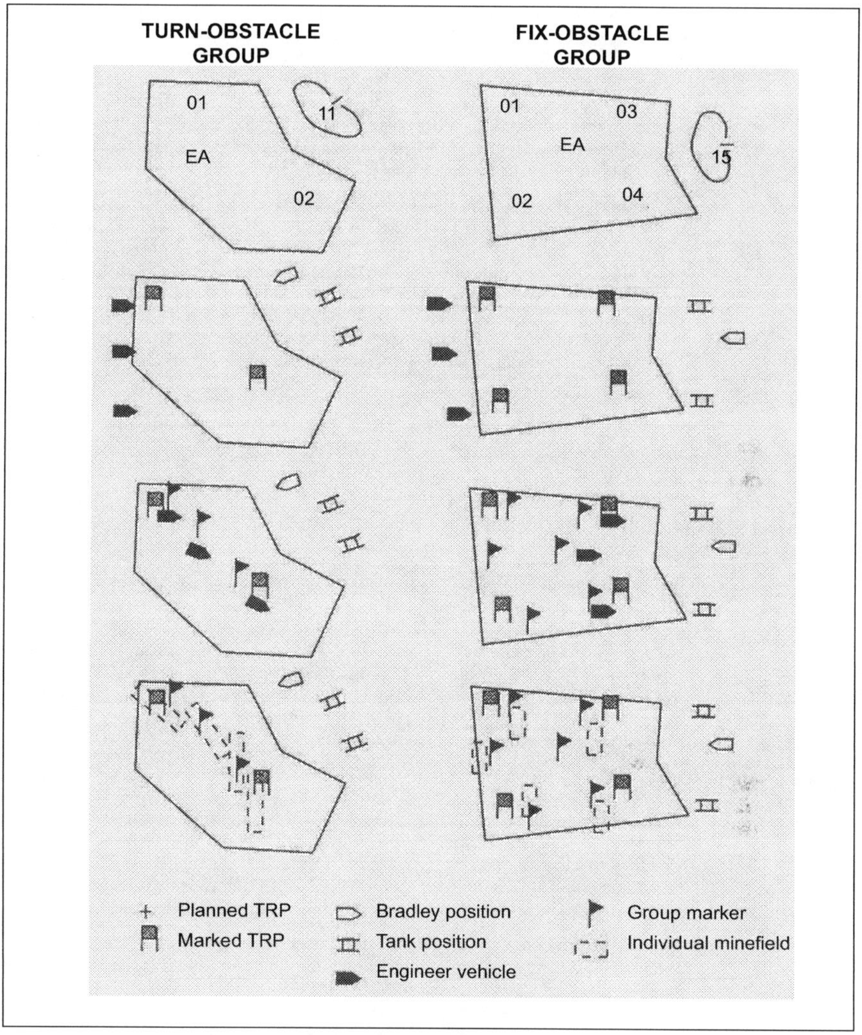

Figure 2-18. Minefield siting

Once the company/team marks the general limits and the orientation of the obstacle group, the engineers can begin marking individual obstacles (if not already done). To mark individual obstacles, the engineer platoon uses the group markers as guides. As shown in Figure 2-18, the group markers may lend themselves well as the start and end points of individual obstacles; however, this is not always the case. As the engineer platoon refines the group limits into the site of individual obstacles, the platoon can then begin the necessary site layout based on the method of obstacle emplacement.

Siting is not the last thing done during preparations. The time and resources involved in emplacing tactical obstacles require that siting begins concurrently with establishing the defensive position. It is imperative that the unit sites the obstacles as soon as the company/team commander has established the EA and identified tentative positions for key weapons. It is not necessary that all weapons are in place and dug in before siting. Normally, well-marked fire control measures and one known position per maneuver platoon (not dug in) are all that is required to site the obstacles effectively.

Emplacing Minefields

Based on the group effect, resources allocated, and the engineer plan, the platoon leader determines the method of emplacement for individual minefields. The procedures for emplacing scatterable, row, and standard-pattern minefields are contained in Chapters 3, 6, and 7, respectively.

Determining Resource Requirements

The engineer must determine the number of individual minefields needed to make up the group and ensure the allocation of required resources. The amount of linear obstacle effort for a group is equal to the width of the AA, multiplied by the resource factor. In Figure 2-19, page 2-40, the AA is 1,500 meters wide, the tactical-obstacle effect is to turn the enemy, and the resource factor is 1.2. The linear minefield requirement is 1,800 meters. One turn-effect minefield has a front of 500 meters (1,800/500 = 4 minefields [round up]). The number of mines and the time required to emplace each minefield depends on the emplacement method.

MINEFIELD SUPPLY OPERATIONS

Requests for Class IV/V obstacle material that are originated at TF level or below go to the TF S4. The TF S4 processes the requests and sends them to the forward support battalion (FSB). The FSB processes the requests and forwards them to the brigade-level FSB, the division material-management center (DMMC), and the corps material-management center (CMMC).

The issue of Class IV obstacle material normally involves large quantities of material; therefore, corps support elements normally use their transportation assets to deliver the material directly to the emplacement sites or to the Class IV/V supply point.

Units request Class V obstacle material somewhat differently. The TF S4 notifies the brigade S4 of Class V requirements. The brigade S4 notifies the division ammunition officer (DAO) in the DMMC, who authorizes Class V issue by the ammunition transfer point (ATP). Class V obstacle material,

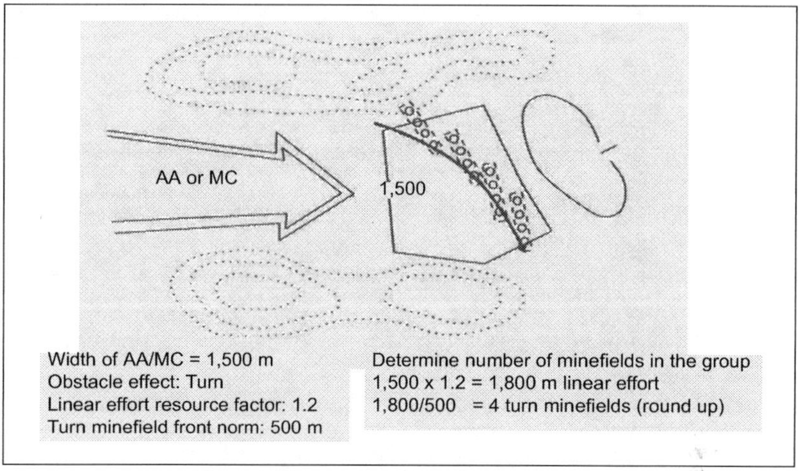

Figure 2-19. Example of minefield resourcing

unlike most ammunition, is delivered to the user at the obstacle emplacement site or the Class IV/V supply point.

A supply request includes the quantity, the required delivery time, the transportation responsibilities, and the desired delivery location. The quantity includes the total for each type of obstacle. There may be several Department of Defense identification codes (DODICs) and national stock numbers (NSNs) involved, depending on the types of obstacles required. The required delivery time is very important for ensuring an early start on the preparation of the battlefield because the lack of material could adversely affect the mission. Transportation responsibilities must be clearly understood. Material handling equipment (MHE) is required to ensure a rapid turnaround of haul assets.

In addition, the brigade staff coordinates with the TF staff to identify the location of Class IV/V supply points in the TF sectors. Prompt identification of TF Class IV/V supply points is required if the obstacle material is forwarded from the corps into the TF sector. If the material is not forwarded into the TF sector, the brigade is responsible for delivering it to the TF.

At the TF level, sustaining obstacle operations is an extremely difficult task. Centralized throughput operations by the corps or the division stop at the TF level. Mass quantities of obstacle material, especially mines, are centrally received, broken down into minefield packages, and then distributed throughout the sector based on the obstacle plan. At some point in the distribution plan, the TF turns over control of the obstacle material to engineers who then emplace the obstacles. Obstacle logistics (especially for mine warfare) at the TF level can be complex, require prudent use of scarce haul assets and MHE, and demand positive C^2.

The supply of obstacle material may vary slightly for obstacle groups developed at corps, division, and brigade levels. The staff at the level where the obstacle group is planned determines the resources required for the obstacle. It also determines how the emplacing unit will get the material. For example, if the corps staff plans a reserve obstacle group but the detailed planning is done at TF level, the TF resources the obstacle group. However, if the corps staff plans the obstacle group in detail, it determines the resources required. The corps staff also plans the delivery of obstacle material to the emplacing unit. Alternately, the corps staff could direct the emplacing unit to pick up the obstacle material from a location such as the corps storage area.

This section describes some of the underlying principles in mine supply operations. It concentrates on the flow of Class IV/V material (mines) through the battalion sector. The maneuver unit is responsible for the flow of obstacle material within the maneuver battalion sector; however, it is effectively a shared responsibility between the engineer and the maneuver unit.

Resupply Nodes

There are two critical mine resupply nodes within the TF sector—the Class IV/V supply point and the mine dump. The relative locations of the Class IV/V supply point and the mine dump are shown in Figure 2-20.

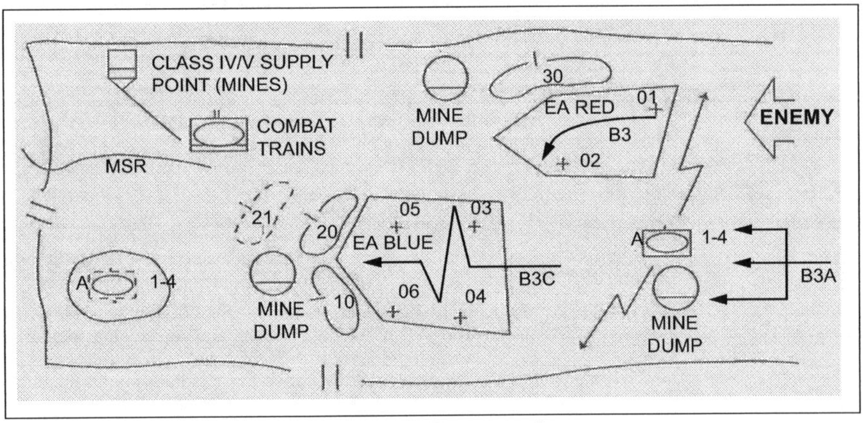

Figure 2-20. Mine resupply

Class IV/V Supply Point

A Class IV/V supply point is a central receiving point for obstacle material in a TF sector. It is where the TF receives and transfers control of obstacle material pushed forward by higher levels. The supply point is established and operated by the TF and is centrally located to support all planned obstacles within the TF sector. Where the tactical-obstacle plan allows, the supply point should be located near the TF combat trains to better facilitate C^2 and the availability of equipment. The TF combat trains command post (CTCP) should provide C^2 of operations at the Class IV/V supply point. The TF engineer will normally furnish a representative that provides technical assistance to the CTCP.

The main purpose of a Class IV/V supply point operation is to receive obstacle material and reconfigure it based on the requirements for each obstacle group. Each supply point must have a dedicated S4 representative to track the flow of obstacle material in and out of the supply point. The supply point should have dedicated MHE to off-load the bulk quantities of obstacle material and reconfigure them into minefield packages as required. Obstacle material is normally broken down into minefield packages if the material is not already delivered in combat-configured loads. This may require a dedicated engineer representative to ensure that obstacle material is configured properly. Table 2-7 shows personnel requirements for a Class IV/V supply point.

Table 2-7. Personnel requirements for a Class IV/V supply point

Personnel	Responsibilities
TF S4/S4 NCOIC	Provides overall C^2 and Class IV/V accountability Coordinates for MHE Is assisted by the RTO
TF engineer representative	Supervises organization of Class IV/V material into packages to support obstacle groups Provides technical assistance on Class IV/V supply-point setup and mine uncrating and inspection
Infantry squad*	Downloads incoming Class V trucks
Infantry squad*	Downloads incoming Class IV trucks
Infantry squad*	Uploads outgoing Class IV/V trucks
Two infantry squads*	Uncrates and inspects mines
Support PL/PSG	Provides C^2 for cross loading of Class IV/V material
Infantry squad*	Provides local security for the Class IV/V supply point
* or equivalent-sized element from a supported unit	

The most labor-intensive task at the Class IV/V supply point is the uncrating of mines. This requires dedicated manpower from the supported maneuver force and the tools needed to break shipping bands and uncrate the mines from their containers. Another important aspect of uncrating mines is tracking fuses and booster charges. As the mines are uncrated, fuses and booster charges are separated. However, the same number and type of fuses and booster charges must be task-organized with minefield packages. This requires strict supervision; mistakes can quickly lead to confusion and a waste of emplacement time.

Because of the assets involved at a Class IV/V supply point, a TF is normally capable of operating only one supply point at any given time. If the TF sector is extremely wide or deep, several supply points may be planned; however, only one can be operated at a time, based on the commander's priorities for obstacle emplacement.

Mine Dump

A mine dump is the most forward mine resupply node. It is where mines are task-organized into mine-strip packages and then inspected, prepared, and loaded onto emplacing vehicles. A mine dump is not a permanent supply point, and it is not always used. Its use depends on the method of minefield resupply, and these techniques are discussed in more detail later in this chapter. When used, one mine dump supports a single obstacle group. It is activated or deactivated upon initiation and completion of the obstacle group's emplacement. Mine-dump operations are primarily an engineer company or platoon responsibility. However, it is a good technique to augment mine-dump operations with personnel from the company/team overwatching the obstacle group being emplaced. A mine dump may be located in the vicinity of the company/team position or closer to the obstacle group.

There are three critical tasks that must be accomplished at the mine dump:

- The minefield packages are further task-organized into mine-strip packages (complete with the right number, type, and mix of fuses and boosters) as they are transported to the mine dump. For example, if a platoon is emplacing a standardized disrupt row minefield, mines are task-organized into three mine-strip packages. As the engineer platoon moves to the mine dump to resupply, each emplacing vehicle loads a designated mine-strip package.
- The mines are prepared for emplacement; but they are not fused at the mine dump. Preparation includes loosening and greasing fuse and booster wells and ensuring proper functioning of the mine.
- The mines are transloaded onto emplacing vehicles or a mine-delivery system.

Transportation of mines from the Class IV/V supply point to the mine dump is a supported TF responsibility. However, it is normally shared between the engineer company and the TF, since neither one has the haul capability to simultaneously service all active mine dumps.

Resupply Rules

The following rules govern mine resupply:

- Mines should be uncrated at the Class IV/V supply point to preserve transportation assets going forward.
- Mines are task-organized into minefield packages at the Class IV/V supply point.
- Transportation from the Class IV/V supply point to the mine dump is a shared engineer and maneuver unit responsibility when a mine dump is used.
- Mines are inspected and prepared at the last supply node (Class IV/V supply point or mine dump) before they are loaded onto an emplacing vehicle or a mine-delivery system.

- Class IV/V supply points are set up using authorized ammunition procedures and distance requirements.

Supply Location

Consider the following when selecting a location for the Class IV/V supply point or the mine dump:

- Carrying capacity. The location of key supply nodes and the type of resupply method used depends in part on the type, amount, and availability of haul assets. The carrying capacity plays a large role. In short, the more material a vehicle can carry, the more turnaround time you can afford. Table 2-8 provides the mine haul capacity for various types of vehicles.
- Traffic circuit. Vehicles must be able to enter, load, unload, and exit without interfering with the loading and unloading of other vehicles.
- Camouflage and cover. Protection from observation and thermal imaging is desired. Protection from artillery and air attack should be considered. Residue must be removed.
- Defense. The site must be organized for defense against enemy patrols and saboteurs.
- Time. Time factors for handling the obstacle material, to include all unloading, uncrating, inspecting, and loading, must be considered.
- Operators. Leaders and soldiers must be specifically allocated to operate Class IV/V supply points and mine dumps. They will probably remain there until the task is complete. The supply node may have to be collocated with or be near a source of manpower. Table 2-7, page 2-42, provides general guidance on how much manpower is required to sustain mine resupply operations.

Resupply Methods

The methods for obstacle material resupply are—

- Supply point.
- Service station.
- Tailgate.

In each method, corps or division transport delivers Class IV/V supplies forward to a designated Class IV/V supply point in each TF sector. The primary differences in the methods are how the material is delivered from the Class IV/V supply point to the obstacle location and whether or not mine dumps are established.

Supply Point

The supply-point method (Figure 2-21, page 2-46) requires that the emplacing engineer platoon return to the Class IV/V supply point each time it resupplies. This method does not require a separate mine dump. In effect, it moves the normal tasks associated with a mine dump to the supply point. Mines are prepared and inspected at the supply point as they are being loaded onto an emplacing vehicle or a mine-delivery system.

Table 2-8. Class IV/V haul capacity

Vehicle	Concertina Wire[1]	M15 AT Mine	M19 AT Mine	M21 AT Mine	M16 AP Mine	M14 AP Mine	MOPMS Mine	Volcano Mine	MICLIC Reload[2]	Hornet
HMMWV 1,124 kg, 6 cu m	2	51	34	27	55	56	15	1	NA	1
M35 2½-ton truck 2,250 kg, 12.5 cu m	4	102	69	55	111	113	30	2	2	2
M1078 2½-ton truck 2,250 kg, 13.4 cu m	4	102	69	55	111	113	30	2	2	2
M54 5-ton truck 4,500 kg, 13.6 cu m	7	204	138	109	222	227	61	5	3	5
M1083 5-ton truck 4,500 kg, 15.6 cu m	8	204	138	109	222	227	61	5	3	5
M930 5-ton dump truck (without sideboards) 4,500 kg, 3.8 cu m	2	112	64	32	168	71	23	3	2	2
M930 5-ton dump truck (with sideboards) 4,500 kg, 8.2 cu m	4	204	138	70	222	153	51	5	3	4
M1090 5-ton dump truck 4,500 kg, 3.8 cu m	2	112	64	32	168	71	23	3	2	2
HEMTT truck 9,000 kg, 15 cu m	8	408	277	128	444	317	94	10	7	8
12-ton S&T 10,800 kg, 24.5 cu m	13	489	333	208	533	514	148	12	9	13
40-ton lowboy 36,000 kg, 49.3 cu m	27	1,466	1,035	419	1,777	1,035	308	30	27	27
M548 cargo 5,400 kg, 14.9 cu m	8	244	166	125	266	272	74	6	4	6
M1077 PLS flat rack 14,900 kg, 17.6 cu m	9	440	352	164	586	293	110	11	9	9
No of mines per box	NA	1	2	4	4	90	21	240	NA	30
Weight per box (kg)	531	22	33	41	21	20	73	833	1,195	810
Size of box (cu m)	1.8	0.04	0.05	0.12	0.03	0.06	0.16	1.6	1.8	1.8

[1]The number of concertina = bundles; 1 bundle = 40 rolls
[2]Line charge + rocket

Several considerations may drive the use of supply-point resupply. First, if there are no additional haul assets to transport obstacle material forward from the Class IV/V supply point, the supply-point method may be the only viable technique. Secondly, the minefield group may be close enough to the supply point that any other method is less efficient.

- Advantages.
  - — Minimizes unloading and loading of material.
  - — Requires minimal augmentation of haul assets.

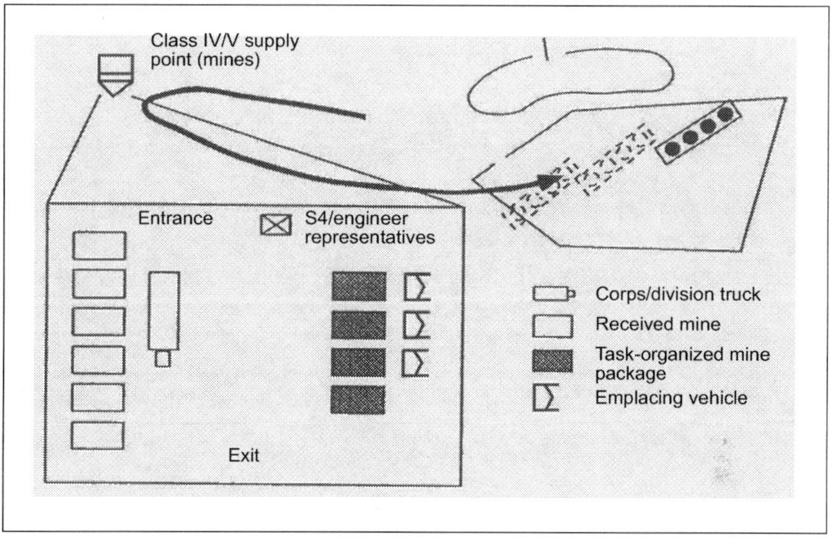

Figure 2-21. Supply-point resupply method

— Allows manpower and equipment to be massed at a single supply point.

— Streamlines C^2 of material.

- Disadvantages.

— Requires more movement of the platoon, which may take away from emplacement time.

— Requires that the platoon move in and out of the area where the minefields are being emplaced, increasing the risk of fratricide.

— May disrupt the emplacement of individual obstacles when emplacing vehicles cannot carry enough material to start and complete the obstacle. This causes emplacing vehicles to stop work, reload, and pick up where they left off.

— Requires a larger Class IV/V supply point that is capable of receiving mass quantities of obstacle material and multiple loading platoons simultaneously.

Service Station

The service-station method (Figure 2-22) centers on the activation of a mine dump forward of the Class IV/V supply point. The mines are transported to a mine dump using a combination of engineer and TF haul assets that are normally under the control of the emplacing engineer. At the mine dump, material is stockpiled and prepared by the mine-dump party. Obstacle material is further task-organized into minefield and mine-strip packages. The emplacing platoon moves to a mine dump to resupply emplacing vehicles or dispensers. Once the obstacle group is emplaced, the mine dump is deactivated or moved to support another obstacle group.

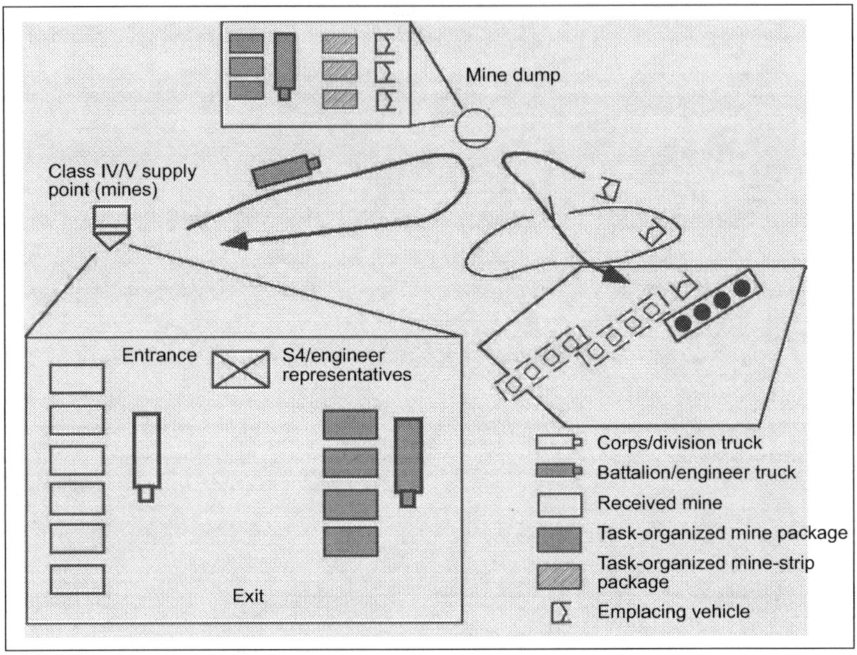

Figure 2-22. Service-station resupply method

There are several considerations for using the service-station method. First, it is used when the obstacle group is located too far from the Class IV/V supply point to allow efficient turnaround. Since this method provides for prestocking obstacle material forward, it is used when available haul assets have a relatively small capacity. This requires making frequent, short-duration resupply trips and stocking mines to keep pace with emplacement. It also streamlines emplacement since there is an opportunity to task-organize the mines into mine-strip packages, based on the emplacement method and the type of minefield. Finally, while it still requires the emplacing platoon to stop laying and resupply, it minimizes the distance and the amount of time the platoon must travel to reload. This requires that a small party be left at the minefield to help pick up where emplacement stopped.

- Advantages.

- Allows prestockage of obstacle material to keep pace with emplacement.
- Minimizes the distance and the amount of time the emplacing platoon must travel to reload.
- May provide additional manpower and security if it is located near a company/team.

- Disadvantages.
  - Requires additional loading and unloading of obstacle material.
  - May require augmentation with haul assets.
  - Disrupts emplacement by requiring the emplacing platoon to stop obstacle emplacement, move to the supply point, reload, and return to the minefield.

Tailgate

The tailgate resupply method (Figure 2-23) transports obstacle material directly from the Class IV/V supply point to the emplacing platoon at the obstacle site. Obstacle material is transported with TF and engineer haul assets. Obstacle material is transloaded by emplacing engineers to emplacing vehicles or dispensers at the obstacle site.

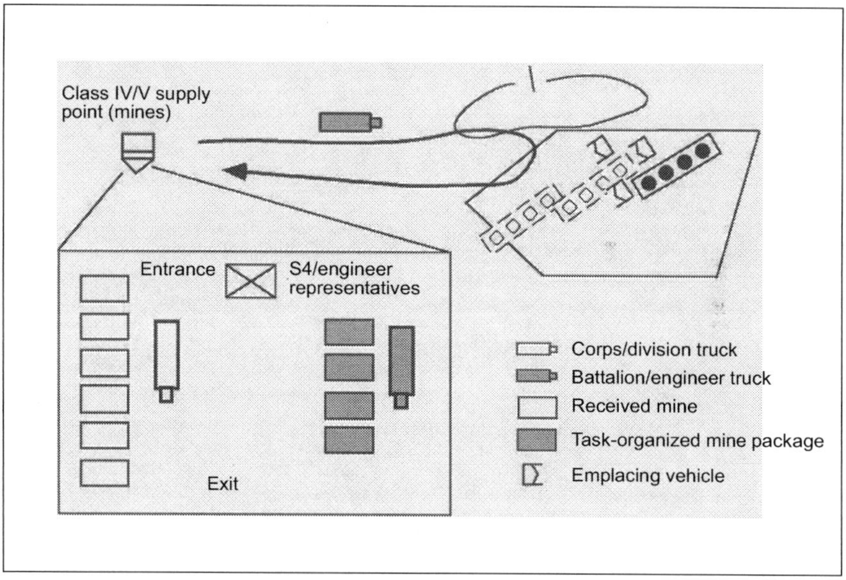

Figure 2-23. Tailgate resupply method

Two overriding considerations drive the decision to use the tailgate resupply method. First, if obstacle emplacement is being conducted during limited visibility, the tailgate method is the primary resupply method. It minimizes disruption of emplacement and the chance of fratricide as engineers move back into the work area after reloading. Secondly, tailgate resupply is the primary method used when establishing a hasty defense or when the situation is unclear and an attack can happen at anytime. Since obstacle material remains loaded until transferred to an emplacing vehicle, the tailgate method enables engineers to quickly break contact without risking a loss of obstacle material to the enemy. The tailgate resupply method is the preferred method for light forces.

- Advantages.
  - Minimizes loading and unloading of obstacle material.
  - Allows engineers to break contact rapidly in the event of enemy attack without losing obstacle material to the enemy.
  - Minimizes the movement of platoons in and out of the minefield (suitable for limited visibility).
- Disadvantages.
  - Requires augmentation by high-capacity transportation assets that are capable of offsetting the loss in turnaround time if the vehicle has to wait at the obstacle site.
  - May result in inefficient use of haul assets.
  - Complicates C^2 in linking mine transport assets with emplacing engineers since the engineers continue emplacement.
  - Causes task-organizing of minefield packages to be conducted concurrently with loading.

MINEFIELD MARKING

CRITERIA

Minefields must be marked to prevent fratricide. Marking ensures that friendly soldiers do not accidentally enter a minefield, and it is a requirement under STANAGs and Geneva Convention agreements. When emplacing minefields behind the forward line of own troops (FLOT) (in the main battle or rear area), mark the minefields on all four sides. This includes air-delivered Volcano minefields that are sited and emplaced before the enemy attacks.

Gator, RAAM, and ADAM minefields are exceptions to the rule. To preserve the system's flexibility and because of the relative inaccuracy of emplacement, these minefields are not normally marked before emplacement unless the tactical situation permits. Marking the area where mines are to be emplaced by artillery or fixed-wing aircraft is not recommended. Mines could likely be emplaced outside the marked area.

Forward of the FLOT, minefields are not generally marked before emplacement. However, commanders must make every attempt to mark these minefields as soon as the tactical situation allows. For scatterable minefields, a commander may choose to remove markings once the self-destruct (SD) time of the mines has expired; but the location of the minefield must still be recorded and forwarded to higher and adjacent units in case some of the mines did not self-destruct.

PERIMETER

Construct a perimeter fence (Figure 2-24) to mark a minefield. Start emplacing the perimeter fence before emplacing mines, preferably during site layout if the tactical situation permits. For conventional minefields, ensure that the perimeter fence is at least 15 meters outside the nearest mine or cluster. For scatterable minefields, the area inside the perimeter fence must include the safety zone. (See Chapter 3 for more details.)

Place warning signs for areas containing emplaced mines. Space the warning signs 10 to 50 meters apart, depending on the terrain. If using pickets and barbwire to mark the minefield, ensure that the wire is waist-high. If using concertina wire, use a one-roll height. Place additional strands of barbwire or rolls of concertina at the discretion of the commander.

TECHNIQUES

A commander may decide to mark individual minefields in an obstacle group or to mark the group as a whole (Figure 2-25, page 2-52). Depending on the size and the location of minefields, either technique may have the advantage of using fewer resources or labor. Normally, marking individual minefields in a fix-obstacle group requires less resources than marking the entire obstacle group. The opposite is usually true for disrupt-, turn-, and block-obstacle groups. The decision to mark individual minefields or obstacle groups should not be based solely on logistical considerations. A commander must consider the amount of tactical and/or sustainment movement required in and around the obstacle groups as well as the capability of the unit's C^2 forces.

The following advantages and disadvantages to marking individual minefields versus marking the obstacle group are provided to help commanders make the best decision.

Marking Individual Minefields

- Advantages.
  - Returning units forward of the minefields have more routes (tactical lanes or bypasses) through the obstacle group.
  - Tactical lanes need only pass through individual minefields.
  - Mine recovery is easier.
- Disadvantages.
  - Obstacle may not provide the desired effect.
  - Enemy units can more easily bypass individually marked minefields in a fix- or block-obstacle group.

Marking Obstacle Groups

- Advantages.

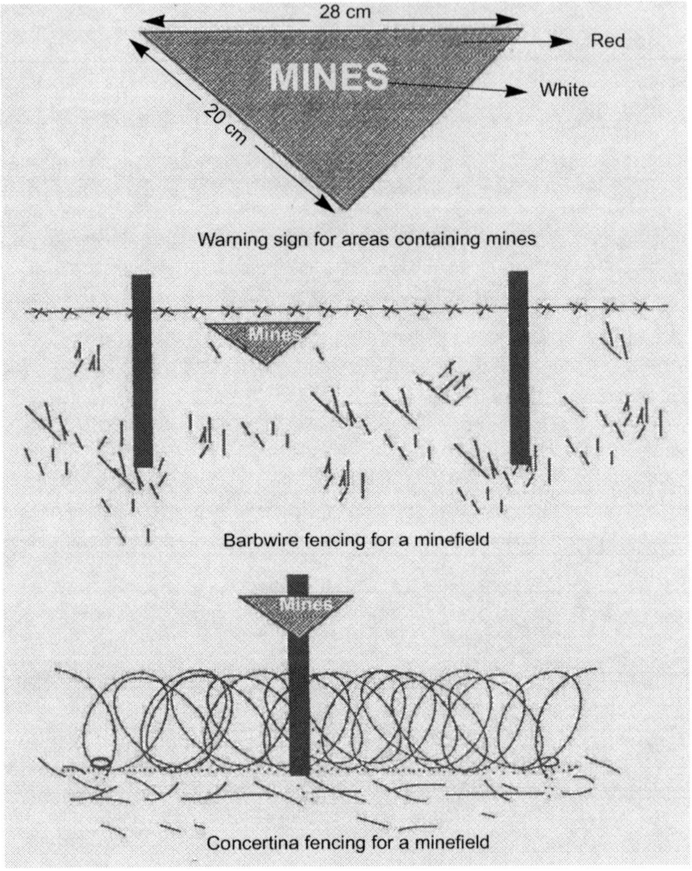

Figure 2-24. Minefield marking

- Obstacle is more likely to provide the desired effect.
- The enemy cannot easily discern individual minefields and decide when to employ breach assets in a fix-or block-effect minefield.

- Disadvantages.
  - Friendly patrols cannot clearly see if the minefield is tampered with unless they are within the perimeter fence.

Obstacle Effect	Individual Minefield	Obstacle Group
Disrupt	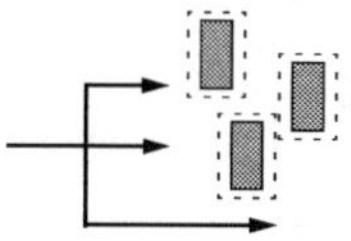	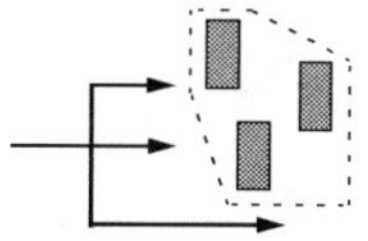
Turn	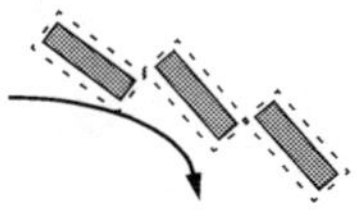	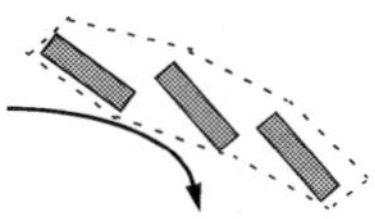
Fix	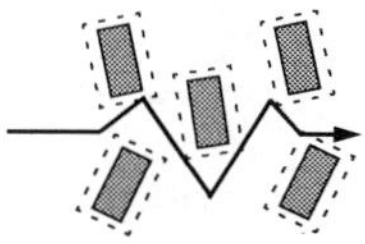	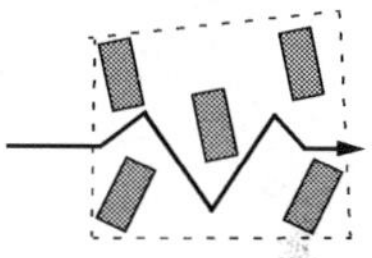
Block	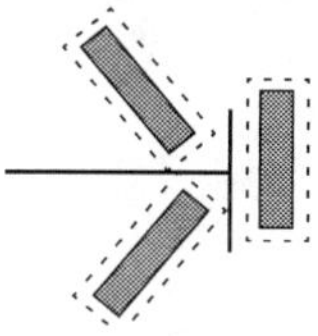	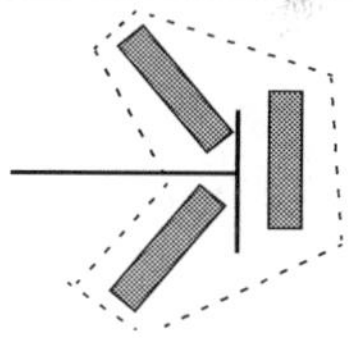

Figure 2-25. Marking of minefields and obstacle groups

— Tactical lanes need to pass through the entire obstacle group. Friendly units passing through the lanes will be slowed considerably.

MINEFIELD TURNOVER

Once an obstacle group is completed, the emplacing unit conducts minefield or obstacle turnover with the owning unit. Occasionally, the owning unit will transfer responsibility for a minefield to another unit. Minefield or obstacle turnover ensures that the commander of the owning unit is familiar with the minefield and understands his responsibilities concerning it. Turnover is conducted whether or not there are lanes/gaps to be closed. Minefield turnover is a must; the time and the location for the turnover is established during the initial siting coordination.

The engineer must verbally address the following items with the overwatching company/team:

- Intelligence.
  — Provide an update on enemy activity forward of the minefield.

- Discuss expected enemy reconnaissance efforts.
- Brief on local, friendly, and enemy situations.

- Maneuver.
  - Discuss obstacle protection against enemy dismounted patrols. Recommend that the maneuver unit conducts security/patrols to protect the minefield during limited visibility.
  - Discuss fire control measures.
- Mobility/survivability.
  - Discuss the obstacle's intended effect on enemy maneuver.
  - Discuss the minefield front and depth and walk/ride the minefield trace. Provide grid coordinates of the minefield trace.
  - Discuss minefield composition.
  - Discuss friendly minefield marking.
  - Discuss lane/gap closure, if applicable. Confirm the signal or the activity that initiates lane closure.
  - Train units on how to close lanes. This may mean training the unit on emplacing conventional mines or using the MOPMS.
- Fire support.
  - Update the company FIST on grid coordinates for the minefield trace.
  - Discuss indirect fires covering the minefield.
- CSS. Provide mines/material required to close lanes/gaps and ensure that all necessary material is available and prepared.
- C^2.
  - Transfer graphics and documentation (minefield records, demolition-target folders, or other written records).
  - Report completion of the turnover to the higher engineer and supported unit headquarters.
  - Complete an obstacle-turnover report (a sample work sheet is shown in Figure 2-26, page 2-54) and submit it to higher headquarters.
  - Forward the written minefield report and record (DA Form 1355 or 1355-1-R) to the next higher commander common to both units.

OBSTACLE-TURNOVER/TRANSFER REPORT

OBSTACLE DATA

Obstacle Number	Obstacle Type	Emplacing Authority
Obstacle Location *(8-digit UTM grid to center of mass)*	Obstacle Effect	Emplacing Unit

TURNOVER DATA

Transferring Unit	Receiving Unit	Date/Time of Transfer
Next Higher Command Common to Both		Date Obstacle Emplaced

TURNOVER CHECKLIST

☐ Intelligence.
- Enemy activity forward of the obstacle *(discuss enemy recon and breaching efforts)*.
- Friendly activity in the vicinity of the obstacle *(discuss LOGPAC, recon, counterrecon maneuver)*.

☐ Maneuver.
- Obstacle protection against enemy recon/breaching *(discuss overwatch, counterrecon patrols, sunrise sweep)*.
- Location of TRPs and indirect-fire targets *(point out location on graphics and terrain)*.
- Rearward/forward passage of lines *(discuss planned and routine activity around the obstacle)*.

☐ Mobility/survivability.
- Obstacle trace *(discuss corner grids on map and point them out on terrain)*.
- Obstacle marking *(rear side, all four sides, fence, signs, HEMMS poles, or no marking)*.
- Obstacle composition.
- Location of gaps/lanes.
- Method of closing gaps/lane *(MOPMS, ADAM, RAAM, conventional mines)*.
- Location of material to close gaps/lanes.

☐ CSS. Planned casualty evacuation and supply routes in the vicinity of the obstacle.

☐ C^2.
- DA Form 1355 or 1355-1-R, dated ________, transferred to receiving unit.
- Obstacle overlay, dated ________, transferred to receiving unit.
- Target folder, dated ________, preparing headquarters ______________, transferred to receiving unit.
- Higher headquarters (transferring, receiving, and emplacing units) notified of transfer.
- DA Form 1355 or 1355-1-R forwarded. (**NOTE: Three copies are required for transfer—one to transferring unit, one to receiving unit, and one to next higher headquarters common to both.)**

CERTIFICATION

The undersigned certify that the receiving unit has assumed full responsibility for the obstacle indicated above. The unit commander understands all the information related to the obstacle, the obstacle intent, and his responsibility to transfer or recover the obstacle upon his departure.

______________________________	______________________________
Signature of Transferring Unit Commander	Signature of Receiving Unit Commander
______________________________	______________________________
Printed Name and Rank	Printed Name and Rank

Figure 2-26. Sample obstacle-turnover work sheet

MINEFIELD INSPECTION AND MAINTENANCE

Mines left in the ground for a long time may deteriorate and malfunction for one or more of the following reasons:

- Moisture may have entered the igniter or the body of the mine and either neutralized the explosive or corroded the metal parts. Such actions may be aggravated by local factors (soil acidity or wide temperature swings).
- Frost or heat may have subjected the mine to mechanical strain and caused distortion.
- Insects or vegetation may have caused obstructions.
- Animals may have turned mines over or detonated them.

Technical inspections should only be made by experienced engineers or explosive ordnance disposal (EOD) personnel. When a minefield deteriorates below the operating level, additional mine strips/rows are added to restore its effectiveness. They are sited to the front or the rear of the existing minefield to increase its depth. New mine strips/rows are treated as separate minefields.

Technical inspections of minefields are normally done at three-month intervals. They are done more frequently during extreme weather conditions. Detailed information on mine inspections can be found in Technical Manual (TM) 9-1345-203-12.

This chapter implements STANAG 2990.

Chapter 3

Scatterable Mines and Mine Delivery Systems

SCATMINEs are laid without regard to a classical pattern. They are designed to be delivered or dispensed remotely by aircraft, artillery, missile, or a ground dispenser. All US SCATMINEs have a limited active life and self-destruct after that life has expired. The duration of the active life varies with the type of mine and the delivery system.

SCATMINE systems enable a tactical commander to emplace minefields rapidly in enemy-held territories, contaminated territories, and in most other areas where it is impossible for engineers to emplace conventional minefields. Some systems allow for rapid emplacement of minefields in friendly areas. As with all minefields and obstacles, scatterable minefields are an engineer responsibility.

Based on the tactical plan, the maneuver commander's staff engineer determines the minefield location, size, density, and emplacement and SD times. With this information and a thorough understanding of the available systems, he can then recommend the type of minefield (conventional or scatterable) to be emplaced. If a scatterable minefield is selected, he recommends the delivery system and coordinates the minefield with appropriate staff officers.

GENERAL CHARACTERISTICS

Most US SCATMINEs have similar characteristics. SCATMINEs are much smaller in size and weight than conventional mines. For example, a standard AT SCATMINE weighs approximately 1.8 kilograms and has 600 grams of explosive; an M15 conventional mine weighs 13.5 kilograms and has 10 kilograms of explosive. Arming mechanisms, arming times, and SD times of SCATMINEs differ based on the dispensing system.

ANTIPERSONNEL MINES

There are two general categories of AP SCATMINEs—wedge-shaped and cylindrical (Figure 3-1, page 3-2). Table 3-1, page 3-2, summarizes the characteristics of each AP SCATMINE.

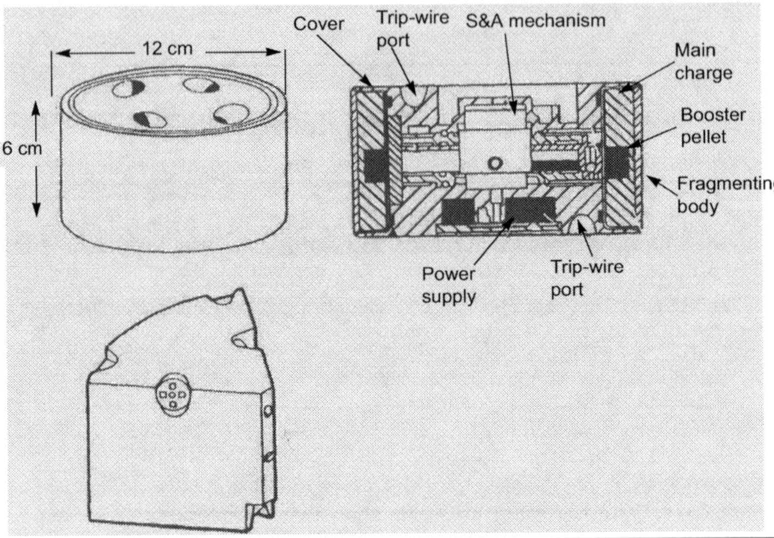

Figure 3-1. AP SCATMINEs

Table 3-1. Characteristics of AP SCATMINEs

Mine	Delivery System	DODIC	Arming Time	Fuse	Warhead	AHD	SD Time	Explosive Weight	Mine Weight	Number of Mines
M67	155-mm artillery (ADAM)	D502	Within 1 min after ground impact	Trip wire	Bounding frag	20%	4 hr	21 g Comp A5	540 g	36 per M731 projectile
M72	155-mm artillery (ADAM)	D501	Within 1 min after ground impact	Trip wire	Bounding frag	20%	48 hr	21 g Comp A5	540 g	36 per M692 projectile
BLU 92/B	USAF (Gator)	K291 K292 K293	2 min	Trip wire	Blast frag	100%	4 hr 48 hr 15 days	540 g Comp B4	1.44 kg	22 per CBU 89/B dispenser
M77	MOPMS	K022	2 min	Trip wire	Blast frag	0%	4 hr (recycle up to 3 times)	540 g Comp B4	1.44 kg	4 per M131 dispenser
Volcano	Ground/ air	K045	2 min	Trip wire	Blast frag	0%	4 hr 48 hr 15 days	540 g Comp B4	1.44 kg	1 per M87 canister

The M67 and M72 AP SCATMINEs are wedge-shaped and dispensed from an ADAM projectile, which is a special 155-millimeter artillery munition. Each mine weighs 540 grams and is 7 centimeters high.

The M74, BLU 92/B, M77, and Volcano AP SCATMINEs are all cylindrical in shape. They are 6 centimeters high and 12 centimeters in diameter. Cylindrical AP SCATMINEs kill enemy soldiers through the combined effects of blast and fragmentation. Each mine contains 540 grams of composition B4 as its main charge. The charge detonates upon actuation and shatters the mine's metal casing to produce shrapnel. Shrapnel is propelled upward and outward from the mine and produces fatal casualties to a distance of 15 meters. Each mine has eight trip wires (four on the top and four on the bottom) that deploy after ground impact up to 12 meters from the mine. Trip wires are similar in appearance to very fine thread; they are olive-drab green in color and weighted at the free end. A tension of 405 grams applied to one trip wire is enough to create a break in the electrical circuit and cause the mine to detonate.

Antitank Mines

All AT SCATMINEs (Figure 3-2) have similar functional characteristics. They are cylindrical in shape, weigh approximately 1.8 kilograms, contain 585 grams of cyclonite (RDX) explosive as the main charge, and have a magnetically induced fuse. The characteristics of each AT SCATMINE are summarized in Table 3-2, page 3-4.

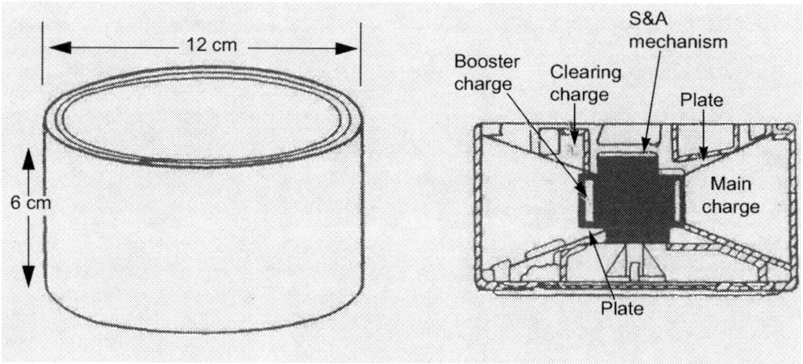

Figure 3-2. AT SCATMINE

Table 3-2. Characteristics of AT SCATMINEs

Mine	Delivery System	DODIC	Arming Time	Fuse	Warhead	AHD	SD Time	Explosive Weight	Mine Weight	Number of Mines
M73	155-mm artillery (RAAM)	D503	Within 1 min after ground impact	Magnetic	M-S plate	20%	48 hr	585 g RDX	1.7 kg	9 per M718 projectile
M70	155-mm artillery (RAAM)	D509	Within 1 min after ground impact	Magnetic	M-S plate	20%	4 hr	585 g RDX	1.7 kg	9 per M741 projectile
BLU 91/B	USAF (Gator)	K291 K292 K293	2 min	Magnetic	M-S plate	NA	4 hr 48 hr 15 days	585 g RDX	1.7 kg	72 per CBU 89/B dispenser
M76	MOPMS	K022	2 min	Magnetic	M-S plate	NA	4 hr (recycle up to 3 times)	585 g RDX	1.7 kg	17 per M131 dispenser
Volcano	Ground/ air	K045	2 min 30 sec	Magnetic	M-S plate	NA	4 hr 48 hr 15 days	585 g RDX	1.7 kg	5 per M87 canister; 6 per M87A1 canister

AT SCATMINEs are designed to produce a K-Kill instead of an M-Kill. They produce a kill by using an SFF warhead (created from an M-S plate). The warhead penetrates the vehicle's belly armor, and spalling metal from the vehicle (caused by the mine blast) kills occupants instantly. Even though the crew is killed, the drive train may be undamaged and the vehicle may continue to move. On enemy tanks with autoloaders, the detonation of rounds in the belly-mounted ammunition carousel is very likely. The mine may not achieve a kill when the track of an armored vehicle runs directly over it.

The magnetic fuse is designed to detonate as the magnetic field changes over the mine. The warhead is bidirectional, meaning that it can fire from the top or the bottom. AHDs are built into 20 percent of M70, M73, and M75 mines. Although Volcano, M76, and BLU 91/B mines do not have AHDs, they may detonate when moved, because the mine may sense a significant change from its original orientation.

Due to their small size, the reduced explosive, and the possibility of landing with an improper orientation (on their side or at an angle), AT SCATMINEs have less chance of destroying a vehicle than a conventional full-width AT mine. An armored vehicle will not always be destroyed after an encounter with an AT SCATMINE. Further, the effectiveness of SCATMINEs in water obstacles is reduced even more, because 5 centimeters of water prevents the formation of the M-S slug. Although the blast wave is accentuated by underwater placement (attacking hatches and covers), mining of banks and approaches is recommended instead.

CAPABILITIES

Faster Response

SCATMINEs can be emplaced more rapidly than conventional mines, so they provide a commander with greater flexibility and more time to react to changes in situations. The commander can use SCATMINEs to maintain or regain the initiative by acting faster than the enemy. Using SCATMINEs also helps preserve scarce mine resources.

Remote Placement

All SCATMINEs are remotely emplaced. This enhances battlefield agility and allows the maneuver commander to emplace mines rapidly to best exploit enemy weaknesses. SCATMINEs can be used as situational obstacles or to attack enemy formations directly through disrupt, fix, turn, and block obstacles. Modern fusing, sensing, and AHDs allow SCATMINEs to better defeat enemy attempts to reduce the minefield.

Increased Tactical Flexibility

Upon expiration of the SD time, the minefield is cleared and the commander can move through an area that was previously denied to enemy or friendly forces. In many cases, the SD period may be set at only a few hours. This feature allows for effective counterattacks to the enemy's flank and rear areas.

Efficiency

SCATMINEs can be emplaced by a variety of delivery methods. They can be deployed by fixed-wing aircraft, helicopters, artillery, manpack, or ground vehicles. They satisfy the high mobility requirements of modern warfare. Manpower, equipment, and tonnage are reduced for their emplacement.

Increased Lethality

AT SCATMINEs utilize an SFF that is created from two M-S plate charges to produce a full-width kill. In simple terms, a metal plate is formed into a high-velocity slug that punches a hole in the belly of a tank. The effect produces an M-Kill against the vehicle's engine, track, or drive train; or it produces a K-Kill when the on-board ammunition is set off and the crew is killed or incapacitated or the vehicle's weapon system is destroyed. AT SCATMINEs are designed to destroy any tank in the world. In order to form an SFF, the mine requires a certain standoff between the vehicle and the target. Mines must also be nearly perpendicular to the target (laying on either side). The M-S plate is actually two plates—one facing the top of the mine and one facing the bottom. This ensures that it will successfully attack the target while lying on either side.

AP SCATMINEs are actuated by a trip wire and utilize a blast-fragmentation warhead.

LIMITATIONS

Extensive Coordination

Because SCATMINEs are a very dynamic weapon system, great care must be taken to ensure that proper coordination is made with higher, adjacent, and subordinate units. To prevent friendly casualties, all affected units must be notified of the location and the duration of scatterable minefields. Recording and reporting procedures for SCATMINEs are discussed in detail in Chapter 8, and they were specifically designed to minimize friendly casualties.

PROLIFERATION OF TARGETS

SCATMINEs may be regarded by some commanders as easy solutions to tactical problems. Target requests must be carefully evaluated, and a priority system must be established because indiscriminate use of weapon systems will result in rapid depletion of a unit's basic load. Controlled supply rates (CSRs) will probably be a constraint in all theaters.

VISIBILITY

SCATMINEs are highly effective, especially when fires and obscurants strain the enemy's C^2. SCATMINEs lay on the surface of the ground, but they are relatively small and have natural coloring.

ACCURACY

SCATMINEs cannot be laid with the same accuracy as conventional mines. Remotely delivered SCATMINE systems are as accurate as conventional artillery-delivered or tactical aircraft-delivered munitions.

ORIENTATION

Between 5 and 15 percent of SCATMINEs will come to rest on their edges; mines with spring fingers will be in the lower percentile. If there is mud or snow more than 10 centimeters deep, the number will be in the higher percentile. When employing ADAMs or RAAMs in more than 10 centimeters of snow or mud, high-angle fire should be used and the number of mines increased. AP mines may be less effective in snow, because the deployment of trip wires is hindered. Melting of the snow may also cause the mines to change positions and activate AHDs.

LIFE CYCLE

All SCATMINEs have a similar life cycle, although specific times vary based on the SD time and the dispensing system.

For safety reasons, SCATMINEs must receive two arming signals at launch. One signal is usually physical (spin, acceleration, or unstacking), and the other is electronic. This same electronic signal activates the mine's SD time.

Mines start their safe-separation countdown (arming time) when they receive arming signals. This allows the mines to come to rest after dispensing and allows the mine dispenser to exit the area safely. Table 3-1, page 3-2, and Table 3-2, page 3-4, show arming times for individual SCATMINEs.

Mines are armed after the arming time expires. The first step in arming is a self-test to ensure proper circuitry. Approximately 0.5 percent of mines fail the self-test and self-destruct immediately.

After the self-test, mines remain active until their SD time expires or until they are encountered. Mines actually self-destruct at 80 to 100 percent of their SD time. The time period from when the mines begin to self-destruct and when they finish is called the SD window (Table 3-3). No mines should remain active after the SD time has been reached. Two to five percent of US SCATMINEs fail to self-destruct as intended. Any mines found after the SD time must be treated as unexploded ordnance (UXO). For example, mines with a 4-hour SD time will actually start self-destructing at 3 hours and 12 minutes. When the 4-hour SD time is reached, no unexploded mines should exist.

Table 3-3. SD windows

SD Time	SD Window Begins
4 hours	3 hours 12 minutes
48 hours	38 hours 24 minutes
5 days	4 days
15 days	12 days

LETHALITY AND DENSITY

LETHALITY AND TACTICAL-OBSTACLE EFFECT

Scatterable minefields are employed to reduce the enemy's ability to maneuver, mass, and reinforce against friendly forces. They increase the enemy's vulnerability to fires by producing specific obstacle effects (disrupt, fix, turn, and block) on the enemy's maneuver. To achieve this aim, individual minefields must be emplaced with varying degrees of lethality. During emplacement, lethality is varied primarily by changing the minefield density. Therefore, there is a direct correlation between the obstacle effect and the minefield density. In order to achieve the tactical-obstacle effect, use the following guidance when selecting minefield density:

- Disrupt.
  - Low density.
  - Probability of encounter: 40 to 50 percent.
  - Linear density: 0.4 to 0.5 mine per meter.
- Fix.
  - Medium density.
  - Probability of encounter: 50 to 60 percent.
  - Linear density: 0.5 to 0.6 mine per meter.
- Turn.
  - High density.
  - Probability of encounter: 75 to 85 percent.

- Linear density: 0.9 to 1.1 mines per meter.

- Block.
  - High density.
  - Probability of encounter: 85+ percent.
  - Linear density: More than 1.1 mines per meter.

DENSITY

Density is normally expressed as linear or area. For conventional mines, linear density is normally used and is expressed in the average number of mines per meter of minefield front. For SCATMINE systems, area density is normally used and is expressed as the average number of mines per square meter. Since SCATMINE systems normally employ a preset combination of AT and AP mines, the area density includes both. For example, a scatterable minefield with an area density of 0.006 mine per square meter may have an AT density of 0.004 AT mine per square meter and an AP density of 0.002 AP mine per square meter. Due to the varying dimensions of scatterable minefields that can be created by the different types of employment devices, the exact density of a scatterable minefield cannot be determined. However, an estimate of the average density can be determined by using the following formulas:

- Linear density equals the number of mines divided by the minefield front.

$$\frac{\textit{number of mines}}{\textit{minefield front}} = \textit{mines per meter}$$

- Area density equals the number of mines divided by the minefield area.

$$\frac{\textit{number of mines}}{\textit{front} \times \textit{depth}} = \textit{mines per square meter}$$

- Area density can be converted to linear density by multiplying the area density by the minefield depth. **(NOTE: Converting area density to linear density is not always accurate due to the space between minefield strips.)**

$$\textit{area density} \times \textit{minefield depth} = \textit{linear density}$$

EXAMPLE: A 650- by 200-meter Gator minefield contains 564 mines (432 AT and 132 AP).

- Area density: 564 ÷ (200 x 650) = 0.004 mine per square meter.
  - AT area density: 432 ÷ (200 x 650) = 0.003 mine per square meter.

— AP area density: 132 ÷ (200 x 650) = 0.001 mine per square meter.

- Linear density: 564 ÷ 650 = 0.87 mine per meter.

  — AT linear density: 432 ÷ 650 = 0.67 mine per meter.

  — AP linear density: 132 ÷ 650 = 0.2 mine per meter.

COMMAND AND CONTROL

Due to the delivery means, C^2 of SCATMINEs is more complex than conventional mines. SCATMINEs are very dynamic weapon systems because they can be rapidly emplaced and then cleared via their SD capability. Also, the physical boundary of a scatterable minefield is not clearly defined. These characteristics require impeccable communications and coordination to ensure that all friendly units know where mines are located, when they will be effective, and when they will self-destruct.

AUTHORITY

The corps commander has emplacement authority for all scatterable minefields within the corps AO. He may delegate this authority to lower echelons according to the guidelines contained in Table 3-4.

Table 3-4. Emplacement authority

System Characteristics	Emplacement Authority
Ground- or artillery-delivered, with SD time greater than 48 hours (long duration)	The corps commander may delegate emplacement authority to division level, who may further delegate it to brigade level.
Ground- or artillery-delivered, with SD time of 48 hours or less (short duration)	The corps commander may delegate emplacement authority to division level, who may further delegate it to brigade level, who may further delegate it to TF level.
Aircraft-delivered (Gator), regardless of SD time	Emplacement authority is normally at corps, theater, or army command level, depending on who has air-tasking authority.
Helicopter-delivered (Volcano), regardless of SD time	Emplacement authority is normally delegated no lower than the commander who has command authority over the emplacing aircraft.
MOPMS, when used strictly for a protective minefield	Emplacement authority is usually granted to the company, team, or base commander. Commanders at higher levels restrict MOPMS use only as necessary to support their operations.

Based on how the commander wants to shape the battlefield, he must specifically delegate or withhold the authority to employ SCATMINE systems. The commander's guidance concerning SCATMINEs is found in the unit's OPORD/operation plan (OPLAN). Additional information is included in their engineer and fire-support annexes, if used.

Due to the complete control a commander has over the MOPMS, emplacement authority guidelines do not apply to the MOPMS. It is impractical for the corps or brigade commander to authorize every MOPMS protective minefield. Therefore, authority to emplace MOPMS minefields is specifically delegated. In general, units can emplace MOPMS protective minefields as required for their own self-defense and report them as they do any protective obstacle. Any MOPMS minefield used as part of an obstacle plan must be reported as a scatterable minefield.

COORDINATION

Basic responsibilities of key commands, staff elements, and units are outlined in Table 3-5. The fire-support coordinator (FSCOORD) is involved in planning artillery-delivered (ADAM and RAAM) SCATMINEs, and the air liaison officer (ALO) is involved in planning air-delivered (Gator and Volcano) SCATMINEs. The engineer has primary responsibility for planning and employing SCATMINE systems. It is vital that coordination be conducted with all units and subunits that will be effected by the employment of SCATMINEs. A scatterable minefield warning (SCATMINWARN) will be sent to all effected units before the emplacement of the minefield (see Chapter 8 for more details).

Table 3-5. Coordination responsibilities

Element	Responsibilities
G3/S3 with Engineer FSCOORD/ALO	Plan and coordinate the minefield location, size, composition, density, SD time, safety zone, and emplacement time
	Designate and brief the emplacing unit
	Incorporate the minefield and the safety zone into the obstacle plan
	Receive and forward the scatterable minefield report and record
	Disseminate information concerning the minefield in the SCATMINWARN to adjacent and subordinate units prior to laying
	Post operation maps with the minefield location, safety zone, and DTG of the SD time; and disseminate the SCATMINWARN 1 hour prior to initiation of the SD sequence
Emplacing Unit	Calculate the logistical requirements
	Calculate the safety zone
	Emplace the minefield
	Report the amount of ammunition expended
	Prepare and forward the scatterable minefield report and record to the authorizing commander via appropriate channels
Maneuver Units	Be aware of the calculated safety-zone boundary and advise subunits of its location

EMPLOYMENT AND EMPLACEMENT

Employment considerations and emplacement techniques and procedures differ for each type of SCATMINE and delivery system. This section discusses the characteristics of each delivery system and provides tactical considerations for the employment of each system on the battlefield. Techniques and procedures for emplacing minefields intended to disrupt, fix, turn, and block are also discussed; and they build on tactical-obstacle design principles discussed in Chapter 2.

Area-Denial Artillery Munitions and Remote Antiarmor Mines

ADAMs and RAAMs are delivered by a 155-millimeter howitzer (Figure 3-3). There are no special modifications or adaptations necessary for the firing system. Mines are contained within a projectile and are dispensed while the projectile is in the air. The effective range for the M109 howitzer is 17,500 meters, and for the M198 howitzer, 17,740 meters.

Figure 3-3. Emplacement of ADAMs and RAAMs

The M692 (long-duration) and the M731 (short-duration) ADAM projectiles deliver AP mines with different SD times. Each ADAM round contains 36 mines. The M731/M731A1 round contains M72 AP mines with 4-hour SD times; the M692/M692A1 round contains M67 AP mines with 48-hour SD times. SD times are preset during the manufacturing process and cannot be changed.

The wedge-shaped ADAM is a bounding-fragmentation mine that deploys up to seven tension-activated trip wires 6 meters away from the mine. After ground impact, trip wires are released and the mine is fully armed. The ADAM contains a metal-jacketed sphere that is filled with 21 grams of composition A5 as its main charge. A liquid-explosive propelling charge positions itself at the bottom of the sphere after impact with the ground. When the mine is jarred or tilted, or when one of its trip wires receives a tension of at least 405 grams, the sphere propels upward 0.6 to 2.4 meters and detonates. The lethal casualty radius is between 6 and 10 meters.

The M741 (short-duration) and the M718 (long-duration) RAAMs are artillery-delivered AT mines. Each RAAM round contains nine mines. The M741/M741A1 round contains M70 AT mines with 4-hour SD times; the M718/M718A1 round contains M73 AT mines with 48-hour SD times. The SD times are preset during the manufacturing process and cannot be changed. The RAAM mine utilizes an SFF warhead, has a magnetic-influence fuse, weighs 1.7 kilograms, and has a small (12 centimeters in diameter by 6 centimeters in height) cylindrical shape.

The new model ADAM and RAAM mines (designated by an A1 suffix) have a 45-second arming time; the older models have a 2-minute arming time. The new model RAAM has a built-in feature that defeats magnetic, signature-duplicating breaching devices.

Employment

The ADAM and RAAM systems were designed to provide a flexible, rapid-response mining capability. These systems provide the maneuver commander with the capability to emplace mines directly on top of, in front of, or behind enemy forces. This is one of their greatest advantages. Their responsiveness allows the mission to be executed quickly and allows the commander to effectively influence a rapidly changing battlefield. They also allow the commander to emplace minefields while maintaining maximum standoff from the target. In short, their emplacement does not require committing any force (ground or air) forward. ADAM and RAAM systems may be used for the following purposes:

- Defense.
  - Develop targets for long-range AT weapons.
  - Close gaps and lanes in other obstacles.
  - Delay or disrupt attacking forces.
  - Deny the enemy unrestricted use of selected areas.
  - Disrupt movement and commitment of second-echelon forces.
  - Disrupt and harass enemy C^2, logistics (excluding medical), and staging areas.
  - Reinforce existing obstacles.
  - Disrupt or delay river crossings.
- Offense.
  - Supplement flank reconnaissance and security forces to protect flanks along AAs.
  - Suppress and disrupt enemy security elements once contact has been made.
  - Hinder the withdrawal of enemy forces.

— Hinder the ability of the enemy to reinforce the objective area.

The time and the number of rounds required to install effective ADAMs and RAAMs limit their use. Their range is limited to 17,500 or 17,740 meters, depending on which howitzer (M109 or M198, respectively) is used. Many of the deep-interdiction missions that support force-projection doctrine require a greater distance. Due to the large footprint created when the minefield is fired, many mines will scatter outside the planned minefield area. It is therefore necessary to plot the safety zone in order to prevent fratricide. The fire-support element (FSE) is responsible for plotting the safety zone, and the staff engineer should be familiar with the process and the expected results. The staff engineer ensures that the safety zone is plotted on the tactical command post (TCP)/TOC operation overlay.

Emplacement

ADAM and RAAM mining missions are requested through normal artillery-support channels. Although the actual numbers vary based on the unit and the mission, a representative basic load for an artillery battalion consists of approximately 32 ADAM and 24 RAAM (short SD time) rounds per artillery piece. **NOTE: The rounds with long SD times are normally used for preplanned targets and are issued from an ammunition supply point (ASP) on a mission-by-mission basis.**

Once the proper authorization has been received to employ the mines, requests for ADAMs and RAAMs are processed in the same way as other requests for fire support, including targets of opportunity. Allocate enough time for processing the request and completing firing procedures. This ensures that the enemy has not moved out of the target area before execution. (FM 90-7 contains more information on this process.) The use of ADAMs and RAAMs for preplanned fires requires close coordination among the Assistant Chief of Staff, G3 (Operations and Plans) (G3)/Operations and Training Officer (US Army) (S3), the staff engineer, and FSE sections. Coordination should also be made with the S2 and the S3 during the development of the decision support template (DST) to identify the proper named areas of interest (NAIs), target areas of interest (TAIs), trigger points, and decision points.

There are two critical aspects when emplacing ADAM and RAAM minefields:

- Designing the minefield to achieve the required effect.
- Ensuring the technical correctness of resourcing and delivering the minefield.

The following discussion provides general guidance for designing the minefield to achieve the desired effect and for determining the safety zone to assess the impact on maneuver. Appendix H of FM 6-20-40 serves as the primary source for technically resourcing and delivering artillery-delivered minefields.

ADAM and RAAM minefields can be emplaced to achieve disrupt, fix, turn, and block effects based on the principles outlined in Chapter 2. The engineer is responsible for deciding the required location, the density, the size, the composition, and the duration of the minefield based on the tactical-obstacle plan and the obstacle restrictions of the higher unit. The engineer provides this information to the FSE. Table 3-6 provides guidance on the minefield density and size necessary to achieve the desired obstacle effect.

Table 3-6. RAAM and ADAM minefield density and size

Obstacle Effect	RAAM		ADAM		Width (meters)	Depth (meters)
	Area[1]	Linear[2]	Area[1]	Linear[2]		
Disrupt	0.001	0.2	0.0005	0.1	200	200
Turn	0.002	0.8	0.001	0.4	400	400
Fix	0.002	0.4	0.0005	0.1	200	200
Block	0.004	0.6	0.002	0.8	400	400

[1]Area density = mines per square meter
[2]Linear density = mines per meter

The FSE determines all the technical aspects for delivering the minefield, such as the number of rounds required per aim point, the number of aim points required, the size of the safety zone, and the time required to emplace mines. There is a wide variety of factors involved in determining the number of rounds, the size of the safety zone, and the emplacement time. These factors are the range-to-target time, the battery-to-minefield angle, the high- or low-angle trajectory, and the method of firing (observer adjust or meteorological data plus velocity error [Met+VE] transfer). The FSE must tell the engineer whether the minefield mission is feasible. Feasibility is based on the number of rounds available, the scheme of indirect fires, and the availability of artillery tubes.

The engineer is primarily concerned with two technical aspects of delivery provided by the FSE—the safety zone and the emplacement time. The engineer uses the safety zone and the minefield duration to assess the impact of the minefield on the mobility requirements of the scheme of maneuver. The engineer depicts the safety zone on the obstacle overlay. He also uses the safety zone to identify requirements for minefield marking if the unit leaves or turns over the area before the SD time. The engineer and the FSE use the emplacement time to synchronize the delivery of the minefield with the tactical plan.

GATOR

The Gator (Figure 3-4) has a longer range than any other SCATMINE system. It provides a means to rapidly emplace minefields anywhere that can be reached by tactical aircraft. The Gator is produced in two versions—the United States Air Force (USAF) CBU-89/B system that contains 94 mines (72 AT and 22 AP) per dispenser and the United States Navy (USN) CBU-78/B system that contains 60 mines (45 AT and 15 AP) per dispenser.

The mines used with the Gator are the BLU-91/B AT mine and the BLU-92/B AP mine. They are similar to the mines used with the Volcano system. The mines are capable of three field-selectable SD times (4 hours, 48 hours, and 15 days). Both types of mines are encased in a plastic, square-shaped protective

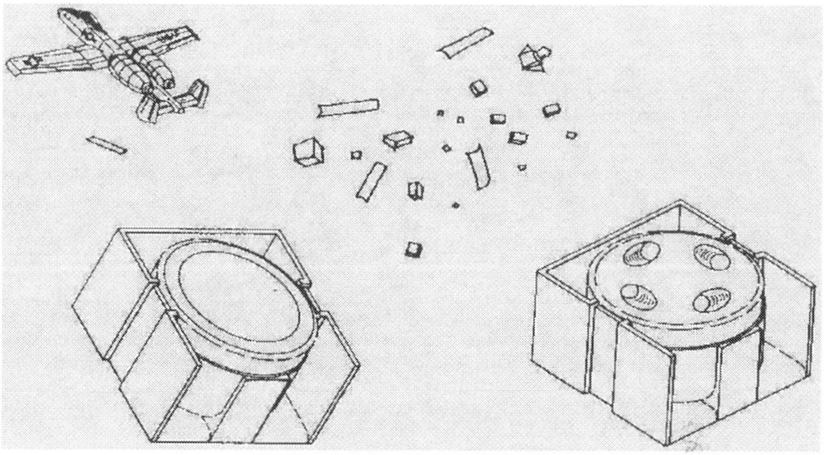

Figure 3-4. Gator SCATMINE system

casing that is designed to aid dispersion and lessen ground impact upon delivery.

The mines are contained inside tactical munition dispensers (TMDs) that are attached under the wings of high-performance, fixed-wing aircraft. The TMD is a USAF dispenser that was designed for common use with cluster munitions. The Gator is compatible with the USAF A-10, F-4, F-15, F-16, B-1, and B-52 aircraft and with the USN A-6, A-7, F-4, FA-18, and AV-8B aircraft.

The TMD is released in the air and allowed to fall free. Four linear charges along the edge of the TMD cut the outer casing, and the mines are aerodynamically dispersed. The maximum delivery speed is 800 knots at altitudes of 75 to 1,500 meters. The area of minefield coverage depends on the number of munitions carried, the aircraft speed and altitude, and the altitude where the fuse functions and opens the dispenser. The average area covered is approximately 200 by 650 meters.

Employment

Gator missions are primarily used at long range to disrupt, fix, turn, or block enemy troop movement beyond the fire-support coordination line (FSCL). For use in interdiction missions beyond the FSCL, submit requests for Gator missions as early as possible to nominate targets for the theater air-tasking order. Gator munitions are well-suited for placing minefields on specific concentrations of forces (artillery, logistic, and C^2) that are out of range of conventional artillery.

While the Gator can provide close combat support, deep-interdiction mining is expected to be its primary mission. Gator minefields are normally employed in conjunction with other deep indirect-fire attacks, such as area of interest (AI), battlefield air interdiction (BAI), or joint air-attack team (JAAT). However, a Gator minefield may be employed in conjunction with close air support (CAS) and covered by close indirect- and direct-fire systems. Typical mining missions include—

- Isolating objectives.
- Countering ADA/artillery fires.
- Denying terrain.
- Disrupting and disorganizing support activities.
- Inflicting personnel and equipment losses.

The extended range of the Gator system, together with its speed and responsiveness, makes it one of the most influential weapon systems on the deep battlefield. The primary limitations of the Gator are the availability of high-performance aircraft to emplace the mines and the system's relative ineffectiveness on units in column. During any conflict, aircraft will be in high demand and will not always be immediately available for a Gator mission when required. Communications may also pose a problem because mission execution is a joint US Army-USAF operation.

The Gator is well suited to support contingency operations and amphibious landing operations in an immature theater when there is no danger to friendly forces or host-nation assets. Gator minefields are one of the light-force commander's few durable, long-range antiarmor weapons.

Emplacement

As an aircraft-delivered munition, the Gator is a corps asset. The Gator is a BAI mission and is controlled by the tactical air control center (TACC). Missions should be requested as early as possible (no later than 36 hours in advance) through fire-support channels to the corps FSE. As a mine system, Gator missions must be approved by corps. The corps FSE passes the mission to the theater or army air headquarters to be included on the theater air-tasking order for execution. In support of BAI or CAS, Gator sorties may be allocated down to battalion level, with final control exercised by the battalion ALO. Immediate Gator missions can also be requested directly from the maneuver unit's TACC. The same records and reports applicable to other SCATMINE systems are used with the Gator mine system. Close cooperation and coordination among the G3/S3, the staff engineer, and the ALO are required for planning and executing Gator missions.

As with artillery-delivered minefields, the engineer is primarily responsible for identifying the minefield location, size, duration, and density. Minefield density is varied by changing the orientation of the minefield with respect to the target AA. Figure 3-5 illustrates how minefield orientation is changed to achieve a fix or block effect. Normally, Gator is employed as a fix obstacle with a front of 650 meters. Emplacing a fix-obstacle group along a battalion AA (1,500 meters) requires two Gator sorties, each delivering one minefield. Each Gator minefield would have a front of 650 meters and a depth of 200 meters.

The minefields would be delivered at different locations so that the group covers the entire AA and affects the entire enemy battalion.

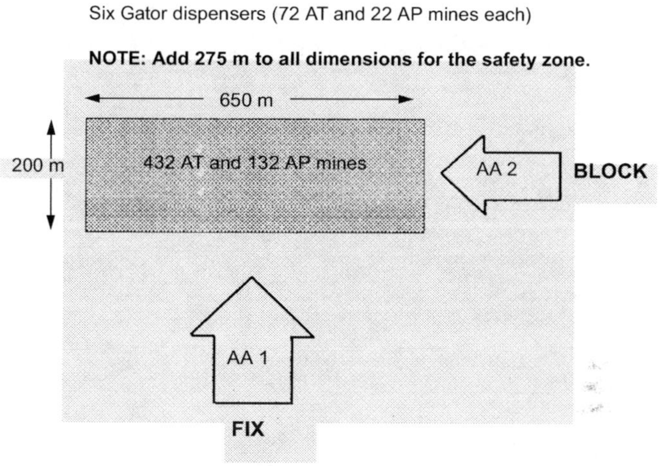

Figure 3-5. Gator minefield

VOLCANO

The Volcano multiple-delivery mine system (Figure 3-6, page 3-18) can be dispensed from the air or on the ground. It can be mounted on any 5-ton truck, an M548 tracked cargo carrier, a heavy expanded mobility tactical truck (HEMTT), a palletized load system (PLS) flat rack, or a UH-60A Blackhawk helicopter. The Volcano uses modified Gator mines and consists of four components (Figure 3-7, page 3-18)—the mine canister, the dispenser, the dispenser control unit (DCU), and the mounting hardware (aircraft also require a jettison kit). The Volcano uses M87 and M87A1 mine canisters. The M87 mine canister is prepackaged with five AT mines, one AP mine, and a propulsion device inside a tube housing. The M87A1 mine canister is prepackaged with six AT mines and a propulsion device. The mixture of mines is fixed and cannot be altered. Mines are electrically connected with a web that functions as a lateral dispersion device as the mines exit the canister. Spring fingers mounted on each mine prevent it from coming to rest on its edge. All canisters are capable of dispensing mines with 4-hour, 48-hour, and 15-day SD times. The SD times are field-selectable prior to dispensing and do not require a change or modification to the mine canister. The arming time is 2 minutes 15 seconds for AT and AP mines. The reload time (not including movement time to the reload site) for an experienced four-man crew is approximately 20 minutes.

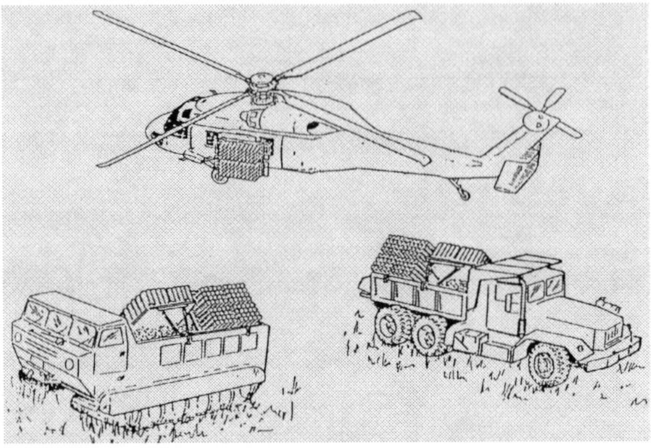

Figure 3-6. Volcano mine system

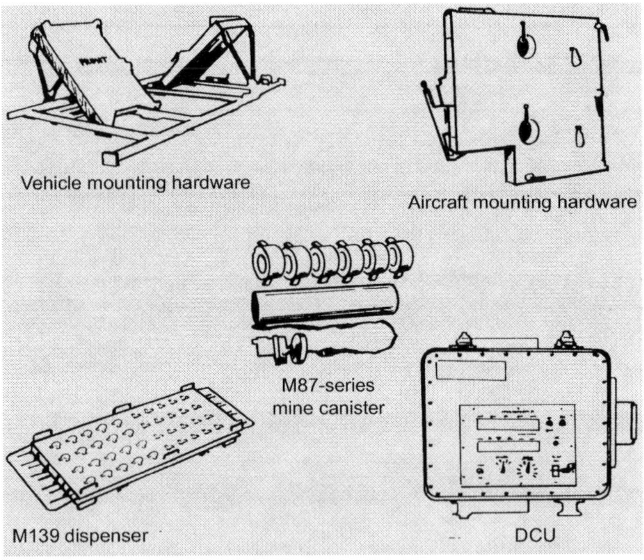

Figure 3-7. Volcano components

The dispenser consists of an electronic DCU and four launcher racks. Four racks can be mounted on a vehicle, and each rack can hold 40 M87-series mine canisters. The racks provide the structural strength and the mechanical support required for launch and provide the electrical interface between the mine canisters and the DCU. Mounting hardware secures the racks to the vehicle or the aircraft. Mounting hardware for the Blackhawk includes a jettison subassembly to propel the Volcano racks and canisters away from the aircraft in the event of an emergency.

The operator uses the DCU to control the dispensing operation electrically from within the carrier vehicle. The DCU provides controls for the arming sequence and the delivery speed and sets mine SD times. The DCU allows the operator to start and stop mine dispensing at anytime. A counter on the DCU indicates the number of remaining loaded canisters on each side of the carrier.

Mines are dispensed from their canisters by an explosive propelling charge. For ground vehicles, the mines are dispensed 25 to 60 meters from the vehicle at ground speeds of 8 to 90 kph. The average time to emplace one ground Volcano load (160 canisters) is 10 minutes.

Employment

The primary mission of the Volcano is to provide US forces with the capability to emplace large minefields rapidly under varied conditions. The Volcano can be rapidly attached to air or ground vehicles. It is used to emplace tactical minefields; reinforce existing obstacles; close lanes, gaps, and defiles; protect flanks; and deny probable enemy air-defense sites. Volcano minefields are ideal for providing flank protection of advancing forces and for operating in concert with air and ground cavalry units on flank guard or screen missions.

The air Volcano is the fastest method for emplacing large tactical minefields. When employed by combat aviation elements in support of maneuver units, close coordination between aviation and ground units assures that Volcano-dispensed mines are emplaced accurately and quickly. Although mine placement is not as precise as it is with ground systems, air Volcano minefields can be placed accurately enough to avoid the danger inherent in minefields delivered by artillery or jet aircraft. Air Volcano minefields can be emplaced in friendly and enemy territory. They should not be planned in areas of enemy observation and fire because the helicopter is extremely vulnerable while flying at the steady altitude, the speed, and the path required to emplace the minefield. The air Volcano is the best form of an obstacle reserve because a minefield can be emplaced in minutes.

The ground Volcano is designed to emplace large minefields in depth. It is normally employed by combat engineer units. These mounted dispensers are primarily used to emplace tactical minefields oriented on enemy forces in support of maneuver operations and friendly AT fires. The system is vulnerable to direct and indirect fires, so it must be protected when close to the FLOT. It is ideal for use as an obstacle reserve, employed when the enemy reaches a decision point that indicates future movement. Obstacles can then be emplaced in depth on the avenues the enemy is using, leaving other avenues open for friendly movement.

Emplacement

The principles and procedures of Volcano emplacement are significantly different for air- and ground-delivery systems. This section outlines the use of the ground Volcano system to emplace disrupt, fix, turn, and block minefields. The air Volcano system is discussed in detail in Appendix D. Both air and ground Volcano systems are capable of emplacing nonstandard minefields. However, the emplacement norms below streamline identifying resource requirements and conducting emplacement drills.

Air and ground Volcano systems emplace a minefield with an average AT linear density of 0.72 mine per meter and an AP linear density of 0.14 mine per meter. These densities may vary slightly since some mines will fail the arming sequence and self-destruct 2 to 4 minutes after dispensing. Additionally, some mines may not orient correctly, will not deliver their full mine effect, and will not produce a K-Kill. The probability of failing the arming sequence and misorienting is relatively small and does not appreciably degrade the minefield's lethality. For tracked vehicles, the AT density yields more than 80 percent probability of encounter. Volcano AT mines do not have AHDs but are highly sensitive to any movement once they are armed. Any attempt to remove the mines will likely result in detonation.

The basic site layout is extremely important, and it is the same for air and ground Volcano minefields. The limits of Volcano minefields are marked before emplacement when the situation (planned targets within the main battle area [MBA] of a defensive operation) allows it. The minefield is not premarked when the situation (offensive operations or situational obstacles) does not allow it. If the mines have not self-destructed, the minefield is marked before the unit leaves the area or turns it over to an adjacent unit. Minefield marking must include the safety zone, which is 40 meters from the start and end points and 80 meters to the left and right of the centerline. The start and end points of the strip centerline are marked based on the minefield front and the number of strips. For a ground Volcano minefield, guide markers are emplaced along the path of the centerline but are offset left to allow the host vehicle to remain on the centerline. When using a ground-delivery system, minefield marking must leave a gap along each centerline for vehicle entrance and exit. The number of guide markers used depends on the terrain and the visibility. Guide markers are not required for an air Volcano minefield because the pilot will use the start and end points of the centerline as reference points.

Figure 3-8 illustrates the emplacement pattern for standard disrupt and fix minefields using the ground or air Volcano. Disrupt and fix minefields use only one centerline to give a minefield depth of 120 meters (ground) or 140 meters (air), not including the safety zone. The strip centerline is 277 meters (ground) or 278 meters (air) long. The host vehicle moves toward the start point, achieving and maintaining the ground or air speed selected on the DCU. The operator depresses the launch switch on the DCU when the vehicle passes the start marker, and he stops dispensing mines when the vehicle passes the end marker. The operator dispenses 40 canisters (20 on each side) along the centerline. One full load of ground or air Volcano emplaces four disrupt or fix minefields. For ground emplacement, the vehicle moves out of the minefield, marks the exit, and waits a minimum of 4 minutes before approaching the minefield. This delay allows faulty mines to self-destruct.

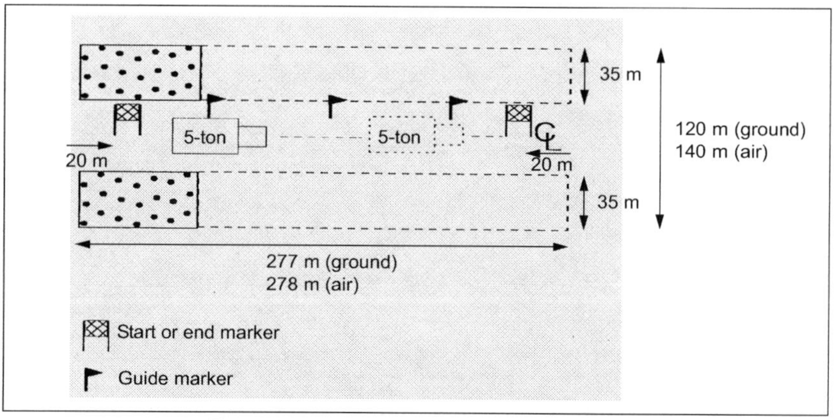

Figure 3-8. Volcano disrupt and fix minefields

Turn and block minefields (Figure 3-9, page 3-22) are emplaced using the same basic procedures as those used for disrupt and fix minefields. However, turn and block minefields use two strip centerlines along a front of 555 meters (ground) or 557 meters (air). During site layout, centerlines are separated by at least 320 meters for both ground and air delivery. This gives a total minefield depth of 440 meters (ground) or 460 meters (air). The operator dispenses 80 canisters along each centerline (40 on each side); therefore, turn and block minefields require a total Volcano load of 160 canisters. One full load of ground or air Volcano emplaces one turn or block minefield. Wherever possible, two ground Volcanoes are employed simultaneously on turn and block minefields. When only one ground delivery system is used, the crew must wait 4 minutes after dispensing the first strip before dispensing the second strip. This allows mines that fail the arming sequence to self-destruct. For air delivery, two sorties are also optimal; but demands for sorties elsewhere in the division may preclude the simultaneous employment of two Blackhawks.

Modular Pack Mine System

The MOPMS (Figure 3-10, page 3-22) is a man-portable, 162-pound, box-shaped mine dispenser that can be emplaced anytime before dispensing mines. The dispenser contains 21 mines (17 AT and 4 AP). The mines have leaf springs along their outer circumference that are designed to push the mines into proper orientation if they land on their side.

Each dispenser contains seven tubes; three mines are located in each tube. When dispensed, an explosive propelling charge at the bottom of each tube expels mines through the container roof. Mines are propelled 35 meters from the container in a 180-degree semicircle (Figure 3-11, page 3-23). The resulting density is 0.01 mine per square meter. The safety zone around one container is 55 meters to the front and sides and 20 meters to the rear.

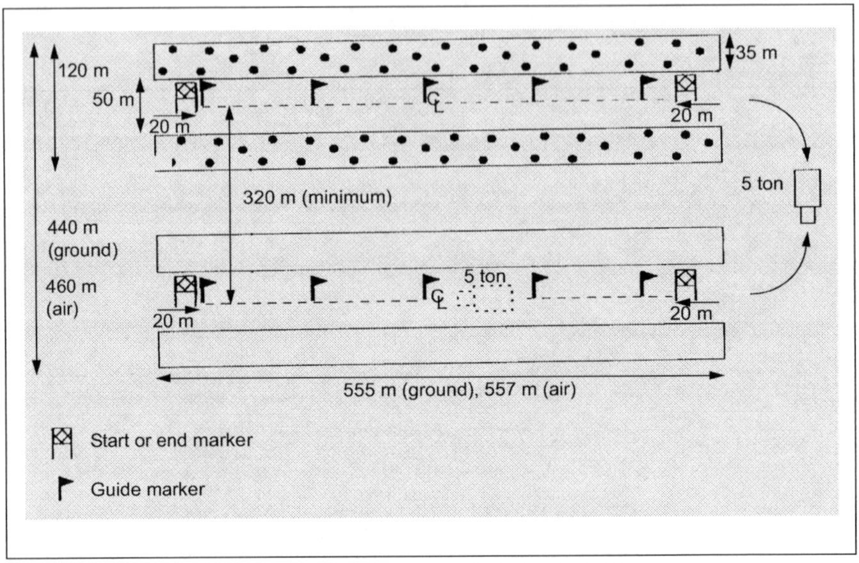

Figure 3-9. Volcano turn and block minefields

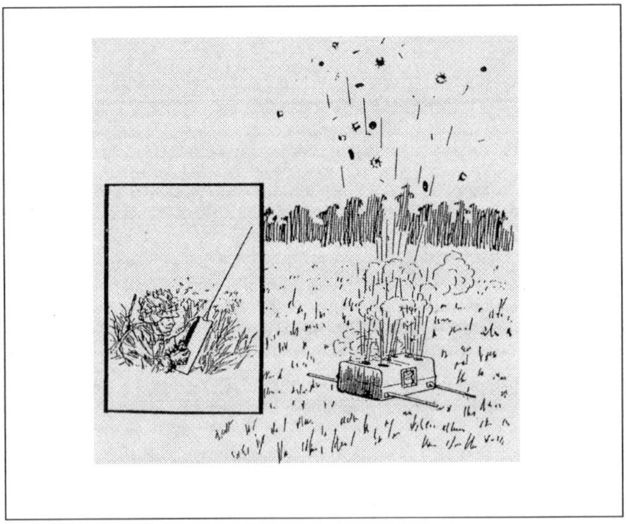

Figure 3-10. MOPMS

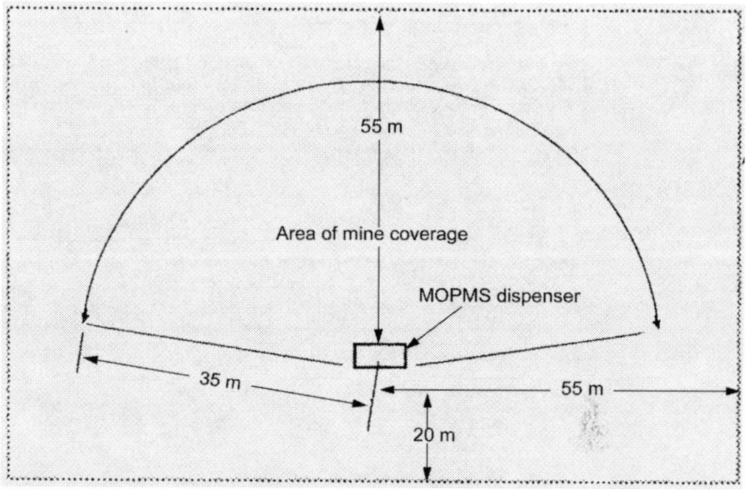

Figure 3-11. MOPMS emplacement and safety zone

Mines are dispensed on command using an M71 remote-control unit (RCU) or an electronic initiating device. Once mines are dispensed, they cannot be recovered or reused. If mines are not dispensed, the container may be disarmed and recovered for later use.

The RCU can recycle the 4-hour SD time of the mines three times, for a total duration of approximately 13 hours. Mines with a 4-hour SD time will begin to self-destruct at 3 hours and 12 minutes. All active mines must be recycled within 3 hours of the initial launch or last recycle. This feature makes it possible to keep the minefield emplaced for longer periods if necessary. The RCU can also self-destruct mines on command, allowing a unit to counterattack or withdraw through the minefield, as necessary, rather than waiting until the SD time has expired. The RCU can control up to 15 MOPMS containers or groups of MOPMS containers from a distance of 300 to 1,000 meters via separate pulse-coded frequencies. Coded frequencies defeat threat electronic countermeasures directed against the system.

If the M71 RCU is unavailable, a direct wire link is used in conjunction with an M32, M34, or M57 blasting machine. By using the M32 10-cap blasting machine, one MOPMS dispenser can be detonated at a maximum range of 1,000 meters. The M34 50-cap blasting machine can detonate one MOPMS at a maximum range of 3,000 meters. (Due to internal resistance, the maximum range is decreased by 400 meters for each additional MOPMS connected in series.) The M57 claymore-type FD can fire only one MOPMS at a maximum range of 100 meters. When controlled by direct wire, MOPMS dispensers cannot be command-detonated, and the SD time cannot be recycled.

WARNING

The MOPMS dispenser has seven launch tubes. If all seven tubes are not visible after deployment, mines are jammed in the tube(s). In this event, clear the area and notify EOD. The dispenser is considered to be UXO; do not attempt to recover the dispenser.

Employment

The MOPMS provides a self-contained, on-call minefield emplacement capability for all forces. It can be command-detonated, reused (if mines are not dispensed), and directly emplaced to provide complete and certain coverage of small or critical targets. The ability to command-detonate mines or extend their SD time provides an added flexibility not currently available with other SCATMINE systems. With its unique characteristics, the MOPMS is ideally suited for the following minefield missions:

- Emplacing hasty protective minefields.
- Emplacing deliberate protective minefields (cases emplaced, but mines not dispensed).
- Emplacing nuisance minefields (trails, crossing sites, landing zones [LZs], drop zones [DZs], and road junctions).
- Emplacing tactical disrupt and fix minefields.
- Closing gaps and lanes in existing minefields.
- Temporarily closing counterattack routes.
- Supporting ambushes.
- Supporting military operations in built-up areas (MOBA) operations.

When the MOPMS is used to close lanes, the container is positioned and dispensed by personnel in an overwatch position from a safe standoff. The MOPMS is ideally suited for creating a small disrupt obstacle in support of engineers executing a reserved demolition target. Engineers prepare the reserved target for demolition and emplace several MOPMS units on the enemy side, just out of target range. When the last forward element passes through the target, the firing party detonates the charges. If something goes wrong or the firing party needs more time, MOPMS mines can be dispensed to disrupt the enemy before it reaches the target.

The MOPMS provides light and special forces with a versatile, compact system for emplacing nuisance minefields. It can be used in low-, mid-, and high-intensity conflicts and in a variety of environments. The MOPMS cannot be transported long distances by hand because of its weight, so its use is limited.

Emplacement

MOPMS dispensers are issued as standard Class V munitions and are drawn from an ASP on a mission-by-mission basis. RCUs are organizational issues of equipment and are assigned to engineer and combat arms units. Due to the weight of the system, it will normally be transported by vehicle, as close as possible to the emplacement site, where it can easily be hand-emplaced by four soldiers using the four foldout carrying handles.

To ensure that the minefield will be dispensed in the proper location, the container should be carefully sited by the noncommissioned officer in charge (NCOIC). Several containers can be used together to provide a greater area of coverage or a higher mine density. If mines are not dispensed immediately, containers should be camouflaged and, if possible, buried. When placed in sand or snow, brace the containers to prevent them from moving during mine dispensing. Designate a firing point that gives the operator clear observation of the area to be mined. Firing systems must be inspected according to MOPMS operating instructions. If mines are dispensed immediately, remove empty containers to avoid revealing the minefield location.

The MOPMS can be employed to emplace disrupt and fix tactical minefields. Emplacement procedures are the same as for protective minefields above. However, MOPMS containers are arranged in a specific pattern to achieve the necessary depth, front, and density. Once the minefield is marked (to include the safety zone), MOPMS containers are arranged as shown in Figure 3-12 for a disrupt minefield. The safety zone is 55 meters from the front and sides and 20 meters from the rear of the container. The disrupt minefield uses four MOPMS containers that are spaced 70 meters apart to give a minefield front of 280 meters. Other MOPMS containers are offset from the baseline by 35 meters to give the minefield a depth of 70 meters. All containers are fired using the same RCU or FD.

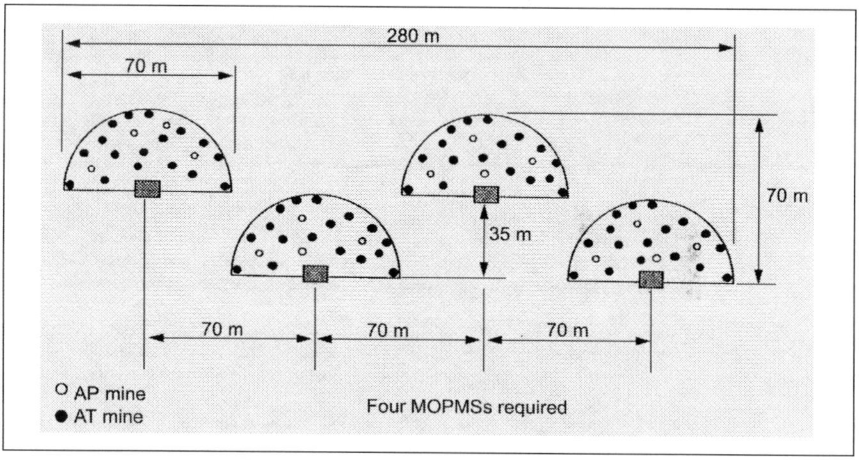

Figure 3-12. MOPMS in a disrupt minefield

Figure 3-13 illustrates the arrangement of MOPMS containers for a fix minefield. The basic layout is the same as the disrupt minefield; however, the fix minefield has one additional MOPMS that is placed 70 meters forward of the baseline to act as an IOE. This gives the same 280-meter minefield front but increases the minefield depth to 115 meters.

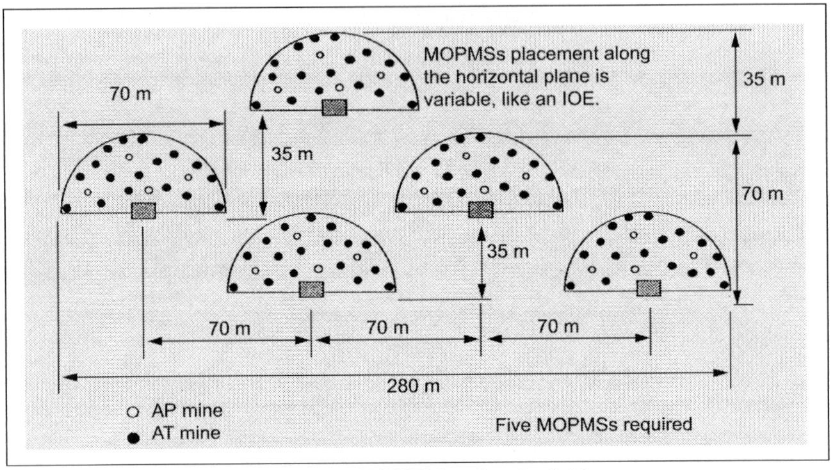

Figure 3-13. MOPMS in a fix minefield

MOPMS can be used to construct turn and block tactical minefields using the principles outlined in Chapter 2; however, turn and block minefields require more containers than are normally available to a unit.

MARKING

The maneuver unit that is responsible for the area of ground in which the minefield is emplaced is also responsible for marking the minefield. This normally requires direct coordination between elements of the maneuver command (usually the engineer) and the delivering/emplacing unit. However, it is unrealistic to expect units to mark artillery-delivered ADAM and RAAM, air-delivered Volcano, or Gator minefields. For this reason, units operating in the vicinity of these minefields must know calculated safety zones and use extreme caution. Scatterable minefields are marked to protect friendly troops as shown in Table 3-7. Ground Volcano minefields are marked according to the guidelines below.

Table 3-7. Marking scatterable minefields

Minefield Location	Marking
Enemy forward area	Unmarked
Friendly forward area	Sides and rear marked
Friendly rear area	All sides marked

SAFETY ZONES

A safety zone is an area where a stray or outlying mine has a chance of landing and laying to rest. The commander must prevent friendly forces from maneuvering into the safety zone during the minefield's life cycle. Depending on its specific location on the battlefield, the safety zone may be marked with a fence.

The safety zone around a Volcano minefield is shown in Figure 3-14.

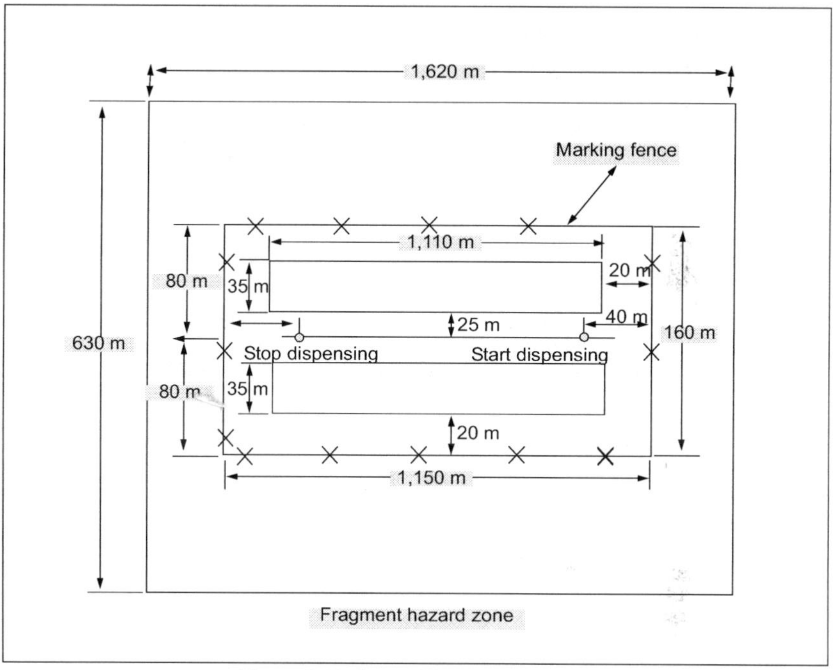

Figure 3-14. Ground Volcano minefield

FRAGMENT HAZARD ZONES

If an AT mine that is oriented on its side self-destructs, the EFP can theoretically travel 640 meters. This is the maximum fragment hazard zone; however, the chances of being struck are negligible at this distance. Tests indicate that the acceptable risk distance is 235 meters from the outer edges of the minefield's safety zone. This fragment hazard zone is also associated with the Gator and MOPMS AT mines. When the MOPMS is used for protective minefield missions, commanders must be made aware of the fragment hazard zone.

Use Table 3-8 to determine safety zones and fragment hazard zones.

Table 3-8. Safety and fragment hazard zones

System	Safety Zone	Fragment Hazard Zone
ADAM/RAAM	500 to 1,500 meters from aim point(s) (depends on delivery factors)	235 meters from the outside dimensions of the safety zone
Gator	925 x 475 meters from aim point(s)	1,395 x 945 meters from aim point(s)
Ground Volcano	1,150 x 160 meters	235 meters from start and stop points and the centerline
Air Volcano	1,315 x 200 meters	235 meters from start and stop points and the centerline
MOPMS	See page 3-28 for specific placement.	235 meters from the outside dimensions of the safety zone

FENCING

Fencing for ground Volcano minefields (Figure 3-14, page 3-27) is emplaced 80 meters beyond the centerline of the minefield and 40 meters from the start and stop points. Fencing should be no closer than 20 meters from the nearest mine.

Air Volcano minefields are not normally marked by fencing. However, if air Volcano minefields are emplaced in friendly areas, they are marked with fencing to protect friendly personnel. Fencing is installed before delivering an air Volcano, and it is located 100 meters from the centerline of the minefield and 100 meters from the start and end points. Appendix D contains detailed information pertaining to air Volcano minefields.

Chapter 4

Special-Purpose Munitions

Special-purpose munitions are hand-emplaced and used to create an expedient obstacle, enhance existing ones, and attack specific types of targets. The commander can employ these munitions to support his scheme of maneuver, to mass firepower, and to disrupt or destroy enemy forces in depth. Special considerations must be made in the planning process to effectively employ special-purpose munitions.

M18A1 CLAYMORE

The M18A1 claymore munition (Figure 4-2) is a fragmentation munition that contains 700 steel balls and 682 grams of composition C4 explosive. It weighs 1.6 kilograms and can be detonated by command **(Korea Only: or trip wire)**. It is activated by electric or nonelectric blasting caps that are inserted into the detonator well. The claymore projects a fan-shaped pattern of steel balls in a 60-degree horizontal arc, at a maximum height of 2 meters, and covers a casualty radius of 100 meters. The forward danger radius for friendly forces is 250 meters. The backblast area is unsafe in unprotected areas 16 meters to the rear and sides of the munition. Friendly personnel within 100 meters to the rear and sides of the munition should be in a covered position to be safe from secondary missiles. If the M18A1 is employed in a minefield for 72 hours or more, the minefield must be fenced on all sides.

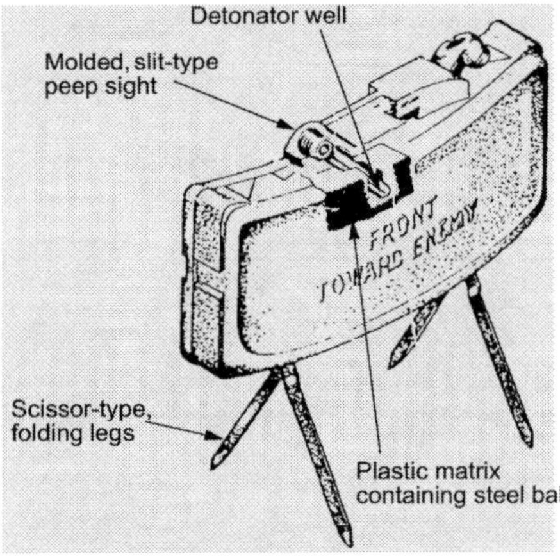

Figure 4-2. M18A1 claymore

When employing the M18A1 claymore with other munitions or mines, separate the munitions by the following minimum distances:

- 50 meters in front of or behind other M18A1s.
- 3 meters between M18A1s that are placed side by side.
- 10 meters from AT or fragmentation AP munitions.
- 2 meters from blast AP munitions.

SELECTABLE LIGHTWEIGHT ATTACK MUNITION

The selectable lightweight attack munition (SLAM) (Figure 4-3) is a multipurpose munition with an antitamper feature. The SLAM is compact and weighs only 1 kilogram, so it is easily portable. The SLAM is intended for use against APCs, parked aircraft, wheeled or tracked vehicles, stationary targets (such as electrical transformers), small fuel-storage tanks (less than 10,000-gallon), and ammunition storage facilities. The EFP warhead can penetrate 40 millimeters of homogeneous steel.

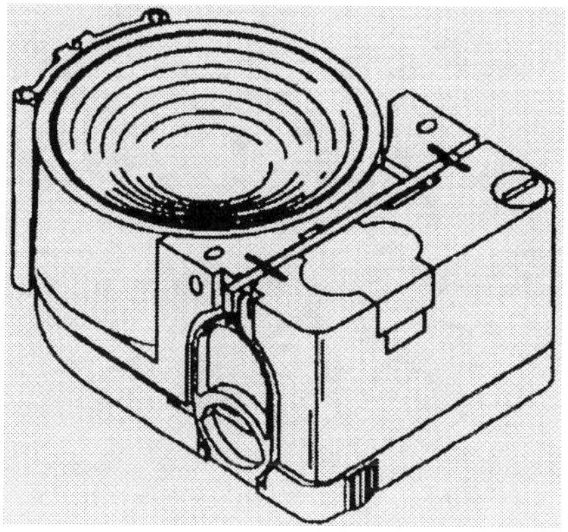

Figure 4-3. SLAM

The SLAM has two models—one is self-neutralizing (M2) and the other is self-destructing (M4):

- The M2 is solid green and has no labels, brands, or other distinguishing marks. This device is used by SOF and is not available to other units.
- The M4 is green with a black warhead (EFP) face. This device is normally used by units designated as light, airborne, air assault, crisis response, and rapid deployment.

See Appendix B for a description of major SLAM components.

OPERATING MODES

The SLAM has four possible employment methods—bottom attack, side attack, timed demolition, and command detonation.

Bottom Attack

The SLAM has a built-in magnetic sensor, so it can be used as a magnetic-influenced munition against trucks and light armored vehicles (Figure 4-4, page 4-4). It can be concealed along trails and roads where target vehicles operate and can be camouflaged with dry leaves, grass, and so forth without affecting EFP performance. Mud, gravel, water, and other debris that fill the EFP cup have minimal impact on EFP formation and effectiveness as long as the debris does not extend beyond the depth of the EFP cup. The magnetic sensor is designed to trigger detonation when it senses a vehicle's overpass. For the EFP to form properly, it needs a minimum of 13 centimeters from the point of emplacement to the target. The bottom-attack mode is active when the selector switch is set to *4, 10, or 24 HOURS* and the passive infrared sensor (PIRS) cover is in place. The SLAM will self-destruct (M4) or self-neutralize (M2) if the selected time expires before the SLAM is detonated by a vehicle.

Figure 4-4. SLAM in bottom-attack mode

Side Attack

The SLAM is equipped with a PIRS that was specifically developed for the side-attack mode (Figure 4-5). The PIRS detects trucks and light armored vehicles by sensing the change in background temperature when vehicles cross in front of the PIRS port. The PIRS is directional and aligned with the EFP when the device is aimed. The side-attack mode is active when the SLAM selector switch is set to *4, 10, or 24 HOURS* and the PIRS cover is removed to expose the PIRS. The SLAM will self-destruct (M4) or self-neutralize (M2) if the selected time expires before it is detonated by a vehicle.

Timed Demolition

The SLAM's built-in timer will trigger detonation at the end of a selected time (Figure 4-6). The timed-demolition mode is active when the SLAM selector switch is set to *15, 30, 45, or 60 MINUTES*. In this mode, the magnetic sensor and the PIRS are inoperable, and the SLAM will detonate after the selected time has expired.

Command Detonation

This mode provides manual warhead initiation using standard military blasting caps and a priming adapter (Figure 4-7). The command-detonation capability bypasses the SLAM's fuse and safing and arming (S&A) assembly.

Figure 4-5. SLAM in side-attack mode

Figure 4-6. SLAM in timed-demolition mode

Figure 4-7. SLAM in command-detonation mode

ANTITAMPER FEATURE

The SLAM has an antitamper feature that is only active in the bottom- and side-attack modes. The SLAM will detonate when an attempt is made to change the selector switch's position after arming.

M93 HORNET

The M93 Hornet (Figure 4-8) is an AT/antivehicular off-route munition made of lightweight material (35 pounds) that one person can carry and employ. The Hornet is a nonrecoverable munition that is capable of destroying vehicles by using sound and motion detection methods. It will automatically search, detect, recognize, and engage moving targets by using top attack at a standoff distance up to 100 meters from the munition. It is employed by combat engineers, rangers, and SOF.

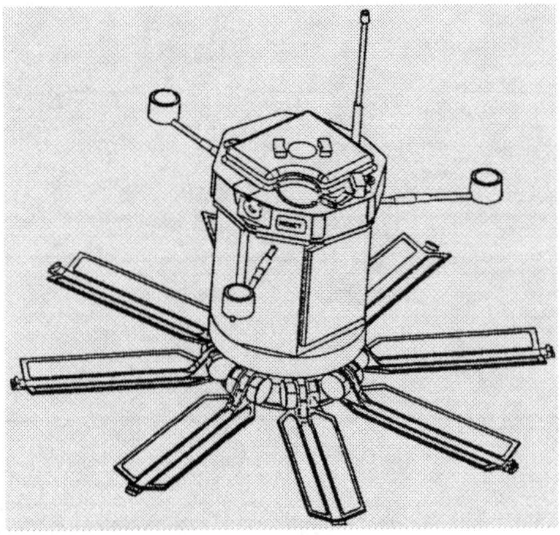

Figure 4-8. M93 Hornet

The RCU is a hand-held encoding unit that interfaces with the Hornet when the remote mode is selected at the time of employment. After encoding, the RCU can be used to arm the Hornet, reset its SD times, or destroy it. The maximum operating distance for the RCU is 2 kilometers.

High winds, heavy rain, snow, ice, extreme cold, and extreme heat reduce the Hornet's ability to detect targets at maximum range. Radio-frequency (RF) jamming devices (such as the hand-emplaced, expandable jammer [HEXJAM]), limit the Hornet's communication capabilities if they are placed in the munition field, but they will not affect the Hornet's ability to engage targets and will not damage the system. RF jamming devices affect the remote arming of current Hornet systems using the MOPMS RCU, and they will affect future Hornet's two-way communications capability with the Centurion remote control device.

See Appendix B for a description of Hornet components.

EMPLOYMENT CONSIDERATIONS

The Hornet's active battery pack is inserted during prearming and has an estimated life of 4 hours. The active battery pack powers the munition from the time it is inserted until the end of the safe-separation time, when the built-in reserve battery is activated. To prevent munitions from becoming duds, do not prearm them too early. Allow adequate time for travelling to the obstacle site, emplacing mines, throwing arming switches, and expiration of safe-separation times.

Once the Hornet is armed and the self-test is performed, the munition will remain active until its SD time expires or until it is encountered. The SD time (4 hours, 48 hours, 5 days, 15 days, or 30 days) is determined by the mission and the commander's intent. The munition will self-detonate after the SD time has expired.

Hornet munitions have an employed life of 60 days in the prearmed mode (remote arming) and 30 days in the armed mode. If the temperature exceeds 100°F, the employed life drops to 15 days in the prearmed mode and 30 days in the armed mode.

EMPLOYMENT ROLES

Combat engineers or maneuver forces under engineer supervision emplace Hornets in close operations; SOF or rangers emplace Hornets in deep operations. Hornets will be employed throughout the entire depth of the battle space to support Army operations.

Close Operations

In close operations, the Hornet can be—

- Used to fix the enemy and weaken it along its AA.
- Emplaced as an offensive-support weapon system because of its quick emplacement time and wide attack area.
- Employed rapidly along exposed flanks during a maneuver as a situational obstacle to disrupt the enemy's counterattacks.

- Used as a stand-alone tactical obstacle or as a reinforcement to conventional obstacles.
- Used to disrupt and delay the enemy, allowing long-range weapons to engage more effectively.

Deep Operations

In deep operations, the Hornet can be—

- Emplaced along key routes in gauntlet obstacles to disrupt and delay threat second-echelon forces, resupply operations, and key lines of communication (LOC).
- Used at C^2 and logistics sites to disrupt enemy operations.

Rear Operations

In rear operations, the Hornet can be emplaced (unarmed) along key routes in preparation for possible retrograde operations.

Early-Entry Operations

In early-entry operations, the Hornet can be—

- Used as an additional antiarmor weapon to supplement light forces.
- Used along high speed AAs in gauntlet obstacles to buy time and space.

TACTICAL EMPLACEMENT

There are four basic emplacement scenarios for the Hornet.

Conventional Minefield Reinforcement

The Hornet can be used to reinforce a conventional turn, block, or fix minefield (Figure 4-9).

Platoon engineers emplace the conventional minefield first, and then they traverse the safe lane that is perpendicular to the minefield. The Hornets are employed in two staggered rows, spaced 100 meters apart, 50 to 100 meters from the front edge (on the enemy side) of the conventional minefield. It is also recommended that a row of Hornets be placed 50 meters behind the minefield to reduce the enemy's breaching capability. (This row will be emplaced after the safe lane is closed.) The emplacing vehicles work toward the safe lane.

Two squads employ Hornets in two rows of ten each. One or more soldiers provide security. Under the supervision of a noncommissioned officer (NCO), four soldiers in each squad vehicle start prearming the Hornets, if necessary. They—

- Rotate the handle.
- Remove the cover.
- Insert the active battery pack and verify functionality via a solid status light.
- Reinstall the active battery-pack cover.

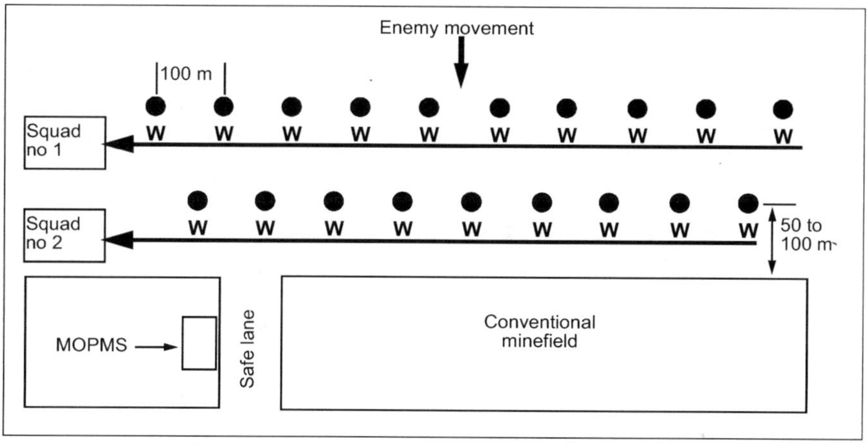

Figure 4-9. Hornet reinforcing a conventional minefield

- Select the SD time.
- Encode the Hornet with the M71 RCU and verify functionality via a flashing status light.
- Reinstall the cover.

Each emplacement vehicle moves to the first Hornet emplacement site in each row. The emplacing soldier and the arming soldier dismount. The emplacing soldier is handed a Hornet from the vehicle. He emplaces the Hornet at the designated site and returns to the vehicle.

The arming soldier rotates the handle on the Hornet, removes the cover and the safety and handling (S&H) band, rotates the SD switch to *U*, and pushes the arm switch to *ARM*. He then returns to the vehicle, taking the cover and the S&H band with him. The vehicle travels to the next Hornet emplacement site.

After all the Hornets in the two leading rows have been emplaced and armed, the emplacing vehicles exit through the safe lane and usually secure it with a MOPMS. The emplacement vehicles must be at least 475 meters (safe standoff distance) from the nearest Hornet within 30 minutes. Before sending a remote arming signal, vehicles must wait at least 36 minutes after the arming switch is thrown on the last Hornet emplaced. If a rear row is required, it is emplaced at this time. The Hornets are then remotely armed with the M71 RCU, when required. They are now capable of covering the minefield by fire and engaging threat tracked vehicles.

Scatterable Minefield Reinforcement

The Hornet can be used to reinforce a Volcano or MOPMS turn, block, or fix minefield (Figure 4-10). Hornet munitions are emplaced, using the same procedures as above, **before** the Volcano or MOPMS minefield is emplaced.

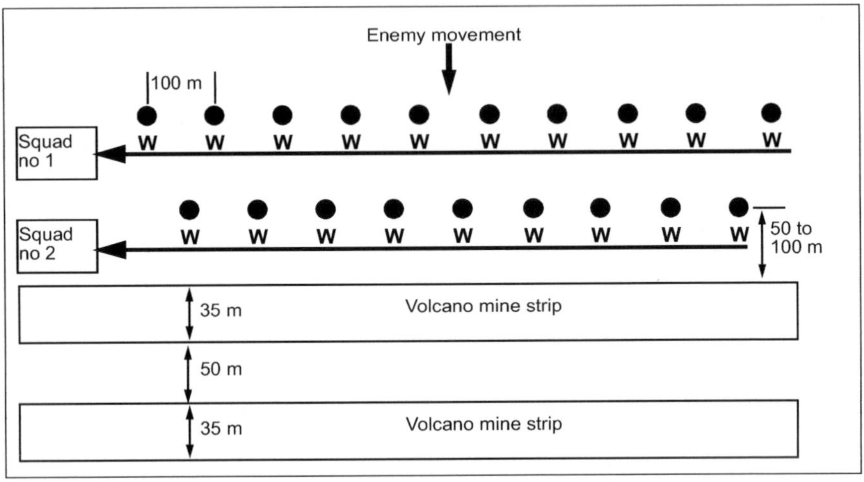

Figure 4-10. Hornet reinforcing a Volcano minefield

To ensure that the Volcano dispensing vehicle has sufficient time to reach the safe standoff distance (475 meters), Volcano dispensing should start no later than 30 minutes (minus the Volcano dispensing time) after the first Hornet is armed. This allows Hornet emplacing squads to be finished or nearly finished before the Volcano dispenser begins emplacing the minefield.

Area-Disruption Obstacle

When the X-pattern is employed, the Hornet is very effective as a disrupting obstacle (Figure 4-11). An area-disruption obstacle is employed to disrupt the enemy's approach prior to the start of the direct-fire battle. It causes disruption and attrition of the advancing threat force and encourages follow-on forces to seek an alternate route. Therefore, multiple area-disruption obstacles will typically be employed to adequately cover the cross-country AA. This requires coordinated action among multiple squads.

An engineer platoon emplaces a Hornet area-disruption obstacle. The obstacle typically consists of 20 Hornets (five clusters of four Hornets each) employed in an X-pattern over a 1- by 1-kilometer area. Individual Hornets are emplaced 100 meters apart. Emplacing this obstacle must be done as a dispense-and-roll operation to ensure that the emplacing vehicles can reach the safe standoff distance (475 meters) from any armed Hornets.

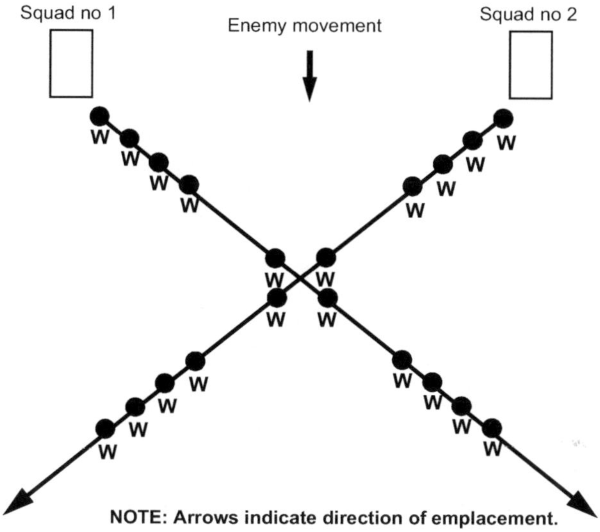

Figure 4-11. Hornet area-disruption obstacle

Area-disruption obstacles are normally armed by remote, but they can be manually armed under the following conditions:

- METT-TC requires rapid emplacement and arming.
- Terrain reconnaissance determines that there are no major impediments (rough terrain, vegetation) to maneuver.
- Emplacement is done during daylight hours (mission-oriented protective posture [MOPP] level 0 only).

Hornets are prearmed the same as above. Two squads lay the Hornets in unison, starting with the two emplacement sites closest to the enemy. Each squad drives in a straight line, crossing paths at the middle of the X, and emplaces ten Hornets.

A soldier in the back of each emplacing vehicle throws the arming switch and sets the Hornet down or drops it off (base down) the back of the vehicle. After all the Hornet clusters are emplaced, squad vehicles quickly travel to the 475-meter safe standoff distance (no further than 2 kilometers) to prepare for remote arming. Hornets can be remotely armed 36 minutes after the arming switch is thrown on the last Hornet emplaced. If manual arming is used, Hornets automatically arm at the end of their safe-separation time (5 to 6 minutes after the arming switch is thrown).

Gauntlet Obstacle

Hornet gauntlet obstacles (Figure 4-12) are emplaced by an engineer platoon and are very effective in constricted terrain along the enemy's AA and at choke points. A Hornet gauntlet typically consists of 40 to 50 Hornets employed in a series of clusters (Figure 4-13). Each cluster contains 3 to 6 Hornets. The Hornets in each cluster are emplaced at 50-meter intervals, perpendicular to the road centerline, on alternating sides of the road/AA, and 25 to 50 meters (depending on the terrain and the vegetation) off the side of the road/AA. The distance between clusters varies from 750 to 2,000 meters so that the advancing threat force is kept guessing about when they will encounter the next cluster.

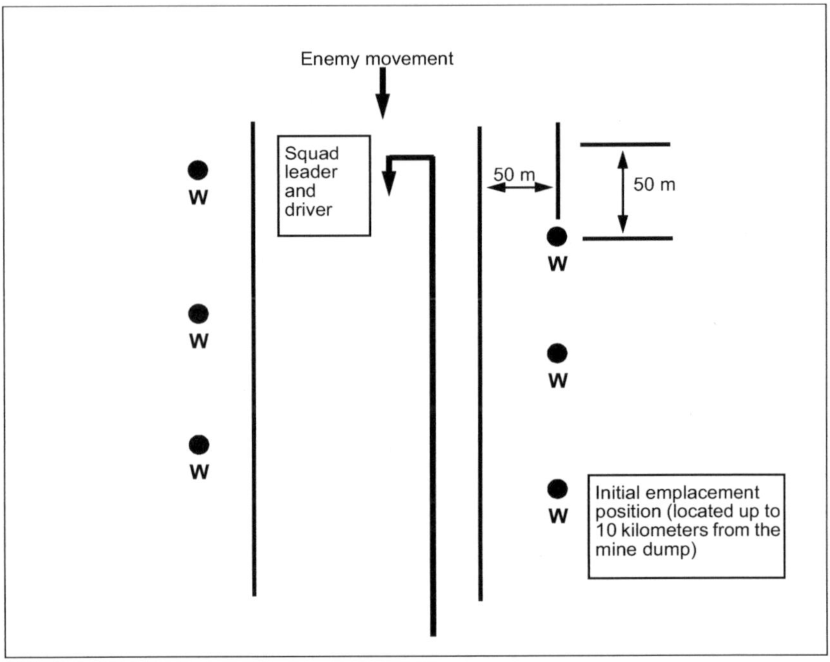

Figure 4-12. Hornet gauntlet obstacle (one cluster)

Before laying any Hornets, the munitions are prearmed as above. Soldiers also set the target switch to *HVY* for clusters closest to the enemy, so that the Hornets will only engage heavy tracked vehicles. The intent is to make threat forces commit to a route they perceive to be clear.

Hornets are emplaced beginning on the friendly side of the cluster. The first engineer squad emplaces Hornet munitions beginning with the cluster closest to the enemy. The emplacement vehicle drives even with the first Hornet

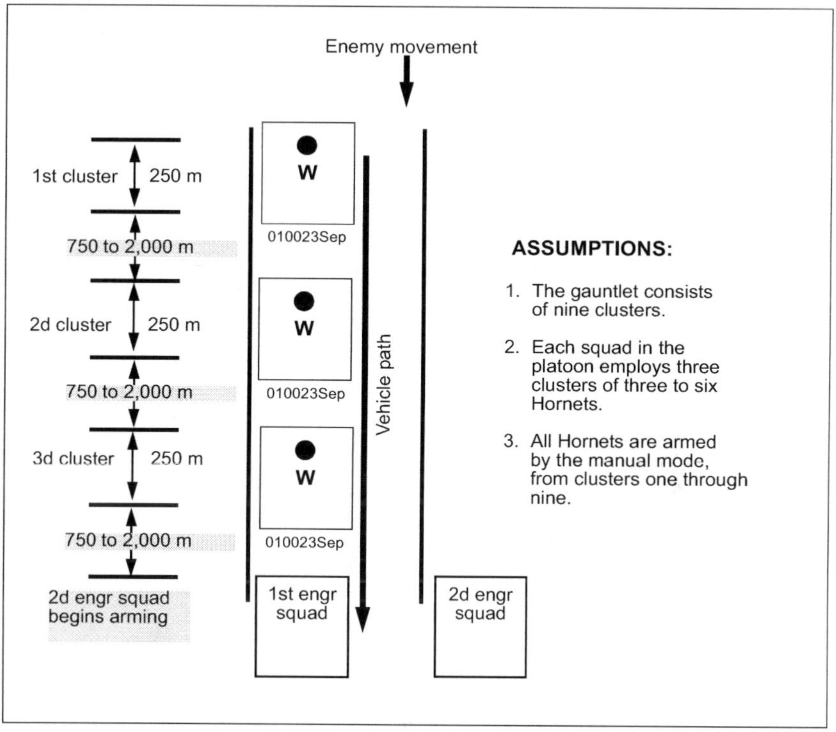

Figure 4-13. Hornet gauntlet obstacle (platoon)

emplacement site. The emplacing soldier dismounts, and a soldier in the vehicle hands him a Hornet. The emplacing soldier then proceeds to the Hornet emplacement site. The vehicle travels to a point even with each subsequent emplacement site. A soldier deploys at each emplacing site to lay one of the remaining Hornet munitions in the cluster. The vehicle then turns around and stops even with the last Hornet (on the enemy side) in the cluster.

Upon reaching the Hornet employment location, each emplacing soldier removes the cover and the S&H band, rotates the SD switch to *U*, and on the command (audible or visual signal) of the NCOIC, pushes the arm switch to *ARM*. Once the Hornets' arming switches are thrown, soldiers return to the road, taking the covers and the S&H bands with them, and wait to be picked up by the emplacement vehicle. After all the soldiers are in the emplacing vehicle, the driver quickly travels to the safe standoff distance (475 meters). The Hornet munitions in the first cluster will arm at the end of the safe-separation time (5 to 6 minutes).

The squad repeats the emplacement process for the next Hornet cluster in the gauntlet, taking care not to emplace any Hornets or drive within 475 meters of the previous cluster. Each squad in the platoon typically emplaces three clusters in the Hornet gauntlet, or 9 to 18 total Hornets.

Figure 4-14 shows Hornet emplacement in the battle space. The example used is in support of a defensive position where Volcano mines are used as tactical obstacles.

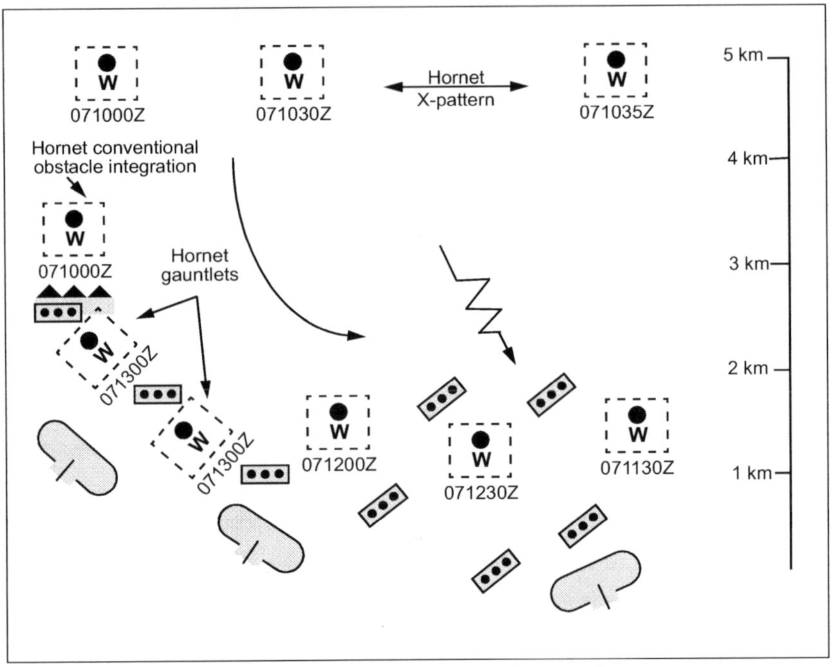

Figure 4-14. Hornet-enhanced turn-and fix-obstacle groups

Deep-Battle Interdiction Weapon

SOF or ranger units emplace Hornet munitions in the deep battle area as interdiction weapons. A typical mission requires a unit ranging in size from a six-man team to an entire company. The number of Hornets carried by the unit depends on the mission and the mode of insertion (vehicle, aircraft, or dismounted troops); a man can normally carry only one Hornet. Hornets are typically used to support a raid against an enemy position or complex and at bridges or choke points along high-speed AAs used by advancing second-echelon forces or for resupply. In these roles, the Hornets are employed similar to the clusters in a gauntlet obstacle.

Camouflage and Concealment

The best camouflage and concealment for the Hornet is tall grass and brush. The Hornet can be partially buried if the terrain or the vegetation does not provide effective natural camouflage and concealment. Placing the Hornet in a hole degrades its performance, so it should only be done when Hornets cannot be covered by fires or protected from tampering by dismounted enemy. The following conditions must be met:

- The depth of the hole must not exceed 4 inches, because the acoustic sensors must be above ground level.
- The hole must not restrict the Hornet's ability to rotate and tilt its body and to fire the sublet. To meet this requirement, the hole must be at least 36 inches wide and flat enough to support the munition. Although the Hornet should be placed on a flat surface if possible, it can operate on slopes up to 15 degrees.

Munitions placed at ground level should be no closer to obstructions than the distances shown in Table 4-1.

Table 4-1. Hornet minimum emplacement distances

Maximum Obstruction Height	Minimum Employment Distance from Obstruction
1 m	3 m
2.4 m	5 m
6.5 m	15 m
25 m	25 m

When the Hornet is emplaced and concealed, remove all indicators of excess soil and camouflage material before performing the arming sequence.

Recording and Marking

When the Hornet munition field is completed, the OIC will identify an NCO to be the recorder. The NCO will collect data from the NCOICs of the emplacing squads and complete DA Form 1355 as outlined in Chapter 8. The OIC will ensure that the DA Form 1355 is completed timely and accurately.

Marking the Hornet munition field will be completed as prescribed in Chapter 2. The fence will be no closer than 150 meters from the nearest Hornet munition. Marking must be completed before emplacing the munitions.

This chapter implements STANAG 2990.

Chapter 5

Conventional Mines

Conventional mines are hand-emplaced mines that require manual arming. This type of mine laying is labor-, resource-, and transport-intensive. Soldiers emplace conventional mines within a defined, marked boundary and lay them individually or in clusters. They record each mine location so that the mines can be recovered. Soldiers can surface lay or bury conventional mines and may place AHDs on AT mines.

ANTITANK MINES

The M15, M19, and M21 AT mines are used by US forces. They are shown in Figure 5-1, and their characteristics are listed in Table 5-1, page 5-2.

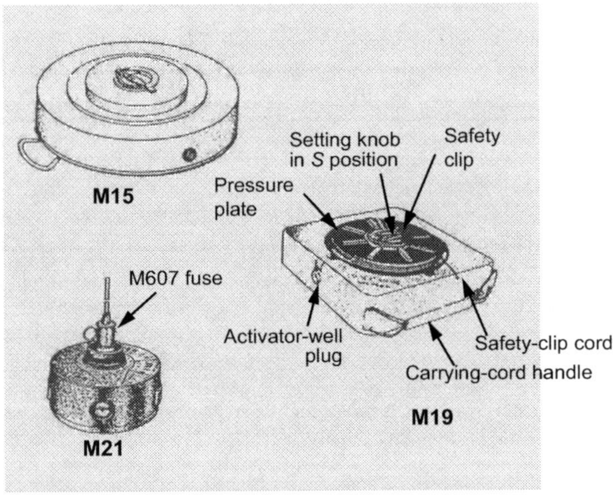

Figure 5-1. AT mines

M15

The M15 AT mine is 337 millimeters in diameter and 125 millimeters high. It weighs 13.5 kilograms and contains 9.9 kilograms of Composition B explosive. The primary fuse well is on the top center of the mine; secondary fuse wells are on the side and bottom. The M15 can contain the following fuses:

Table 5-1. Characteristics of AT mines

Mine	DODIC	Fuse	Warhead	AHD	Explosive Weight	Mine Weight	Mines per Container
M15 w/ M603 fuse	K180	Pressure	Blast	Yes	9.9 kg	13.5 kg	1
M15 w/ M624 fuse	K180 (mine) K068 (fuse)	Tilt rod	Blast	Yes	9.9 kg	13.5 kg	1
M19	K250	Pressure	Blast	Yes	9.45 kg	12.6 kg	4
M21	K181	Tilt rod or pressure	SFF	Yes*	4.95 kg	7.6 kg	4

*Conventional AHDs will not couple with this mine; however, the M142 multipurpose FD can be emplaced under this mine.

- M603 fuse. When the M603 fuse is employed on the primary fuse well, the M15 is a track-width mine that is activated by 158 to 338 kilograms on the pressure plate. This produces an M-Kill.
- M624 fuse. When the M624 fuse (with tilt rod) is employed on the primary fuse well, the M15 is a full-width mine that is activated by a deflection of 20 degrees or 1.7 kilograms of pressure to the tilt rod. Depending on the armor, this produces an M-Kill or a K-Kill.

M19

The M19 AT mine is a low-metallic, square-shaped mine that measures 332 by 332 millimeters and is 94 millimeters high. It weighs 12.6 kilograms and contains 9.45 kilograms of Composition B explosive, a tetryl booster pellet, and an M606 integral fuse. When the setting knob on the pressure plate is in the *S* (safe) position, the mine cannot function by action of the main fuse. After the safety clip has been removed and the setting knob turned to the *A* (armed) position, a force of 157.5 to 225 kilograms on the pressure plate depresses the Belleville spring and begins the firing chain. A standard FD may be used with the M2 activator in any of the secondary fuse wells on the sides or the bottom of the mine. When the M19 is employed, it is difficult to detect because of its plastic construction. It produces an M-Kill with a blast effect.

M21

The M21 AT mine is 230 millimeters in diameter and 206 millimeters high. It weighs 7.6 kilograms and has 4.95 kilograms of Composition H6 explosive. The mine is activated by 1.7 kilograms of pressure against a 61-centimeter-long rod on the end of the M607 fuse. It uses an M-S plate to produce a K-Kill. The M21 with tilt rod must be buried or staked (use three stakes at the 12, 4, and 8 o'clock positions) so that enemy vehicles will not tip the mine over. Without the tilt rod, the mine is activated by 130.5 kilograms of pressure on the M607 fuse and produces an M-Kill by blast effect.

ANTIPERSONNEL MINES

The M14 and M16 AP mines are used by US forces on the Korean peninsula. They are also used by many other countries. The M16 AP mine is likely to be seen in a modified form. These mines are shown in Figure 5-2, and their characteristics are listed in Table 5-2.

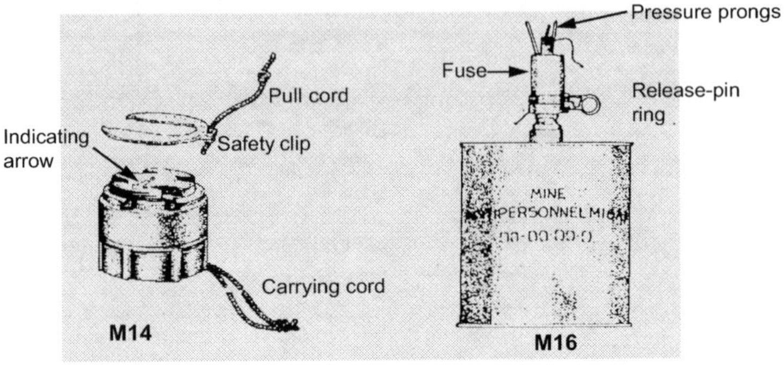

Figure 5-2. AP mines

Table 5-2. Characteristics of AP mines

Mine	DODIC	Fuse	Warhead	AHD	Explosive Weight	Mine Weight	Mines per Container
M14	K121	Pressure	Blast	No	28.4 g	99.4 g	90
M16-series	K092	Pressure or trip wire	Bounding frag	No	450 g	3.5 kg	4

M14

The M14 AP mine is a low-metallic blast mine consisting of a main charge (28.4 grams of tetryl) and a plastic fuse with a steel firing pin. It is cylindrical in shape (56 millimeters in diameter and 40 millimeters high) and weighs 99.4 grams. The pressure plate has an indented, yellow arrow that points to the *A* or *S* position on top of the fuse body. A force of 11.5 to 13.5 kilograms depresses the pressure plate and causes the Belleville spring to drive the firing pin into the detonator. The M14 is not designed to kill, but to incapacitate. The M14 AP mine has been modified by gluing a metal washer to the bottom of the mine. The modification was directed to improve the detectability of the mine. Unmodified mines are not authorized for use by US forces.

M16

The M16 AP mine is a bounding fragmentation mine that consists of a mine fuse (M605), trinitrotoluene (TNT) explosive, a propelling charge, and a projectile that are contained in a sheet-steel case. The mine is 103 millimeters in diameter, 199 millimeters high (including the fuse), and weighs 3.5 kilograms. The principal difference between the M16, M16A1, and M16A2 versions are in the construction of the detonators and boosters. The casualty radius is 27 meters for the M16 and M16A1 and 30 meters for the M16A2. A pressure of 3.6 to 9 kilograms applied on one or more of the three prongs of the M605 fuse or a pull of 1.4 to 4.5 kilograms on the trip wire will activate the mine.

EMPLACING MINES

The method used to lay and conceal each type of mine depends on the method of mine operations, the type of ground in which the mine is to be laid, and the type of ground cover available for camouflage.

Standard-pattern mine laying is laborious and time-consuming, but it is more effective and flexible than row mine laying and allows better mine concealment. Standard-pattern mine laying is well suited for protective minefields, and it can be used in terrain where the nature of the ground makes row mine laying impractical.

To achieve the maximum effect, mines must be laid where they cannot be seen and where a vehicle or a person exerts enough pressure to detonate them. The following rules should be applied to achieve the maximum effects of mines:

Mines With Prongs

Korea Only: If the mine is activated by its prongs, it should be buried flush with the ground so that only the tips of the mechanism are exposed (Figure 5-3). A mine buried in this manner is held firmly upright. The target exerts a direct, downward pressure rather than a sideways thrust. The mine is protected from damage and is difficult to see. If it is buried more deeply, it becomes unreliable because the layer of spoil may prevent the mine mechanism from operating.

If the mine is activated by a trip wire, it should be buried so that the trip wire is at least 2 to 3 centimeters above the ground (Figure 5-4).

Mines With Pressure Plates

Mines with pressure plates will function when completely buried as long as the cushion of earth above them is not too thick. AT mines are normally buried with the top of the mine approximately 5 centimeters below ground level.

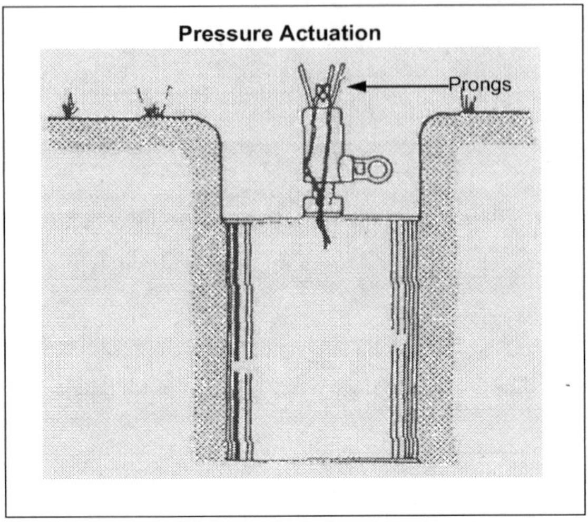

Figure 5-3. Prong-activated AP mine

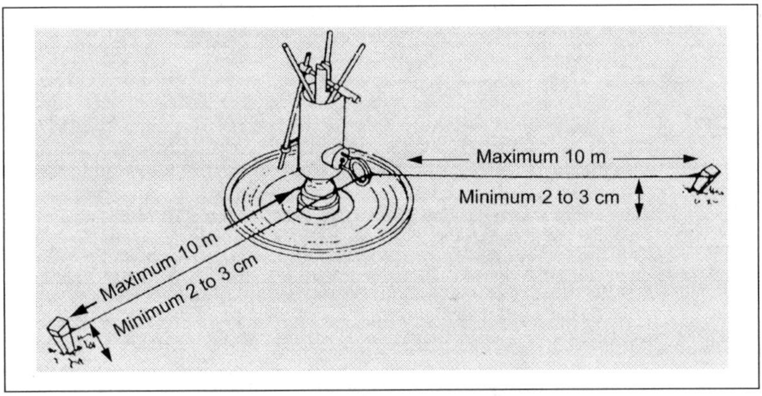

Figure 5-4. Trip-wire-activated AP mine

Korea Only: AP mines are usually placed in a hole and covered with camouflage material. If the hole is only slightly larger than the mine, the weight of the target may be supported by the shoulder of the hole, and the mine will fail to activate. Such bridging action can be avoided if the hole is dug much wider than the mine (Figure 5-5).

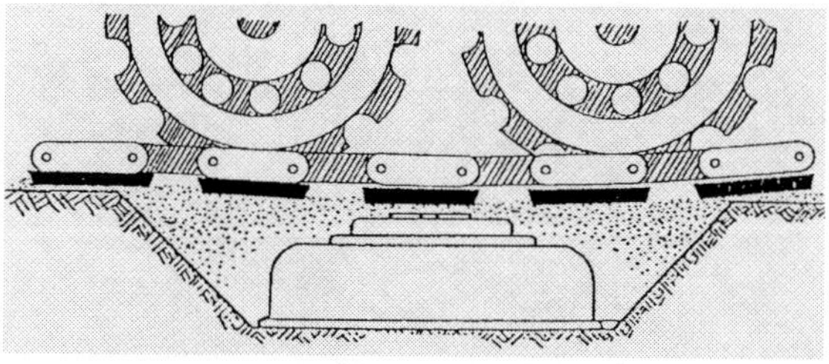

Figure 5-5. Buried mine with pressure plate

MINES WITH TILT RODS

Tilt-rod fuses normally require the body of the mine to be buried and the tilt-rod assembly to be clear of the ground (Figure 5-6). A tilt-rod fuse is preferred in areas where vegetation is sufficient to conceal the extension rod. Camouflage materials are carefully used to prevent premature detonation or interference with the normal functioning of the fuse. Extension rods are camouflaged before the mine is armed. If tilt rod mines are surface-laid, they must be staked.

BEARING BOARDS

High pressure is required to activate AT mines. When burying a mine in soil that has a low bearing pressure (such as soft sand or clayey soil), it may be necessary to place a board or another bearing plate under the mine. Otherwise, the mine may not detonate when it is forced down.

CONCEALMENT

After digging the hole for a mine, place the spoil in a sandbag to reduce the evidence of mining. If a sandbag is not available, heap the spoil. Camouflage all traces of digging after the mine is laid. If the ground cover is turf or other matted, root material, remove spoil that cannot be hidden. Cut the sod in an X, I, or U shape in the area where the mine is to be placed; lay the mine; and then roll the sod back in place to camouflage the mine. Loose earth over a mine will eventually consolidate, so the mine location should have a small mound immediately after laying (Figure 5-7). Ensure that the mound is inconspicuous and that it blends with the surrounding area. It is very important that you make a final check after concealing each mine so that you can correct faults progressively, because they cannot be corrected later.

AT mines in standard-pattern minefields should be buried. However if conditions dictate, mines with a single-impulse fuse may be laid on the surface. Mines with double-impulse fuses should always be buried, because if

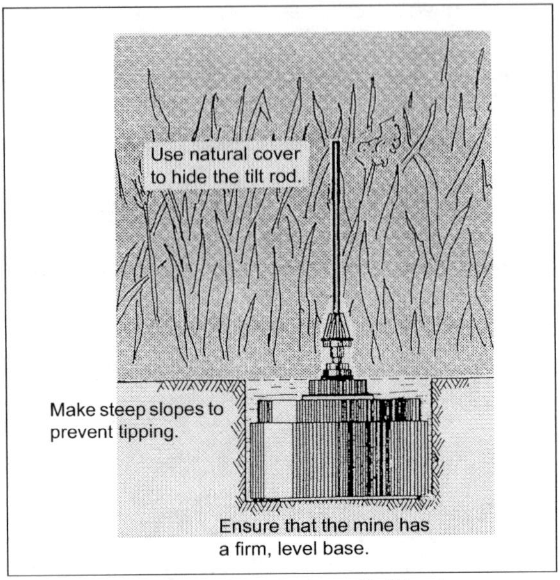

Figure 5-6. Buried mine with tilt rod

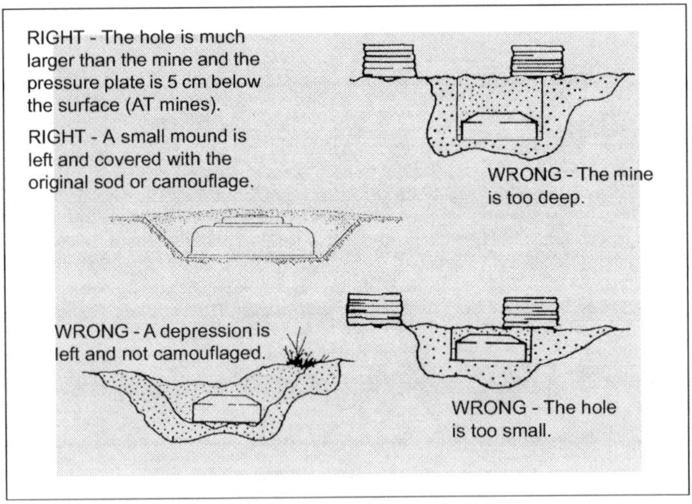

Figure 5-7. Buried and concealed mines

they are surface-laid, they may be physically damaged when pressure is exerted by a tracked vehicle. Buried mines also have some resistance to countermeasures, but surface-laid mines have none. Consideration must also be given to sympathetic detonation of AT mines (Table 5-3). US conventional mines do not have integral AHDs, so allow extra time to lay mines with AHDs.

Table 5-3. Sympathetic detonation chart

Type	M16	M15	M19
Surface-laid	NA	4.0 m	4.0 m
Buried flush	1.5 m	2.4 m	5.5 m
Buried 5 cm	NA	1.5 m	4.8 m

The difficulty of burying mines in very rocky ground and the necessity for surface laying will have a bearing on which mines are suitable. For example, small, blast-type AP mines are hard to detect and easy to camouflage. They are much easier to camouflage than larger fragmentation mines. The type of AT mine used will make little difference, because the mine's size will always make camouflage very difficult.

MANEUVER ASSISTANCE

During large mine-laying operations, engineers seldom have sufficient manpower to carry out all minefield tasks. Other combat arms units must often provide work parties. Engineers must be capable of organizing, controlling, and supervising combined arms work parties. They must also instruct them in new equipment and techniques. Work parties may be integrated with engineers or given certain tasks that are within their capabilities.

When laying a standard-pattern minefield, consider supplementing work parties with other combat arms soldiers to perform the following:

- Executing Class IV/V supply point or mine dump missions. Soldiers uncrate and prepare mines and remove empty boxes and residue.
- Laying. Soldiers position mines within strips and dig holes.
- Marking. Soldiers construct the perimeter fence and emplace mine signs.

Unpacking, preparing, and loading mines are the most time-consuming tasks when laying a row minefield; and they are ideal tasks for other combat arms soldiers.

This chapter implements STANAG 2036.

Chapter 6

Row Mining

Row mining is a means of emplacement for tactical minefields. For example, a typical tactical minefield could contain several rows of regularly spaced mines.

USE

Row mining is not a new idea. It has been used since the beginning of modern mine warfare and is very effective. It is especially effective in support of maneuver-oriented doctrine. Row mining is faster than standard-pattern mining and improves the maneuver commander's flexibility by providing him an obstacle that requires less manpower effort.

Mines may be surface-laid or buried, and they are often laid directly from a slow-moving vehicle. This reduces the time and the personnel required to emplace a minefield. Row mining can be used as a tactical or situational obstacle. Minefields are usually emplaced at or near the FLOT, along flank AAs, to support security operations. Speed and efficiency make row mining a desirable option, and row mining supports current doctrine.

RULES

Rules governing authority, reporting, recording, and marking are generally the same for row minefields as they are for other minefields. Row mining is simply a method of laying mines.

The most important factor in row mining is the requirement for strict C^2. Row mining is potentially the most hazardous form of mine laying. It entails vehicles and personnel moving in and around mines without the safety of a centerline strip. Leaders must place extreme emphasis on safety because the laying procedure is very rapid.

Most of the rules governing row mining are defined in STANAG 2036. A summary of those rules and some additional rules that apply are shown below.

- Rows.
  - There are two types of mine rows—regular and short. Short rows are described under IOE rules below.
  - Regular rows are marked and recorded. They are designated by letters (A, B, and so forth), with Row A being closest to the enemy.
  - The minimum distance between rows of AT mines is 8 meters.
  - **Korea Only: The minimum distance between any row and a row containing AP mines is 15 meters.**

- The distance between the start row marker and the first mine in a row is the mine spacing for that row.
- Start and end row markers are permanent markers and must be made of detectable material.

- Clusters.
  - Clusters are placed on the row centerline and directed toward the enemy side.
  - A cluster in row mining usually consists of one AT mine **(Korea Only: but it may also contain AP mines)**.
  - Cluster composition must remain the same throughout the row.
  - **Korea Only: Different types of AP mines may be used in a cluster.**
  - **Korea Only: The total number of mines in one cluster will not exceed five; no more than one will be an AT mine.**
  - The type of AT mine may vary from one cluster to another.
  - **Korea Only: A cluster of AP mines can be laid in a 2-meter semicircle on the enemy side of the baseline.**
  - When a cluster contains a mine that is equipped with an AHD, the mine is armed before the AHD is armed. The cluster is not armed until all personnel are at least 60 meters away.
  - Omitted clusters do not contain mines. They are recorded on DA Form 1355 (see Chapter 8).
  - Clusters are omitted within lanes, within gaps, in areas less than 2 meters from boundaries and lanes, and in areas where the terrain (trees, rocks) prohibits emplacement.
- IOE.
  - The IOE is located on the enemy side of the minefield.
  - The IOE baseline must be at least 15 meters from Row A.
  - IOE mines are buried.
  - IOE short rows are labeled at start (I1) and end (I1E) points.
  - IOE short rows must be at least 15 meters apart.
- **Korea Only: Trip wires.**
  - **Trip wires can be used in regular rows, but only one mine per cluster can be actuated by a trip wire.**
  - **No more than two trip wires can be used on a mine.**
  - **Trip wires are not considered AHDs.**
  - **Trip wires must be at least 2 meters from a minefield lane, a cluster, another trip wire, an IOE short row, or a minefield perimeter fence.**

— **Trip wires can only be used with AP fragmentation mines.**

- Lanes.
  - — Lanes are sited before laying begins. Lane locations should not be obvious.
  - — Clusters must be at least 2 meters from lane edges.
  - — The number of lanes must be sufficient to ensure that no one lane is overused and turned into an obvious track.
  - — Sufficient mines must be stockpiled so that the responsible unit can seal lanes suspected of being located by the enemy.
- General.
  - — The spacing between mines or clusters can vary from 4 to 10 meters but must remain constant within the row.
  - — Mines and clusters must be at least 15 meters from the perimeter fence.
  - — If the distance between a mine or a cluster and any turning point is less than the spacing for that row, omit that mine or cluster. The mine immediately following a turning point is always located at the mine spacing for that row.
  - — The minefield has two landmarks located to the rear, never to the extreme side or front.
  - — Global-positioning systems (GPSs) can only be used to determine the coordinates for minefield landmarks and reference points (RPs).

WARNING
Do not use GPSs to chart or record minefield perimeter coordinates or to determine safe routes through or around existing minefields.

  - — If landmarks are more than 200 meters away from the last row or are out of the direct line of sight, intermediate row markers or landmarks are placed at least 75 meters from the last end row marker.
  - — Landmarks can be used for more than one minefield. This must be recorded in the remarks block of DA Form 1355.
  - — Back azimuths are not used to record the minefield.
  - — Measurements are in meters.

LOGISTICS

CALCULATIONS

To simplify the calculation process, a minefield requirements computation work sheet (Figure 6-1, pages 6-5 through 6-8) has been developed. This work sheet is provided to the platoon leader or sergeant as a step-by-step guide to the mathematics involved in the logistical computation process. Properly completed, the work sheet provides the number of mines to order (by type), the number of regular strips to be emplaced, cluster composition, the estimated man-hours required to install the minefield, the amount of fencing and marking material required, the number of truckloads required to carry the mines, and the number of rolls of engineer tape required.

Step-by-step procedures for completing the work sheet are shown in Figure 6-2, pages 6-9 through 6-14. Each step is explained in the example to facilitate the understanding of the logic behind the calculations.

Use the following steps to determine the number of AT mines required for a row minefield when not using the standard row minefields discussed later in this chapter. Round the resulting numbers up to the nearest whole number.

Step 1. Determine the number of mines required.

$$density \times front = number\ of\ mines$$

Step 2. Determine the number of mines per row.

$$front \div mine\ Spacing = number\ of\ mines\ per\ row$$

Step 3. Determine the number of rows.

$$number\ of\ mines \div number\ of\ mines\ per\ row = number\ of\ rows$$

Step 4. Determine the actual number of mines.

$$number\ of\ mines\ per\ row \times number\ of\ rows = number\ of\ mines$$

Step 5. Determine the number of mines to request (includes a 10 percent resource factor).

$$number\ of\ mines \times 1.1 = number\ of\ mines\ to\ request$$

Step 6. Determine the number of vehicle loads by using Table 2-8, page 2-45.

Step 7. Determine the fencing and marking material required.

Sample Problem: Your platoon has been tasked to emplace a 400-meter row minefield with a density of 0.5-0-0 (AT-AP fragmentation-AP blast). You have decided to space the mines 6 meters apart. Determine the number of M15 mines to order and the number of 5-ton dump trucks (with sideboards) required to deliver the mines.

- Step 1. 0.5 × 400 meters = 200 mines
- Step 2. 400 ÷ 6 = 66.6 = 67 mines per row
- Step 3. 200 ÷ 67 = 2.98 = 3 rows
- Step 4. 67 × 3 = 201 mines
- Step 5. 201 × 1.1 = 221.1 = 222 mines
- Step 6. 222 ÷ 204 = 1.08 = 2 5-ton trucks
- Step 7.
  - — Concertina: ([400 × 2] + [200 × 2] + 160) × 1.4 = 1,904 meters of concertina
  - — Pickets: 1,904 ÷ 15 = 126.9 = 127 pickets
  - — Signs: 127 pickets = 127 signs

MINEFIELD REQUIREMENTS COMPUTATION WORK SHEET

GIVEN			
Desired density	AT ______	APF ______	APB _____
IOE representative cluster	AT ______	APF ______	APB _____
Front	______ meters		
Depth	______ meters		
Percentage of AHDs	______%		
Type of mines	AT ______	APF ______	APB _____
Type of truck/trailer	__________		
Lanes/gaps/traffic tapes	______		
Trip-wire safety tapes	______		
PART 1. NUMBER OF MINES			
A. IOE live clusters = front ÷ 9	______ ÷ 9 = _____ (round up)		
	AT	APF	APB
B. IOE representative cluster × Number of IOE clusters = Number of mines in IOE	______ × _____ ______	______ × _____ ______	______ × _____ ______
C. Desired density × Minefield front = Mines in regular strips	______ × _____ ______	______ × _____ ______	______ × _____ ______
D. Subtotal of mines (line B + line C)	______	______	______
E. 10% excess factor =	× 1.10	× 1.10	× 1.10
Total number of mines to order (round up for each)	______	______	______
PART 2. NUMBER OF REGULAR STRIPS			
A. Add desired density	AT _____	+ APF _____	+ APB ____ = _____
B. 0.6 × line A above	0.6 × _____ = _____ (round up)		
C. 3 × AT desired density	3 × _____ = _____		

Figure 6-1. Minefield requirements computation work sheet

D. Number of regular strips required = highest number of line B or C _____

PART 3. NUMBER OF AHDs

% AHDs × total number of AT mines _____

PART 4. STRIP CLUSTER COMPOSITION

A. Desired density

AT: 3 × _____ = _____ APF: 3 × _____ = _____ APB: 3 × _____ = _____

B. Cluster composition table

STRIP	AT	APF	APB	STRIP TOTAL (cannot exceed 5)
A	_____	_____	_____	_____
B	_____	_____	_____	_____
C	_____	_____	_____	_____
D	_____	_____	_____	_____
E	_____	_____	_____	_____
F	_____	_____	_____	_____
G	_____	_____	_____	_____
H	_____	_____	_____	_____
I	_____	_____	_____	_____
TOTAL (Cannot exceed desired density × 3 as computed in A above)	_____	_____	_____	_____

PART 5. NUMBER OF MAN-HOURS FOR INSTALLATION

Number of mines ÷ emplacement rate = mines per man-hour

Number of AT mines:______ ÷ 4 = _____ (round up)

Number of APF mines:______ ÷ 8 = _____ (round up)

Number of APB mines:______ ÷ 16 = _____ (round up)

_____ + _____ + _____ × 1.2 = _____ man-hours (round up)

Figure 6-1. Minefield requirements computation work sheet (continued)

PART 6. AMOUNT OF FENCING AND MARKING MATERIAL

A. Concertina wire or single-strand barbwire

([front × 2] + [depth × 2] + 160) × 1.4 = meters of concertina or single-strand barbwire required

([_____ × 2] + [_____ × 2] + 160) × 1.4 = _____ (round up)

Number of pickets = amount of concertina or single-strand barbwire ÷ 15

_____ ÷ 15 = _____ (round up)

- OR -

B. Double-strand barbwire

([front × 2] + [depth × 2] + 160) × 2.8 = meters of double-strand barbwire required

([_____ × 2] = [_____ × 2] + 160) × 2.8 = _____ (round up)

Number of pickets = amount of double-strand barbwire ÷ 30

_____ ÷ 30 = _____ (round up)

C. Number of signs = number of pickets = _____

PART 7. NUMBER OF TRUCKLOADS

AT mines

_____ cases per truck × _____ mines per case = _____ mines per truck

_____ mines required ÷ _____ mines per truck = _____ truckloads of AT mines

APF mines

_____ cases per truck × _____ mines per case = _____ mines per truck

_____ mines required ÷ _____ mines per truck = _____ truckloads of APF mines

APB mines

_____ cases per truck × _____ mines per case = _____ mines per truck

_____ mines required ÷ _____ mines per truck = _____ truckloads of APB mines

Total truckloads

_____ AT truckloads + _____ APF truckloads + _____ APB truckloads =

_____ total truckloads required (round up)

Figure 6-1. Minefield requirements computation work sheet (continued)

PART 8. AMOUNT OF ENGINEER TAPE

A. Minefield boundaries depth × 2 = _____ × 2 = _____

B. Regular strips front × number of regular strips = _____ × _____ = _____

C. IOE front + (number of IOE clusters × 3) = _____ + (____ × 3) = _____

D. Lanes and gaps depth × 2 × number of lanes and gaps = _____ × 2 × _____ = _____

E. Traffic tapes depth × number of traffic tapes _____ × _____ = _____

F. Trip-wire safety tape front × number of regular strips with trip wire _____ × _____ = ______

G. Subtotal (lines A + B + C + D + E + F)
_____ + _____ + _____ + ______ + _____ + _____ = _____ meters (round up)

H. Number of rolls to order (line G × 1.2)
______ × 1.2 = _____ meters

_____ meters ÷ 170 meters per roll = _____ rolls of engineer tape (round up)

PART 9. SANDBAGS

A. Number of clusters in IOE (from 1A) = _____

B. Number of clusters in minefield = number of clusters in IOE × 3 × number of regular strips (from 2D) = _____

C. Total number of clusters (line A + line B) = _____

D. Number of sandbags = number of clusters × 3 sandbags per cluster (line C × 3) = _____

Figure 6-1. Minefield requirements computation work sheet (continued)

Basic information pertaining to the minefield is normally determined by the engineer company commander or the staff engineer. It is provided to the OIC or NCOIC of the emplacing unit during the mission briefing. In this example, the following guidance is given to the emplacing unit:

Desired density	AT 1	APF 4	APB 8
IOE representative cluster	AT 1	APF 2	APB 2
Front	200 meters		
Depth	300 meters		
Percentage of AHDs	10%		
Type of mines	AT M15	APF M16A2	APB M14
Type of truck/trailer	5-ton dump (with sideboards)		
Lanes/gaps/traffic tapes	1 lane, 1 traffic tape (foot troops)		
Trip-wire safety tapes	3		

The rest of this work sheet is completed by using the above information.

The regular strip has a cluster density of one cluster every 3 meters. The IOE has a cluster density of one-third that of a regular strip, or one cluster every 9 meters. Therefore, to obtain the number of clusters in the IOE, the length of the strip is divided by 9. Decimals are rounded up to the next higher whole number.

PART 1. NUMBER OF MINES

Step 1.

IOE live clusters 200 ÷ 9 = 23 (rounded up)

The representative cluster composition for the IOE clusters is established and provided by the commander based on METT-TC factors. The number of clusters in the IOE is multiplied by the cluster composition to determine the number of mines, by type, in the entire IOE.

Step 2.

	AT	APF	APB
IOE representative cluster ×	1	2	2
Number of IOE clusters =	23	23	23
Number of mines in IOE	23	46	46

The minefield front multiplied by the desired density determines the number of mines in the minefield.

NOTE: The desired density pertains only to the regular strips and does not take into account the number of mines in the IOE which were calculated in Step 2.

Figure 6-2. Step-by-step procedures for completing the minefield requirements computation work sheet

Step 3.

Desired density ×	1	4	8
Minefield front =	200	200	200
Mines in regular strips	200	800	1,600

The number of mines required for the IOE (Step 2) is added to the number of mines in the regular strips (Step 3).

Step 4. Subtotal of mines

(Step 2 + 3)	223	846	1,646

Ten percent is added to the total number of mines required to allow for damaged items and irregularities in terrain and strip length. This is accomplished by multiplying the total number of mines (Step 4) by 1.1. Decimals are rounded up to the next higher whole number.

Step 5.

10% excess factor =	1.1	1.1	1.1
Total number of mines to order	246	931	1,811

These figures represent the total number of mines, by type, required for the entire minefield. When ordering by the case rather than by individual mines, the total should be divided by the number of mines per case and rounded up to the next whole case. (See Table 2-8, page 2-45.)

PART 2. NUMBER OF REGULAR STRIPS

Step 1.

Add desired density	AT 1	+ APF 4	+ APB 8 = 13

Each regular mine strip has a cluster every 3 meters; therefore, its density is one-third cluster per meter of front. A total density of 13 mines per meter of front in the previous example would equal 3 × 13 or 39 mines per 3 meters of front. Clusters may contain a maximum of five mines, so the resulting figure must be divided by 5. In short, to determine the minimum number of regular strips required, the total density must be multiplied by three-fifths (3 meters between clusters and five mines per cluster). For ease of calculation, three-fifths is converted to the decimal 0.6. Decimals are rounded up to the next highest whole number.

Step 2.

0.6 × Step 1	0.6 × 13 = 8 (rounded up)

The calculations to determine the minimum number of regular strips previously described are not suitable when the ratio of AT to AP mines is greater than 1:4. For example, if the desired density is 1-1-1, the total density is 3. The minimum number of strips would be 3 × 3/5 = 1.8, rounded up to 2 strips. However, because of the restriction on the number of AT mines per cluster, it is impossible to obtain a density of 1 AT mine per meter of front with only 2 strips. A minimum of 3 regular strips is required. The alternative means of determining the number of regular strips is founded by multiplying the AT desired density by 3.

Figure 6-2. Step-by-step procedures for completing the minefield requirements computation work sheet (continued)

Step 3.

3 × AT desired density 3 × 1 = 3

The number of regular strips calculated by the first method and the alternative method are compared, and the higher figure is used as the minimum number of regular strips. The 8 determined by the 3/5 rule is larger than the 3 determined by the alternative method. Therefore, the minimum number of regular strips in the example is 8.

Step 4.

Number of regular strips required = highest number of Step 2 or 3 = 8

PART 3. NUMBER OF AHDs

0.1 × 223 = 23 (rounded up)

PART 4. STRIP CLUSTER COMPOSITION

The cluster composition table is prepared by the OIC of the laying unit to control the allocation of mines to a regular strip. The cluster composition remains constant within a particular strip, but it may vary among different strips. As the mines are allocated by strip, no more than 1 AT mine can be placed in each representative cluster, and each cluster can have a maximum of 5 mines.

A tabular format is prepared (see Figure 6-1, page 6-6) to facilitate the distribution of mines by emplacement personnel. Note that each component of the desired density is multiplied by 3. The number 3 is always used regardless of the minimum number of regular strips because it is the number of mine strips required to give a minefield density of 1 mine per meter of front when a cluster contains only 1 mine of each type. Each mine strip has a cluster every 3 meters; therefore, it has a density of one-third mine per meter when a cluster contains 1 of each type of mine.

Step 1. Desired density

AT: 3 × 1 = 3 APF: 3 × 4 = 12 APB: 3 × 8 = 24

The total of each column in the table cannot exceed the number of mines above. For example, with an APF desired density of 4, 3 × 4 = 12, so the total APF mines in the representative cluster composition for each of the regular strips cannot exceed 12.

PART 5. NUMBER OF MAN-HOURS FOR INSTALLATION

Remember, the total number of mines includes the mines in regular strips and the mines in IOE short strips. The laying rates are—

AT mines: 4 per man-hour.

APF mines: 8 per man-hour.

APB mines: 16 per man-hour.

Figure 6-2. Step-by-step procedures for completing the minefield requirements computation work sheet (continued)

The number of man-hours required for each mine type is computed and rounded up. These amounts are totalled and a 20 percent excess factor is included by multiplying the total by 1.2. The resulting figure is the total number of man-hours required for emplacement and represents straight work time only. It does not take into account the time for transportation to and from the emplacement site, meals, and breaks; limited visibility; or NBC conditions. The commander should use his judgment and past experience to determine the time required for transportation, meals, and breaks. When working under limited visibility or NBC conditions, the total man-hours (after the excess factor has been included) should be multiplied by 1.5.

In this example, a total of 357 man-hours is required as determined below. Note that each decimal is rounded to the next higher whole number.

Number of mines ÷ emplacement rate = mines per man-hour

Number of AT mines	246 ÷ 4 = 62 (rounded up)
Number of APF mines	961 ÷ 8 = 121 (rounded up)
Number of APB mines	1,811 ÷ 16 = 114 (rounded up)
Total	62 + 121 + 114 = 297 × 1.2 = 357 (rounded up)

PART 6. AMOUNT OF FENCING AND MARKING MATERIAL

Standard-pattern minefields must be marked and fenced. The amount of fencing required depends on whether barbwire (single- or double-strand) or concertina is used.

The amount of wire for a single-strand barbwire or a single-strand concertina fence is calculated with the following formula:

([front × 2] + [depth × 2] + 160) × 1.4

The amount of wire for a double-strand barbwire fence is calculated with the following formula:

([front × 2] + [depth × 2] + 160) × 2.8

Step 1.

Concertina Wire or Single-Strand Barbwire

([200 × 2] + [300 × 2] + 160) × 1.4 = 1,624

The number of pickets required is determined by dividing the total amount of fence by 15.

Number of pickets = amount of fence ÷ 15   1,624 ÷ 15 = 109 (rounded up)

Step 2.

Double-Strand Barbwire

([200 × 2] + [300 × 2] + 160) × 2.8 = 3,248 (rounded up)

The number of pickets required is determined by dividing the total amount of fence by 30.

Number of pickets = amount of fence ÷ 30   3,248 ÷ 30 = 109 (rounded up)

Figure 6-2. Step-by-step procedures for completing the minefield requirements computation work sheet (continued)

The number of minefield marking signs is equal to the number of pickets.

NOTE: These calculations determine the marking and fencing materials required for the minefield perimeter only. Additional materials may be required for lanes and gaps.

PART 7. NUMBER OF TRUCKLOADS

The number of vehicles required depends on the type and amount of mines as well as the type of vehicles available. The total mines, by type, required is divided by the haul capacity of available vehicles to determine the number of truckloads required to transport the mines.

In this example, crated M15 AT mines, M16A2 APF mines, and M14 APB mines are hauled in M930 5-ton dump trucks (with sideboards). (See Table 2-8, page 2-45).

AT mines:

246 mines required ÷ 204 mines per truck = 1.2 truckloads of AT mines

APF mines:

931 mines required ÷ 888 mines per truck = 1.05 truckloads of APF mines

APB mines:

1,811 mines required ÷ 13,770 mines per truck = 0.13 truckloads of APB mines

Total truckloads:

1.2 AT truckloads + 1.05 APF truckloads + 0.13 APB truckloads = 2.38 truckloads = 3 truckloads (rounded up) required

PART 8. AMOUNT OF ENGINEER TAPE

An extensive amount of engineer tape is used to mark the initial layout of a standard-pattern minefield. Engineer tape comes in 170-meter rolls and is used to mark several portions of the minefield.

NOTE: In this example, only one lane and one roll of traffic tape is required.

Step 1.		
Minefield boundaries	depth × 2 300 × 2 = 600	
Step 2.		
Regular strips	front × number of regular strips	200 × 8 = 1,600
Step 3.		
IOE	front + (number of IOE clusters × 3)	200 + (23 × 3) = 269
Step 4.		
Lanes and gaps	depth × 2 × number of lanes and gaps	300 × 2 × 1 = 600
Step 5.		
Traffic tape	depth × number of traffic tapes	300 × 1 = 300

Figure 6-2. Step-by-step procedures for completing the minefield requirements computation work sheet (continued)

Step 6.

Trip-wire safety tape — front × number of regular strips with trip wire — 200 × 3 = 600

Step 7.

Total of Steps 1 through 6 = 600 + 1,600 + 269 + 600 + 300 + 600 = 3,969 meters

Step 8.

Add 20% excess — total amount of engineer tape required for the minefield × 1.2 = 3,969 × 1.2 = 4,762.8 = 4,763 (rounded up)

Step 9.

Total number of rolls — total amount of engineer tape, in meters, from Step 8 ÷ 170 meters per roll = 4,763 meters ÷ 170 = 28.02 = 29 rolls

PART 9. SANDBAGS

To determine the number of sandbags for the removal of spoil.

Step 1.

Number of clusters in IOE (Part 1, Step 1) = 23

Step 2.

Number of clusters in minefield = number of clusters in IOE × 3 × number of regular strips (Part 1, Step 4)

23 × 3 × 8 = 552

Step 3.

Total number of clusters (add Steps 1 and 2) = 575

Step 4.

Number of sandbags = number of clusters × 3 sandbags per cluster

575 x 3 = 1,725

Figure 6-2. Step-by-step procedures for completing the minefield requirements computation work sheet (continued)

Task Organization

To maximize the efficiency of the row-mining process, the platoon leader must task-organize his platoon. The organization of the task, as a whole, is intricate and places great demands on the leader. Leave nothing to chance when planning and executing a row minefield, because each situation is different. Make allowances for transporting, handling, and controlling the mines. The officer in charge (OIC) and the squad leaders must be able to exercise control throughout the task under all conditions. Always observe safety.

Organize the platoon into four parties—siting and recording, marking, mine dump, and laying.

Siting-and-Recording Party

The platoon leader directs the party and is responsible for siting, recording, and reporting the minefield. This party consists of one or two soldiers and a vehicle to carry material. (If a vehicle is not available, increase the party to three soldiers.) Because siting is usually done in daylight, the party must take appropriate physical-security measures. The party starts well ahead of the actual laying, sets out control markers, and avoids using sharp turns. The party marks the vehicle traffic routes to and from the minefield rows.

When siting is complete, the OIC identifies one member of the party to be the recorder and assigns the remaining soldiers to other tasks. The recorder collects data from the laying party NCOIC and completes DA Form 1355 as outlined in Chapter 8. The OIC ensures that the DA Form 1355 is completed timely and accurately.

Marking Party

This party is composed of an NCOIC and personnel who are not working as members of other teams. After the minefield is sited, the marking party emplaces fence posts, wire, and marking signs.

Mine-Dump Party

This party is controlled by the platoon sergeant (PSG) and is composed of personnel who are not working as members of other teams. The mine-dump party accounts for all strip packages that arrive from other sources, sets up vehicle mine sets at the mine dump, and hauls supplies as required. The PSG places row packages in a location that maximizes speed and provides concealment for minefield emplacement, and he also ensures that the mine dump is prepared for night operations. The party marks the mine dump's entrance and exit and the routes to them. The PSG verifies the strip feeder reports with the squad leaders upon the completion of each row and passes the reports to the recording party. The PSG is not required to stay at the mine dump continuously; he has the flexibility to move around the area to perform other activities.

The mine-dump party creates vehicle sets by setting aside the number of mines and fuses that are required by each laying vehicle. The party loosens and then hand-tightens arming and shipping plugs, helps load the mines onto laying vehicles, and disposes of residue. Soldiers may also assist the marking party and provide local security. For initial vehicle loads, the mine-dump party may be assisted by the laying party.

Laying Party

This party consists of an NCOIC, four soldiers, and a vehicle to carry the mines. The NCOIC controls the movement of each laying vehicle. He directs each vehicle to start and stop laying and controls immediate-action drills. The NCOIC initiates a strip feeder report with the PSG or the mine-dump NCOIC, receives azimuths from the siting party, and directs his element to the correct row. He is responsible for replacing the temporary row markers with permanent markers and for ensuring that mines are laid according to the azimuths, mines are spaced correctly, and the strip feeder report is accurate.

The NCOIC ensures that the end row marker is emplaced at the completion of each row, and he closes the strip feeder report with the PSG.

NOTE: Using tilt-rod fuses requires additional soldiers to stake mines, insert fuses, and arm mines.

When laying three rows at once, each laying party consists of an APC or an organic squad vehicle and a carrier, sapper, and digging team.

Carrier Team

This team is comprised of the APC driver and the track commander (TC). They ensure that the APC maintains the proper speed and stays on the proper course.

Sapper Team

This team is composed of the squad leader and the remaining squad members. It provides personnel to lay and arm mines. Each soldier carries wrenches and fuses. The squad leader supervises laying and tasks personnel who are not needed for laying to other parties.

Digging Team

The digging team buries mines. It consists of an NCOIC and several soldiers (may be soldiers from supported maneuver units) who are equipped with suitable digging tools. Increase the arming party by two to speed up the laying process or task personnel who are not needed to other parties. **NOTE: If mines are surface-laid, there is no digging team.**

SITE LAYOUT

Once the platoon leader has coordinated the location of the minefield(s) with the maneuver commander, siting in the minefield can begin. Siting is the first step in the actual laying process and is done for safety and control. Although the minefield may be emplaced at night or during limited visibility, the siting party should site the minefield under favorable conditions, preferably during daylight. Siting consists of identifying landmarks; establishing routes; and emplacing start, end, and intermediate row markers. Actual control measures (stakes or pickets) should not stand out to such an extent that they give away the minefield orientation, but they must be easily discernible to the laying party.

Certain features, like thick woods and deep, wide streams, are natural obstacles. Mine rows should be laid to reinforce terrain and increase the effectiveness of the minefield.

Mine Rows

Mine rows are labeled with a letter and should be laid in order. Row A is nearest the enemy, followed by rows B, C, D, E, and so forth. When laying tactical minefields, each row has permanent start and end row markers. Intermediate markers may be required, depending on the row length and the terrain. Platoon leaders determine the number of laying vehicles to be employed. The preferred technique is to use three vehicles so that three rows can be laid simultaneously. Using more than three vehicles is beyond the platoon's C^2 capabilities and is not considered. The distance between rows is determined by the following factors:

- Depth and density of the minefield.
- Terrain.
- Suitability of the ground.
- Desired obstacle intent.

NOTE: Rows are spaced 50 meters apart in standardized row minefields (discussed later in this chapter).

Mine Spacing

The minefield OIC determines the mine spacing. The desired density, the availability of laying vehicles, the number of rows, and the possibility of sympathetic detonation (Table 5-3, page 5-8) affect the distance between mines. **NOTE: Mines are spaced 6 meters apart in standardized row minefields (discussed later in this chapter).**

Control Measures

Control measures are temporary markers that are used to guide vehicles and troops during row-mining operations. Markers are constructed of different materials for different uses. For example, use VS17 panels on poles for start and end row markers, and use M133 hand-emplaced minefield marking set (HEMMS) poles with flags for intermediate markers. Use the following temporary markers:

- Start row (does not replace the mandatory permanent marker).
- Start laying (first intermediate marker after the start row marker).
- Intermediate (used between the last row marker and the next visible point, not more than 100 meters away).
- Change of direction or turning point (actually consists of three markers—warning, turning point, and new direction).
- Stop laying.
- End row (does not replace the mandatory permanent marker).

The following materials may be used to construct temporary markers:

- U-shaped pickets.
- HEMMS poles.
- Wooden posts.
- Steel rods.
- Engineer tape.
- VS17 panels.

Control measures for laying mines at night require lights or infrared (IR) equipment as follows:

- Chem-lights placed in U-shaped pickets or hand-held.

- Directional flashlights taped in U-shaped pickets or hand-held.
- HEMMS lights used with U-shaped pickets or poles.
- Lights from a minefield marking set number 2.
- IR reflectors.

NOTE: The use of control measures should be incorporated into unit standard operating procedures (SOPs).

Procedures

The minefield OIC arrives on the site with the siting-and-recording party. He selects Landmark 1 and then sites the left (or right) boundary fence and start row markers (all start and end row markers are permanent markers). The siting-and-recording party takes distances and azimuths to be used in preparing the recording form. If the tactical situation permits and the marking party is ready, emplacement of the fence should begin.

If the minefield is to have an IOE row, the siting-and-recording party proceeds across the IOE and establishes I1, I1E, I2, I2E, and so on until it reaches the end. Personnel proceed down the right (or left) boundary and emplace start row marker A1. Proceeding from A1 to A2, they place intermediate markers as required. Personnel use different colored markers to identify each row (for example: Row A, red light; Row B, green light; Row C, blue light). For IR markings, they use multiple horizontal IR light sources that are spaced at least 6 inches apart (for example: Row A, one light; Row B, two lights; Row C, three lights). When they reach A2, they emplace an end row marker and repeat the procedure from B1 to B2, C1 to C2, and so on until they emplace all the required control measures (Figure 6-3). The siting-and-recording party establishes Landmark 2 and the left (or right) rear fence location. Personnel also assist the PSG in siting mine dumps near the minefield.

MINE-LAYING VEHICLES

Soldiers normally lay row minefields from a tactical vehicle. Consider vulnerability, capacity, and trafficability when selecting a vehicle. Before emplacing the minefield and preparing the vehicle for mine laying, drive it in a random pattern across the minefield site. The random pattern deceives the enemy by masking the actual laying pattern. Load enough mines so that each vehicle can complete at least one entire row before reloading, but do not stack fused mines more than two-high.

LAYING A ROW MINEFIELD

The following drills demonstrate how to lay a minefield:

Drill 1

Squad vehicles arrive on the site and proceed down the left (or right) boundary of the minefield to their assigned row. (A separate party must be detailed to install the IOE.) At the start row marker, the squad vehicle moves into position and prepares to lay mines. The squad leader for Row A directs Vehicle 1 to move out.

Mines are laid on the ground at the required spacing, along the temporary markers positioned by the siting-and-recording party.

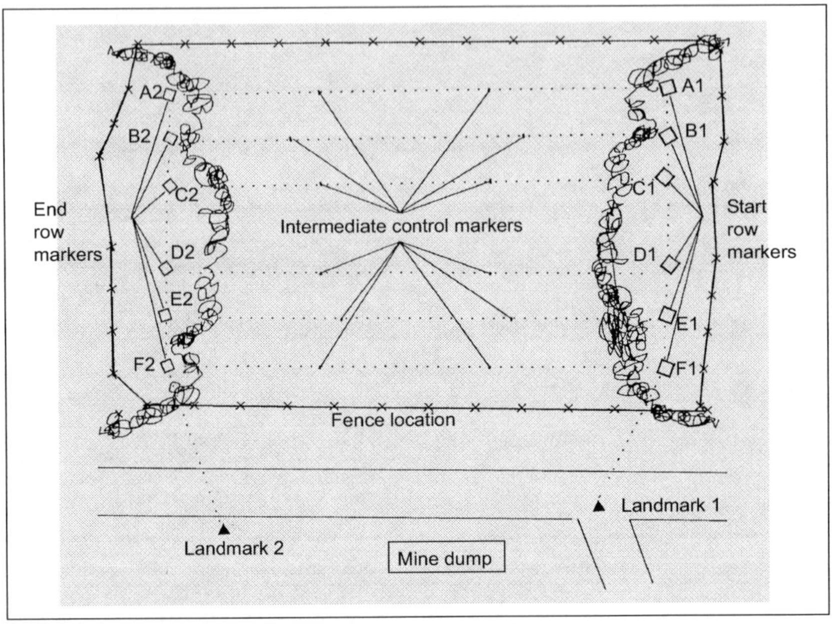

Figure 6-3. Site layout

As mines are laid, the arming party moves behind the vehicle and arms the mines. Personnel remove temporary markers installed by the siting-and-recording party and replace the end row markers with permanent markers.

When Vehicle 1 moves a safe distance (approximately 25 meters) along Row A, Vehicle 2 begins to lay mines on Row B. When Vehicle 2 moves a safe distance along Row B, Vehicle 3 begins to lay mines on Row C.

The marking party continues to emplace the left and right boundary fences (Figure 6-4a, page 6-20).

The IOE party exits the minefield outside the left (or right) boundary after it completes the IOE.

When Vehicles 1 and 2 finish their assigned rows, they move past the end row marker and execute a left (or right) turn and wait for Vehicle 3 to complete its row. All vehicles move in column down the left (or right) boundary to the mine dump, load the next row's mines, and then move to their next assigned row. The process of laying and arming mines is repeated (Figure 6-4b, page 6-21).

After the minefield is laid, all the vehicles exit down the left (or right) boundary and out the rear. The marking party completes the rear boundary fence, and the recording party completes DA Form 1355. The OIC or PSG ensures that the DA Form 1355 is complete and accurate and signs it.

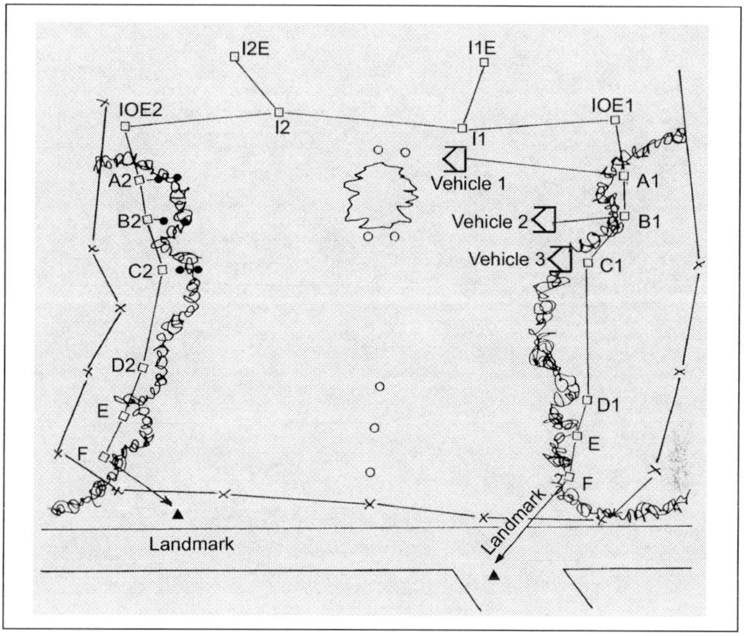

Figure 6-4a. Laying a minefield

Drill 2

This drill may be used to speed up mine laying; however, strict C^2 is vital to ensure security and safety. This method is difficult to use when the terrain is rugged or when weather or visibility is subject to change.

The drill is conducted by three squad vehicles, each laying one row. Row B has turning points and Rows A and C do not. If the minefield has six rows, Row E has turning points and Rows D and F do not. The squad leader (laying leader) in Row B (and Row E, if required) is in charge of the overall laying.

Squad vehicles arrive on the site and proceed down the left (or right) boundary of the minefield to their assigned row. Squad vehicles move into position at start row markers and prepare to lay mines.

The laying leader directs Vehicle 1 to move out on Row A. The sapper team lays mines on the ground at the required spacing. If an IOE is required, the Row A team emplaces the IOE concurrently with Row A and at the same spacing. When Vehicle 1 reaches the IOE short-row start marker, the laying party lays mines along an azimuth designated by the laying leader (Figure 6-5). After the IOE short row is laid, Vehicle 1 returns to Row A and continues laying mines.

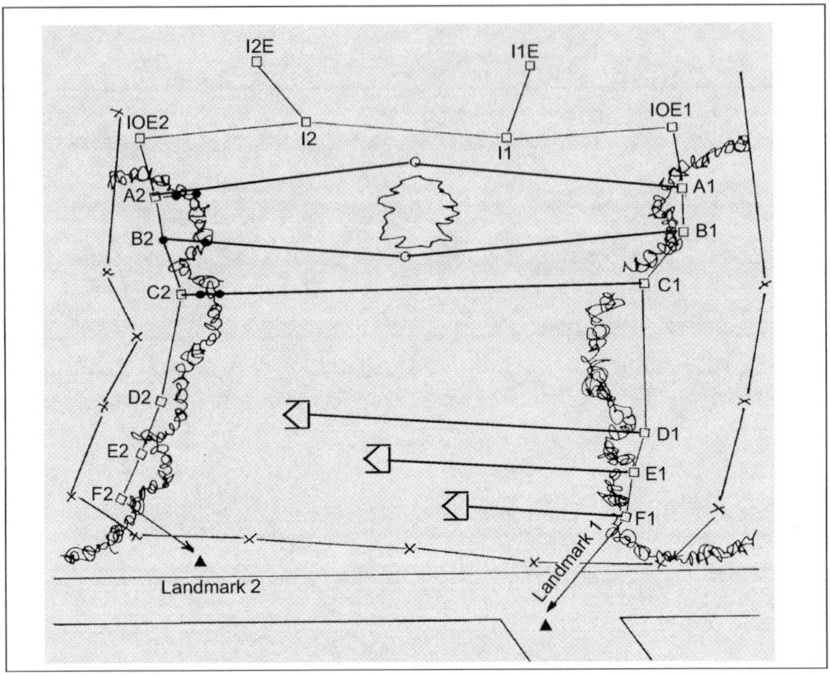

Figure 6-4b. Laying a minefield (continued)

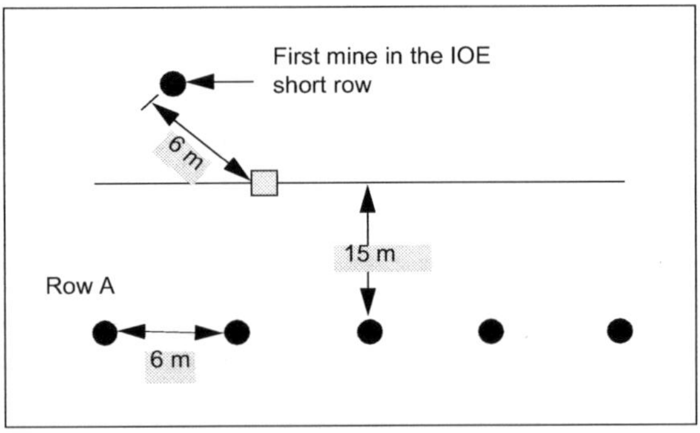

Figure 6-5. Laying an IOE short row

After all the mines are laid, the arming party moves behind the vehicle and arms the mines. Personnel remove temporary markers and replace start and end row markers with permanent markers. The arming party must be distinguishable from everyone else. The last member of the arming party should wear a colored vest or carry a specific colored chem-light. No one is allowed behind the last member of the arming party.

The NCOIC completes a strip feeder report (Figure 6-6) and gives it to the recording party. The strip feeder report includes the number of mines laid, the type of mines laid, azimuths of IOE strips and turning points, AHDs emplaced (by cluster number), and any other information (such as omitted mines) the platoon leader requires.

STRIP FEEDER REPORT FOR STRIP/ROW ________					
Type of Mine	Number of Mines	AHDs by Cluster	IOE-Strip Azimuth	Turning-Point Azimuth	Remarks

Figure 6-6. Sample strip feeder report

Vehicle 1 moves down Row A and lays mines until the laying leader directs it to stop. (The laying leader chooses vehicle stops to coincide with the locations of turning points.) The laying leader then directs Vehicle 3 to begin laying mines along Row C. Vehicle 3 lays mines on Row C until the laying leader directs it to stop (somewhere well past Vehicle 1).

Vehicle 2 moves toward Vehicle 1 and begins to lay mines on Row B. He lays mines to within 15 meters of Vehicle 1. Vehicle 2 then turns toward Vehicle 3 and lays mines on Row B to within 15 meters of Vehicle 3. Vehicle 2 then turns back toward Vehicle 1 and continues to lay mines in this pattern until Row B is laid.

Figure 6-7 shows vehicle positions when using the above method to lay a row minefield.

NOTES:

1. At night or during low visibility, Vehicle 1 has two red flashlights and Vehicle 3 has two green flashlights. The flashlights are held side by side, and pointed toward Vehicle 2. The driver of Vehicle 2 moves forward until he is within 15 meters of the lights or until the light holder turns the lights off.

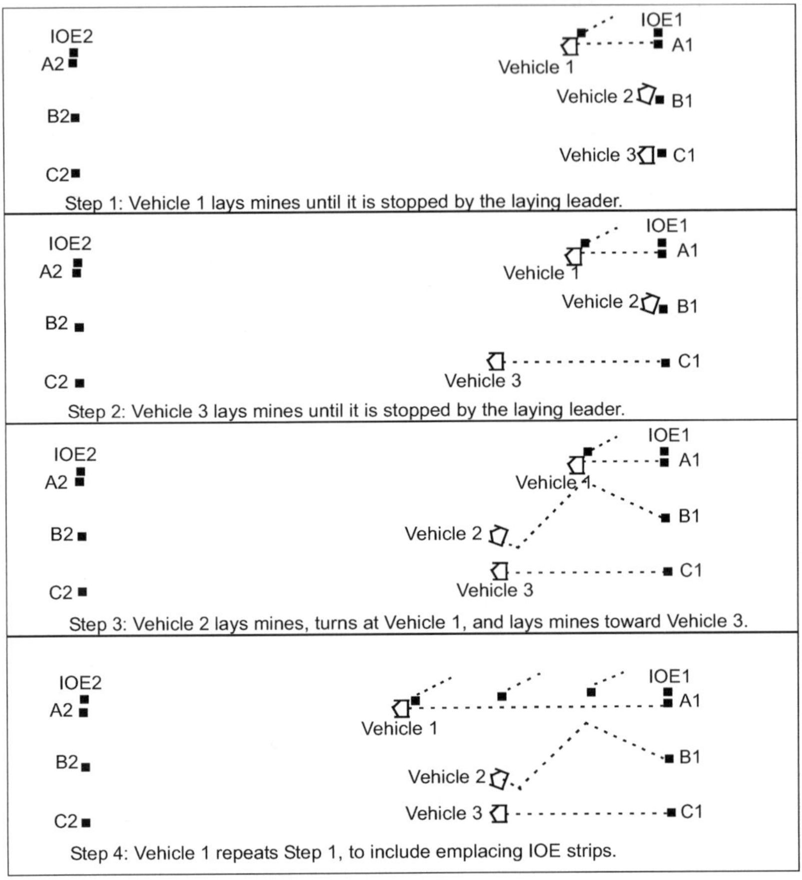

Figure 6-7. Laying a row minefield

2. If the platoon leader feels that low visibility or other reasons preclude the use of vehicle positions as turning points, he may have the siting party emplace turning-point markers (three intermediate markers) for Vehicle 2 to use as a guide. In this event, the three vehicles emplace mines simultaneously.

After Vehicles 1 and 2 finish their assigned rows, they move past the end row marker, execute a left (or right) turn, and wait for Vehicle 3 to complete its row. All the vehicles move in column down the left (or right) boundary to their next assigned row, if there is one, and continue to lay and arm mines. This process is repeated until the entire minefield is laid. All the vehicles then exit the minefield down the left (or right) boundary and out the rear. The marking party completes the rear boundary fence, and the recording party completes DA Form 1355. The OIC or PSG ensures that the DA Form 1355 is complete and accurate and signs it.

Immediate-Action Drill

If the enemy attacks the platoon during minefield emplacement, the laying party should execute the following actions:

- Sapper teams enter vehicles and recover spacing sandbags.
- Vehicle 1 exits the minefield by making a wide turn around the front of the other two vehicles.
- Vehicle 2 follows by making a wide turn around the front of Vehicle 3.
- Vehicle 3 exits the minefield.
- The three squads conduct the immediate-action drill as ordered by the platoon leader.

Squad Drill

During row mining, the squad in each laying vehicle performs the following actions:

- Squad leader.
  - Directs the squad to start laying mines.
  - Supervises mine arming and placing.
  - Allocates a vehicle, if possible, to help remove spoil from the site.
- Carrier team.
  - Moves the APC to the row start point.
  - Lowers the APC ramp until it is horizontal or opens the rear door. (If using the APC ramp to distribute mines, chains the ramp open to support the weight.)
  - Moves the APC at a low speed (3 to 5 kph) in a straight line toward the row end point.
- Sapper team.
  - Soldier 1 ties the rope to the end of the lowered ramp or the tow pintle.
  - Soldier 2 ties the partially filled sandbag on the other end of the rope. (The rope length from the end of the ramp door to the sandbag is the correct spacing between mines [Figure 6-8]).
  - Soldier 3 (squad leader) positions the team members. Soldier 1 is at the rear of the compartment, Soldier 2 sits on the edge of the APC ramp or open door, and Soldier 4 walks behind the APC.
  - Soldier 1 fuses a mine and passes it to Soldier 2. **(Korea Only: If AP mines are also laid, they are given out simultaneously.)**
  - Soldier 2 records all the mines issued.
  - Soldier 2 places the fused mine on the ground when the sandbag tied to the rope is even with the previously laid mine.
  - Soldier 3 (squad leader) walks behind the vehicle and supervises mine laying.

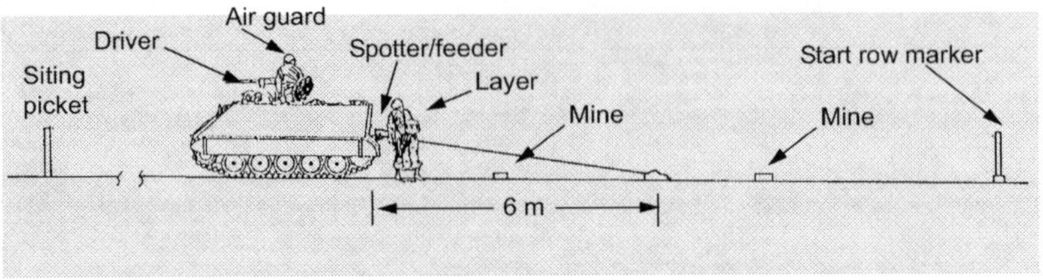

Figure 6-8. Measuring distances between mines with sandbags

- Soldier 4 walks behind the vehicle and arms mines.
- After the mine row is armed and camouflaged, Soldier 4 buries pins, clips, and shipping plugs 30 centimeters to the rear of the start row marker.
- The sapper team repeats the above steps until the end of the row is reached.

- Digging team, if needed. (The NCOIC selects the mine to be buried by each soldier and supervises the operation.)
  - Follows the laying party along the friendly side of the row.
  - Digs in mines but leaves them exposed until arming is complete.
  - **Korea Only: Arms AP mines in a cluster before arming AT mines.**

MARKING, RECORDING, AND REPORTING ROW MINEFIELDS

Marking procedures for row minefields are the same as those for other minefields (see Chapter 2).

Row minefields are recorded on DA Form 1355 (Figures 6-9a and 6-9b, pages 6-26 and 6-27). Reporting procedures for intent, initiation, status, and completion reports are detailed in Chapter 8.

STANDARDIZED TACTICAL ROW MINEFIELDS

The specific composition of a tactical row minefield depends on METT-TC factors and available resources. To aid in standardization of platoon techniques, four compositions have been developed to match desired obstacle effects. Using standardized minefields facilitates planning the obstacle type, size, and logistical requirements. It is imperative that the design and the effect of these minefields are well understood. They are an integral part of combined arms obstacle doctrine and form the cornerstone of engineer obstacle operations.

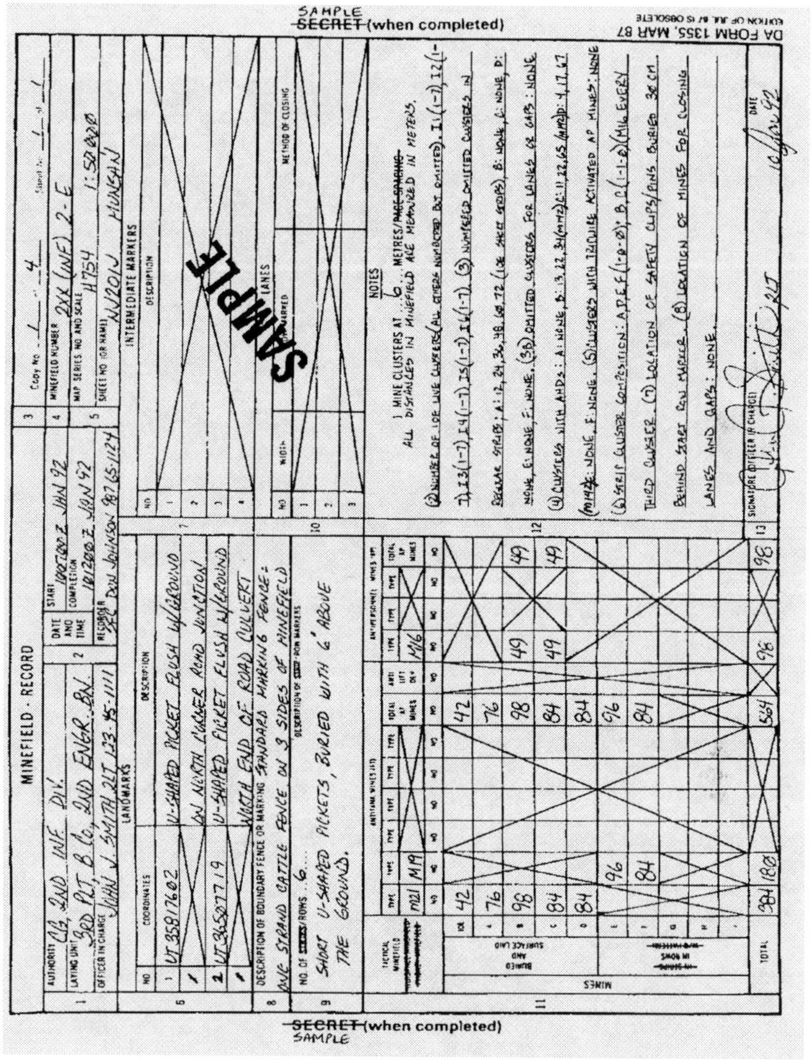

SAMPLE
~~SECRET~~ (when completed)

MINEFIELD - RECORD

1 AUTHORITY: CG, 2ND INF. DIV.
LAYING UNIT: 3RD PLT, B Co., 2ND ENGR. BN.
OFFICER IN CHARGE: JOHN J. SMITH, 2LT, 123-45-1111

2 DATE AND TIME: START 100700Z JAN 92; COMPLETION 101200Z JAN 92
RECORDER: SFC DON JOHNSON 987-65-1124

3 Copy No. 1 of 4 — Sheet No. 1 of 1
4 MINEFIELD NUMBER: 2XX (INF) 2-E
5 MAP SERIES NO AND SCALE: H754 1:50,000
SHEET NO (OR NAME): NJ201J MUNSAN

6 LANDMARKS

NO	COORDINATES	DESCRIPTION
1	UT 35817602	U-SHAPED PICKET FLUSH W/GROUND
✓		ON NORTH CORNER ROAD JUNCTION
2	UT 36507719	U-SHAPED PICKET FLUSH W/GROUND
✓		NORTH END OF ROAD CULVERT

7 INTERMEDIATE MARKERS

NO	DESCRIPTION
1	
2	
3	
4	

8 DESCRIPTION OF BOUNDARY FENCE OR MARKING: STANDARD MARKING FENCE: ONE STRAND CATTLE FENCE ON 3 SIDES OF MINEFIELD

9 NO. OF ~~STRIPS~~/ROWS: 6
DESCRIPTION OF ~~STRIP~~-ROW MARKERS: SHORT U-SHAPED PICKETS, BURIED WITH 6" ABOVE THE GROUND.

10 LANES

NO	WIDTH	HOW MARKED	METHOD OF CLOSING
1			
2			
3			

SAMPLE

11 MINES

TACTICAL MINEFIELD		AT TYPE m21 NO	AT TYPE M19 NO	TYPE NO	TYPE NO	TYPE NO	TYPE NO	TOTAL AT MINES NO	ANTI LIFT DEV NO	AP TYPE M16 NO	TYPE NO	TYPE NO	TOTAL AP MINES NO
BURIED AND SURFACE LAID	IOE	42						42					
	A	76						76					
	B	98						98		49			49
	C	84						84		49			49
	D	84						84					
SNOW MI	E		96					96					
	F		84					84					
	G												
	H												
	I												
TOTAL		384	180					564		98			98

12 NOTES
1 MINE CLUSTERS AT 6 METRES/~~PACE SPACING~~
ALL DISTANCES IN MINEFIELD ARE MEASURED IN METERS.
(2) NUMBER OF IOE LIVE CLUSTERS (ALL OTHERS NUMBERED BUT OMITTED). I1(1-7), I2(1-7), I3(1-7), I4(1-7), I5(1-7), I6(1-7). (3) NUMBERED OMITTED CLUSTERS IN REGULAR STRIPS; A: 12, 24, 36, 48, 60, 72 (IOE SHORT STRIPS), B: NONE, C: NONE, D: NONE, E: NONE, F: NONE. (3B) OMITTED CLUSTERS FOR LANES OR GAPS: NONE. (4) CLUSTERS WITH AHDs: A: NONE, B: 13, 22, 34 (M1A2) C: 11, 22, 65 (M1A2) D: 4, 17, 67 (M1A2) E: NONE, F: NONE. (5) CLUSTERS WITH TRIPWIRE ACTIVATED AP MINES: NONE (6) STRIP CLUSTER COMPOSITION: A, D, E, F (1-0-0); B, C (1-1-0) (M16 EVERY THIRD CLUSTER. (7) LOCATION OF SAFETY CLIPS/PINS. BURIED 30 CM BEHIND START ROW MARKER. (8) LOCATION OF MINES FOR CLOSING LANES AND GAPS: NONE.

13 SIGNATURE (OFFICER IN CHARGE): [signature] 2LT — DATE: 10 Jan 92

DA FORM 1355, MAR 87 — EDITION OF JUL 67 IS OBSOLETE

~~SECRET~~ (when completed)
SAMPLE

Figure 6-9a. Sample DA Form 1355 for a row minefield (front)

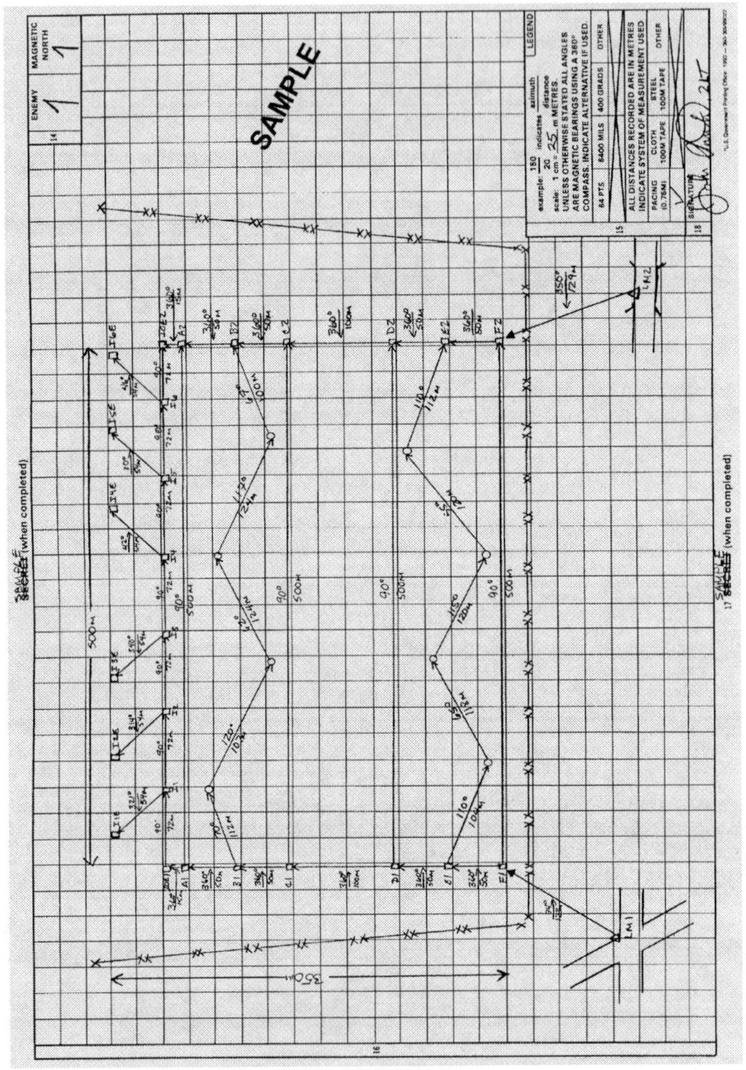

Figure 6-9b. Sample DA Form 1355 for a row minefield (back)

Disrupt and Fix

Disrupt and fix row minefields (Figure 6-10) are similarly constructed, but a fix minefield has an IOE and does not have AHDs.

DISRUPT

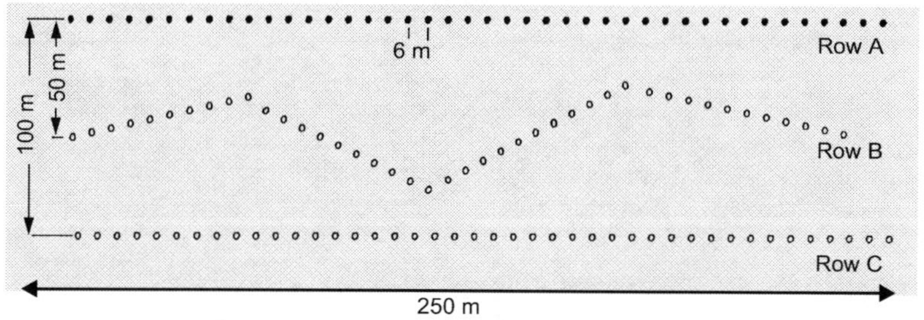

	Disrupt	Fix
Front	250 m	250 m
Depth	100 m	120 m
Rows of AT full-width mines	1	1
Rows of AT track-width mines	2	2
IOE	No	Yes
AHD	No	No
Platoon hours required	1.5	1.5
Full-width mines	42	63
Track-width mines	84	84
Density	0.5	0.6

FIX

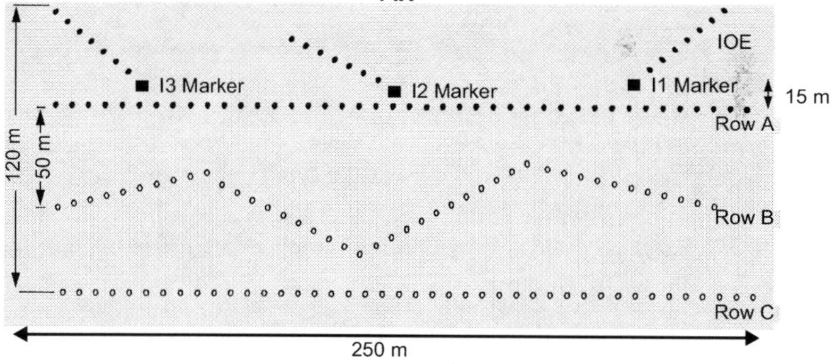

Figure 6-10. Standardized disrupt and fix row minefields

Disrupt and fix row minefields are emplaced as follows:

- Row A.
  - — 42 full-width AT mines (tilt-rod) are placed 6 meters apart.
  - — No turning points.
  - — Mines are surface-laid (staked) or buried.
- Row B.
  - — Start and end row markers are emplaced 50 meters behind Row A.
  - — 42 track-width AT mines are placed 6 meters apart.
  - — Preferably no more than 3 turning points.
  - — Mines are surface-laid or buried.
- Row C.
  - — Emplaced 100 meters behind Row A.
  - — 42 track-width AT mines are placed 6 meters apart.
  - — No turning points.
  - — Mines are surface-laid or buried.
- IOE (fix minefield only).
  - — 3 IOE short rows.
  - — A separate IOE baseline.
  - — The IOE baseline is on the enemy side, 15 meters from Row A.
  - — 7 full-width AT mines are placed 6 meters apart on each IOE short row.
  - — Mines are buried.
  - — The first IOE short row is placed 48 meters from the IOE end marker, the next short row is 84 meters from the first short row, and the third short row is 84 meters from the second short row.

TURN

A turn minefield (Figure 6-11, page 6-30) consists of four rows of full-width mines and two rows of track-width mines.

A turn row minefield is emplaced as follows:

- Row A.
  - — 84 full-width AT mines (tilt-rod) are placed 6 meters apart.
  - — No turning points.
  - — Mines are surface-laid (staked) or buried.
- Row B.
  - — Start and end row markers are emplaced 50 meters behind Row A.
  - — 84 full-width AT mines (tilt-rod) are placed 6 meters apart.

NOTE: The center of the minefield shown in this figure is omitted for clarity.

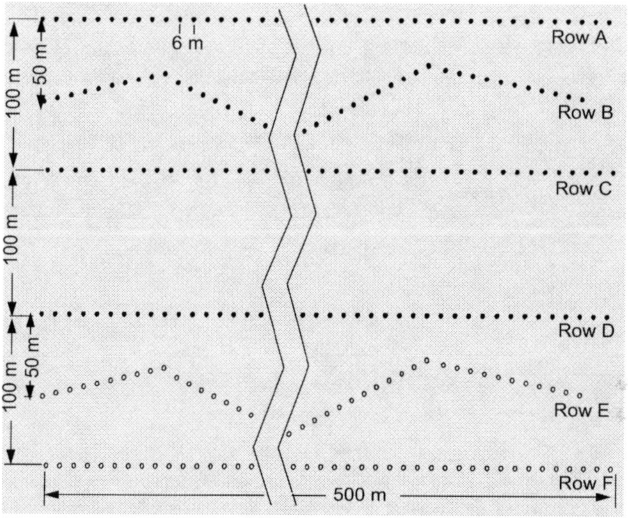

Front	500 m
Depth	300 m
Rows of AT full-width mines	4
Rows of AT track-width mines	2
IOE	No
AHD	No
Platoon hours required	3.5
Full-width mines	336
Track-width mines	168
Density	1.0

Figure 6-11. Standardized turn row minefield

— Preferably no more than 5 turning points.

— Mines are surface-laid (staked) or buried.

- Row C.

— Emplaced 100 meters behind Row A.

— 84 full-width AT mines (tilt-rod) are placed 6 meters apart.

— No turning points.

— Mines are surface-laid (staked) or buried.

- Row D.
  - Emplaced 100 meters behind Row C.
  - 84 full-width AT mines (tilt-rod) are placed 6 meters apart.
  - No turning points.
  - Mines are surface-laid (staked) or buried.
- Row E.
  - Start and end row markers are emplaced 50 meters behind Row D.
  - 84 track-width AT mines are placed 6 meters apart.
  - Preferably no more than 5 turning points.
  - Mines are surface-laid or buried.
- Row F.
  - Emplaced 100 meters behind Row D.
  - 84 track-width AT mines are placed 6 meters apart.
  - No turning points.
  - Mines are surface-laid or buried.

BLOCK

A block minefield (Figure 6-12, page 6-32) has an IOE **(Korea Only: and AP mines)** and has 20 percent AHDs in two of its rows of full-width mines. AHDs are placed in Rows B and C for the best effect. **(Korea Only: A block minefield also requires 84 M16 or M14 AP mines and has a density of 0.17 M16/M14 AP mine per linear meter.)**

A block row minefield is emplaced as follows:

- Row A.
  - 84 full-width AT mines (tilt-rod) are placed 6 meters apart.
  - No turning points.
  - Mines are surface-laid (staked) or buried.
- Row B.
  - Start and end row markers are emplaced 50 meters behind Row A.
  - 84 full-width AT mines (tilt-rod) are placed 6 meters apart.
  - Preferably no more than 5 turning points.
  - Mines are surface-laid (staked) or buried.
- Row C.
  - Emplaced 100 meters behind Row A.
  - 84 full-width AT mines (tilt-rod) are placed 6 meters apart.
  - No turning points.

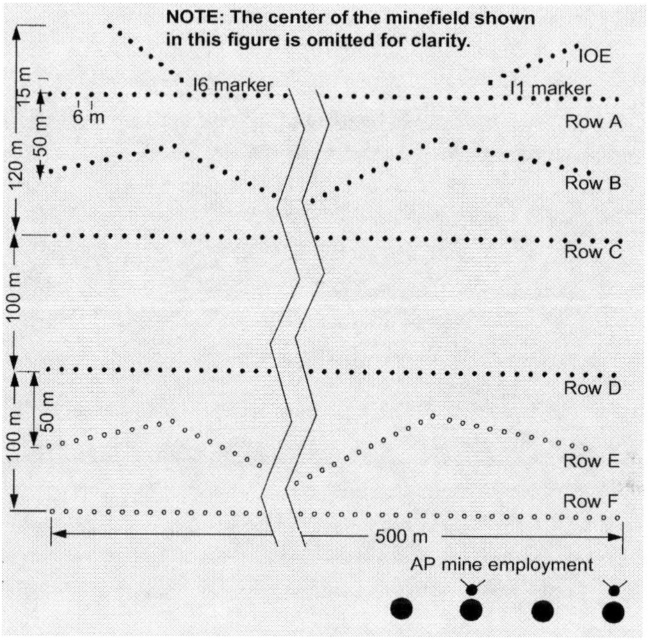

Front	500 m
Depth	320 m
Rows of AT full-width mines	4
Rows of AT track-width mines	2
IOE	Yes
AHD	Yes
Platoon hours required	5
Full-width mines	378
Track-width mines	168
AT mine density	1.1
Korea Only: **AP frag mines**	**84**
Korea Only: **AP mine density**	**0.17**

Figure 6-12. Standardized block row minefield

  - Mines are surface-laid (staked) or buried.
- Row D.
  - Emplaced 100 meters behind Row C.
  - 84 full-width AT mines (tilt-rod) are placed 6 meters apart.
  - No turning points.
  - Mines are surface-laid (staked) or buried.
- Row E.
  - Start and end row markers are emplaced 50 meters behind Row D.
  - 84 track-width AT mines are placed 6 meters apart.
  - Preferably no more than 5 turning points.
  - Mines are surface-laid or buried.
- Row F.
  - Emplaced 100 meters behind Row D.
  - 84 track-width AT mines are placed 6 meters apart.
  - No turning points.
  - Mines are surface-laid or buried.
- IOE.
  - 6 IOE short rows.
  - A separate IOE baseline.
  - 7 full-width AT mines are placed 6 meters apart on each IOE short row.
  - Mines are buried.
  - The first IOE short row is placed 72 meters from the IOE end marker, and the five subsequent IOE short rows are placed at 72-meter intervals on the IOE baseline.
- **Korea Only: AP mines.**
  - **AP mines are placed on two rows of full-width AT mines.**
  - **AP mines are placed in a cluster around AT mines.**
  - **One AP mine is placed in front of every other AT mine.**

HASTY PROTECTIVE ROW MINEFIELDS

Hasty protective row minefields are temporary in nature and are used as part of a unit's defensive perimeter. Units usually use mines from their basic load. If time permits, mines should be buried to increase their effectiveness, but they can be surface-laid. The minefield can be easily recovered because AHDs, nonmetallic mines, or low-metallic mines are not used. Mines are employed outside the hand-grenade range but within the range of small-caliber weapons. All mines are picked up by the emplacing unit upon leaving the area, unless enemy pressure prevents mine retrieval or the minefield is being transferred to a relieving commander.

RULES

The brigade commander has the initial authority to employ hasty protective row minefields. This authority may be delegated to the battalion or company commander on a mission-by-mission basis.

Most of the rules governing hasty protective row mining are defined in STANAG 2036. A summary of those rules and some additional rules that apply are shown below.

- Rows.
  - Rows are marked and recorded. They are designated by letters (A, B, and so forth) with Row A being closest to the enemy.
  - The minimum distance between rows of AT mines is 8 meters.
  - **Korea Only: The minimum distance between any row and a row containing AP mines is 15 meters.**
  - The distance between the start row marker and the first mine in a row is the mine spacing for that row.
  - An IOE row is not used.
  - Start and end row markers are permanent markers and must be made of detectable material.
- General.
  - The spacing between mines is at least 4 meters to prevent sympathetic detonation. There is no maximum distance between mines within the row.
  - Mines are at least 15 meters from the perimeter fence.
  - The minefield must be fenced on all sides if M18A1 AP mines are employed and the minefield will be in place for more than 72 hours.
  - The minefield has at least one landmark that is located to the rear, never to the extreme side or front.
  - The minefield has an easily identifiable RP (tree, stump, stake).
  - Back azimuths are not used to record the minefield.
  - Measurements are in meters.
  - AHDs, nonmetallic mines, or low-metallic mines are not used.
  - Minefields are recorded on DA Form 1355-1-R (see Chapter 8).

SITE LAYOUT

Requesting and receiving permission to lay mines is the first step when emplacing a hasty protective row minefield. The next step is conducting a thorough leader's reconnaissance of the proposed minefield area. Identify mine locations that cover likely AAs, enhance key weapon systems, and cover dead space. Establish an easily identifiable RP between the minefield and the unit position. From the RP, visualize mines as running in rows parallel to the unit position (Figure 6-13). After the RP is established and the minefield is visualized, complete recording and emplace the mines, but do not arm them. This procedure simplifies recording and makes retrieval quicker and safer.

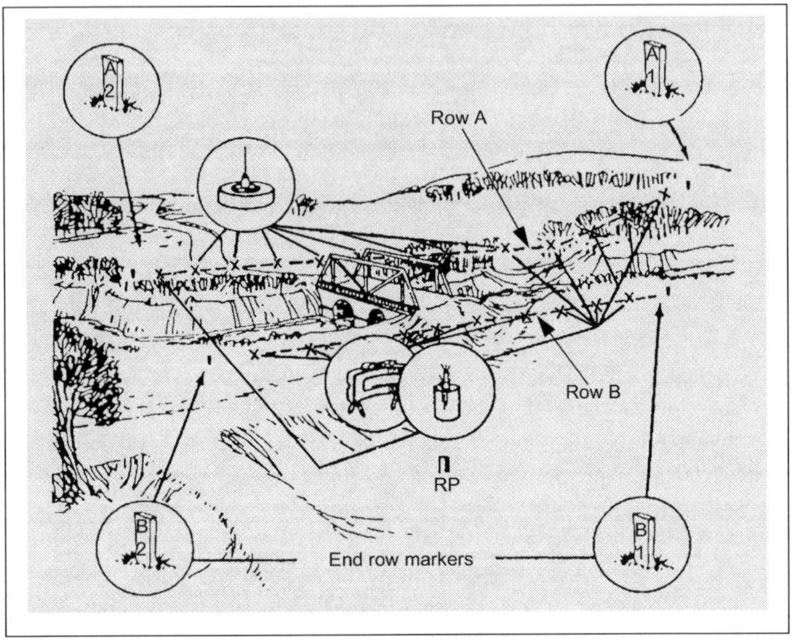

Figure 6-13. Site layout

Mine Rows

The row closest to the enemy is designated as Row A; succeeding rows are designated B, C, D, and so on. The ends of rows are shown by two end row markers. They are labeled with the letter of the row and the number 1 for the right end of the row or the number 2 for the left end of the row. The rows are numbered from right to left, facing the enemy. The marker should be an easily identifiable object, such as a wooden stake with a nail or a steel picket so that it can be found with an AN/PSS-12 mine detector.

Laying Procedures

From the RP, the leader measures the magnetic azimuth, in degrees, to a selected point on the right side (facing the enemy) of the tentative minefield. He paces off the distance and records it in meters. This point (B1) marks the beginning of the second row. The leader places a marker at B1 and records the azimuth and the distance on DA Form 1355-1-R.

From B1, the leader measures the azimuth and distance to a second point on the right side of the minefield (facing the enemy). He places a marker at this point (A1), and records the information.

The leader measures the distance and the azimuth from A1 to the location of the first mine in that row. The distance (or spacing) from the end row marker to the first mine is the mine spacing for that row. After the leader records the location, the mine is emplaced, but it is not armed.

The distance and azimuth are measured from the first mine to the second mine, and so on, until all the mines are emplaced and the locations are recorded. This procedure is repeated for the second row. As each mine is recorded, it is assigned a number to identify it in the minefield record.

When the last mine location is recorded for a row, the distance and the azimuth are measured from that point to another arbitrary point, A2 or B2. A marker is placed here in the same manner as A1 and B1. Next, the distance and azimuth from the RP to B2 and from B2 to A2 are measured and recorded.

When all the mines have been placed and recorded, the leader measures the distance and the azimuth between the RP and a permanent landmark that can be found on the map. He records the information on DA Form 1355-1-R. The landmark is used to assist others in locating the minefield if it is transferred or unexpectedly abandoned.

Mines can be armed after recording is complete. Mines nearest the enemy are armed first, allowing soldiers to safely work their way back to the unit position. Pins and clips are buried 30 centimeters behind row markers, the RP, or any easily identifiable, accessible location. Record the location of the pins and clips in the remarks section of DA Form 1355-1-R. The leader then reports the completion of the minefield to higher headquarters.

If the minefield is transferred to another unit, the transferring unit leader briefs the gaining unit leader. The gaining unit leader signs and dates the mines-transferred block on the DA Form 1355-1-R. The form is destroyed when the minefield is removed. If the minefield is abandoned unexpectedly, the DA Form 1355-1-R is forwarded to higher headquarters.

Mine Removal

When removing mines from a hasty protective row minefield, the leader first determines the best method to use:

- If the minefield has been under constant observation from the time it was laid and has not been tampered with, the squad leader directs the personnel who laid the mines to pick up the same mines. He uses DA Form 1355-1-R preceded by a mine detector to determine the types of mines to be removed and where they are located.
- If the minefield has not been under constant observation, may have been tampered with, or the personnel who laid the mines are not available or do not remember the location of the mines, the squad leader uses DA Form 1355-1-R and a clearance team as outlined in Chapter 11 to locate and remove mines.

The leader retrieves safety devices, shipping plugs, and other items that accompanied the emplaced mines. Using the azimuths and distances provided on the DA Form 1355-1-R, the removal team starts at the RP and moves to B1. They then move from B1 to the mine and remove the mine. If B1 is destroyed, the team moves from the RP to B2. The team then shoots a back azimuth (subtract 180 degrees) from the recorded azimuth from B2 to the first mine and removes the mine. Personnel continue this process until all the mines have been removed. The stakes at Al, B1, A2, and B2 are necessary because it is safer to find a stake than to find an armed mine.

The removal team observes basic safety precautions by maintaining 30 meters between personnel, not running, and moving only in cleared areas. The team starts with the row closest to the defender and works toward the enemy. Personnel—

- Check the sides and bottoms of the mines for AHDs, and disarm or mark the mines as they are found.
- Replace all pins, clips, and other safety devices before the mines are removed from the ground.
- Turn arming dials to *SAFE* or *UNARMED;* or if mines have screw-type fuses, remove the fuses and take them away from the mines.
- Lift the mines from the holes after they have been rendered safe.
  - If a mine was put in place and kept in sight by the individual who removes it, he lifts it directly from the hole after rendering it safe.
  - If a mine has not been kept in sight, the individual attaches a 60-meter rope or wire to the mine, takes cover, and pulls the mine from the hole.
- As each mine is removed, place a tick mark beside it on the DA Form 1355-1-R.
- Assemble all the mines in one location for accountability.

NOTE: AHDs are not used in hasty protective row minefields. However, as a safety precaution, consider all mines to be equipped with AHDs until proven otherwise.

The leader confirms the removal of the mines and accounts for the number of mines, by type, as recorded on the DA Form 1355-1-R. The leader may find it necessary to confirm an exploded mine to account for all the mines.

To confirm a mine explosion that was not witnessed, identify the crater or traces of burnt soil made by the detonated mine and place a tick mark beside the mine number on the DA Form 1355-1-R. Ensure that the crater found in the vicinity of the mine was caused by a land mine and not by artillery. A mine crater is normally circular, and it shows traces of burnt soil. The impact and the soil dispersion of artillery is normally elongated.

The squad leader confirms that each mine is disarmed and safe. The removal team cleans and repacks serviceable mines for future use and destroys the others; they repack serviceable mines in their original containers and store them according to the unit SOP and local regulations. The removal team removes and stores the row markers. The leader submits a report to his higher headquarters stating that the minefield has been removed and that the area is clear.

This chapter implements STANAG 2036.

Chapter 7

Standard-Pattern Minefields

Emplacing standard-pattern minefields is laborious and time-consuming, but it allows better mine concealment than row mining. Standard-pattern laying is well suited for protective and nuisance minefields. It can be used in terrain where the nature of the ground makes row mining impracticable.

To achieve their maximum effect, mines must be laid so that they cannot be seen and so that a vehicle's wheel or track or a person's foot exerts enough pressure to detonate them.

The method used to lay mines depends on the mine operation, the type of ground in which the mine is to be laid, and the type of ground cover that is available for camouflage.

COMPONENTS

MINE STRIPS

The mine strip is the foundation of a standard-pattern minefield. If a mine strip was laid in one straight line, the enemy could easily locate the mines; therefore, mine strips are laid in several segments as shown in Figure 7-1, page 7-2.

When siting, laying, and recording mine strips, all measurements are expressed in meters. Directions are recorded as magnetic azimuths, in degrees.

MINE CLUSTERS

The cluster is the basic unit of a minefield. It consists of one to five mines that are laid within a 2-meter-radius semicircle (Figure 7-2, page 7-3). When clusters are placed in a mine strip, they are numbered progressively and may consist of—

- One AT mine.
- **Korea Only: One AT mine and one to four AP mines.**
- **Korea Only: One to five AP mines.**

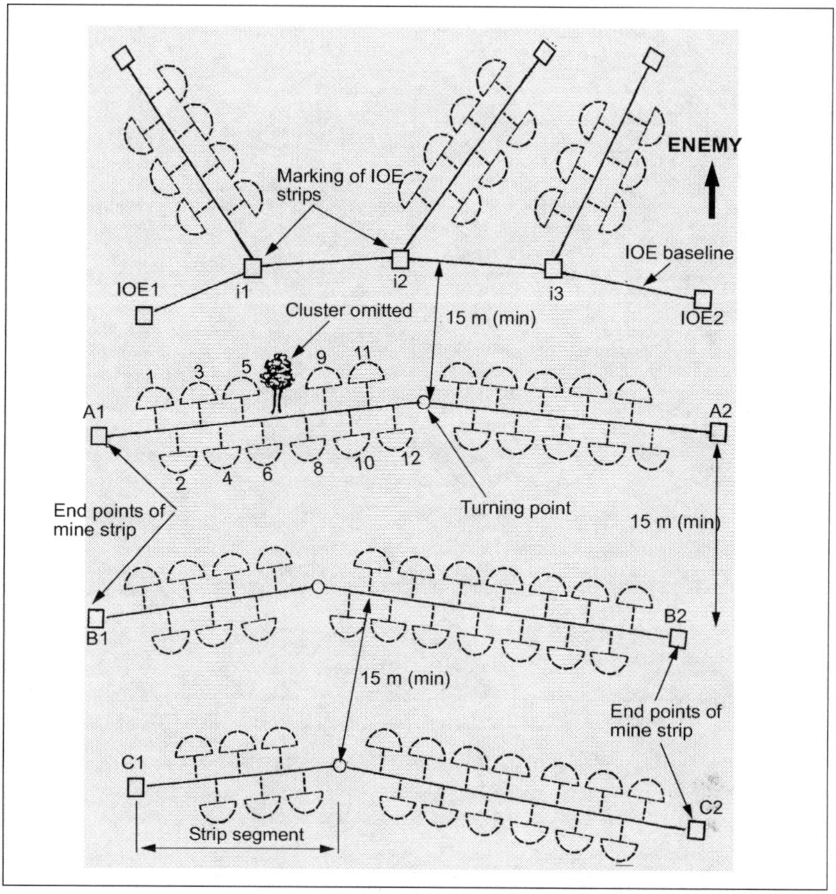

Figure 7-1. Minefield layout

Clusters are placed at 6-meter intervals, center to center, to form rows. Two parallel rows, 6 meters apart, form a mine strip. The arrangement of clusters in a mine strip is shown in Figure 7-3.

RULES FOR POSITIONING CLUSTERS WITHIN A STRIP

The first cluster is placed on the enemy side of the strip centerline, 6 meters from the beginning-of-strip marker. The following clusters are numbered consecutively. Odd-numbered clusters are always on the enemy side of the

Individual laying of AT mine (to be at base of cluster)	
Korea Only: Cluster with one AT mine plus several AP mines within or on a 2-meter semicircle of the AT mine (the AT mine must be the base mine)	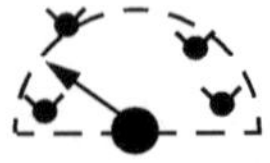
Korea Only: Individual laying of AP mine (to be at base of cluster)	
Korea Only: Cluster with several AP mines within or on a 2-meter semicircle of the center AP mine (the most easily detected mine is the base mine)	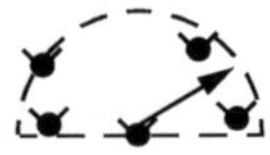

Figure 7-2. Cluster compositions

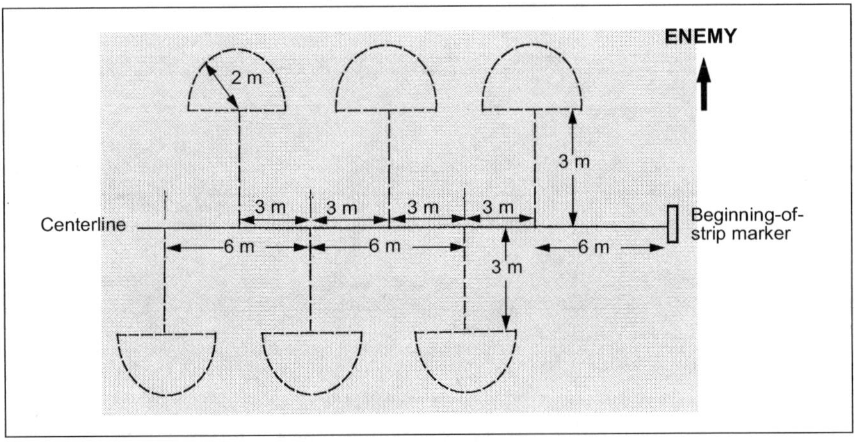

Figure 7-3. Arrangement of clusters in a mine strip

strip centerline. The direction of laying follows the numbering (A1 to A2, B1 to B2, and so on).

The IOE is normally the first part of the minefield encountered by the enemy. It consists of a baseline from which short strips are extended (Figure 7-4, page 7-4). Short strips along the IOE deceive the enemy on the minefield's pattern, spacing, and size. IOE placement and composition are largely dictated by the time allowed for laying the minefield, terrain conditions at the laying site, and the tactical situation.

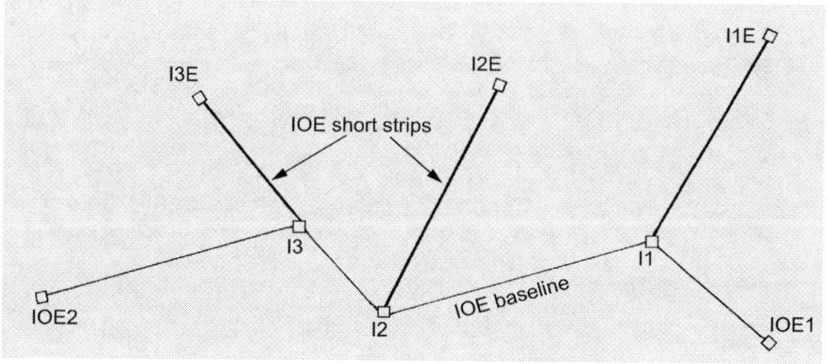

Figure 7-4. IOE baseline with short strips

The IOE baseline extends from one end point (IOE1) to another end point (IOE2). The laying direction is indicated by end-of-strip markers. Laying always begins at IOE1. Intermediate or turning points are marked in consecutive order beginning with I1. On the enemy side of the IOE baseline, short strips are extended from turning points at irregular angles. They are identified by turning-point markers.

Turning points should be no more than 45 degrees from the last azimuth. The length of short strips is not standard. A marker is emplaced at the end of each short strip. Markers are numbered in consecutive order beginning with I1E for recording purposes. No trip wires are used in the IOE, but AHDs may be employed.

Korea Only: AP mines actuated by trip wires are placed on the enemy side of each regular strip. No more than one mine per cluster uses trip wires, and no more than two trip wires extend from the mine. Trip wires are angled toward the enemy and should be at least 2 meters from the cluster, the lane border, and the minefield boundary. Trip wires are only used with AP fragmentation mines; they are not considered to be AHDs.

STANDARD-PATTERN MINEFIELD RULES

The following rules apply to tactical and protective standard-pattern minefields. They do not apply to nuisance minefields.

Clusters

- A cluster is a 2-meter-radius semicircle located 3 meters off the strip centerline.
- There are two types of clusters—live and omitted.

- — A live cluster contains one AT mine. **(Korea Only: A live cluster contains as many as five mines, only one can be an AT mine).**
- — Omitted clusters do not contain mines, but they are numbered and recorded on DA Form 1355 (see Chapter 8).

- Clusters are omitted within lanes and gaps; in areas less than 2 meters from boundaries, lanes, or another cluster (including the IOE); and in areas where the terrain (trees, rocks) prohibits emplacement.
- The base mine in a live cluster is the first mine laid. It is 3 meters from the strip centerline.
- When a live cluster contains an AT mine, the mine is always used as the base mine. **(Korea Only: If an AT mine is not present, the largest metallic AP mine is the base mine.)**
- The first cluster in a mine strip is located on the enemy side, 6 meters from the beginning-of-strip marker.
- The minimum distance between a cluster and a lane, a gap, a boundary, or another cluster is 2 meters (measured from the edge of the cluster).
- Cluster composition is the number of mines, by type, in any cluster in a specific group.
- Cluster composition remains the same through the entire mine strip and is recorded on DA Form 1355.
- **Korea Only: The types of AP mines may vary within a cluster.**
- The cluster boundary must at least 15 meters from the minefield perimeter fence.
- Clusters are numbered, beginning with the first cluster on the enemy side. Odd-numbered clusters are always on the enemy side of the strip, and even-numbered clusters are always on the friendly side.

Regular Strips

- A regular strip (sometimes referred to as a *lettered strip*) consists of a strip centerline and two rows of clusters (Row 1, enemy side; Row 2, friendly side).
- Regular strips are marked and recorded. They are designated by letters (A, B, and so forth), with Strip A being closest to the enemy.
- A standard-pattern minefield contains at least three regular strips.
- The minimum distance between strip centerlines is 15 meters; there is no maximum distance.
- Safety tapes are used to ensure that personnel installing trip wires do not move forward into armed clusters. A safety tape is used behind each regular strip. Safety tapes are 8 meters from the strip centerline (3 meters from the outer edge of the cluster).

- The marking of end points indicates the direction of laying (for example, A1 to A2).

IOE

- The IOE consists of a baseline from which short strips are extended.
- The IOE is located on the enemy side of the minefield.
- The number of clusters in an IOE is approximately one-third the number used in a regular strip.
- The first cluster along a short strip is placed on the enemy side and must be at least 6 meters from the IOE baseline; the cluster boundary must be at least 2 meters from the IOE baseline (Figure 7-5). If the short strip is exactly parallel to the enemy direction of travel, the NCOIC designates the enemy side of the strip.
- The IOE baseline is labeled at the beginning (IOE1) and end (IOE2) according to the direction mines are emplaced.
- Short strips are labeled at turning points (I1) and at the end (I1E).
- The IOE contains a safety tape that is 2 meters behind the IOE baseline and runs parallel with it.
- The IOE baseline is at least 15 meters from any point of the strip centerline of a regular strip; there is no maximum distance.
- An IOE short strip is at least 15 meters from another IOE short strip; there is no maximum distance.
- AHDs may be employed.

IOE Short Strips

- Short strips originate from turning points along the IOE baseline.
- The number and length of short strips depend on the tactical situation and the resources available.

Trip Wires (Korea Only)

- **Trip wires are not used in an IOE.**
- **Trip wires may be used in regular strips, but only one mine per cluster may be actuated by a trip wire.**
- **Trip wires are employed no closer than every third cluster.**
- **No more than two trip wires can be used on one mine.**
- **Trip wires are used only on the enemy side of the strip.**
- **Trip wires are not considered AHDs.**
- **Trip wires are located at least 2 meters from a lane, a safety tape, a cluster, another trip wire, the IOE baseline, and the minefield perimeter fence.**
- **Trip wires can only be used with AP fragmentation mines.**

Turning Points

- Clusters must be at least 3 meters from turning points.

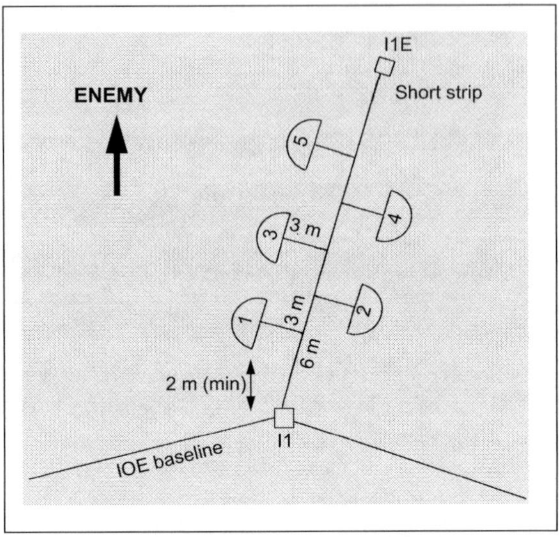

Figure 7-5. Clusters on an IOE short strip

- The first cluster after a turning point is laid on the opposite side of the strip centerline from the last cluster before the turning point.
- The angle of any given turning point cannot exceed 45 degrees from the last azimuth. (This ensures a minimum distance of 2 meters between cluster boundaries in the same row.)

Lanes

- Minefield lanes (Figure 7-6, page 7-8) are used by dismounted patrols and vehicles.
- Lanes are sited before laying begins.
- Lane locations should not be obvious.
- Clusters are not laid within 2 meters of lane edges.
- Lanes are zigzagged, not straight.
- Lanes cross the strip centerline at approximately right angles.
- Direction changes will not exceed 45 degrees. (This ensures that long vehicles will be able to negotiate turns.)
- The number of lanes must be sufficient to ensure that no one lane is overused and turned into an obvious track.
- Sufficient mines are stockpiled so that the responsible unit can close lanes suspected of being located by the enemy.
- Recommended minefield lane widths are—
  - Footpath: 1 meter.

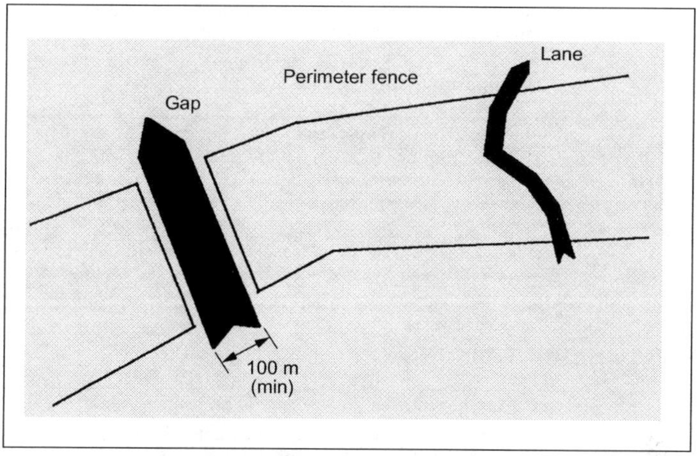

Figure 7-6. Minefield lanes and gaps

— One-way vehicle lane: 8 meters.

— Two-way vehicle lane: 16 meters.

Gaps

- Minefield gaps (Figure 7-6) are left so that friendly forces can pass through the minefield in tactical formation.
- Gaps must be at least 100 meters wide.
- Gaps are sited before laying begins.
- Gaps are located along recognizable features (fences, tracks, creeks).
- Gaps should run straight through a minefield and not contain bends.
- Sufficient mines must be stockpiled so that the responsible unit can close gaps when necessary.
- Gaps should closely resemble the rest of the minefield so that they will not be discovered by the enemy.
- The ground within a gap should be disturbed with tracks to represent the passage of a mine-carrying vehicle.
- Signs of mine laying (digging, scattered spoil, crates) should be visible in a gap.

General

- The farthest extremities of a regular strip determine the minefield front.
- Minefield depth is measured from the first to the last strip and includes the IOE (if applicable).
- Strips can be laid left to right or right to left.
- Back azimuths are not used to record the minefield.

- The minefield will have two landmarks located to the rear, never to the extreme side or front.
- GPSs can be used to determine the coordinates for minefield landmarks and reference points.

> **WARNING**
> **Do not use GPSs to chart or record minefield perimeter coordinates or to determine safe routes through or around existing minefields.**

- If landmarks are more than 200 meters away from the last regular strip or are out of the direct line of sight, intermediate markers are placed at least 75 meters from the last end-of-strip marker.
- Landmarks can be used for more than one minefield. They are recorded on DA Form 1355.

LOGISTICAL CALCULATIONS

Emplacing standard-pattern minefields allows the unit to calculate the number of mines required for a minefield accurately.

To simplify the calculation process, a minefield requirements computation work sheet (Figure 6-1, pages 6-5 through 6-8) has been developed. Step-by-step procedures for completing the work sheet are shown in Figure 6-2, pages 6-9 through 6-14.

CLUSTER CALCULATION

Accuracy is essential when emplacing a standard-pattern minefield. The following check system has been designed so that the minefield OIC can accurately record the number of mines laid. Compare the strip feeder report with the—

- Number of pins and clips returned by the laying party.
- Mine tally sheet.
- Cluster computation.

EXAMPLE: You receive a strip feeder report for Strip A. There are 26 clusters in the strip. The strip has three segments (30, 42, and 21 meters) and two turning points.

Step 1. Add the total length of the strip as determined by the recording party.

$30 + 42 + 21 = 93$ meters

Step 2. The first and last cluster in the strip are located 6 meters in from the end-of-strip markers. Subtract 12 from Step 1.

$93 - 12 = 81$ meters

Step 3. Clusters are not located on turning points. Multiply the number of turning points by 3 and subtract it from Step 2.

$81 - (3 \times 2) = 75$ meters

Step 4. Divide Step 3 by the cluster spacing.

$75 \div 3 = 25$ clusters

Step 5. Add one cluster, because when a line is divided, there is one more interval than spacing.

$25 + 1 = 26$ clusters

From these computations, the minefield OIC is able to cross-check the information on the strip feeder report. In this example, Strip A should have a total of 26 clusters, less any omitted clusters.

PLATOON ORGANIZATION

The platoon is the basic unit used to install a standard-pattern minefield. Orders to the laying unit OIC specify the proposed location, the length, and the mine type and density. The platoon organization and equipment are shown in Table 7-1.

Table 7-1. Platoon organization and equipment

Personnel	Officer	NCO	EM	Equipment
Supervisory	1	1	0	Officer: Map, lensatic compass, notebook, and minefield record forms NCO: Map, lensatic compass, and notebook
Siting party	0	1	3	Stakes or pickets, sledgehammers, engineer tape on reels, and nails to peg tape
Marking party	0	1	2	Rolls of barbwire or concertina, marking signs, lane signs, wire cutters, gloves, sledgehammers, pickets, and picket pounder
Recording party	0	1	2	Sketching equipment, lensatic compass, minefield record forms, map, and metric tape
First laying party	0	1	6-8	Notebook for squad leader, picks, shovels, and sandbags
Second laying party	0	1	6-8	Same as first laying party
Third laying party	0	1	6-8	Same as first laying party
Totals	1	7	25-31	

The OIC performs a map study, and if the situation permits, conducts a ground reconnaissance of the site. He determines locations for mine strips, landmarks, fences, mine dumps, and approaches. Using the minefield requirements computation work sheet (Figure 6-1, pages 6-5 through 6-8), the OIC determines the required number of mines and other materials. He arranges for mines to be delivered and organizes the platoon into siting, laying, recording, marking, and mine-dump parties.

Siting Party

The siting party places boundary stakes or pickets as strip markers at the beginning and end of each mine strip and at the points where strips change direction. It lays tape on the centerlines of each strip, lane, and traffic path. After siting is complete, the siting party augments other parties.

Laying Parties

One laying party is responsible for installing, arming, and camouflaging all the mines on a strip or a portion of a strip. Each laying party is then assigned additional strips.

Recording Party

The recording party obtains the necessary reference data, prepares DA Form 1355 (see Chapter 8), and when needed, installs intermediate markers.

Marking Party

The marking party erects fences and signs to mark minefield boundaries and lanes. After marking is complete, the marking party augments other parties.

Mine-Dump Party

The mine-dump party, controlled by the PSG, is composed of personnel who are not working as members of other teams. It accounts for all Class IV and V supplies that arrive from other sources, organizes mine sets at the mine dump, and hauls additional supplies as required. The PSG is not required to stay at the mine dump continuously; he has the flexibility to move around the area to perform other activities.

Mine-Emplacement Procedures

Mine-emplacement procedures are shown in Figure 7-7. In the illustration, the minefield is laid from right to left.

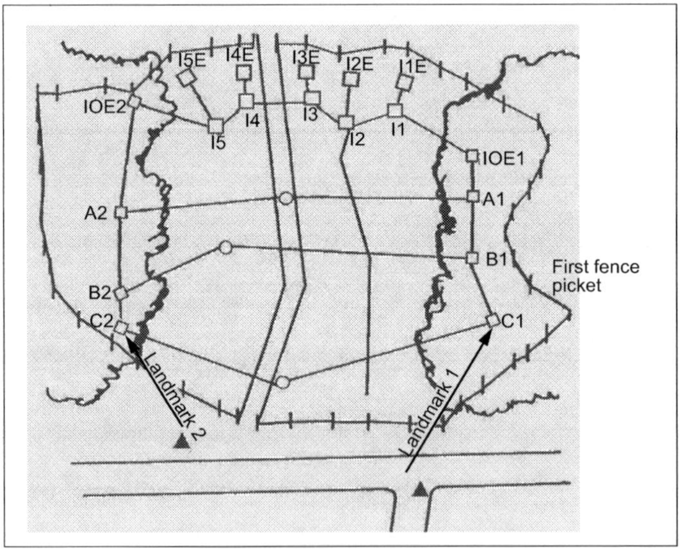

Figure 7-7. Mine-emplacement procedures

The OIC arrives at the site with the siting and marking parties. He goes to the right or left (depending on the laying direction) rear boundary of the minefield. This part of the minefield is the farthest from the enemy. The OIC indicates the starting point of the rear strip (this is Strip C in a three-strip minefield), and the siting party emplaces a beginning-of-strip marker.

The OIC designates a starting point for the marking party that is at least 15 meters to the right of the boundary stake. He indicates where the minefield marking fence should be placed. The marking party immediately begins to install fence pickets, working in a counterclockwise direction.

When all the pickets are installed, the marking party encircles the field with a single strand of barbwire (at waist height) and emplaces mine signs. Personnel emplace a second strand of barbwire if required.

From the boundary stake of Strip C, the OIC moves in the direction of the enemy and establishes the starting point of Strip B. Strip centerlines should not be parallel nor less than 15 meters apart. Two members of the siting party emplace a beginning-of-strip marker at the starting point of Strip B. The remaining two members lay tape between the two strips and fasten the tape to the ground at frequent intervals to prevent movement. This procedure is followed until the beginning-of-strip markers of the three regular strips (C, B, and A) and the IOE on the right-hand side of the minefield have been installed.

At the IOE beginning-of-strip marker (IOE1), the OIC gives the siting party a sketch of the minefield and instructions on siting the IOE baseline and strip centerlines. The NCOIC and one other member of the siting party immediately begin setting stakes to indicate the IOE baseline. The centerline laying team lays tape on the IOE baseline, leaving tape reels where tapes run out. Short strips extending from the IOE baseline are established at the same time. Each short strip starts with a marker that is designated as I1, I2, and so forth and ends with a marker that is designated as I1E, I2E, and so forth. Turning points are not used on short strips.

Upon reaching the IOE end-of-strip marker (IOE2), the NCO moves away from the enemy side, establishes the left boundary stake of Strip A, stakes out Strip A, and repeats the procedure until all strip centerlines are taped. All stakes are driven flush with the ground.

While the IOE is being taped, the recording party obtains reference data for the DA Form 1355. Working behind the siting party, the recording party starts from Landmark 1 (designated by the OIC) and proceeds to C1. After C2 has been sited, the recording party proceeds from Landmark 2 to C2 to establish the distance and the azimuth. Finally, the recording party ties C1 and C2 to both landmarks in case one of the landmarks is removed or destroyed. The amount of detail obtained by the recording party depends on the tactical classification of the minefield and any special orders. Aerial photographs taken of the minefield before the tracing tape is removed become valuable supplements to the DA Form 1355.

As soon as laying parties arrive at the site with mines, they establish mine dumps a minimum of 150 meters apart and 50 meters behind the field. They uncrate and stack AT mines and place fuses and detonators in separate boxes.

Fuse types are not mixed. **(Korea Only: AP mines are left in their crates, and the crate lids are removed.)**

When the siting party completes the centerline staking, it installs lane tapes and traffic tapes, respectively. Lane tapes are used by tactical vehicles and patrols. Traffic tapes are used by laying personnel to assist in camouflage and to reduce the amount of traffic on strip centerlines. Traffic tapes are laid perpendicular to the minefield trace at 100-meter intervals.

Mine Emplacement

The laying party must know the cluster composition of the strip, the location of any omitted cluster, and future lane locations. When the centerline tape for a regular strip has been installed, the NCOIC designates all but two members of the laying party to emplace mines in the ground. The remaining two soldiers, usually the most experienced, are designated as fusers and are responsible for arming mines. Layers carry the maximum load of mines to be used as base mines in the clusters. Fusers carry the fuses and the detonators.

The NCOIC moves to the right or left (depending on the laying direction) beginning-of-strip marker of the strip and organizes the layers into one column to his rear, directly on the centerline. He measures 6 meters along the centerline for the first cluster and, pointing perpendicular from the centerline and in the direction of the enemy, indicates the placement of the base mine. The first layer on the enemy side places a mine on the ground, 3 meters from the centerline.

The NCO measures 3 more meters and indicates the placement of the second base mine on the opposite (friendly) side of the strip. The first layer on that side places a base mine on the ground. As the initial load of mines is laid, each layer returns to the nearest mine dump for another load. Fusers follow behind layers and insert mine fuses, but they do not arm the mines. This procedure is followed until the end-of-strip marker on the far side of the minefield is reached.

Korea Only: The NCO tells layers the number and type of mines to be placed next to the base mine in each cluster. As AP mines are being placed, the NCO proceeds along the strip and ensures that the proper number of AP mines is placed in each cluster. The NCO places a spool of trip wire next to the mines that are to be activated by trip wire.

When all the mines are positioned in clusters, one layer is assigned to dig the holes for all the mines in a cluster. He places the spoil from the holes in sandbags and leaves the sandbags beside the base mine in each cluster. The layer checks the positioning of the mines in the holes, removes the mines from the holes, and places the mines beside the holes. **(Korea Only: The layers anchor trip wires with nails or stakes and wrap the loose ends of trip wires around the fuses.)**

When digging has progressed at least 25 meters from the first mine laid, the arming procedure begins. Fusers arm all the mines in a cluster, beginning with the mine farthest from the centerline and work backward. They place all the mines in the holes **(Korea Only: attach trip wires)** and arm and camouflage the mines. They place filled sandbags on the centerline of the strip, opposite the base mine. Fusers keep their back toward the centerline. Other personnel must remain at least 25 meters from the fusers (Figure 7-8).

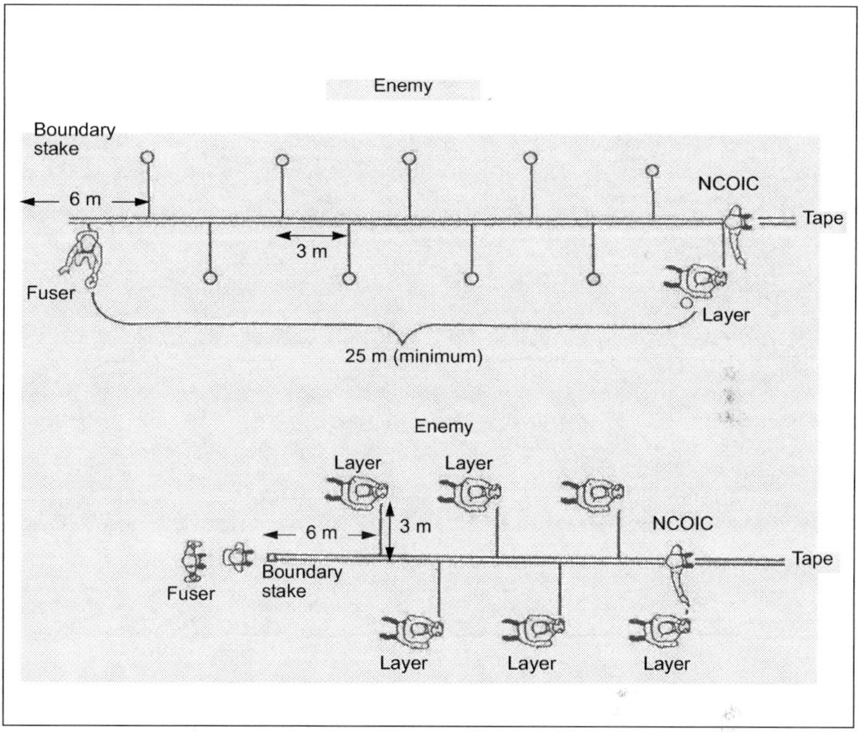

Figure 7-8. Laying and fusing mines

Mines located in lanes are not initially buried. They are placed aside to prevent confusion when counting clusters. The mines can be buried after the lane is closed. Upon completing the arming operation, fusers give the safety clips to the NCO, who verifies that all the mines have been armed and camouflaged. The NCO checks the strip and ensures that sandbags, tape, and debris have been picked up. The NCO gives the safety clips to the PSG, who buries them 30 centimeters to the rear of the beginning-of-strip marker.

All mines and other explosive items are recorded upon issue. They are summarized on a mines tally sheet (see Table 7-2). If more than one mine dump is established, a mines tally sheet is kept at each dump and the information is later transferred to a master tally sheet.

Table 7-2. Sample mines tally sheet

Strip/Laying Party	Movement of Mines	M15	M16	Trip Wire	AHD Type
IOE strip/ Party 1	No forecasted	23	46	—	—
	No issued	23	46	—	—
	No returned	—	—	—	—
	No used	23	46	—	—
Strip A/ Party 2	No forecasted	81	154	35	—
	No issued	81	154	35	—
	No returned	7	6	9	—
	No used	74	148	26	—
Strips B, C, D, and so forth	No forecasted				
	No issued				
	No returned				
	No used				
Total	No forecasted				
	No issued				
	No returned				
	No used				
Date _______ Rank _____ Name _____________________ Signature _________________________					

The PSG ensures that the number of mines used per tally sheet is entered on the DA Form 1355.

When a lane is no longer required through a minefield, it is closed by a lane-closure team that consists of one NCO and two other soldiers. Before closure starts, the NCO checks the minefield record to ascertain the—

- Width of the lane.
- Cluster composition of each strip.
- Total number of mines required.
- Number of strips that intersect the lane.
- Azimuth of each strip.
- Distance between strips, along the lane centerline.
- Location of the mine dump.

The procedures for lane closure is as follows (see Figure 7-9, page 7-16):

- The lane-closure team moves along the safe-lane centerline until it arrives at the strip safe-lane marking picket.
- The NCO lays out a strip centerline tape and a tape along both sides of the safe lane to mark its boundaries.
- The team lays clusters.
- The team recovers the tapes.

- The above steps are repeated at successive strips.
- The NCO amends the DA Form 1355.

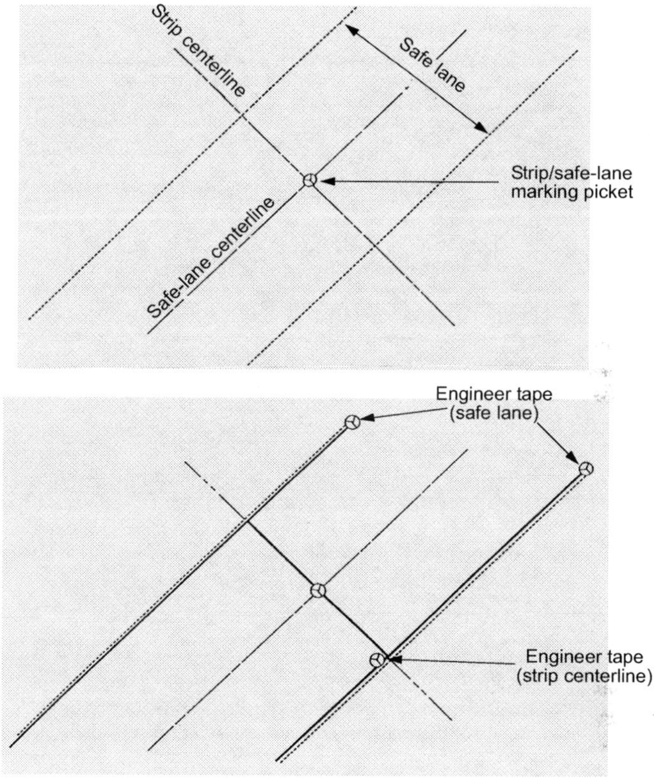

Figure 7-9. Lane closure

NOTE: When gaps must be closed, fences are temporarily erected along the side boundaries. They are removed later to avoid the indication of a passage through the minefield.

NCOICs do not act as working members of their parties; they ensure that—

- No one moves back into a mined area.
- Any irregularity, such as an omitted cluster, is recorded.

- All safety devices are recovered and checked against the minefield record.

Safety tapes are used to create a network of safe routes through a minefield. They may be removed progressively but are normally left in place until the minefield is complete.

Korea Only: When trip wires are used, safety tapes are laid between strips where the trip wire will be positioned (including Strip A and the IOE baseline). If trip wires are not used, safety tapes are recommended but are not mandatory.

NUISANCE MINEFIELDS

SITING

Consider the following factors when siting a nuisance minefield:

- The effort needed by the enemy to bypass a mined area.
- The importance of an area or a route to the enemy.
- Achievement of the goal (use the minimum amount of effort needed to maximize casualties and the effect on enemy morale).

Remember, the more ingenuous the methods of concealment, the longer it will take to lay mines.

Observation and covering fires are not essential and are seldom feasible for nuisance minefields. Their value depends on effective siting and concealment to cause surprise.

The minefield OIC is responsible for detailed siting and the design of a nuisance minefield. He must consider the minefield from the enemy's point of view and assess the courses open to the enemy when he encounters it. Such considerations may expose weaknesses in the initial plan and bring about a change to the proposed minefield layout or may lead to a decision to site the minefield elsewhere.

LOCATION

In wooded or hilly terrain, the enemy's logistics transport will normally be confined to existing routes. Nuisance mines at selected sites along roads can impose considerable delay on the enemy and have a cumulative effect on his resources and morale. The best sites for axial mining are—

- Natural defiles and constructed areas that are difficult to bypass (cuts, embankments, causeways, fords, forest tracks, built-up areas).
- In the vicinity of road craters, AT ditches, and any obstacles that have to be cleared.
- Around culverts.
- Demolished bridges, particularly the home bank, including likely adjacent crossing places and alternative building sites.
- Likely assembly areas.
- Covered approaches or dead space.

- In the vicinity of enemy-needed fuel, supplies, and engineering materials that cannot be destroyed or removed.
- Railroads (in or near culverts, bridges, sharp turns, tunnels, steep grades). Mines should be laid where enemy trains cannot bypass the mined area on branches or spurs.

Laying

There is no requirement for recording the precise location of individual nuisance mines. Recording mine positions that are laid to a pattern is easy and quick. Pattern laying should be used when it can be done without prejudicing concealment.

When the number of mines to be laid on the site makes it impracticable or undesirable to lay mines in a pattern, they may be scatter-laid unless otherwise directed. Scatter-laying by hand is useful in road blocks, bridge abutments, and craters when it would be difficult and wasteful to lay mines in a pattern. Again, scatter-laying along routes to be denied to the enemy will add considerably to the delay imposed. All available types of AT mines **(Korea Only: and AP mines)** are used to make nuisance minefields complex and difficult to remove. Combinations of mine types should be varied constantly so that each minefield presents a clearance problem. Deeply buried mines can be included; however, they take much longer to lay. These mines may be worthwhile around craters, where the enemy is likely to need earthmoving equipment.

If a nuisance minefield is laid in a standard pattern, standard procedures are followed. If mines are selectively positioned, procedures must be tailored to suit the situation. In all occasions, however, the following rules should be observed:

- The intended position for each mine is clearly marked on the ground before laying begins.
- Laying parties work in pairs, and each pair is detailed to lay specific mines.

Inspection and Maintenance

Mines that are left in the ground for an extended period of time may deteriorate and malfunction for one or more of the following reasons:

- Moisture may have entered the igniter or body of the mine and either neutralized the explosive or corroded the metal parts. Such actions may be aggravated by local factors (soil acidity, wide temperature swings).
- Frost or heat may have subjected the mine to mechanical strain and caused distortion.
- Insects or vegetation may have caused obstructions.
- Animals may have turned mines over or detonated them.

Technical inspections should only be made by experienced engineers or EOD personnel. When a minefield deteriorates below the operating level, additional mine strips and/or rows must be added to restore its effectiveness. They are sited to the front or rear of the existing minefield to increase its depth. New mine strips and/or rows are treated as a separate, additional minefield.

Technical inspections of minefields are normally done at three-month intervals, and personnel work in pairs. They are done more frequently during extreme weather conditions.

HANDOVER

Minefield handover is an extremely important task. The following items need to be addressed between emplacing and overwatching units:

- Discussing minefield compositions.
- Discussing the minefield configuration and walking and/or riding the minefield trace.
- Discussing lane closure, if applicable.
- Training the unit on how to the close lane, if applicable.
- Discussing obstacle protection against enemy dismounted patrols.
- Signing over the written report (Figure 2-26, page 2-54).
- Discussing indirect fires.
- Reporting completion of handover to higher headquarters.
- Forwarding a copy of the written report and the DA Form 1355 to higher headquarters.

This chapter implements STANAG 2036.

Chapter 8

Reporting and Recording

Reporting and recording mine and minefield/munition field information is critical to the success of a unit's mission. It not only provides tactical data for the commander, but it also provides force-protection information to subordinate and adjacent units.

MINEFIELD/MUNITION FIELD REPORTS

A minefield/munition field report is an oral, electronic, or written communication that concerns friendly or enemy mining activities. The report format is specified by the local command. It is submitted by the emplacing unit commander through operational channels to the G3/S3 of the authorized headquarters. The headquarters integrates the report with terrain intelligence and disseminates it along with tactical intelligence. The report is sent by the fastest, most secure means available. Figure 8-1, page 8-2, summarizes the minefield/munition field report flow at the division level and below.

REPORT OF INTENTION

When planning to emplace a minefield/munition field, the unit must submit a report of intention to notify their higher headquarters. The report doubles as a request when it is initiated at levels below emplacement authority. The report includes—

- Tactical purpose of the minefield/munition field.
- Estimated number and type of mines to be emplaced.
- Location.
- Proposed start and completion times.
- Type of minefield/munition field.
- Whether mines are surface-laid or buried.
- Whether AHDs are used.
- Location and width of lanes and gaps.

Conventional minefields/munition fields that are part of an OPLAN or a general defense plan (GDP) approved by the authorizing commander do not require a report of intention because inclusion in an OPLAN or a GDP implies an intention to lay.

REPORT OF INITIATION

A report of initiation is mandatory. It informs higher headquarters that emplacement has begun and that the area is no longer safe for friendly movement and maneuver. The report specifies the time emplacement began

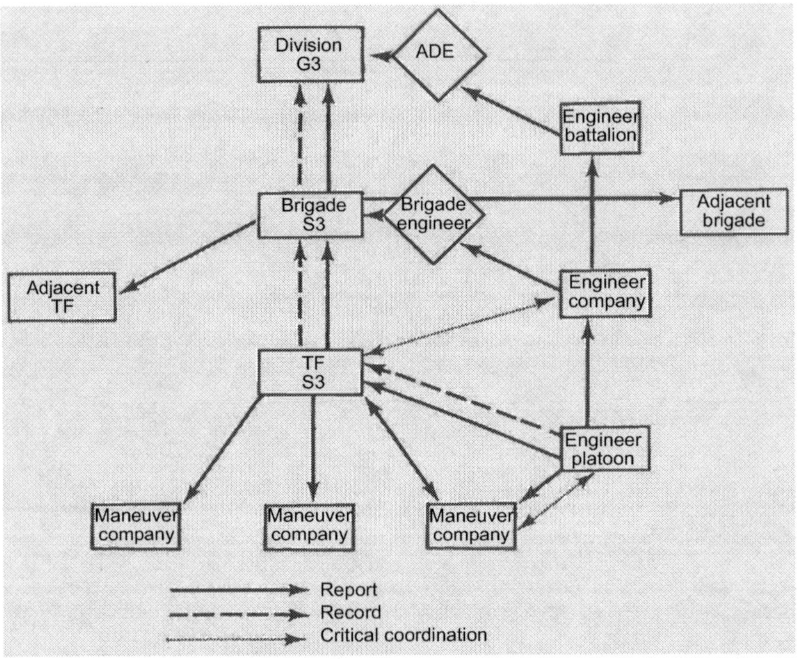

Figure 8-1. Conventional minefield/munition field reporting chain

and identifies the location and the target number of minefields/munition fields.

Report of Completion

A report of completion is usually an oral report to the authorizing commander. It indicates that the minefield/munition field is complete and functional. A report of completion is immediately followed by a completed DA Form 1355 or 1355-1-R.

Report of Transfer

Minefield/munition field responsibility is transferred from one commander to another in a report of transfer (Figure 2-26, page 2-54). This report is signed by the transferring and receiving commanders. It includes a certificate stating that the receiving commander was shown or otherwise informed of all the mines within the transferring commander's zone of responsibility. The report states that the receiving commander assumes responsibility for those mines.

The report of transfer is sent to the next higher commander who has authority over the transferring and receiving commanders.

REPORT OF CHANGE

A report of change is made immediately upon any change or alteration to a previously reported minefield/munition field. It is sent to the next higher commander and through channels to the headquarters that keeps the written minefield/munition field record. A report of change is made by the commander responsible for surveillance and maintenance of the minefield/munition field.

PROGRESS REPORTS

During the emplacing process, the commander may require periodic reports on the amount of work completed.

MINEFIELD/MUNITION FIELD RECORDS

Most conventional minefields/munition fields are recorded on DA Form 1355; hasty protective row minefields/munition fields are recorded on DA Form 1355-1-R.

The laying unit prepares the minefield/munition field record. The OIC signs the form and forwards it to the next higher command as soon as possible. Once the information is entered on the form, the form is classified SECRET, NATO SECRET, or SECRET-ROKUS. (When used for training, the record is marked SAMPLE.) The number of copies prepared depends on the type of minefield/munition field emplaced and the local procedures. Unit SOPs should provide advanced guidance on how minefield/munition field information is to be passed to higher, lower, and adjacent commands. Minefield/munition field records are circulated on a need-to-know basis. Large minefields/munition fields are recorded on two or more DA Forms 1355.

Once the minefield/munition field is removed, the date of removal, the time removal was complete, and the name of the unit that removed the minefield/munition field will be placed in the remarks block of the form. The form will then be forwarded to the unit's central control cell for mine clearance information. The central control cell will retain the form for two years and then destroy it.

When changes are made to an existing minefield/munition field, a new record must be prepared. This record is marked REVISED. It shows the minefield/munition field as it is after changes. The original minefield/munition field number remains unchanged. Some changes which require a new record are—

- Relocation of mines in safe lanes.
- Relocation of safe lanes.
- Changed lane or minefield/munition field marking.
- Inclusion of the minefield/munition field into a larger minefield/munition field system.
- Removal or detonation of mines.
- Addition of mines to the minefield/munition field.

Conventional minefield/munition field records are forwarded through operational channels to theater army headquarters (TAHQ) where they will be maintained on file by the theater engineer. If a TAHQ has not been established, minefield/munition field records are maintained on file with the assistant corps engineer of the AO where the minefield/munition field is located.

Minefield Record

DA Form 1355 consists of a single sheet that is printed on both sides and folded in half. The front side contains blocks of tabular data, the inside is a graph of 1-centimeter squares for a scaled sketch of the field, and the back side includes a computation formula and instructions. The scale for plotting minefields/munition fields depends on the size of the field. To avoid using two sheets for the sketch, adjust the scale so that one form will support the sketch. For very large minefields/munition fields, two sheets may be required. The system of measurement and the scale size must be indicated in the legend block. A second form may be used to support any additional information in the notes block. Any blocks or lines not used on the form must be crossed out to avoid unauthorized entries on the form.

An example of a completed DA Form 1355 is shown in Figures 8-2a through 8-2c, pages 8-5 through 8-7.

The following step-by-step instructions are provided for completing DA Form 1355:

Block 1. Enter complete data on the laying authority and the laying unit. The OIC block should include rank, name, and social security number (SSN).

1	AUTHORITY: CG 2nd INF DIV
	LAYING UNIT: B Co 2nd ENGR BN 1st PLT
	OFFICER IN CHARGE: 1LT R. YOUNG 762-01-1352

Block 2. Enter the date-time group (DTG) for start and completion times. The recorder block includes rank, name, and SSN.

2	DATE AND TIME	START 090630Z JAN 90
		COMPLETION 091800Z JAN 90
	RECORDER: SFC F. LING 550-72-1332	

Block 3. Enter the copy and the sheet number. The number of copies prepared depends on the unit SOP and the minefield/munition field classification.

3	Copy No. 1 of 4	Sheet No. 1 of 1

NOTE: The minefield/munition field record is forwarded by the laying unit. One copy is retained by the overwatching unit, one copy by the next higher command, one copy by the unit central control cell (for mine clearance information), and one copy by the proper national territorial authority.

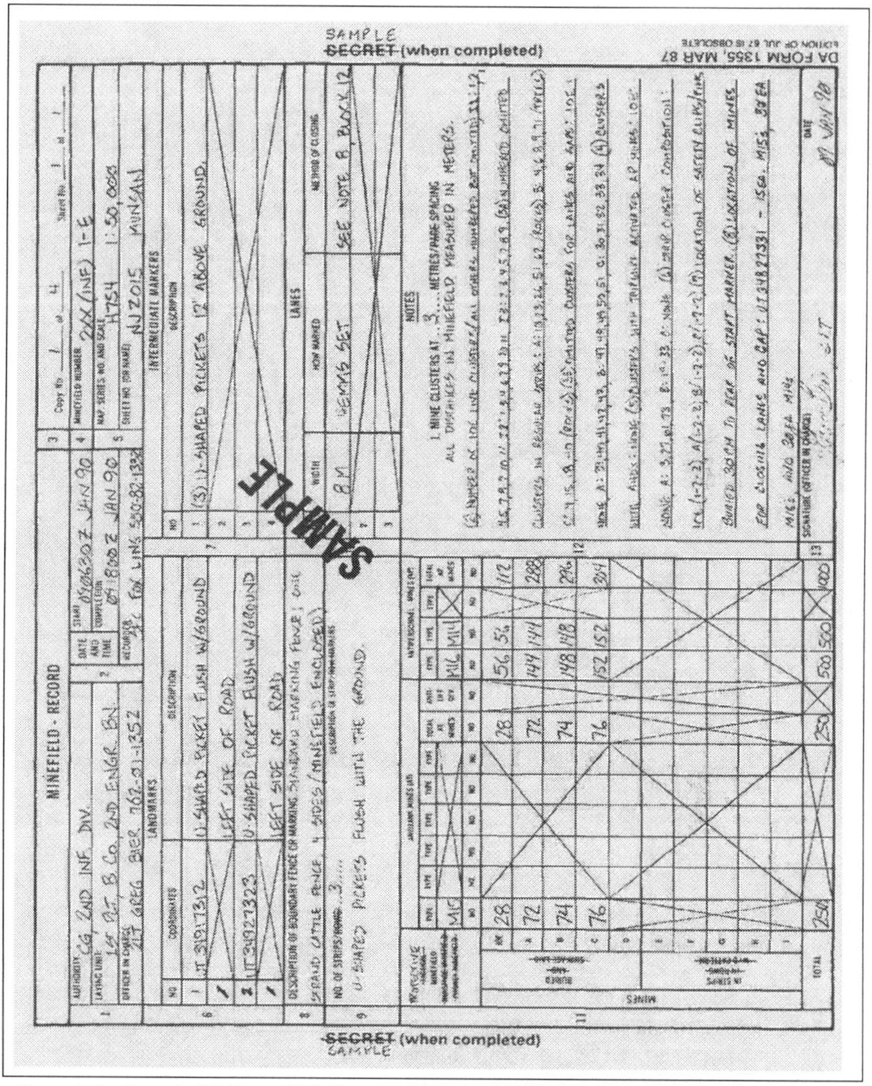

Figure 8-2a. Sample DA Form 1355 (front side) for a standard-pattern minefield/munition field

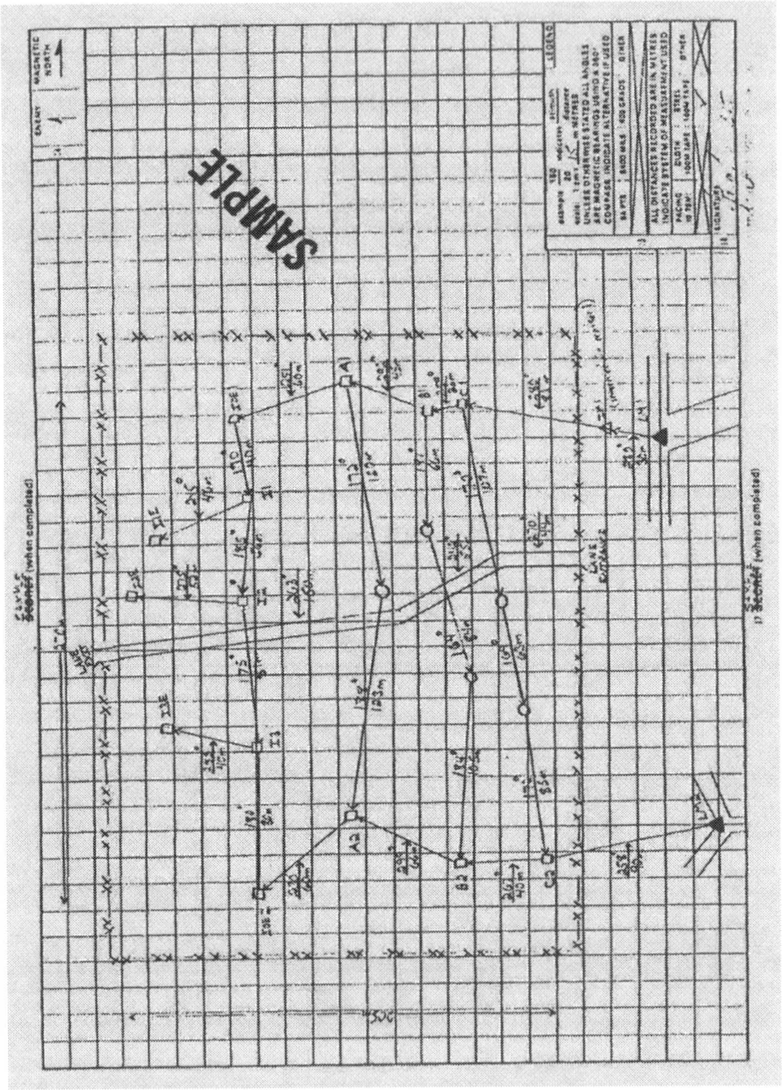

Figure 8-2b. Sample DA Form 1355 (inside) for a standard-pattern minefield/munition field

~~SECRET~~ SAMPLE (when completed)

MINEFIELD REQUIREMENTS COMPUTATION FORMULA

		AT	APF	APB
Desired Density		a____	b____	c____
IOE Representative Cluster		a____	b____	c____
Front ____ meters				
Depth ____ meters				
AHD ____ %				
*1. No. of IOE Clusters = Front ÷ 9	____			
2. No. of IOE clusters x IOE Representative Cluster		a____	b____	c____
3. Mines in Minefield = Front x Desired Density		a____	b____	c____
4. Subtotal of Mines Required = Line 2 + Line 3		a____	b____	c____
*5. Mine Rejections, Strip Length Variances = Line 4 x .1		a____	b____	c____
6. Total AT Mines Required = Line 4 + Line 5		a____	b____	c____
7. Add a + b + c of "Desired Density" =	____			
*8. Line 7 x .6 =	____			
9. AT Mine "Desired Density" x 3 =	____			
10. No. of Regular Strips = Highest No. of Line 8 or 9	____			
*11. No. of AHD = % AHD x Total AT Mines (Line 6)	____			
12. Strip cluster Composition = Desired Density x 3		a____	b____	c____

* NOTE: Round up to the next whole number.

	STRIP	AT	APF	APB	ROW TOTAL (Cannot exceed 5)
"CLUSTER	A	a____	b____	c____	____
	B	a____	b____	c____	____
	C	a____	b____	c____	____
	D	a____	b____	c____	____
COMPOSITION	E	a____	b____	c____	____
	F	a____	b____	c____	____
	G	a____	b____	c____	____
	H	a____	b____	c____	____
TABLE"	I	a____	b____	c____	____
	J	a____	b____	c____	____
COLUMN TOTAL (Totals cannot exceed Line 12).		a____	b____	c____	____

TABULAR DATA (Numbers correspond to numbered blocks on front of form).

1. Enter complete data on authority of laying and on the laying unit. OIC blanks will include name, rank, and SSN.
2. Enter date-time groups for starting and completion times. Recorder blanks will include name, rank, and SSN.
3. Enter copy and sheet numbers. Number of copies will depend upon unit SOP and the classification of the minefield. The number of sheets will depend upon the length and the depth of the minefield versus scale.

4. Enter minefield number as follows:

Designation of unit authorizing installation
Number of obstacle
Status of obstacle
(E=Executed, P=Proposed, U=Under Construction). 3/147-Inf-2-E

5. Enter map data as stated on map(s) used.
6. Enter complete data on at least two landmarks with 8 digit grid coordinates. Cross out unused blocks.
7. Enter description(s) of any intermediate markers used. When a landmark is more than 200 meters from the minefield or the strip/row reference stake cannot be seen from the landmark, [illegible] intermediate marker must be used. If possible, the intermediate marker should not be closer than 75 meters to [illegible] reference stake. Cross out unused blocks.
8. Enter the word "Standard" when the standard marking fence [illegible] the boundary marking if other than the standard marking fence is used.
9. Enter the number of strips/rows laid other than IOE. Describe the [illegible] (Line out words not applicable).
10. Enter the width, marking, and closing provisions for each lane; when [illegible] give the type and number of mines for closing. The location of these mines is described in the "Notes" (Line 12). [illegible] are 1 meter wide, one-way vehicular lanes are 8 meters and two-way vehicular lanes are 16 meters. Cross [illegible] blocks.
11. Enter type of minefield by crossing out lines not needed. Indicate method of laying [illegible] marking out incorrect descriptions. Enter types of mines as AT, APF, APB. (Enter chemical mines under AT mines). For each type of mine, enter number of mines and antihandling devices installed in the IOE and in each Strip or Row. Strips or Rows will be lettered serially, starting with the first one laid. Enter totals. Cross out unused blocks.
12. Enter under Notes information which would be useful to personnel clearing the minefield. Appropriate items include location of chemical mines, location of AT mines with antihandling devices, location AP mines with tripwires, clusters in IOE which contain mines, where safety devices are buried, strip cluster composition and numbered omitted clusters in regular strips.
13. OIC enters signature, rank and date.
14. Enter arrows for the direction of the enemy and magnetic north. The enemy arrow should always point within the top 180 degrees of the paper; the north arrow should follow one of the lines of the graph.
15. Enter scale of sketch for standard pattern minefields the sketch should be drawn to a scale of about 1cm=10 meters.
16. Sketch in the following, as applicable.
    a. Show directional arrows as follows:
        (1) Landmarks (or intermediate markers) to strip markers at starting and finishing points of the last strip laid or to the nearest or farthest mine in a group.
        (2) From landmarks (or intermediate markers) to fence or boundary markers.
        (3) From landmarks to intermediate markers, if used.
        (4) For each straight line section of a lane centerline.
        (5) Between markers of starting points of adjacent strips, including IOE, and between finishing points of adjacent strips, including (IOE).
        (6) For each segment of a strip or of the IOE, label all directional arrows with magnetic azimuth in degrees and distance in meters. Express as a fraction (247 degrees/90 meters). Recorded from friendly to enemy side and from right to left/or left to right.
    b. Show approximate location of protective fence or boundary markers.
    c. Show length and depth of minefield in meters. These dimensions indicate the extremities of the minefield.
    d. Show a grid intersection and give grid coordinates.
    e. Show trace of shoreline and direction and approximate rate in meters per second of water current, for mines laid underwater.
17. Enter security classification of the form. (If the form is used for training, enter the word SAMPLE.)
18. OIC enters signature and rank.

SAMPLE

~~SECRET~~ SAMPLE (when completed)

Figure 8-2c. Sample DA Form 1355 (back side) for a standard-pattern minefield/munition field

Block 4. The minefield/munition field obstacle numbering system (Table 8-1) consists of 11 characters and 1 obstacle status symbol. It shows the type of obstacle, the belt and the zone in which it is located, and the headquarters that established the zone. Abbreviations for obstacle types are shown in Table 8-2.

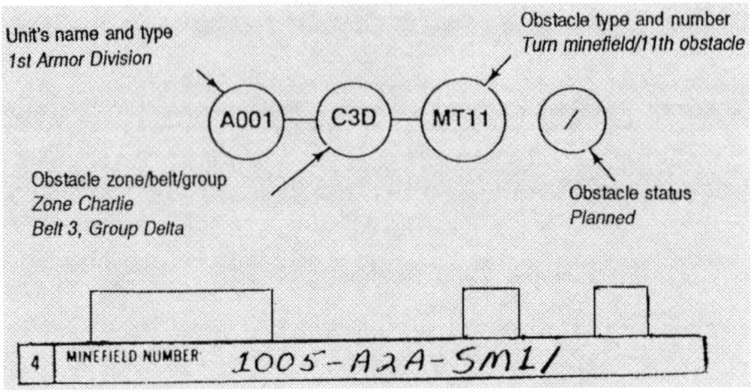

Example: Obstacle number I005-A2A-SM21/ indicates that the 5th Infantry Division planned the obstacle in Zone A. It is the 21st obstacle in Group A, Belt 2, and has not been executed. The obstacle is a MOPMS.

Table 8-1. Minefield/munition field obstacle numbering system

Character	Description
1 through 4	Alphanumeric description of the headquarters type and numerical designation that established the obstacle zone; character 1 designates the unit type with a letter: • A = armor division or brigade • I = infantry division or brigade • C = cavalry division • R = cavalry regiment • Z = corps
5	Letter indicating the obstacle zone
6	Number indicating the belt number in the obstacle zone
7	Letter indicating the group in the obstacle belt
8 and 9	Letters indicating the obstacle type (see Table 8-2)
10 and 11	Two numbers indicating the obstacle number in the group
12	One of four characters indicating the obstacle status— • / = planned obstacle • — = obstacle being prepared • + = prepared obstacle (This sign is for reserve demolition targets and may indicate a readiness state of safe or armed.) • X = completed obstacle

Table 8-2. Abbreviations for obstacle types

B—Bridge Demolition		**W—Wire Obstacle**	
BA	Abutment	WA	Double apron
BS	Span	WB	Booby-trapped
BC	Abutment and span	WF	Tanglefoot
M—Minefield/Munition Field		WG	General-purpose, barbed tape
MD	Disrupt	WN	Nonstandard
MT	Turn	WR	Road block
MF	Fix	WT	Triple standard
MB	Block	**R—Road Crater**	
MN	Nonstandard	RH	Hasty
MP	Protective	RD	Deliberate
MQ	Nuisance	RM	Mined
MS	Standard pattern	**M—Miscellaneous**	
S—Scatterable Minefield/Munition Field		AD	AT ditch
SA	ADAM	AR	Rubble by CEV gun
SB	Gator	AB	Rubble by blade
SR	RAAM	AT	Abatis
SF	ADAM and RAAM	AE	Rubble by explosives
SM	MOPMS	AM	Movable MOBA obstacle (car, bus)
SV	Volcano	AN	Expedient nonstandard
H—Hand-Emplaced Munitions		AL	Log crib, log obstacle
HH	Hornet	AP	Post obstacle (hedgehog, tetrahedron)
HS	SLAM	AH	Log hurdle

Block 5. Enter the map data as stated on the map(s) used.

5	MAP SERIES NO AND SCALE H754 1: 50 000
	SHEET NO (OR NAME) NJ 2015 MUNSAN

Block 6. Enter the grid coordinates and a description of at least two landmarks. If the landmarks are roads, trails, or routes, enter their name or number. This makes identification easier when removing the minefield/munition field. When recording minefields/munition fields, GPSs can only be used to determine the coordinates for minefield/munition field landmarks and RPs.

> **WARNING**
> **Do not use GPSs to chart or record minefield/munition field perimeter coordinates or to determine safe routes through or around existing minefields/munition fields.**

6	LANDMARKS		
	NO	COORDINATES	DESCRIPTION
	1	UT 34917312	U-SHAPED PICKET FLUSH WITH GROUND
	2		NEXT TO ROAD
	2	UT 34927323	U-SHAPED PICKET FLUSH WITH GROUND
	4		NEXT TO ROAD

Block 7. Enter the description of intermediate markers, if applicable. When a landmark is more than 200 meters from the minefield/munition field, or a strip or row reference stake cannot be seen from the landmark, an intermediate marker must be used. If possible, the intermediate marker is at least 75 meters from the strip or row reference stake.

7	INTERMEDIATE MARKERS	
	NO	DESCRIPTION
	1	3 U-SHAPED PICKETS 12" ABOVE GROUND
	2	
	3	
	4	

Block 8. Enter the word STANDARD when a standard marking fence is used. Describe the boundary marking if a standard marking fence is not used. (Use two sides and the rear for a tactical minefield/munition field; use four sides for a protective minefield/munition field.)

8	DESCRIPTION OF BOUNDARY FENCE OR MARKING STANDARD 4 SIDES MINEFIELD FULLY ENCLOSED

Block 9. Enter the number of strips or rows laid. (Do not include the IOE.) Describe the strip or row markers. Cross out words that do not apply.

9	NO. OF STRIPS/~~ROWS~~: 3 DESCRIPTION OF STRIP/~~ROW~~ MARKERS U-SHAPED PICKETS FLUSH WITH GROUND

Block 10. Enter the width, the marking, and the provisions for each lane. When appropriate, give the types of mines and the number of each type of mine for closing. (The location of these mines is described in Block 12.)

10	LANES			
	NO	WIDTH	HOW MARKED	METHOD OF CLOSING
	1	8m	HEMMS SET	15xM15, 30xM16, 30xM14
	2			
	3			

Block 11. Enter the type of minefield/munition field by crossing out the lines that are not needed. Indicate the method of laying by crossing out incorrect descriptions. Enter the types of mines and the number of each type of mine. Also enter the number of AHDs installed in the IOE and in each row. Letter the strips or rows sequentially, starting with the first one laid. Enter totals.

11	Protective ~~TACTICAL~~ MINEFIELD ~~NUISANCE MINEFIELD~~ ~~PHONEY MINEFIELD~~		ANTITANK MINES (AT)								ANTIPERSONNEL MINES (AP)			
			TYPE	TYPE	TYPE	TYPE	TYPE	TYPE	TOTAL AT MINES	ANTI LIFT DEV	TYPE	TYPE	TYPE	TOTAL AP MINES
			M15								M16	M14		
			NO	NO	NO	NO	NO	NO	NO	NO	NO	NO	NO	NO
	MINES		IOE	28						28		56	56		112
		BURIED ~~AND~~ ~~SURFACE LAID~~	A	72						72		144	144		288
			B	74						74		148	148		296
			C	76						76		152	152		304
			D												
		IN STRIPS ~~IN ROWS~~ ~~W/O PATTERN~~	E												
			F												
			G												
			H												
			J												
	TOTAL			250						250		500	500		1000

If the type of munition field is Hornet, enter "Hornet" above the word "tactical" and line out "nuisance minefield" and "phoney minefield." (See Figures 8-3a and 8-3b, pages 8-12 and 8-13.)

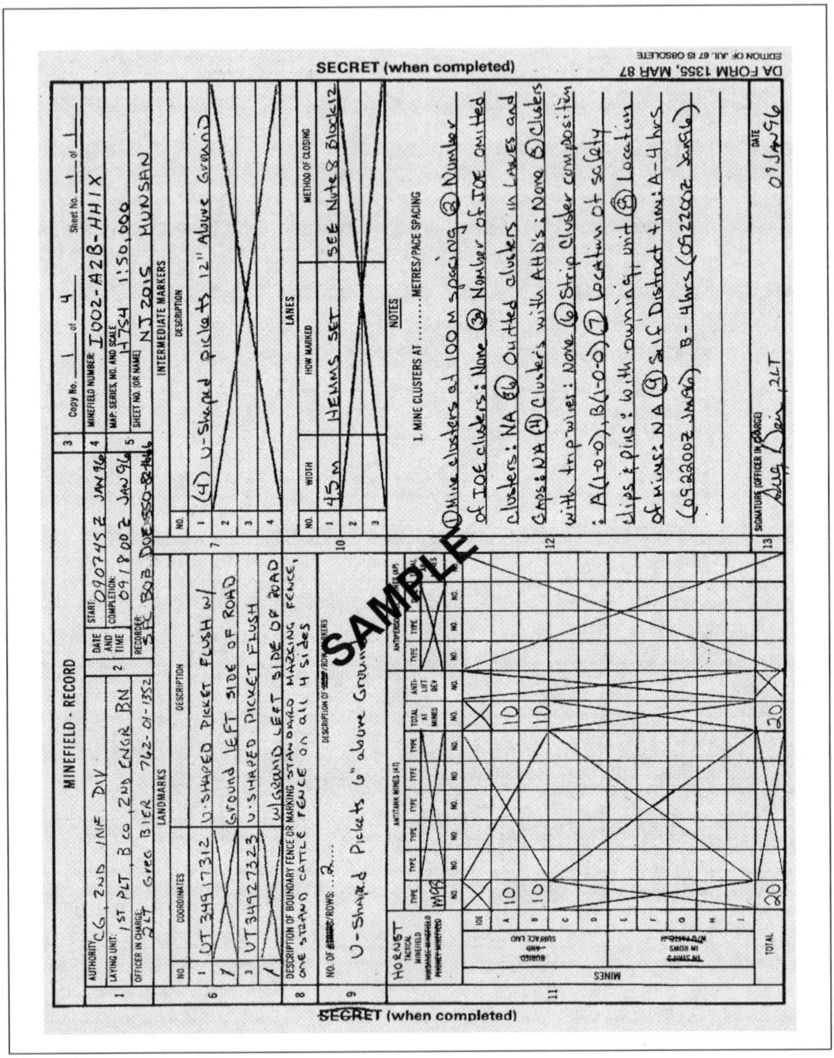

Figure 8-3a. Sample DA Form 1355 (front side) for a Hornet minefield/munition field

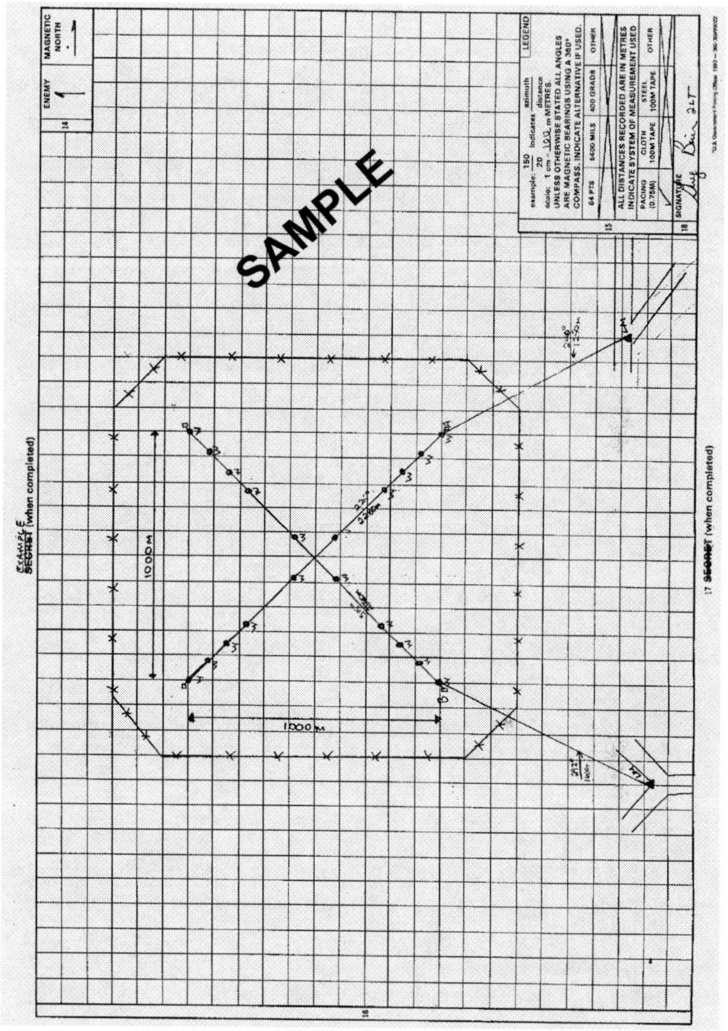

Figure 8-3b. Sample DA Form 1355 (back side) for a Hornet minefield/munition field

Block 12. As a minimum, enter the following items in Block 12. Also enter any additional information that would be useful when removing the minefield/ munition field.

1. Mine clusters at ____ meters spacing.

2. Number of IOE live clusters (all others numbered but omitted).

3a. Number of omitted clusters in the IOE and in regular strips and why they were omitted.

3b. Omitted clusters in lanes and gaps.

4. Clusters with AHDs (identify the mine and the type of device used).

5. Clusters with trip-wire-actuated AP mines.

6. Strip cluster composition.

7. Location of safety clips and pins (buried 30 centimeters to the rear of each start strip or row marker).

8. Location of mines for closing lanes and gaps.

9. SD time for special munitions.

NOTES

12 1. MINE CLUSTERS AT ..3.... METRES/[redacted] SPACING

ALL DISTANCES IN MINEFIELD MEASURED IN METERS

(2) NUMBER OF IOE LIVE CLUSTERS (ALL OTHERS NUMBERED BUT OMITTED); I1: 1-12; I2: 1-13, I3: 1-12 (3A) NUMBER OF OMITTED CLUSTERS IN IOE AND REGULAR LETTERED STRIPS; IOE: NONE, A: NONE B: NONE, C: NONE (3B) OMITTED CLUSTERS FOR LANES AND GAPS; IOE: NONE, A: NONE, B: NONE, C: NONE (4) CLUSTERS WITH ANTI HANDLING DEVICE ON ANTITANK MINE; IOE: NONE, A: NONE, B: NONE, C: NONE (5) CLUSTERS WITH TRIPWIRE ACTIVATED ANTI-PERSONNEL MINES; IOE: NONE, A: NONE, B: NONE, C: NONE (6) STRIP CLUSTER COMPOSITION; IOE: 1:2:2, STRIP ABC: 1:2:1 (7) LOCATION OF SAFETY CLIPS/PINS: BURIED 30cm TO REAR OF START STRIP MARKER (8) LOCATION OF MINES FOR CLOSING LANES OR GAPS: NONE

Block 13. The emplacing unit OIC signs and dates the form.

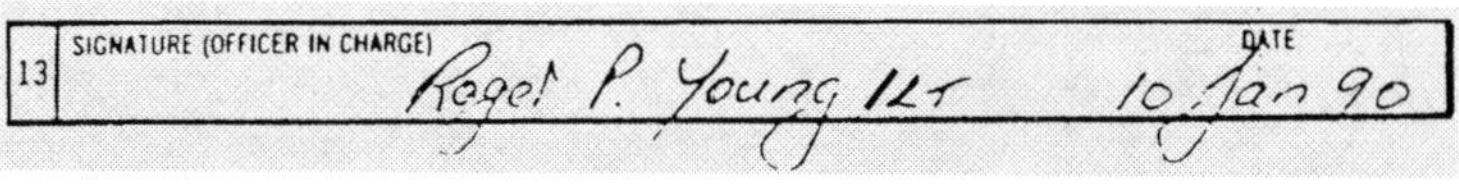

13	SIGNATURE (OFFICER IN CHARGE) Roger P. Young 1LT	DATE 10 Jan 90

The front side of DA Form 1355 is now complete. Continue on the inside as follows:

Block 14. When completing the sketch, enter arrows for the direction of the enemy and the magnetic north. The enemy arrow will always point within the top 180 degrees of the form; the north arrow will follow one of the graph lines.

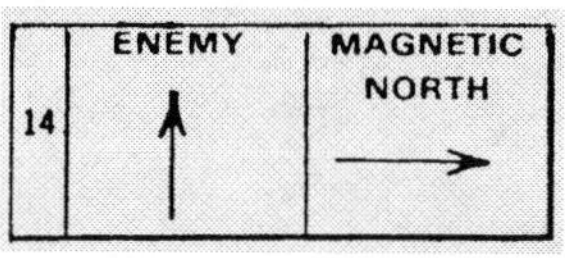

Block 15. If a compass was not available, enter what was used in the information block. Indicate the system of measurement and the scale used.

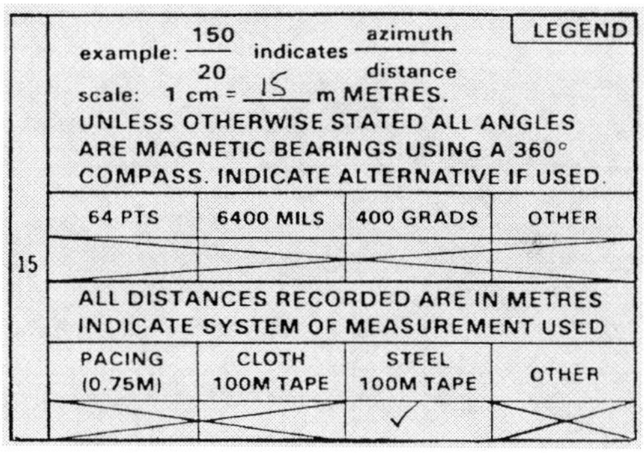

LEGEND

example: 150/20 indicates azimuth/distance

scale: 1 cm = 15 m METRES.

UNLESS OTHERWISE STATED ALL ANGLES ARE MAGNETIC BEARINGS USING A 360° COMPASS. INDICATE ALTERNATIVE IF USED.

64 PTS	6400 MILS	400 GRADS	OTHER
X	X	X	X

ALL DISTANCES RECORDED ARE IN METRES INDICATE SYSTEM OF MEASUREMENT USED

PACING (0.75M)	CLOTH 100M TAPE	STEEL 100M TAPE	OTHER
X	X	✓	X

Block 16. Enter the information shown below on the sketch:

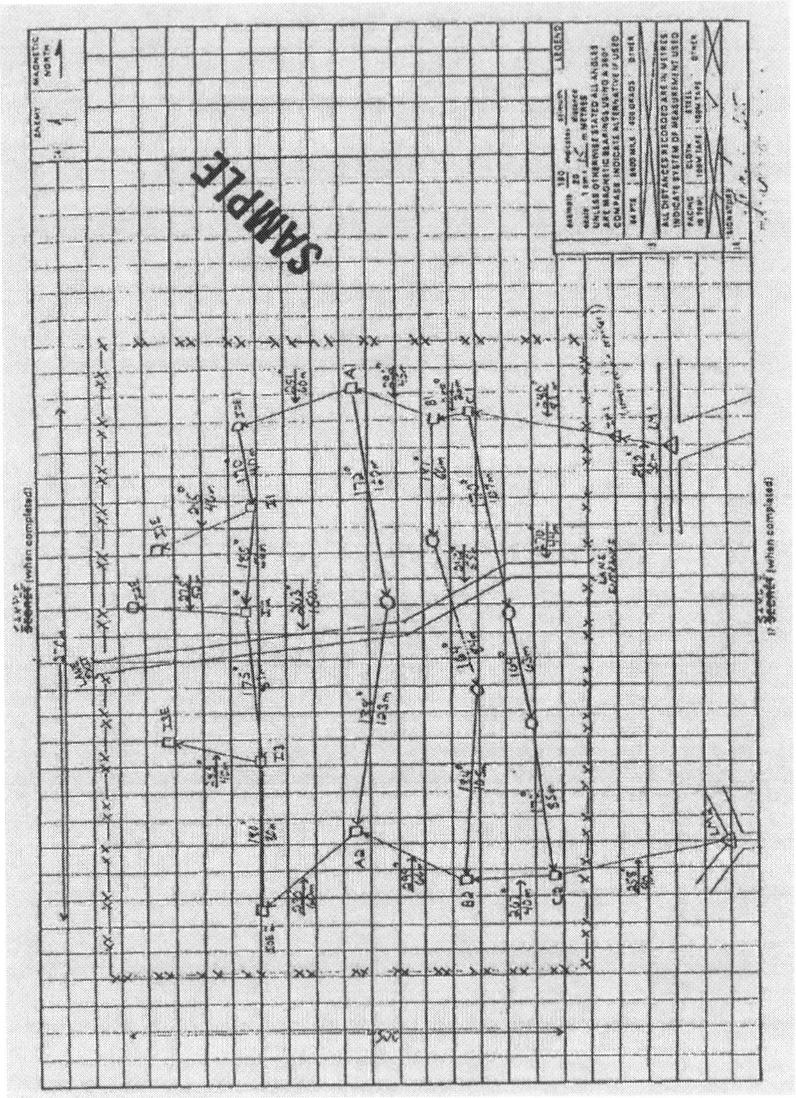

Block 17. Enter the security classification of the form. If the form was used for training, enter the word SAMPLE.

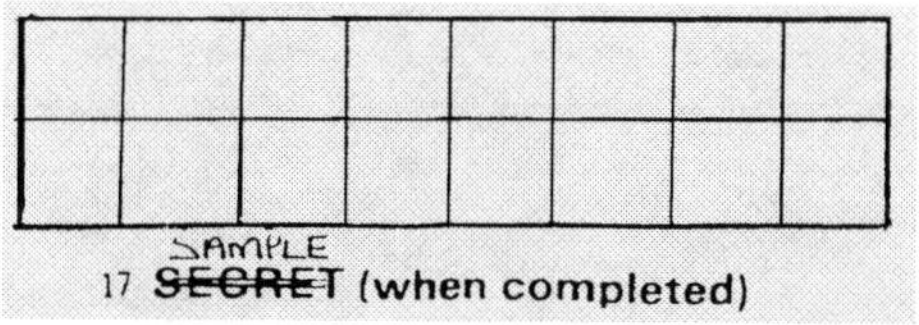
SAMPLE
17 ~~SECRET~~ (when completed)

Block 18. The emplacing unit OIC signs the signature block.

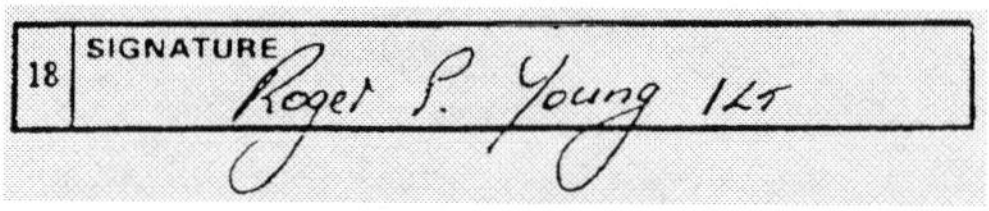
18 SIGNATURE Roger P. Young 1LT

HASTY PROTECTIVE ROW MINEFIELD RECORD

Hasty protective row minefields/munition fields are recorded on DA Form 1355-1-R (Figure 8-4, page 8-18). DA Form 1355-1-R will be printed on 8½- by 11-inch paper. A copy for local reproduction is available in the back of this FM. DA Form 1355-1-R is also available on the Army Electronic Library (AEL) CD-ROM (EM 0001) and the USAPA web site (www.usapa.army.mil).

Use the following formula to determine the scale used on DA Form 1355-1-R:

Distance from RP to the farthest point in the minefield + 10 meters ÷ 4 = scale

Example: 90 meters + 10 meters = 100 meters ÷ 4 = 25 meters

The number 4 is a constant and represents the four concentric rings on DA Form 1355-1-R. Ten is added as a safety margin to ensure that the minefield/munition field sketch is entirely contained within the largest ring. The distance between rings is 2 centimeters; therefore, the scale used in this example is 2 centimeters = 25 meters.

HASTY PROTECTIVE ROW MINEFIELD RECORD

For use of this form, see FM 20-32; the proponent agency is TRADOC.

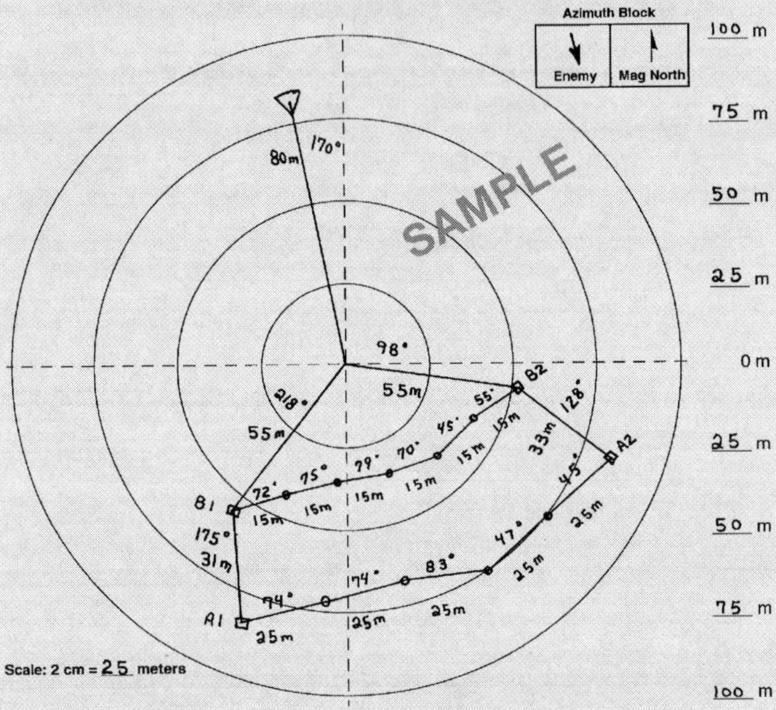

Tabular Block

Row	Type	Actuation	Mine Number
A	M-15	Pressure	1, 4
A	M-21	Tilt-Rod	2, 3
B	M-15	Pressure	1, 3, 5
B	M-15	Tilt-Rod	2, 4

Remarks Points: A1, A2, B1, B2 are marked with orange tent pegs flush to the ground.

Identification Block

Unit 2nd PLT. C. Co. 317TH Eng. BN.	
Reference Point Tree stump right side of the road w/white eng. tape at NA 14503505	
Remarks Landmark: Center of road junction at NA 13403409	
Map and Sheet No. Talbot #5568	
Name of OIC 2nd LT. Roy Bean 123-45-7611	
Signature Roy Bean	Time and Date 1400Z/1 JUL01
Mines Removed	
Mines Transferred	

Figure 8-4. Sample DA Form 1355-1-R

The following step-by-step instructions are provided for completing DA Form 1355-1-R (see Figure 8-5):

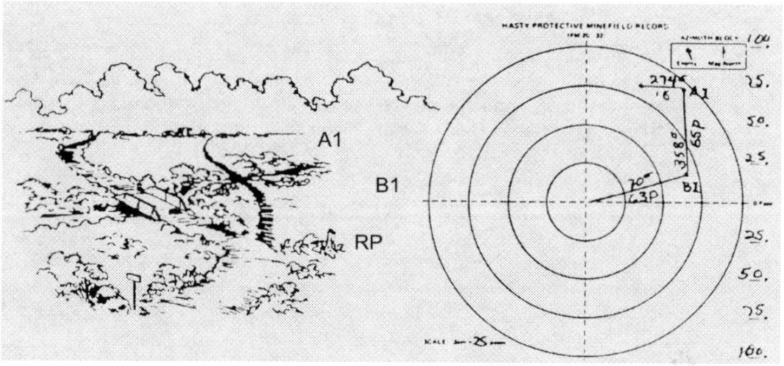

Figure 8-5. Hasty protective row minefield/munition field record

From the RP, the leader measures the magnetic azimuth, in degrees, to a specific point on the right side (facing the enemy) of the tentative minefield/munition field. He determines the distance and records it in meters. He identifies the method of measurement in the remarks block of DA Form 1355-1-R (for example, distances were paced out and multiplied by 0.75 to convert the measurement to meters). This point, called B1 (if there are two rows), marks the beginning of the second row. A marker is placed at B1, and the leader records the azimuth and the distance.

The leader measures the azimuth and the distance from B1 to a second point on the right side of the minefield/munition field (facing the enemy). A marker is placed at this point and recorded as A1.

The leader measures the distance and the azimuth from A1 to the location of the first mine in that row and records the information.

The distance (or spacing) from the end row marker to the first mine is the spacing between all the mines in that row. After the location is recorded, the mine is emplaced, but it is not armed.

The leader measures the distance and the azimuth from the first mine to the second, and so on, until all the mines are emplaced and their locations are recorded. This procedure is repeated for the second row. As each mine is recorded, it is assigned a number to identify it in the minefield/munition field record. When the last mine location is recorded for a row, the distance and the azimuth are measured from that point to another arbitrary point, A2 or B2. A marker is placed here in the same manner as A1 and B1. Next, the distance and the azimuth from the RP to B2 and from B2 to A2 are measured and recorded.

When all the mines have been emplaced and recorded, the leader measures the distance and the azimuth between the RP and a permanent landmark that can be found on the map. He records the information on DA Form 1355-1-R. The landmark is used to help locate the minefield/munition field if it is transferred or unexpectedly abandoned. Finally, the tabular and identification blocks are completed.

Mines can be armed after recording is complete. The mines nearest the enemy are armed first, allowing soldiers to safely work their way back to the unit's position. Pins and clips can be buried 30 centimeters behind row markers, the RP, or any easily identifiable, accessible location. Note the location in the remarks section of DA Form 1355-1-R. The leader then reports the completion of the minefield/munition field to higher headquarters.

If the minefield/munition field is transferred to another unit, the receiving unit leader is briefed by the transferring unit leader. The receiving unit leader signs and dates the mines transferred block on the DA Form 1355-1-R. When the minefield/munition field is removed, the form is destroyed. If the minefield/munition field is abandoned unexpectedly, the DA Form 1355-1-R must be forwarded to higher headquarters.

Nuisance Minefield

It is unnecessary to record the precise location of individual nuisance mines that are laid in a pattern. Recording the position of a minefield/munition field that is laid in a pattern (either a standard pattern or one adopted for the situation) is easy and quick. However, the number of mines to be laid on site may make it impracticable or undesirable to lay mines in a pattern. In this case, they may be laid individually and their positions accurately recorded.

Figures 8-6a and 8-6b, pages 8-21 and 8-22 provide an example of a completed nuisance minefield/munition field record.

SCATTERABLE MINEFIELD/MUNITION FIELD REPORTING AND RECORDING

The speed and responsiveness of SCATMINE employment require accurate, uniform, and timely reports. All known information on scatterable minefields/munition fields is simply and rapidly reported to all affected units. Although SCATMINES have SD capability, they are still recorded; and the information is disseminated to prevent casualties to friendly forces.

Since the locations of individual SCATMINEs are unknown, they cannot be recorded in detail. For most systems, a safety zone is calculated from one or more aim points. For example, a RAAM minefield/munition field is recorded based on the target location (the grid coordinates given to the firing battery). The size of the minefield/munition field depends on the number of rounds fired, the number of aim points, and the angle of fire. An artillery-delivered minefield/munition field is recorded by plotting it on a map, based on the aim point and the safety zone specified in the scatterable minefield/munition field report and record that was prepared by the emplacing unit. A ground-Volcano minefield/munition field can be recorded more accurately by plotting each of the minefield/munition field corner points rather than an aim point.

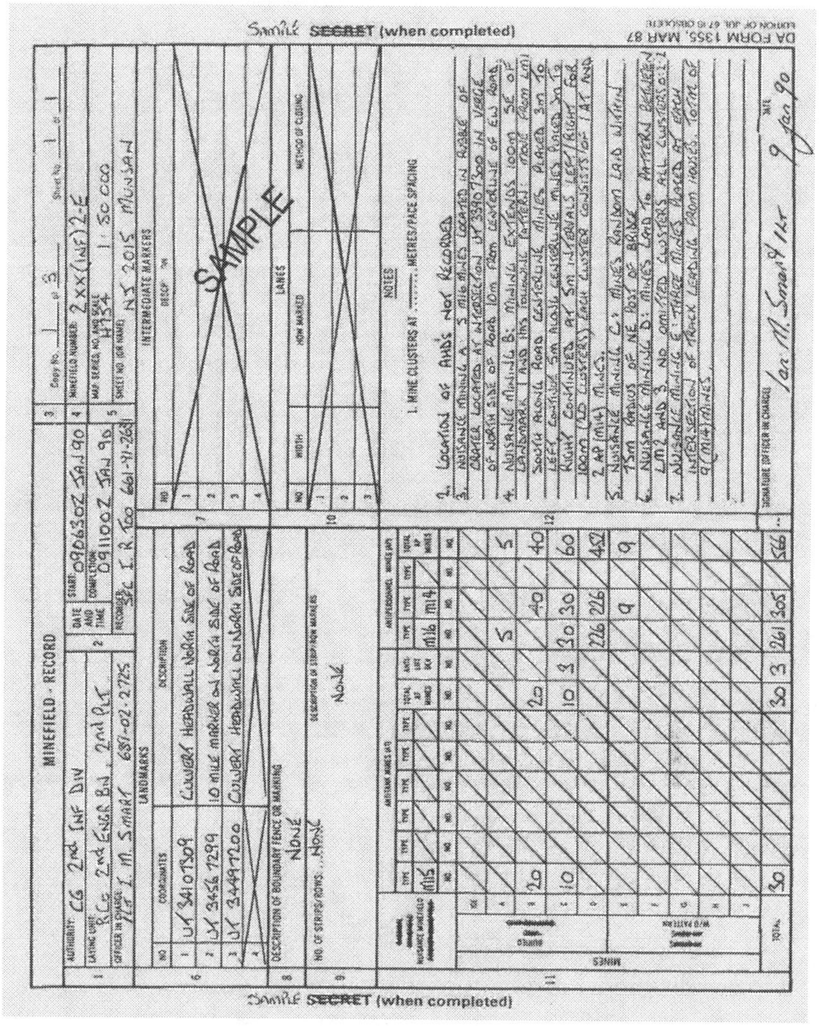

Figure 8-6a. Sample DA Form 1355 (front side) for a nuisance minefield/munition field

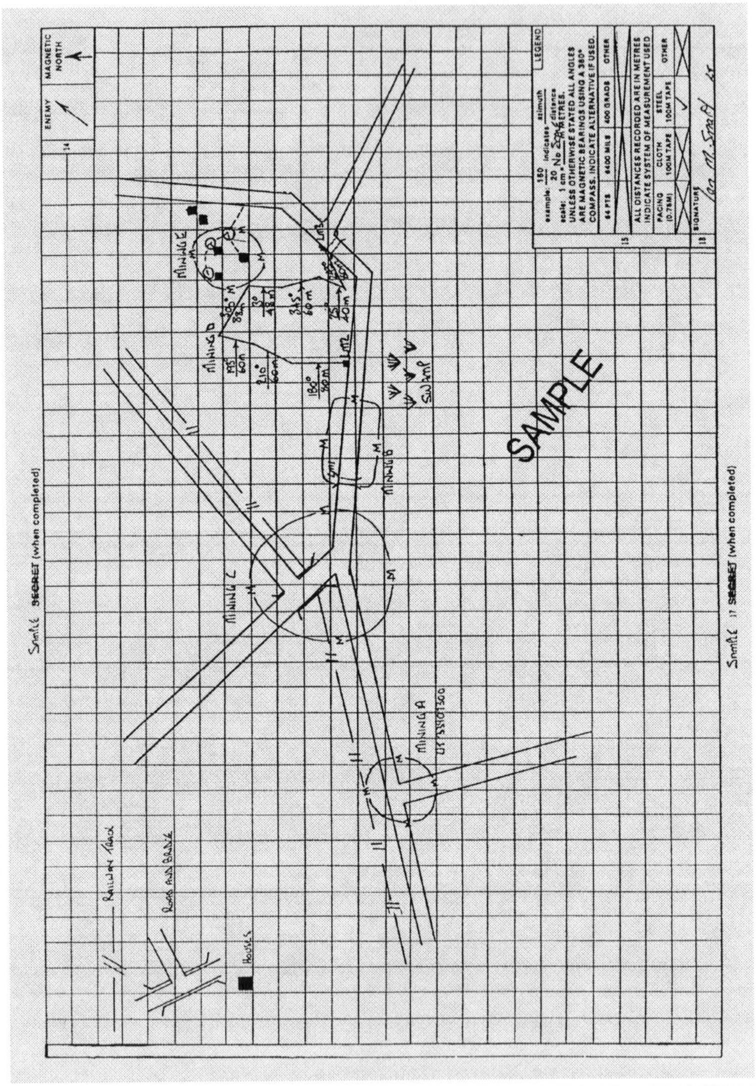

Figure 8-6b. Sample DA Form 1355 (inside) for a nuisance minefield/munition field

To facilitate reporting and recording of scatterable minefields/munition fields, a simple, uniform procedure is used. This procedure combines the report and the record into one document (Figure 8-7) that is applicable for all delivery systems.

Line No	Information Required	Data
1	Approving authority	Enter the approving authority, such as CDR 3AD.
2	Target/obstacle number	If the minefield/munition field is part of an obstacle plan, enter the obstacle number, such as 2XXX0157. This number represents II Corps, target number 157. If the minefield/munition field is not a part of an obstacle plan or does not have a number, then leave this line blank or enter NA.
3	Type of emplacing system	Enter the type system that emplaced the minefield/munition field, such as artillery or Volcano.
4	Type of mines	Enter AP for antipersonnel mines and AT for antitank mines. If both types of mines are used, enter AP/AT.
5	Life cycle	Enter the DTG the minefield/munition field was emplaced and the DTG the last mine SDs.
6-14	Aim point/corner points of minefield/munition field	If the system used to emplace the minefield/munition field uses a single aim point to deliver the mines, enter that aim point, such as MB 10102935. If the system has distinct corner points (Volcano), enter those corner points, such as MB 17954790, MB 18604860, MB 18504890, and MB 18054895.
15	Size of safety zone from aim point	If an aim point is given in Line 6, enter the size of the safety zone from that aim point. **Example:** Artillery emplaces a minefield/ munition field from aim point MB 10102935, and the safety zone is 1,000 x 1,000 m. Enter 500 m so that personnel plotting or receiving the information can plot the coordinates and go 500 m in each direction from the aim point to plot the safety zone.
16	Unit emplacing mines/ report number	Enter the unit emplacing mines and the report number, such as BCO 23 ENGR BN 4. Reports should be numbered consecutively. This would be the fourth minefield/munition field that B Company has emplaced.
17	Person completing report	Enter the person's name completing the report, such as SFC Jones.
18	DTG of report	Enter the DTG of the report, such as 160735ZOCT90.
19	Remarks	Include any other items the reporting unit may feel are important.

Figure 8-7. Scatterable minefield/munition field report and record work sheet

In addition to the scatterable minefield/munition field report and record, the SCATMINWARN (a sample is shown in Figure 8-8, page 8-24) notifies effected units that SCATMINEs will be emplaced. These two reports are the only reports used with scatterable mines.

A completed scatterable minefield/munition field report and record for an ADAM/RAAM artillery mission is shown in Figure 8-9, page 8-24. Note that on line 6, only one grid coordinate is given. It is the aim point used when the mission was fired. Also note that the 500-meter distance from the aim point (line 15) designates a safety zone that is 1,000 by 1,000 meters.

Line	Message
Alpha	Emplacing system
Bravo	AT (Yes or No)
Charlie	AP (Yes or No)
Delta	4 aim or corner points
Echo	Grid coordinates of aim points/corner points and size of the safety zone
Foxtrot	DTG of the life cycle

Figure 8-8. Sample SCATMINWARN

Line No	Information Required	Data
1	Approving authority	CDR 3AD
2	Target/obstacle number	2XXX0157
3	Type of emplacing system	Artillery
4	Type of mines	AT/AP
5	Life cycle	0816102-082020OCT90
6	Aim point/corner points of minefield/munition field	MB 10102935
7		
8		
9		
10		
11		
12		
13		
14		
15	Size safety zone from aim point	500 m
16	Unit emplacing mines/report number	2/48FA/2
17	Person completing report	SFC Hollins
18	DTG of report	061645ZOCT90
19	Remarks	NA

Figure 8-9. Scatterable minefield/munition field report and record for an ADAM/RAAM artillery mission

The SCATMINWARN provides affected units with the necessary warning to plan and execute their operations. The information is kept to a minimum to ensure rapid dissemination. The report may be sent orally, digitally, or hard copy. It is sent before or immediately after the mines have been emplaced. A completed SCATMINWARN for an artillery mission is shown in Figure 8-10.

Line	Message
Alpha	Artillery
Bravo	Yes
Charlie	Yes
Delta	One
Echo	MB 10102935 500 m
Foxtrot	081610Z-081920ZOCT90

Figure 8-10. Sample SCATMINWARN for an artillery mission

MINEFIELD/MUNITION FIELD OVERLAY SYMBOLS

The symbols contained in Figure 8-11, pages 8-26 through 8-30, are extracted from FM 101-5-1 and are provided for posting mine data on maps and overlays.

Description	Symbol
Minefield/Munition Fields	
Korea Only: AP mine	
AT mine	
AT mine with AHD	
Directional mine (arrow points in direction of main effect)	
Mine cluster	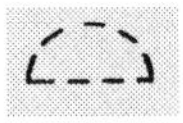
Mine, type unspecified	
Trip wire	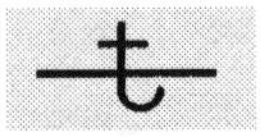
Control Measures	
Zone	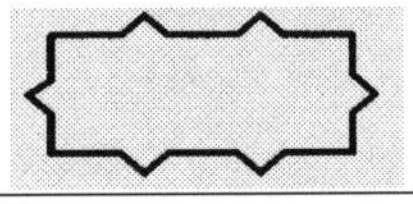
Belt	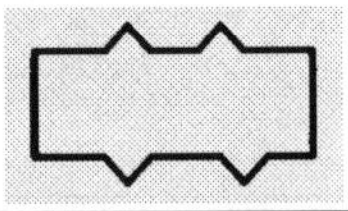
Restrictions	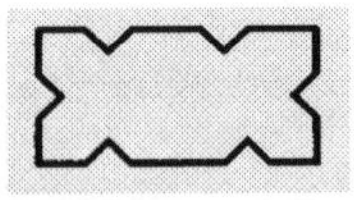

Figure 8-11. Minefield/munition field overlay symbols

Description	Symbol
Block effect	
Turn effect	
Disrupt effect	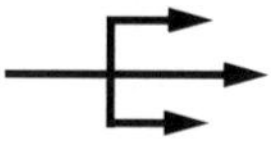
Fix effect	
Conventional	
A planned minefield/munition field consisting of unspecified mines	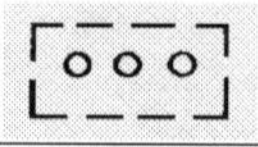
A completed minefield/munition field consisting of unspecified mines	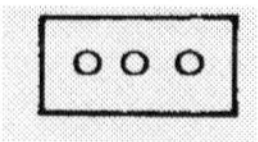
Scatterable minefield/munition field (DTGs used for SD times)	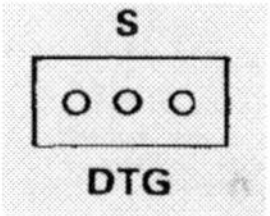
Conventional AP minefield/ munition field reinforced with SCATMINEs	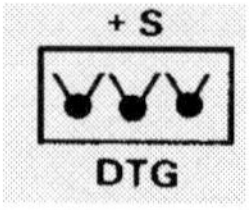
Tactical AP row minefield/ munition field (outline drawn to scale)	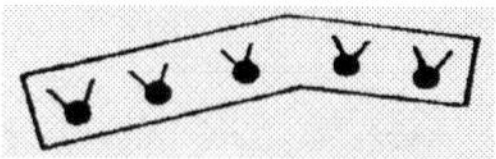

Figure 8-11. Minefield/munition field overlay symbols (continued)

Description	Symbol
Tactical minefield/munition field of scatterable AT mines, effective until 101200Z	
Completed AT minefield/munition field (drawn away from the location and connected by a vector)	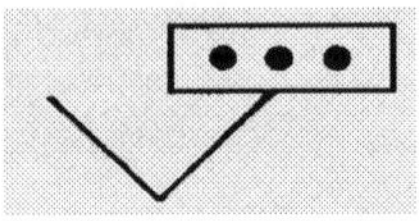
Executed Volcano minefield/munition field (DTG used for SD time)	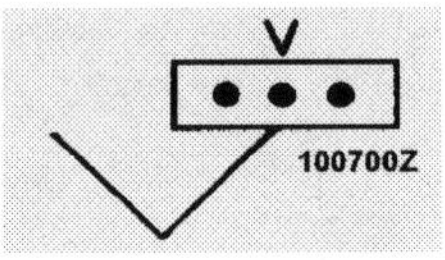
Lane in conventionally laid AT minefield/munition field	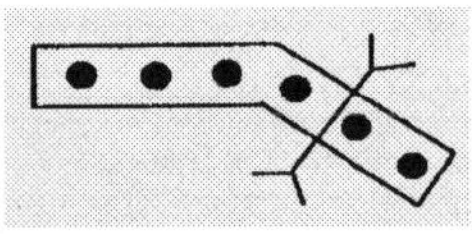
Gap in conventionally laid AT minefield/munition field (DTG opened to DTG closed)	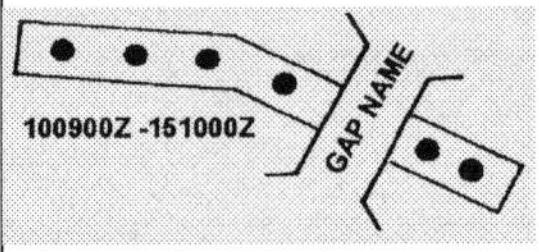

Figure 8-11. Minefield/munition field overlay symbols (continued)

Description	Symbol
AT ditch reinforced with AT mines	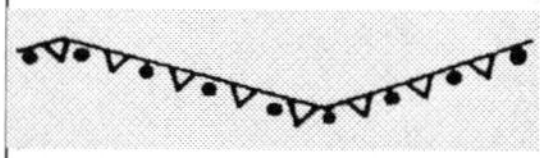
UXO	
UXO area	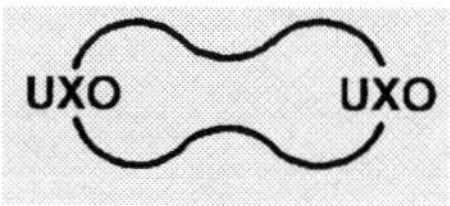
Nuisance	
Nuisance minefield/munition field	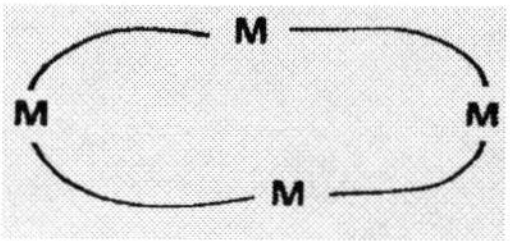
Demolished crossroads with nuisance mines	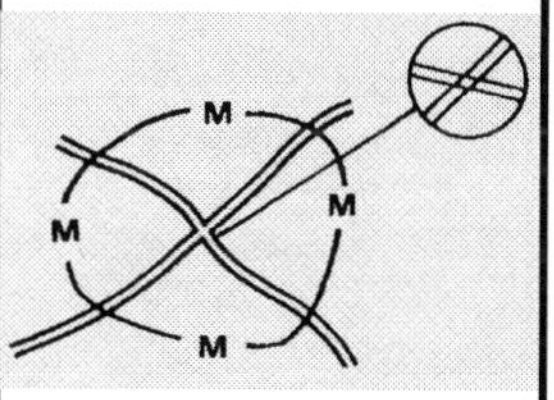
Phony	
Phony minefield/munition field	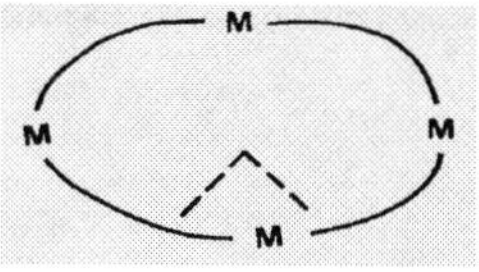

Figure 8-11. Minefield/munition field overlay symbols (continued)

Description	Symbol
Phony minefield/munition field, fenced	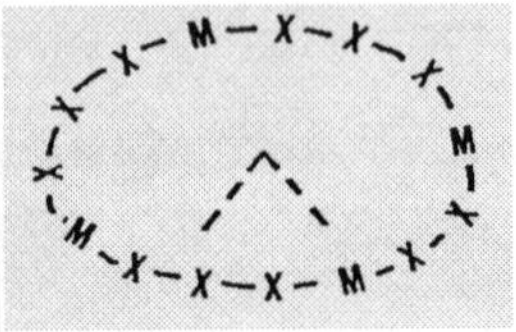
Hornet Symbology	
Planned	
Unarmed	
Armed	
Expended	
Armed munition field (DTG used for SD time)	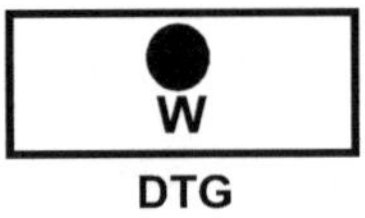

Figure 8-11. Minefield/munition field overlay symbols (continued)

PART TWO

Counteroperations

This part of the manual provides overall guidance for conducting counteroperations by US forces. The types of breaching and clearing operations conducted, the tasks performed, and the equipment required are described in detail. Responsibilities and planning considerations are outlined for each operation.

Chapter 9

Countermine Operations

Countermine operations are undertaken to breach or clear a minefield. All the tasks fall under breaching or clearing operations and include detecting, reporting, reducing, proofing, and marking.

DEFINITIONS

OBSTACLE

The term *obstacle* is used often in this chapter because the same breaching and clearing operations are used for minefields and other obstacles. For the purpose of this manual, breaching and clearing tactics, techniques, and procedures (TTP) focus solely on minefields.

REDUCTION

Reduction is the act or actions taken against an obstacle that diminishes its original effect. For example, creating a lane in a minefield would yield a reduction of the minefield obstacle.

BREACHING

Breaching is the employment of TTP to project combat power to the far side of an obstacle. It is a synchronized combined arms operation that is under the control of the maneuver commander.

AREA CLEARANCE

Area clearance is the total elimination or neutralization of an obstacle or portions of an obstacle. Clearing operations are not conducted under fire. They are usually performed by follow-on engineer forces after a breaching operation or anytime in a friendly AO where an obstacle is a hazard or hinders movement.

ROUTE CLEARANCE

Route clearance is the removal of mines along preexisting roads and trails.

MINE NEUTRALIZATION

Mine neutralization occurs when a mine is rendered incapable of firing on passage of a target. The mine may still be dangerous to handle.

PROOFING

Proofing is done by passing a mine roller or other mine-resistant vehicle through a lane as the lead vehicle. It verifies that a lane is free of mines.

DEMINING

Demining is the complete removal of all mines and UXO within a geopolitical boundary after hostilities cease.

BREACHING OPERATIONS

Breaching is a synchronized combined arms operation that is under the control of the maneuver commander. FM 3-34.2 provides combined arms commanders and staffs with doctrine TTP that are needed to successfully overcome obstacles. Breaching operations make maneuver possible in the face of enemy obstacle efforts. Since obstacles may be encountered anywhere, maneuver forces integrate breaching operations into all movement plans.

When possible, enemy minefields are bypassed to maintain the momentum and conserve critical countermobility assets. However, when making the decision to bypass rather than breach, consider the likelihood of friendly units being channelized into kill zones. Bypassing is done by maneuvering around a minefield or, if aviation assets are available, moving over the minefield. When maneuvering around an obstacle, attempt to locate a portion of the force in overwatch positions to cover the bypass of the main element. Even when the decision is made to conduct a breach, scouts should continue to reconnoiter for bypass routes.

The first step in understanding breaching operations is to know the obstacle breaching theory. Knowing the theory behind breaching operations equips the engineer and the maneuver commander with fundamentals that are needed to integrate breach into the tactical planning, preparation, and execution of an operation.

Successful breaching operations are characterized by the application of the following tenets of breaching:

INTELLIGENCE

In any operation where enemy obstacles interfere with friendly maneuver, obstacle intelligence (OBSTINTEL) becomes a priority intelligence requirement (PIR). Finding enemy obstacles or seeing enemy obstacle activity validates and refines the S2's picture of the battlefield. OBSTINTEL helps determine enemy intentions, plans, and strength. The force engineer is the unit's expert on enemy countermobility, and he assists the S2 in templating enemy obstacles and analyzing OBSTINTEL.

When collecting OBSTINTEL, reconnaissance is a combined arms activity that includes engineers. An engineer squad moves with scouts or the patrol and conducts dismounted reconnaissance of templated or discovered obstacles. Additional information on reconnaissance can be found in FM 5-170. Reconnaissance teams gather the following OBSTINTEL information from the reconnaissance:

- Minefield location. Plot the perimeter location on a large-scale map and refer to recognizable landmarks.

- Perimeter description. Describe how the perimeter is fenced. If it is unfenced, describe how it is marked. If it is unmarked, show how it was recognized.
- Nuisance mines. If you discover a nuisance mine forward of the minefield's outer edge, remember, there may be others. Assembly areas might also be mined.
- Types of mines. Indicate whether mines are AT or AP or have unknown fuses (self-neutralized or SD). If possible, recover specimens of unknown or new mines and note the details.
- Details of any other devices. Describe booby traps, trip wires, and flares.
- Laying method. Indicate whether mines are buried or surface-laid.
- Density and pattern. Include the mine spacing and the number of mine rows; estimate the mine density based on this information.
- Minefield depth. Provide the distance between strips or rows and describe markers.
- Safe lanes and gaps. Plot the location of suspected safe lanes and gaps and describe their markings.
- Ground conditions. Include information on general ground conditions.
- Other obstacles. Plot the location and the construction of other obstacles.
- Enemy defenses. Describe the enemy's location and size. Include the location of enemy direct-fire weapons.

NOTE: Engineers engaged in reconnaissance for OBSTINTEL should rarely, if ever, be used to reduce obstacles during the reconnaissance.

Each reconnaissance team commander submits a detailed OBSTINTEL report to the next higher headquarters when the reconnaissance is complete. A sample OBSTINTEL report is shown in Figure 9-1.

Letter Designation	Explanation
A	Map sheet(s).
B	Date and time the information was collected.
C	Type of minefield (AT, AP, or mixed).
D	Grid references of minefield extremities, if known.
E	Depth of minefield.
F	Estimated time required to clear the minefield.
G	Estimated material and equipment required to clear the minefield.
H	Routes for bypassing the minefield, if any.
I-Y	Grid references of lanes (entry and exit) and width of lanes, in meters.
Z	Additional information such as types of mines used, unknown mines, or types of booby traps.

Figure 9-1. Sample OBSTINTEL report

FUNDAMENTALS

Suppress, obscure, secure, and reduce (SOSR) are the breaching fundamentals that must be applied to ensure success when breaching against a defending enemy. This TTP will always apply but may vary based on specific situations.

Suppress

Suppression is the focus of all available fires on enemy personnel, weapons, and equipment to prevent effective fires on friendly forces. The purpose of suppression is to protect forces who are reducing the obstacle and maneuvering through it and to soften the enemy's initial foothold.

Obscure

Obscuration hampers enemy observation and target acquisition, and it conceals friendly activities and movement. Obscuration smoke deployed on or near the enemy position minimizes the enemy's vision.

Secure

The force secures the breaching site to prevent the enemy from interfering with reduction and to prevent the enemy's passage through lanes created during reduction. In general, enemy tactical obstacles are secured by fires, and protective obstacles are secured by force.

Reduce

Reduction means creating lanes through or over the obstacle to allow passage of the attacking force. The lanes must be sufficient to allow the force to cross and accomplish the mission. (Table 9-1 provides information on lane widths.) The unit reducing the minefield marks the minefield and lane locations and reports their conditions to higher headquarters.

Table 9-1. Lane widths

Lane	Width	Remarks
Assault footpath	1 meter	Allows passage of dismounted troops so that they may continue an attack or secure the far side of a minefield while vehicle lanes are being breached
Initial lane	4.5 meters	Allows passage of breaching and assault forces (initial lanes are widened and marked as soon as the tactical situation allows)
Intermediate lane	4.5 meters	Upgrades sign posting and further defines exit markers
Full lane	10 meters	Allows passage of two-way traffic

ORGANIZATION

The commander organizes the force to accomplish SOSR breaching fundamentals quickly and effectively. This requires him to organize support, breach, and assault forces with the necessary assets to accomplish their missions.

Support Force

The support force is primarily responsible for eliminating the enemy's ability to interfere with the breaching operation. It must isolate the battlefield with fires and suppress enemy fires that are covering the breach location. This involves massive direct and indirect fire to destroy enemy vehicles and personnel who are able to bring fire on the breach force.

Breach Force

The breach force creates lanes that enable the assault force to pass through the obstacle and continue the attack. It is also responsible for marking lanes (length and entry points) to speed passage of assault and follow-on forces. The breach force is a combined arms force. It includes engineers, breaching assets, and enough maneuver force to provide local security. The breach force must be able to deploy and begin reducing the obstacle as soon as enemy fires have been suppressed. It must be capable of creating a minimum of one lane for each assaulting company or two lanes for an assaulting TF. At a minimum, the lanes must be marked and their locations and conditions reported to higher headquarters and follow-on units as prescribed in the unit's SOP. The commander should expect a 50 percent loss of mobility assets in close combat.

Assault Force

The assault force destroys or dislodges the enemy on the far side of the obstacle. It secures the far side by physical occupation in most breaching operations. The assault force may be tasked to assist the support force with suppression while the breach force reduces the obstacle. The assault force must be sufficient in size to seize objectives that eliminate fires on the breaching site.

MASS

Breaching is conducted by rapidly applying concentrated force at a designated point to crack the obstacle and rupture the defense. Massed combat power is directed against an enemy weakness. Smoke and terrain help isolate the enemy force that is under attack. The commander also masses engineers and breaching equipment to reduce the obstacle. The breach force is organized and equipped to use several different reduction techniques in case the primary technique fails (a key vehicle is destroyed or casualties render dismounted engineers ineffective). Additional reduction assets are present to handle the unexpected (50 percent over the requirement are normally positioned with the breach force).

SYNCHRONIZATION

Breaching operations require precise synchronization of the SOSR breaching fundamentals by support, breach, and assault forces. Failure to synchronize effective suppression and obscuration with reduction and assault can result in rapid, devastating losses of friendly troops in the obstacle or the enemy's fire sack. The commander cannot adequately synchronize his force's application of combat power in the short time available to him when he encounters an obstacle. The number of decisions that he must make while under fire in an unclear situation will rapidly overwhelm him. Even with a force trained to execute a combined arms breach, synchronizing all the necessary tasks remains a complex, difficult process. The commander uses the following principles to ensure synchronization through proper planning and force preparation:

- Detailed reverse planning.
- Clear subunit instructions.
- Effective C^2.
- A well-rehearsed force.

CLEARING OPERATIONS

Clearing is the total elimination or neutralization of mines from an area. It is not usually conducted under enemy fire, but it can be conducted by engineers during war or after hostilities as part of nation assistance.

A limited clearing operation can be conducted by follow-on engineers after the breaching force has reduced the minefield and secured the area. Engineers initially improve existing breach lanes by widening and marking them, and they also clear and mark new lanes through the minefield. The clearing operation supports the continued passage of forces.

A clearing operation is also conducted to eliminate all the mines in a minefield (previously identified, reported, and marked in a friendly AO) that hinders mobility or is a hazard to friendly forces.

Upgrade of Breach Lanes

Lane clearance is more deliberate than lane breaching and normally takes longer. Follow-on engineers upgrade breach lanes to improve existing lanes through minefields and to create new lanes. This clearing operation is intended to further reduce the minefield so that follow-on units can pass through it quickly.

The breach force that initially reduced the obstacle and marked the lanes turns over the lanes to follow-on engineers. Follow-on engineers can expect lane widths of 4.5 meters. The total number of lanes depends on the size of the lead assault force. Two to four assault lanes are normal if the lead unit was brigade-size.

If forces continue to pass through existing lanes while further reduction and clearance is conducted, follow-on engineers first begin reducing new lanes. At a minimum, two lanes are required for an assaulting TF and one lane is required for an assaulting company/team.

A limited amount of mechanical breaching assets is available for clearing operations. Follow-on engineers will probably not have tank-mounted mine-clearing blades (MCBs) or mine-clearing rollers (MCRs). The main mechanical clearing asset is an armored dozer with a mine rake. Mine-clearing line charges (MICLICs) are used if available. Engineers conducting clearing operations—

- Ensure that lanes are a minimum of 100 meters apart.
- Reduce additional lanes by using the equipment and techniques outlined in Chapter 10.
- Widen lanes to 10 meters to allow two-way traffic.
- Mark breach lanes by using the original marking system or the division SOP. (Marking procedures are outlined in Chapter 10.)
- Emplace entrance, exit, and left and right lane markers to provide day and night capability.

Traffic control is critical during lane reduction and when shifting lanes to improve existing lanes. Engineers conducting reduction and clearance may also provide guides at the lanes. Control procedures are outlined in FM 90-13-1.

To eliminate the danger of forces entering the minefield adjacent to lanes, the minefield is marked with fencing (barbwire or concertina) and mine markers.

Marking is emplaced across the front, on both sides, between lanes, and to the left and right of the crossing site as far out as practical.

Engineers may also help remove damaged vehicles from minefield lanes. Recovery vehicles should be available near lanes for this purpose.

AREA CLEARANCE

Clearing operations are done when engineers receive a mission to clear an area of mines or to clear a specific minefield in a friendly AO. The minefield was reported and may already be marked on all sides. The worst case would be if the minefield was reported but not marked and its limits were unknown. The engineer unit receiving the mission bases plans on available information and prepares equipment based on the estimate. Detailed techniques and procedures for area and route clearance operations are outlined in Chapter 11.

Actions at the minefield begin with a thorough reconnaissance to identify the minefield limits and the types of mines. This is a time-consuming process that is hazardous to shortcut. Identified limits are marked with an expedient system of single-strand barbwire or concertina. In this situation, since all mines must be destroyed, the unit takes a systematic approach to clearing mines. The procedure depends on the types of mines and whether the mines are buried or surface-laid.

If mines are magnetic- or seismic-fused, mechanical assets are used. Pressure mines can be destroyed by using hand-emplaced explosives. When a manual procedure is used, eliminate trip wires on AP mines with grapnel hooks before moving forward to detect mines.

Using the manual procedure, engineers visually detect mines or detect them with mine detectors and probes. They also mark mines for destruction by explosives. Chapter 11 contains information on minesweeping procedures.

After the mines are destroyed, engineers proof used lanes and routes to ensure that all the mines were eliminated. This is done by using a mine roller or another blast-resistant device. Proofing is discussed further in Chapter 10.

DEMINING

Demining is the complete removal of all mines and UXO to safeguard the civilian population within a geopolitical boundary after hostilities cease. It is an extremely manpower- and time-intensive operation and is sometimes contracted. Although not a formal Army mission or function, SOFs may provide special expertise in training demining organizations, acting as advisors, and taking the lead in providing clearance equipment or techniques that can be useful in demining operations. Demining TTP are outlined in TC 31-34.

Chapter 10

Minefield Reduction

Reduction is the physical creation of a lane through a minefield. It is a fundamental of breaching operations as discussed in Chapter 9 and in FM 3-34.2. A number of tasks (detecting, reporting, reducing, proofing, and marking) directly support or are included in minefield reduction.

DETECTING

Detection is the actual confirmation and location of mines. It may be accomplished through reconnaissance, or it may be unintentional (such as a vehicle running into a mine). Mine detection is used in conjunction with intelligence-gathering operations, minefield bypass reconnaissance, and breaching and clearing operations. There are four types of detection methods—visual, physical (probing), electronic, and mechanical.

VISUAL

Visual detection is part of all combat operations. Personnel visually inspect the terrain for the following minefield indicators:

- Trip wires.
- Signs of road repair (such as new fill or paving, road patches, ditching, culvert work).
- Signs placed on trees, posts, or stakes. Threat forces mark their minefields to protect their own forces.
- Dead animals.
- Damaged vehicles.
- Disturbances in previous tire tracks or tracks that stop unexplainably.
- Wires leading away from the side of the road. They may be firing wires that are partially buried.
- Odd features in the ground or patterns that are not present in nature. Plant growth may wilt or change color, rain may wash away some of the cover, the cover may sink or crack around the edges, or the material covering the mines may look like mounds of dirt.
- Civilians. They may know where mines or booby traps are located in the residential area. Civilians staying away from certain places or out of certain buildings are good indications of the presence of mines or booby traps. Question civilians to determine the exact locations.
- Pieces of wood or other debris on a road. They may be indicative of pressure or pressure-release FDs. These devices may be on the surface or partially buried.

- Patterns of objects that could be used as a sighting line. The enemy can use mines that are fired by command, so road shoulders and areas close to the objects should be searched.

PHYSICAL

Physical detection (probing) is very time-consuming and is used primarily for clearing operations, self-extraction, and covert breaching operations. Detection of mines by visual or electronic methods should be confirmed by probing. Use the following procedures and techniques when probing for mines:

- Roll up your sleeves and remove your jewelry to increase sensitivity. Wear a Kevlar helmet, with the chin strap buckled, and a protective fragmentation vest.
- Stay close to the ground and move in a prone position to reduce the effects of an accidental blast. When moving into a prone position—
  - Squat down without touching your knees to the ground.
  - Scan forward up to 2 meters and to the sides up to 3 meters for mine indicators.
  - Probe the area around your feet and as far forward as possible.
  - Kneel on the ground after the area is found to be clear, and continue probing forward until you are in a prone position.
- Use sight and touch to detect trip wires, fuses, and pressure prongs.
- Use a slender, nonmetallic object as a probe.
- Probe every 5 centimeters across a 1-meter front.
- Gently push the probe into the ground at an angle that is less than 45 degrees.

DANGER
Use extreme caution when probing. If the probe is pushed straight down, its tip may detonate a pressure fuse.

- Apply just enough pressure on the probe to sink it slowly into the ground.
- If the probe encounters resistance and does not go into the ground freely, carefully pick the soil away with the tip of the probe and remove the loose dirt by hand. Care must be taken to prevent functioning the mine.
- When you touch a solid object, stop probing and use two fingers from each hand to carefully remove the surrounding soil and identify the object.
- If the object is a mine, remove enough soil to show the mine type and mark its location. Do not attempt to remove or disarm the mine. Use explosives to destroy detected mines in place, or use a grappling hook and rope to cause mines to self-detonate. Do not use metal grappling hooks on magnetic-fused mines.

Probing is extremely stressful and tedious. The senior leader must set a limit to the time a prober can actually probe in the minefield. To determine a reasonable time, the leader must consider METT-TC factors, weather conditions, the threat level, the unit's stress level, and the prober's fatigue level and state of mind. As a rule, 20 to 30 minutes is the maximum amount of time that an individual can probe effectively.

ELECTRONIC

Electronic detection is effective for locating mines, but this method is time-consuming and exposes personnel to enemy fire. In addition, the suspected mines must be confirmed by probing.

AN/PSS-12 Mine Detector

The AN/PSS-12 mine detector (Figure 10-1) can only detect metal, but most mines have metal components in their design. The detector can locate and identify plastic or wooden mines by a slight metallic signature. Employment and operation procedures for the AN/PSS-12 are discussed in Appendix F, and technical data is available in TM 5-6665-298-10. The detector is hand-held and identifies suspected mines by an audio signal in the headphones.

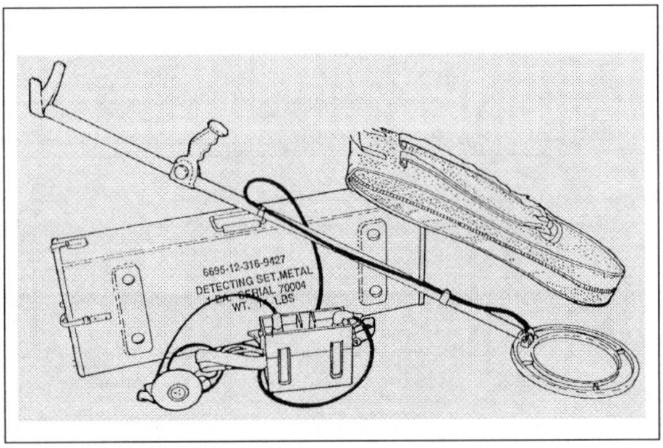

Figure 10-1. AN/PSS-12 mine detector

As in probing, consideration must be taken for the maximum amount time an individual can operate the detector. The leader considers METT-TC factors, weather conditions, the threat level, the unit's stress level, and the individual's fatigue level and state of mind. As a rule, 20 to 30 minutes is the maximum amount of time an individual can use the detector effectively.

Airborne Standoff Minefield Detection System

The Airborne Standoff Minefield Detection System (ASTAMIDS) (Figure 10-2, page 10-4) provides US forces with the capability to detect minefields rapidly. Environmental conditions must be favorable for aircraft and ASTAMIDS operations. ASTAMIDS can be mounted on a UH-60 Blackhawk helicopter, an unmanned aerial vehicle (UAV), or a fixed-wing aircraft. The system detects and classifies thermal and other anomalies as suspected minefields along routes or in areas of interest. ASTAMIDS can be used to protect advancing forces and can operate in concert with air and ground units in reconnaissance missions.

Figure 10-2. ASTAMIDS

System Components

ASTAMIDS hardware and software components consist of a sensor with associated electronics and the minefield-detection algorithm and processor (MIDAP). Surrogate equipment includes an air-data package (GPS, radar altimeter, inertial measurement unit [IMU]), a power supply, a work station(s), a digital data recorder, mounting racks, and a modified floor for the specific aircraft.

Operators view the data displayed on the monitors, communicate with the aircrew, and perform other functions (such as changing data tapes and producing reports). The aircrew must maintain an altitude of 300 feet and an airspeed of approximately 70 knots for the system to detect mines accurately within the sensor's ground swath (approximately 215 feet wide). The system has a 2-hour operational capability, based on standard flight time for the mission profile.

Employment Concept

ASTAMIDS is a fast method for detecting tactical minefields. When it is employed by aviation elements in support of maneuver units, close coordination between aviation and ground units assures that minefield detection is reported accurately and quickly. ASTAMIDS is not as precise as ground detection systems, but it is accurate enough to help mitigate the dangers inherent with minefields. It can be used in both friendly and enemy territories. The use of a Blackhawk ASTAMIDS in areas of threat observation and fire must be planned and coordinated very carefully, because a helicopter is extremely vulnerable while flying the mission profile required for detection (steady altitude, speed, and path).

Once airborne and at its start point, the ASTAMIDS system is placed in the correct detection mode, based on the intended mission (route or area reconnaissance). When the system indicates an initial detection, the operator communicates it to the pilot. The pilot then flies a verification pass over the indicated area. If the system again indicates a detection, the pilot resumes the mission (route reconnaissance) or continues the survey pattern to determine the minefield borders (area reconnaissance). If no detection is indicated on the verification pass, the operator instructs the pilot to resume the flight plan.

Interim Vehicle-Mounted Mine Detector

The interim vehicle-mounted mine detector (IVMMD) is used in all levels of conflict and OOTW. The IVMMD is mounted on a blast- and fragmentation-survivable vehicle; it is designed to detect and mark buried and surface-laid, metallic AT mines. The primary mission of the IVMMD is to detect mines during route clearance. The system should not be used when operating in an environment where the enemy employs mines that are not pressure-fused.

System Components

A complete IVMMD (Figure 10-3) consists of one mine-detection vehicle (MDV), one towing/mine-detection vehicle (T/MDV), three mine detonation trailers, a spare-wheel module for the MDV, a spare-wheel module for the T/MDV, and a container of spare parts.

Figure 10-3. IVMMD components

The MDV's only mission is to detect mines. It can negotiate vertical slopes up to a 20 percent grade. The MDV employs a 4-cylinder engine and a manual transmission to propel the 4.8-ton vehicle with a 3-meter-wide detection array. The detection array consists of two separate induction coils (one for the left side and one for the right side) that detect magnetic fields below the vehicle.

The detection array is suspended between the two axles of the vehicle. When the detector encounters a metallic object, the operator is notified by an audible signal in the earphone. A visual signal appears on the dashboard that denotes which side of the array detected the object. The operator then stops the vehicle, backs it up, and reencounters the metallic object. (The MDV has two detection modes—the locate mode is used to identify the metal object, and the pinpoint mode is used to find the center of the object.) When the operator encounters the strongest signal, he activates the marking system (a nozzle mounted on the rear frame and centered on each detection array) that deploys a water-based ink onto the roadway.

The T/MDV has a 6-cylinder engine and the same detection and marking system as the MDV. The T/MDV tows three detonation trailers. The recommended maximum operating speed while towing the trailers is 20 kph. The T/MDV (with trailers) can negotiate vertical slopes up to a 20 percent grade; however, going down such slopes is difficult. The T/MDV must be in first gear, and the trailer brakes must be deployed to decrease the speed of the vehicle when going down a slope.

The mine detonation trailers are very heavy, and they are specifically designed to apply heavy ground pressure that initiates pressure-activated mines. Each trailer has two axles of different lengths so that the three trailers provide a full 3-meter-wide proofing capability behind the T/MDV. If a mine detonates underneath the trailers, the wheel bolts are designed to sheer so that repair is limited to replacing a single wheel.

The detection array is suspended between the two axles of the MDV. Although the vehicle is designed to produce very little ground pressure, it will detonate most pressure-fused mines, depending on the sensitivity of their fuses.

- The MDV produces 27.9 pounds per square inch (psi) of ground pressure when the tires are inflated to 14.5 psi and 21.8 psi of ground pressure when the tires are inflated to 8.7 psi.
- The T/MDV produces 49.8 psi of ground pressure when the tires are inflated to 29 psi and 23.4 psi of ground pressure when the tires are inflated to 8.7 psi.

Employment Concept

The IVMMD is used to support route-clearance operations. Clearance operations ensure that LOC are safe for the passage of personnel and equipment. The IVMMD should not be used during hours of limited visiblity, because it hampers the operator's ability to see surface-laid mines and visual signatures that indicate mining activities.

MECHANICAL

The track-width mine roller is a mechanical minefield-detection system. It is most effectively deployed to lead columns on route movement, but it can be used to precede tactical formations. In column movement, unit vehicles travel a narrow path, and one or two mine rollers can effectively detect mines in the path. Mine rollers can also be used to detect minefields in front of deployed tactical formations; however, more than one roller is required for a good probability of detection.

REPORTING

Intelligence concerning enemy minefields is reported by the fastest means available. Spot reports (SPOTREPs) are the tactical commander's most common source of minefield intelligence. They originate from patrols that have been sent on specific minefield reconnaissance missions or from units that have discovered minefield information in the course of their normal operations. The information is transmitted to higher headquarters.

REDUCING

Minefield reduction and clearing equipment is broken down into explosive, mechanical, electronic, and manual. Combat engineers and the operators of breach assets practice and become proficient in these reduction means. They integrate them into the breach drills of the units they support. The team applies different TTP to breach drills and prepares and rehearses them as part of the TF plan.

EXPLOSIVE

M58A4 Mine-Clearing Line Charge

The MICLIC (Figure 10-4, page 10-8) is a rocket-propelled, explosive line charge. It is used to reduce minefields that contain single-impulse, pressure-activated AT mines and mechanically activated AP mines. It clears a 14- by 100-meter path. The MICLIC has a 62-meter standoff distance from the launcher to the detonation point. The MICLIC's effectiveness is limited against prong AP mines, magnetically activated mines (including some SCATMINEs), top-attack mines, side-attack mines, and mines containing multiple-impulse or delay-time fuses. It also has little effect on other obstacles, such as log and concrete barriers, antivehicular ditches, and walls. The shock effect and the psychological impact of the detonation make the MICLIC a useful weapon in a close fight or in MOBA.

The MICLIC is mounted on a rubber-tired trailer, or two MICLICs can be mounted on an armored vehicle-launched bridge (AVLB), with the bridge downloaded, using a fabricated I-beam frame (procedures for mounting the MICLIC on the AVLB are outlined in TM 9-1375-215-14&P). This is called an armored vehicle-launched MICLIC (AVLM) (Figure 10-5, page 10-8), and it is the preferred system because no trailer is involved to hinder the mobility of the vehicle.

Towing vehicles for the trailer-mounted MICLIC are a combat engineer vehicle (CEV), an M113 APC, M2 and M3 Bradleys, an M9 armored combat earthmover (ACE), a 5-ton wheeled vehicle, and a 2½-ton wheeled vehicle. The trailer limits the MICLIC's mobility in rough terrain and degrades the maneuverability of the towing vehicle, thereby increasing vulnerability. Since the MICLIC is critical to the breach, it is kept under the protection of the force and is moved to the breach site along easily trafficable, covered, and concealed routes. This effectively prevents the towing vehicle from performing any other task (firing or maneuvering) or serving as an engineer squad vehicle unless MICLIC employment is the squad's only mission. This is an important consideration when selecting the towing vehicle because this vehicle must be solely dedicated to the mission.

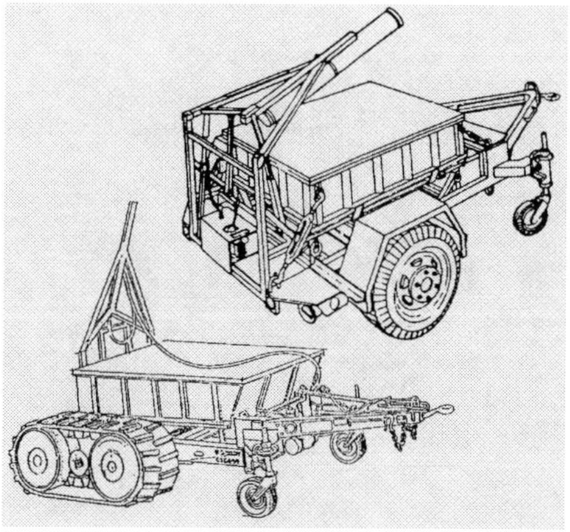

Figure 10-4. MICLIC

Figure 10-5. AVLM

The MICLIC can be fired from within an armored vehicle without exposing soldiers to fires; however, the prefiring preparations must be done in advance at a covered and concealed location near the breach site. The initiating cable is brought into the vehicle through the hatch, which must be left ajar, or through the portal of the periscope, which has been removed. Therefore, the crew is not afforded nuclear, biological, chemical (NBC) protection. When the MICLIC is fired from a wheeled vehicle, however, the crew must move to a covered position outside the backblast area. The special-purpose cable on the firing control switch is long enough to allow adequate standoff.

The vehicle operator must be proficient in all aspects of preparing and deploying the MICLIC, including the critical aspect of selecting the optimum breach site. Although the operator will be directed to the breach site by the engineer platoon leader or the breach force commander, ensuring that he can independently accomplish the task will simplify the operation and greatly enhance its likelihood of success. The towing vehicle and the operator must be selected well in advance and be dedicated solely to the task. The operator must be included in all rehearsals and planning sessions and, if possible, during leaders' reconnaissances.

Each MICLIC trailer transports and fires one charge, and then it must be reloaded. The AVLM can fire both MICLICs before reloading. The loaded charge container weighs 1,283 kilograms, so a lifting device such as a 5-ton wrecker or a HEMTT is needed. Reloading, which can be done by an experienced crew in about 20 minutes, entails loading a rocket on the rail and lifting a new charge container onto the launcher. The reloading operation must be done in a covered and concealed location.

The exact limits and depth of an enemy minefield are seldom known before the breach. This is particularly true when the situation is unclear, and the minefield is encountered simultaneously with enemy contact. The first and only indication that a unit is in a minefield may be when a vehicle encounters a mine. The leading edge of the minefield still may be an uncertainty, because the vehicle could have hit a mine in an interior row. The number of MICLICs needed to clear a single lane through a minefield depends on the minefield depth:

- Clearing a lane through a minefield less than 100 meters deep requires one MICLIC (Figure 10-6, page 10-10). The leading edge of the minefield is identified and, if possible, confirmed by reconnaissance. The MICLIC is deployed from a minimum standoff distance of 62 meters from the leading edge of the minefield.
- Clearing a lane through a minefield more than 100 meters deep or of uncertain depth requires two or more MICLICs (Figure 10-7, page 10-10). If the leading edge cannot be identified, the MICLIC is deployed 100 meters from the possible edge or stricken vehicle. When the first MICLIC is detonated, a second MICLIC moves 25 meters into the first MICLIC's path and fires its charge. This extends the lane an additional 87 meters. Additional MICLICs are used for minefields of extreme depth, and each one moves down the lane 25 meters into the path created by the previous charge.

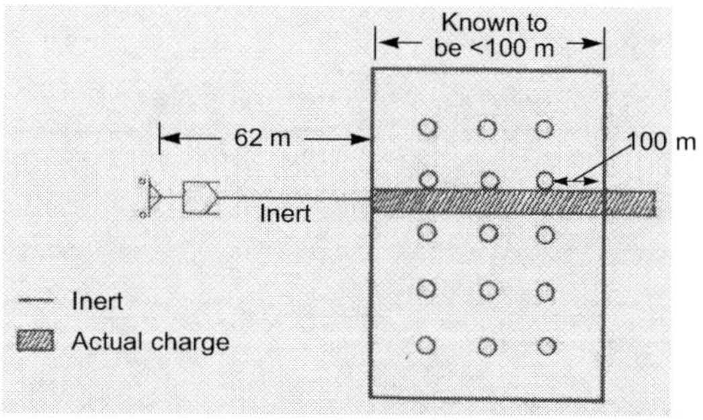

Figure 10-6. MICLIC employment in a minefield less than 100 meters deep

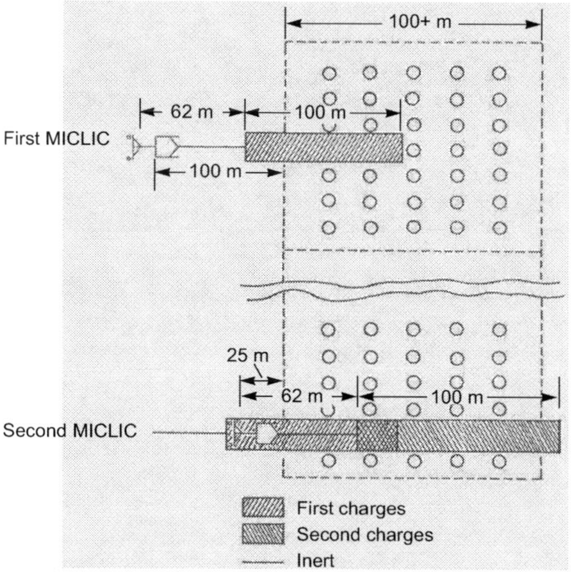

Figure 10-7. MICLIC employment in a minefield of uncertain depth or greater than 100 meters

The neutralization of mines by blast depends on the peak pressure and the impulse. For the MICLIC, the impulse is at a maximum of 3 meters from the line charge (on both sides) and decreases the closer it gets toward the line charge, to a minimum of 1 meter from the line charge. This decrease on impulse causes a *skip zone* (Figure 10-8). This does not mean that neutralization is equal to zero percent; it means that it is not equal to 100 percent. Mines that are buried deeper than 10 centimeters and located 1 to 2 meters from the line charge have a high probability of not being neutralized.

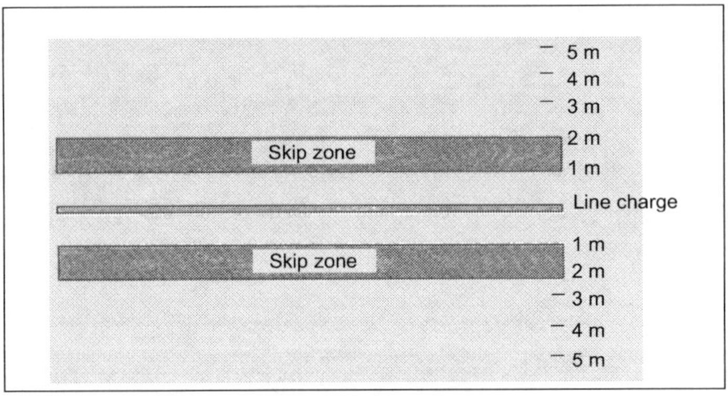

Figure 10-8. Skip zone

Antipersonnel Obstacle Breaching System

The Antipersonnel Obstacle Breaching System (APOBS) (Figure 10-10) is a man-portable device that is capable of quickly creating a footpath through AP mines and wire entanglements. The APOBS is normally employed by combat engineers, infantry soldiers, or dismounted armored cavalry personnel. The APOBS provides a lightweight, self-contained, two-man, portable line charge that is rocket-propelled over AP obstacles from a standoff position away from the edge of the obstacle.

For dismounted operations, the APOBS is carried in 25-kilogram backpacks by no more than two soldiers for a maximum of 2 kilometers. One backpack assembly consists of the rocket-motor launch mechanism, containing a 25-

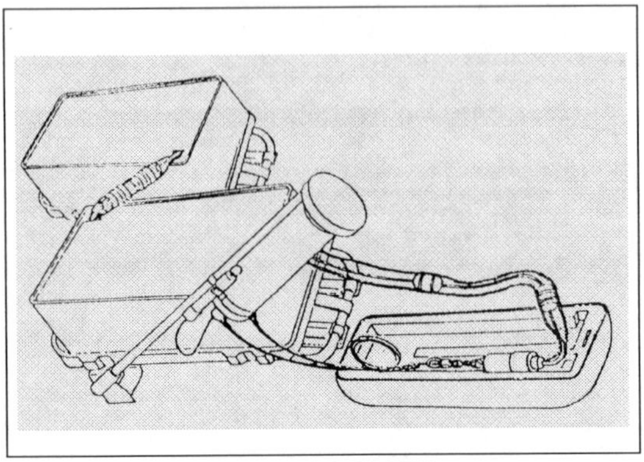

Figure 10-9. APOBS

meter line-charge segment and 60 attached grenades. The other backpack assembly contains a 20-meter line-charge segment and 48 attached grenades. The total weight of the APOBS is approximately 54 kilograms. It is capable of breaching a footpath that is approximately 0.6 by 45 meters and is fired from a 25-meter standoff.

M1A1/M1A2 Bangalore Torpedo

The bangalore torpedo (Figure 10-11) is a manually emplaced, explosive-filled pipe that was designed as a wire breaching device, but it is also effective against simple pressure-activated AP mines. It is issued as a demolition kit and consists of ten 1.5-meter tubes. Each tube contains 4 kilograms of high explosives and weighs 6 kilograms. The kit clears a 1- by 15-meter lane.

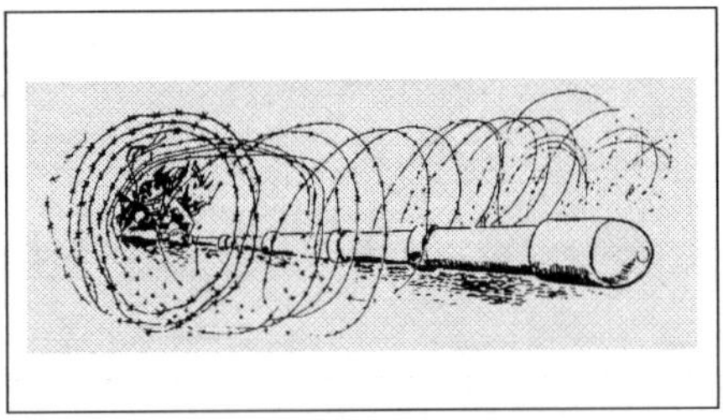

Figure 10-10. Bangalore torpedo

The bangalore torpedo is used by dismounted infantry and engineer troops. An individual soldier or a pair of soldiers connects the number of sections needed and pushes the torpedo through the AP minefield before priming it. A detailed reconnaissance is conducted before employing the bangalore torpedo to ensure that trip wires have not been used.

The bangalore torpedo generates one short impulse and is not effective against pronged, double-impulse, or pressure-resistant AP and AT mines.

> **WARNING**
> **Do not modify the bangalore torpedo. Cutting the bangalore in half or performing any other modification could cause the device to explode.**

MECHANICAL

MCBs and MCRs are fielded as armor battalion sets that contain 12 MCBs and 4 MCRs. Blades clear lanes through minefields, while rollers are used to detect minefields and proof lanes created by other means. Rollers are not a good primary system for lane reduction because multiple mine detonations destroy the roller system and the vehicle pushing it. (The roller is designed to resist two conventional-mine or three scatterable-mine strikes, depending on the mine type.)

The CEV, the ACE, and dozer blades were not designed for breaching minefields and should be employed only as a last resort, because using them is extremely hazardous to the crew and equipment. However, CEVs or ACEs can effectively clear a lane through AP scatterable minefields because they sustain little or no damage and offer protection to the crew. When using a dozer to clear a path through minefields, the operator is exposed to mine effects. Before clearing begins, the operator's cabin should be sandbagged or *up-armored* and the lane should be cleared of trip wires. When using an engineer blade to clear a path through a scatterable minefield, use the skim technique (Figure 10-12). Start skimming 100 meters from the suspected minefield leading edge.

Mine-Clearing Blade

The MCB (Figure 10-13) is used to remove land mines from the minefield. It consists of a blade arrangement with scarifying teeth to extract mines, a moldboard to cast mines aside, and leveling skids to control the depth of the blade.

The MCB lifts and pushes mines, which are surface-laid or buried up to 31 centimeters deep, to the side of the track-width lanes. The blade has three depth settings—21, 25, and 31 centimeters. The blade creates a 1.5-meter cleared path in front of each track. Figure 10-14, page 10-16, shows inside clearance distances between tracks of common track vehicles in relationship to the uncleared area left by the MCB.

Mines armed with AHDs, magnetic fuses, or seismic fuses may be activated when they are lifted by the blade; and they may disable the blade. Mines lifted by the blade are left in the spoil on each side of the furrowed path and remain a hazard until they are removed. Double-impulse mines that are lifted into the spoil on the side have a probability of functioning into the hull of the plowing vehicle. The skid shoe for each blade exerts adequate pressure to activate

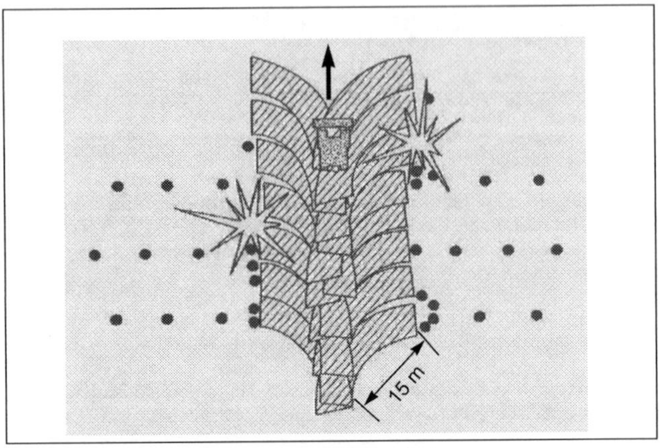

Figure 10-11. Skim technique

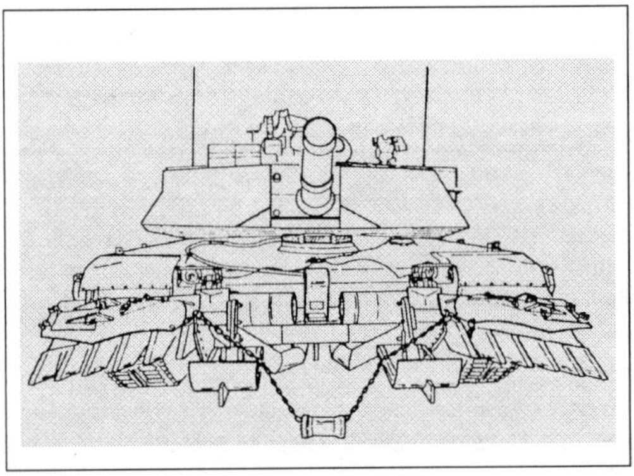

Figure 10-12. MCB

most single-impulse mines, which effectively clears a section of the centerline by explosive detonation. This action may disable the blade. Multiple-impulse pressure fuses encountered by the skid shoe are not defeated. A dog-bone assembly between the blades defeats tilt-rod mines. The improved dog-bone assembly (IDA) projects a magnetic signature and defeats tilt-rod and magnetic mines.

The MCB weighs approximately 3,150 kilograms and can be mounted on an M1 tank without special preparation or modification. Mounting requires lift capability and takes up to an hour, so it must be mounted well in advance of the mission. It is not easy to mount or transfer the MCB to another tank under battlefield conditions.

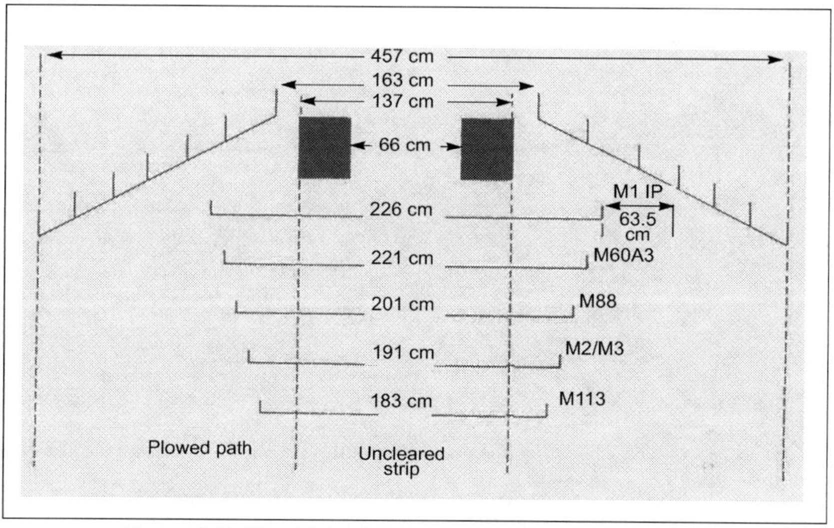

Figure 10-13. Mine-blade width compared to track-vehicle widths

Once mounted, an electric motor raises and lowers the blade. When it is in the raised position, it minimally effects the M1's maneuverability and speed. This will not greatly impact the employment of the weapon system except when the blade is in operation. The MCB is also equipped with an emergency, quick-disconnect feature.

The M1 should perform plowing operations from 8 to 10 kph, depending on soil conditions. It cannot maneuver but must continue in a straight path through the minefield to avoid damaging the blade. The main gun must be traversed to the side during plowing because mine detonation under the blade may cause the gun to be thrown violently into the air, damaging the tube. The area selected for the lane must be relatively flat and free of rocks or other obstacles.

The operator begins plowing approximately 100 meters from the estimated minefield leading edge. He creates a lane extending another 100 meters beyond the estimated minefield far edge to ensure that the lane extends through the entire minefield. Multiple vehicles crossing the breach will deepen the cut made by the MCB, and pressure-fused mines left in the uncleared strip will be dangerous. The uncleared strip should be cleared as soon as possible.

Mine-Clearing Roller

The MCR (Figure 10-15) consists of a roller assembly, a mounting kit, and a hand winch kit. The roller assembly weighs approximately 9,072 kilograms and consists of two push beams mounted to the front of the tank. The rollers are designed to defeat most single-pulse, pressure-activated AT and AP mines. The roller creates a 1.1-meter-wide cleared path in front of each track.

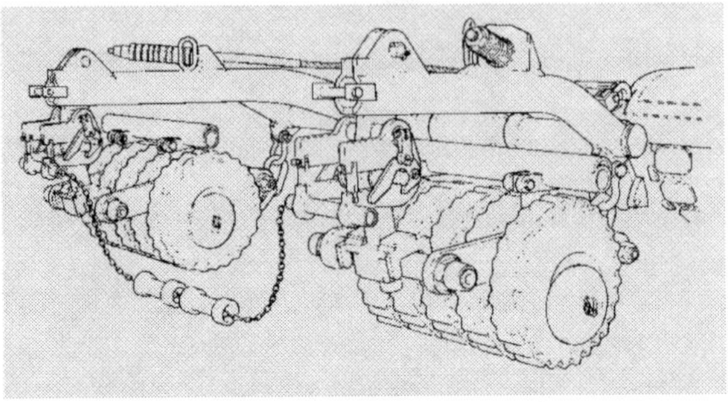

Figure 10-14. MCR

Figure 10-16 shows inside clearance distances between tracks of common track vehicles in relationship to the uncleared area left by the MCR. A dog-bone-and-chain assembly between the rollers defeats tilt-rod mines. The IDA can be fitted to the roller. The roller is designed to withstand multiple mine explosions before damage; however, this depends on the size of the mines. Large blasts may destroy the roller or the vehicle or injure the crew.

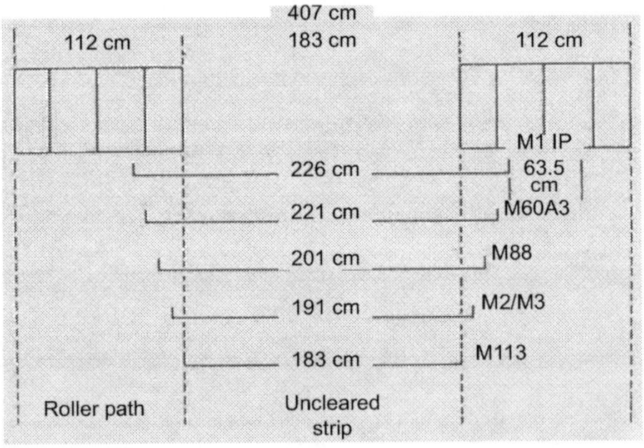

Figure 10-15. Mine-roller width compared to track-vehicle widths

The roller can be mounted on an M1 or M60 tank that is modified with a permanently attached mine-roller mounting kit. Mounting the roller to a tank is a cumbersome, time-consuming operation because it is very difficult under battlefield conditions and requires lift capability. The roller tank is limited to a speed of 5 to 15 kph. When employed in a suspected minefield, the MCR must travel in a relatively straight path, because tight turns may cause the roller to deviate from the path of the track and leave the tank vulnerable to mines. Ground fluctuations, bumps, and berms may cause the roller to lift from the ground and miss mines.

The MCR is not designed to negotiate gaps on its own; however, it can be used on AVLB caution crossings. In this situation, the curbing from the bridge is removed. To prevent damage to the bridge's hydraulic line, the tank driver uses a strap to lift the dog bone and chain when crossing the bridge. The main gun must be traversed to the side when a mine encounter is possible or imminent, because a mine blast can throw the roller or parts of the roller violently into the air and damage the tube. The main gun should only be fired from a temporary halt.

When the situation and the mission permit, MCRs may be employed as lead vehicles to detect minefields. This is most viable when the supported element is traveling in a column. The roller may also be used to lead a supported element traveling in a tactical formation other than a column, but it is less effective than other methods because—

- Vehicles not directly behind the roller may encounter mines passed by the roller.
- The roller may travel well into or completely through a widely spaced minefield without encountering a mine, thus giving the formation a false sense of security.
- A mine encountered by the roller may not be on the leading edge of the minefield.
- The roller vehicle is extremely vulnerable because it can only use its weapon system from a temporary halt.

Rollers are best used to proof lanes in obstacles that are breached by other means, such as a MICLIC or an MCB. A roller pulling a trailer-mounted MICLIC can proof a lane created by a MICLIC that was launched by another vehicle. The roller then fires the second MICLIC and proofs its own lane.

If rollers participate in a deliberate breach operation or if the force incorporates rollers into a hasty breach plan, rollers must be mounted before rehearsals. Unmounted rollers that not being used for the mission are carried in the TF formation on M916 tractor trailers. Rollers require lift capability (such as an M88), a secure location, and 30 to 60 minutes to mount on a tank that is fitted with a mounting kit.

Panther

The M60 Panther (Figure 10-17) is one of several developmental countermine systems used by US forces during operations *Joint Endeavor* and *Joint Task Force Eagle*. The Panther is a remotely controlled vehicle with mine rollers, and it is used to proof lanes and assembly areas. The system consists of a turretless M60 tank, Israeli mine rollers, an antimagnetic actuating device, and an RCU that is mounted in a separate vehicle. Additionally, a remote video camera allows the operator to see the road ahead.

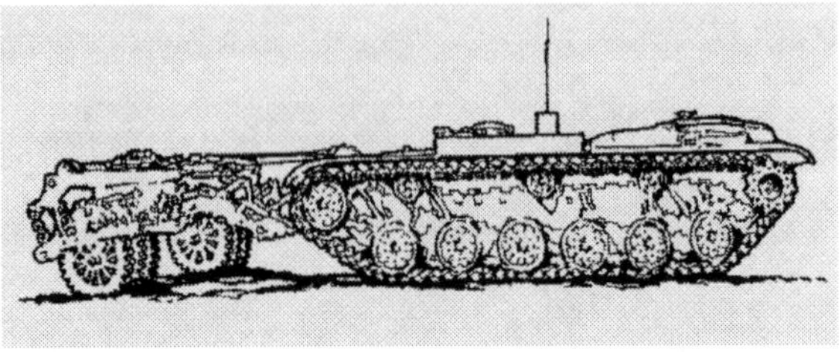

Figure 10-16. Panther

During route clearance or proofing operations, the Panther is the lead vehicle on the route. It is followed closely by an armored control vehicle, usually an M113. The control vehicle contains the Panther operator, the RCU, and the monitor. The monitor displays the route being proofed or cleared through a camera mounted on the Panther. The Panther is controlled from the commander copula or troop hatch of the control vehicle. The control vehicle should be approximately 200 to 300 meters behind the Panther, and its hatches should be secured open. Crew members in the control vehicle should be wearing Improved Body Armor System, Individual Countermine (IBASIC) protective garments.

Mine rollers can be raised for limited travel while mounted on the Panther. If the distance is excessive, the rollers must be transported on a cargo carrier. Rollers must be adjusted before every mission to ensure that they have contact with the ground and that their weight is uniformly distributed. To ensure proper coverage and overlap of rollers, at least three passes should be conducted. Passes should have a minimum of 30 centimeters overlap. Inside roller distances are the same as the MCR.

MiniFlail

The MiniFlail (Figure 10-18. page 10-20) is a remotely operated, line-of-sight, AP-mine and UXO neutralization system that was developed for use by US light forces. It can clear at a rate of 1,200 square meters per hour. The MiniFlail detonates or disables AP mines from a safe operating distance. The MiniFlail neutralizes by striking objects with a rotating chain assembly, called a *flail*, and clears a foot path approximately 1.1 meters wide. The system neutralizes AP mines and UXO by detonation, mechanical destruction, or displacement from the cleared lane. The MiniFlail is approximately 1.3 meters wide, 1.3 meters high, 3 meters long, and weighs 1,100 kilograms. The system is operated by a hand-held controller that has a maximum range of 300 meters. It is fully armored with a material similar to Kevlar, and the tires are filled with foam. The flail is a self-articulating, hydraulically powered shaft with 84 chains; each chain is 0.5 meter long.

Figure 10-17. MiniFlail

M1 Grizzly

The Grizzly (Figure 10-19) provides a hasty capability for breaching complex obstacles of mines, wire, posts, rubble, and tank ditches to create a lane for other vehicles to follow. The Grizzly's primary features are a full-width, 4.2-meter MCB and a power arm. The power arm has a reach of 9 meters and a bucket capacity of 1.2 cubic meters. Its primary missions are to reduce berms and fill AT ditches.

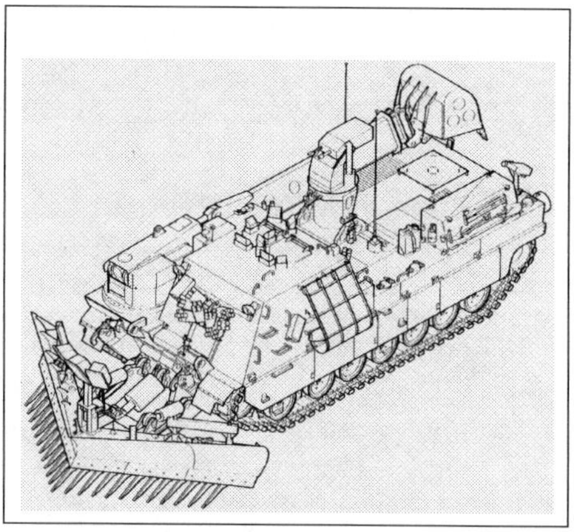

Figure 10-18. Grizzly

The system is armed with an M240 7.62-millimeter machine gun and an MK19 40-millimeter grenade launcher. The Grizzly has an M1-series tank chassis.

The Grizzly lifts and pushes mines, which are surface-laid or buried up to 31 centimeters deep, to the side of full-width lanes. The blade has multiple depth settings, depending on the mission, and it creates a 4.2-meter-wide cleared path. When plowing, the Grizzly is restricted to less than 10 kph, depending on soil conditions. The operator begins plowing approximately 100 meters from the estimated minefield leading edge. He creates a lane extending another 100 meters beyond the estimated minefield far edge to ensure that the lane extends through the entire minefield.

The Grizzly has integrated digital features to enhance battlefield awareness. Some of the digital features are thermal and video cameras, ground-speed sensors, terrain-mapping sensors, and an integrated commander's control station.

Combat Engineer Vehicle with Full-Width Mine Rake

NOTE: The CEV with full-width mine rake will be used in Korea only.

This system consists of a wedge-shaped rake that is mounted to a CEV blade (Figure 10-20).

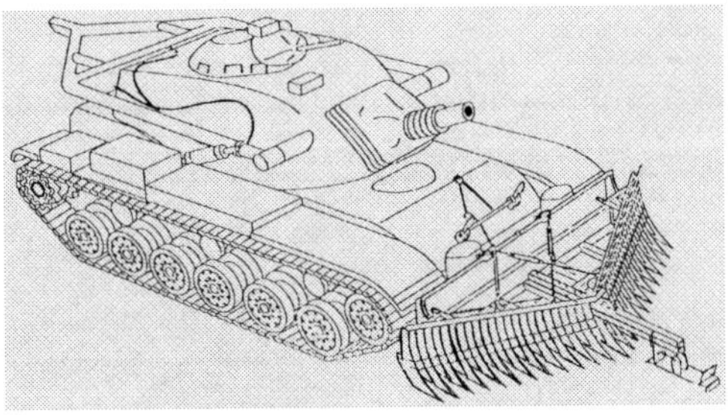

Figure 10-19. CEV with mine rake

The rake weighs 2,025 kilograms and is lifted off its transport vehicle with a HEMTT, a wrecker, an M88, or a CEV boom. The CEV crew uses basic-issue items to install the rake, and installation takes approximately 30 minutes. The rake has a skid shoe to maintain a raking depth of 31 centimeters. The CEV with mine rake provides a vehicle-width clearance (4.5 meters) at 5 to 10 kph. The rake has a quick-disconnect feature. It lifts surface-laid and buried mines (up to 31 centimeters deep) and pushes them off to both sides.

The CEV with mine rake is used to clear lanes during minefield breaching. While it can be employed as the first breaching asset into a minefield, a MICLIC should be used first to eliminate as many mines as possible. The rake is then used to proof the lane. The system can pull a MICLIC and fire it before proofing. Raking begins 100 meters before the minefield and continues 100 meters beyond the suspected limit. The CEV maintains a straight course through the minefield. If the skid shoe is damaged, the operator reduces speed and manually controls the blade depth. This is very difficult and risky.

Mine-Clearing/Armor-Protection Kit

The mine-clearing/armor-protection kit (MCAP) consists of two parts—the mine-clearing rake and the armor protection.

The rake uses a tine that is mounted on a diagonal beam. The rake assembly is designed to sift through the soil, lift out mines, and windrow buried and surface-laid mines to the right of the vehicle. The system clears a 30-centimeter-deep path through a minefield. The rake has a skid shoe that acts as a depth control guide for the operator.

The armor protects the crew against mine blast, small-arms fire, and artillery fire. Protection is also provided for the engine, the fuel tank, and exposed hydraulic cylinders and lines. Ballistic glass blocks are provided at each vision port to permit unrestricted view and operation of the vehicle and the equipment.

The MCAP is mounted on a D7 dozer to perform minefield breaching and lane widening. Proofing the lane must be conducted after the dozer has cleared the lane. Some AP mines may still be left in the lane.

Electronic

The Field-Expedient Countermine System (FECS) is a series of copper coils that fit over the front of tracked and wheeled vehicles. Power is supplied by the vehicle's battery. The coil emits a large magnetic signature that detonates magnetically fused mines located 2 to 5 meters in front of the vehicle. The FECS is designed to defeat magnetically influenced mines only and must be used with other countermine systems.

Manual

When stealth is required or advanced mechanical equipment is unavailable, manual breaching procedures can be used. Engineers use hand-emplaced explosives, grapnel hooks attached to ropes, probes, mine detectors, and hand-emplaced marking equipment to manually breach obstacles. This is the only method that works in all situations and under all conditions because certain types of terrain, weather, and sophisticated fuses can severely degrade the effectiveness of rollers, plows, and line charges.

Surface-Laid Minefields

The enemy possesses a significant mechanical, mine-burying capability. It has the capacity and the propensity for the labor-intensive effort required to bury mines by hand; however, the enemy often lays mines on the surface. Buried mines are usually found in a highly prepared defense that requires a deliberate breach operation. Training and execution of surface and buried minefield breaches should always assume the presence of AHDs and trip wires until proven otherwise.

From covered positions, the engineers first use grapnel hooks to check for trip wires in the lane. The limited range of the tossed hook requires the procedure to be repeated through the estimated width of the obstacle. A demolition team then moves through the lane. The team places a *line main* down the center of the lane, ties the line from the explosives into the line main, and places blocks of explosives next to surface-laid mines. After the mines are detonated, the team makes a visual check to ensure that all of the mines were cleared before directing a proofing roller and other traffic through the lane.

Manual procedures must be well-practiced. Members of the demolition team are assigned special tasks, such as grappler, detonating-cord man, and demolitions man. All of the members should be cross-trained on all the procedures. Demolitions are prepared for use before arriving at the breach site. An engineer platoon uses squads in series through the minefield to clear a lane for a company team. The platoon must rehearse reduction procedures until execution is flawless, quick, and technically safe. During the breach, the engineer platoon will be exposed in the lane for 5 minutes or more depending on the mission, the minefield depth, and the platoon's level of training.

Buried Minefields

Manually reducing a buried minefield is extremely difficult to perform as part of a breaching operation. It is usually part of a clearance operation. If the mine burrows are not easily seen, mine detectors and probes must be used to locate the mines. The mines are then destroyed by hand-emplaced charges. As an alternative, the mines can be removed by using a grappling hook and, if necessary, a tripod (Figure 10-21). Using a tripod provides a vertical lift on the mine, making it easier to pull the mine out of the hole.

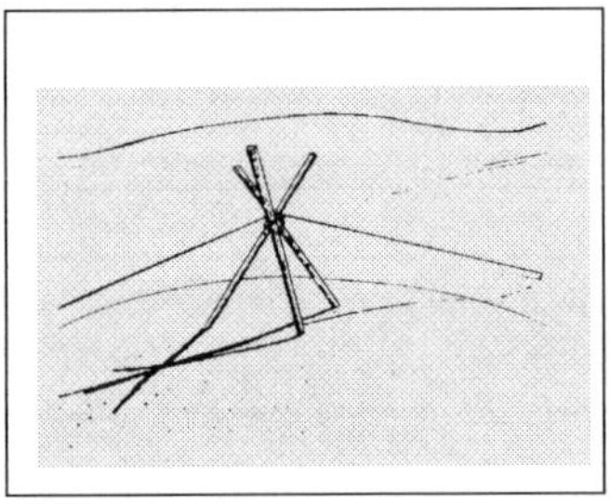

Figure 10-20. Tripod

The platoon leader organizes soldiers into teams with distinct, rehearsed missions including grappling, detecting, marking, probing, and emplacing demolitions and detonating cord. The platoon is exposed in the obstacle for long periods of time.

Grappling Hook

The grappling hook (grapnel) is a multipurpose tool that is used for manual obstacle reduction. Soldiers use it to detonate mines from a standoff position by activating trip wires and AHDs. After the grapnel is used to clear the trip wires in a lane, dismounted engineers can move through the minefield, visually locate surface-laid mines, and prepare the mines for demolition. In buried minefields, soldiers grapple, then enter the minefield with mine detectors and probes.

There two types of grapnel hooks—hand-thrown and weapon-launched.

Hand-Thrown

A 60+-meter light rope is attached to the grapnel for hand-throwing. The throwing range is usually no more than 25 meters. The excess rope is used for the standoff distance when the thrower begins grappling. The thrower tosses the grapnel and seeks cover before the grapnel and rope touch the ground in case their impact detonates a mine. He then moves backward, reaches the end of the excess rope, takes cover, and begins grappling. Once the grapnel is recovered, the thrower moves forward to the original position, tosses the grapnel, and repeats the procedure at least twice. He then moves to the end of the grappled area and repeats this sequence through the depth of the minefield.

Weapon-Launched

A 150-meter light rope is attached to a lightweight grapnel that is designed to be fired from an M16A1 or M16A2 rifle using an M855 cartridge. The grapnel is pushed onto the rifle muzzle, with the opening of the retrieval-rope bag oriented toward the minefield. The firer is located 25 meters from the leading edge of the minefield, and he aims the rifle muzzle at a 30- to 40-degree angle for maximum range. Once fired, the grapnel will travel 75 to 100 meters from the firer's position. After the weapon-launched grapnel hook (WLGH) has been fired, the firer secures the rope, moves 60 meters from the minefield, moves into a prone position, and begins to grapnel. The WLGH can be used only once, but it can be reused up to 20 times for training (blanks are used to fire the grapnel for training).

Multiple grapplers can clear a lane of trip wires quickly and thoroughly, but they must time their efforts and follow procedures simultaneously, if possible. A hit on a trip wire or a pressure fuse can destroy the hook and the cord, so engineers should carry extras.

PROOFING

Proofing is done by passing a mine roller or another mine-resistant vehicle through the minefield as the lead vehicle to verify that a lane is free of mines. An MCB, a Panther, a MiniFlail, or an MCR can be used to proof lanes. If the risk of live mines remaining in the lane does not exceed the risk of loss to enemy fires while waiting, proofing may not be practical. Some mines are resistant to some breaching techniques (for example, magnetically fused mines may be resistant to the MICLIC blast), so proofing should be done when the time available, the threat, and the mission allow.

During a limited clearing operation, proof upgraded breach lanes following a breach. After the minefield is completely cleared, proof the routes used through the area.

MARKING

This section implements STANAGs 2036 and 2889.

This section provides commanders with a standard system for marking breach lanes and bypasses. It centers around a systematic, phased upgrade of lane marking. Each upgrade conforms to the tactical requirements for that phase of the attack, from initial reduction of the obstacle to the passage of larger follow-on forces, as well as the return traffic necessary to sustain the force. Additional guidelines are discussed in FM 3-34.2.

Marking breach lanes and bypasses is critical to obstacle reduction. Effective lane marking allows the commander to project forces through the obstacle quickly, with combat power and C^2. It gives the assaulting force confidence in the safety of the lane and helps prevent unnecessary minefield casualties.

There are two critical components of the lane-marking system:

- Lane-marking pattern (location of markers indicating the entrance, the lane, and the exit).
- Marking device (type of hardware emplaced to mark the entrance, the lane, and the exit).

The lane-marking system outlined in this section centers around standardized marking patterns rather than the marking device. Standardizing the marking pattern is critical to offensive operations. A common lane pattern—

- Enables cross attachments and adjacent units to recognize breach lanes easily with minimal knowledge of a particular unit's tactical SOP.
- Gives all forces a standardized set of visual cues that are needed to pass through a lane safely while maintaining their momentum.
- Facilitates quick conversion to the lane-marking requirements of STANAGs 2889 and 2036 (discussed later in this chapter).

The standard lane-marking hardware is decided by unit commanders. This gives units greater flexibility and allows them to adopt marking devices that are tailor-made for their type of unit and operational focus (such as an armored or light force, a mounted or dismounted attack, limited visibility, thermal capability). However, regardless of the type of device used, it must support the standard lane-marking pattern outlined in the following paragraphs. Therefore, commanders should consider these guidelines and examples before developing or adopting their own marking system.

LANE-MARKING TERMS

The definitions in the following paragraphs provide a common basis for discussing lane marking.

Entrance Markers

Entrance markers indicate the start of a reduced lane through an obstacle. They signify the friendly-side limit of the obstacle and the point at which movement is restricted by the lane width and path. Entrance markers are placed to the left and the right of the entrance point and spaced the width of the reduced lane. They must be visually different from handrail markers to help the force distinguish this critical point in the lane.

Handrail Markers

Handrail markers define the lane path through the obstacle and indicate the limits of the lane width. As a minimum, mounted and dismounted lanes will have a left handrail marker. Mounted and dismounted forces moving through the lane should keep the left handrail marker immediately to their left. As the operation progresses, lane marking may be upgraded to include left and right handrail markers.

Exit Markers

Exit markers indicate the far side of the reduced lane through an obstacle. Like entrance markers, exit markers must be distinguishably different from handrail markers; however, the exit may be marked the same as the entrance. Exit markers are placed to the left and the right of the exit point and spaced the width of the reduced lane. This visual reference is critical when only the left handrail is marked. The combination of entrance markers, left handrail markers, and exit markers provide the driver and the tank commander with visual cues so that they can safely pass through a reduced lane.

Entrance Funnel Markers

Entrance funnel markers augment entrance marking. The V formed by a funnel marker forces the platoon into a column and helps drivers and tank commanders make last-minute adjustments before entering a lane.

Final-Approach Markers

Final-approach markers are highly visible, robust markers that augment the visual signature of entrance funnel markers. They are critical when initial assault forces must maneuver to the breaching site. Normally, the initial assault force can observe the breaching area but cannot clearly distinguish entrance funnel markers. Final-approach markers provide the assault force commander with a highly visible RP toward which to maneuver his formation. They also signal company team commanders to begin changing from combat column to column formation, with platoons in combat column.

Far Recognition Markers

Far recognition markers are highly visible markers that are located between the final-approach marker and the friendly unit. They are primarily used when passing forces are denied direct observation of the final-approach marker due to distance, visibility, or terrain. When possible, far recognition markers should be different from the final-approach marker. Far recognition markers indicate the point at which forces begin changing their formation to posture for the passage. A single far recognition marker may serve up to two initial breach lanes. Once lanes are upgraded to two-way traffic, far recognition markers are required for each two-way lane. When a far recognition marker serves more than one lane, a guide or a traffic-control post (TCP) is collocated with the far recognition marker that is nearest to the breach.

Guides and Traffic-Control Posts

A TCP or a guide consists of a two-man team with communications means. The team assists the commander in controlling the movement of forces. When possible, military police (MP) should man TCPs. However, the commander may initially use other personnel as guides to man critical far recognition markers until the MP establish full TCPs. TCPs and guides provide the commander with a man on the ground who controls traffic flow to the appropriate lanes. When there are multiple lanes branching off a single far recognition marker, the TCP can assist in breaking parts of the formation off into various lanes. The TCP can also help modify the traffic flow when lanes have been closed for maintenance, for lane expansion, or by enemy SCATMINEs. The guide or TCP must give the assault force commander the azimuth and distance to the final-approach marker, identify the device used for the final-approach marker, and provide the level of the lane-marking pattern. For light forces, guides may physically escort passing units from the far recognition marker to the lane entrance.

Levels of Lane Marking and Patterns

The three standard levels of marking for breach lanes and bypasses are initial, intermediate, and full.

Each lane-marking level provides an increase in lane signature and capability. Lane requirements change as a breaching operation matures from an initial breach to the forward passage of large combat forces.

Initial lane-marking requirements are driven by the nature of the fight through the obstacle. Marking must be rapid, providing only the bare minimum signature needed to pass small units who make up the initial assault force. This contrasts with the lane requirements of later phases of an offense where larger units are passed to subsequent objectives. Here, the lane signature must be more extensive and more visible, because it must guide larger forces over a greater distance to the lane's entrance without interruption. Two-way traffic becomes a priority for the simultaneous forward passage of combat units as well as the return traffic (such as ambulances and empty supply vehicles) that is necessary to sustain the force. Lane-marking limits must be absolutely clear to the most inexperienced driver or crewman. A fully developed lane must support two-way traffic and be completely marked.

Bypasses are not marked the same as lanes. They are marked with directional panels indicating the direction of the bypass. The limits of the mine threat must be marked to prevent friendly forces from entering the minefield. Marking the direction of the bypass and the minefield limits will enable the maneuvering element to bypass the minefield without having to unnecessarily defile through a marked lane. Further information on bypass marking can be found in FM 3-34.2.

Commanders must be aware of how the needs of the force change with the operation so that they can anticipate lane-marking and lane-capability requirements. Integrating the levels of lane marking into the overall breaching plan ensures that the unit's needs are satisfied. Forces necessary to mark, maintain, and upgrade lanes must be allocated and tasked with the mission. The phases of the scheme of maneuver and the service-support plan are the basis for analyzing lane requirements. The following paragraphs describe lane-marking patterns in detail and provide guidelines on when the commander should upgrade lane marking and lane capability.

Initial Lane Marking

Initial lane marking (Figure 10-22, page 10-28) is emplaced by the breach force immediately after the lane is reduced and proofed. It provides a signal to the assault force commander that the lane is ready for traffic. Initial lane marking is kept to a minimum, centering on markings needed to pass immediate assault forces through the lane to seize the initial foothold on the objective. Normally, the assault force can observe the breach and does not need the more visual signature of a mature lane marking. The initial lane-marking pattern has the following markers:

- Entrance.
- Exit.
- Left handrail.
- Entrance funnel.
- Final-approach.

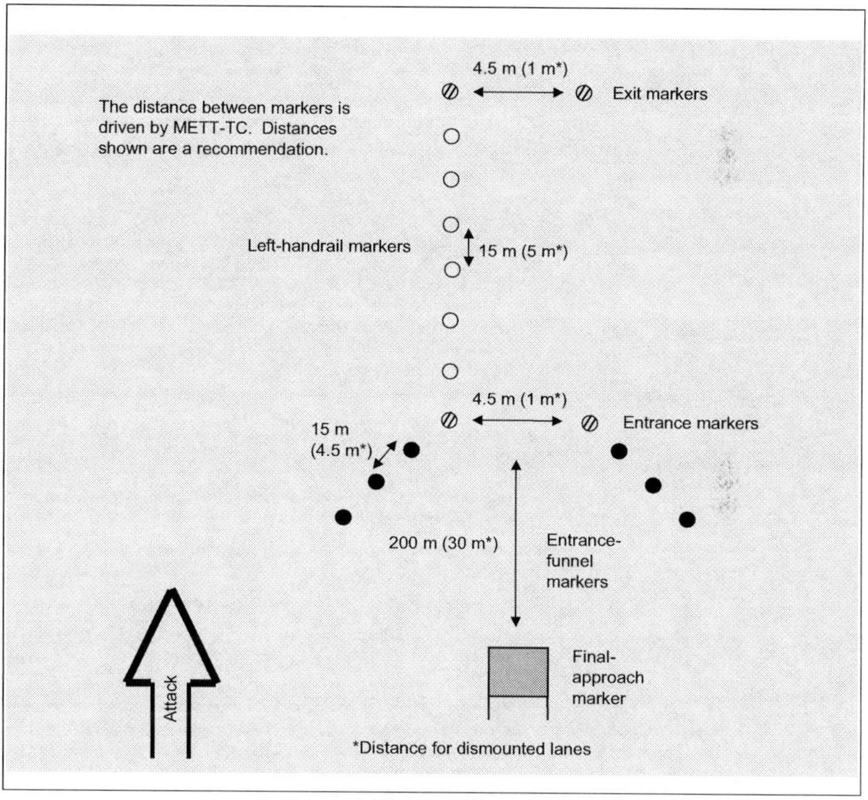

Figure 10-21. Initial lane marking

The entrance, left handrail, and exit markers are the first markers emplaced by the breach force because they define the location and the limits of the reduced lane.

- Entrance markers are placed to the left and the right of the reduced lane's entrance point, and they are spaced the width of the lane (4.5 meters for mounted lanes, 1 meter for dismounted lanes).
- Left handrail markers are placed at the left limit of the lane, along the entire path. Handrail markers are placed at 15-meter intervals for mounted forces and at 5-meter intervals for dismounted forces. Commanders may have to modify the intervals based on the terrain, the visibility, the lane length, and the lane path.
- Exit markers are placed to the left and the right of the reduced lane's exit point, and they are spaced the width of the lane (4.5 meters for mounted lanes, 1 meter for dismounted lanes).

Once the entrance, left handrail, and exit markers are emplaced, the breach force emplaces the entrance funnel markers and the final-approach marker.

- Entrance funnel markers are placed at 15-meter intervals for mounted forces and at 5-meter intervals for dismounted forces. They are placed diagonal to the lane entrance and form a 45-degree V (Figure 10-22).
- The final-approach marker is centered on the lane and placed at least 200 meters from the lane entrance for mounted forces. For dismounted forces, the nature of the attack may initially preclude using a final-approach marker; however, as soon as the mission allows, a final-approach marker is placed 30 meters from the entrance. Final-approach markers for mounted and dismounted forces must be placed on high ground to ensure that they are clearly visible. The commander may modify the recommended distance for the final-approach marker, based on the terrain and the visibility.

Intermediate Lane Marking

Upgrading initial lane marking to intermediate lane marking (Figure 10-23, page 10-30) is triggered by one of two key events—the commitment of larger combat forces who are unable to directly observe the breach or the rearward passage of sustainment traffic (casualty evacuation and vehicle recovery). Intermediate lane marking has two goals:

- Increasing the lane signature to help the passage of larger, more distant combat forces.
- Providing sufficient marking for two-way, single-lane traffic.

Intermediate lane marking builds on initial lane marking by adding right handrail markers, exit funnel markers, far recognition markers, and a farside final-approach marker.

The commander sets the priority of marker emplacement based on the situation. If the scheme of maneuver requires the immediate passage of larger combat forces, the right handrail markers and the far recognition marker may be the priority. On the other hand, if it is necessary to ground evacuate casualties or to recover vehicles, emplacing right handrail markers, exit funnel markers, and a farside final-approach marker may be required first.

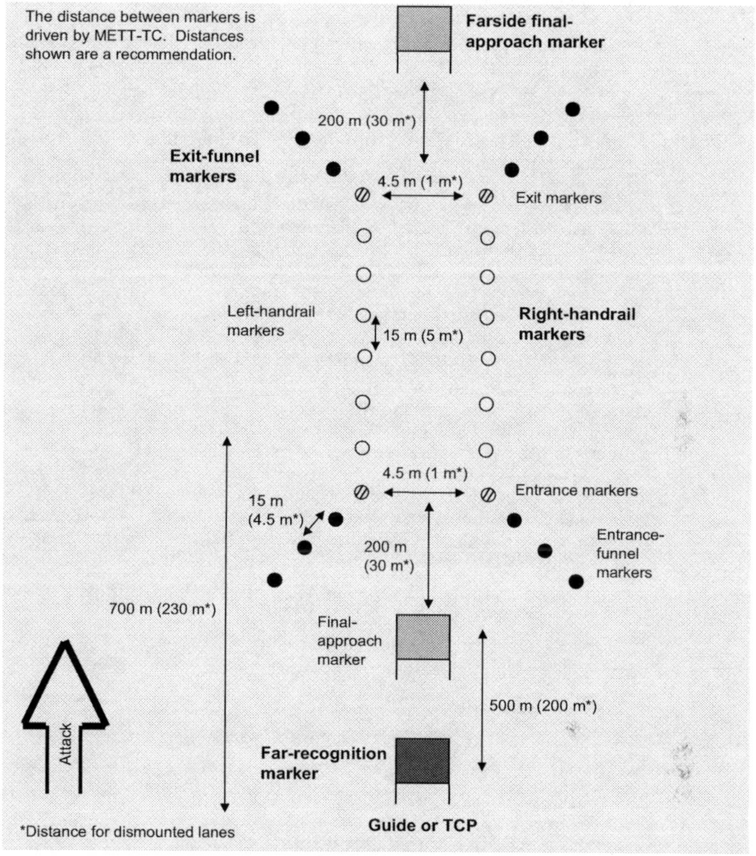

Figure 10-22. Intermediate lane marking

When upgrading to intermediate marking, the first step is to emplace the right handrail markers. Right handrail markers define the rightmost limit of the lane. They are placed the width of the lane as defined by the entrance and exit markers. The right handrail follows a path parallel to the left handrail through the obstacle. Right handrail markers are placed at the same interval as left handrail markers.

Exit funnel markers and a farside final-approach marker are emplaced to mirror the entrance markers. Exit funnel markers prevent the premature deployment of the passing force into combat formation before it is safely outside the obstacle. They also become the entrance funnel markers for rearward passing traffic, giving these forces the visual cues needed to line themselves up on the lane. The exit funnel markers are augmented by a farside final-approach marker to help rearward passing forces clearly identify the lane from their side. The farside final-approach marker is centered on the lane and placed 200 meters (mounted forces) or 30 meters (dismounted forces) from the exit.

A far recognition marker completes intermediate lane marking. It provides commanders with a visual signature or a series of signatures for guiding their movement toward the lane. For mounted forces, the far recognition marker nearest to the breach lane is placed 500 meters from the lane entrance or on the nearest terrain feature. Dismounted forces may require a system of guides instead of far recognition markers for passing combat forces; however, far recognition markers must be emplaced as soon as possible to reduce guide requirements for passing mounted sustainment traffic. This gives the assault force commander the space needed to transition his formation to companies in combat column. Far recognition markers may be emplaced before or concurrent with exit markers, based on the mission and the situation.

The commander collocates guides or TCPs at the far recognition marker when he feels the situation requires more positive control over traffic flow. Commanders should plan for the use of full-time guides once they have upgraded to intermediate marking. TCPs become mission-critical during limited visibility or in restrictive terrain. They should also be used when a single far recognition marker feeds more than one breach lane. TCPs must be manned with a minimum of two soldiers and must have FM communications with the controlling headquarters. It is essential that soldiers acting as guides or TCPs know the—

- Azimuth and distance to the breach lane and the 8-digit grid coordinate of the lane.
- Level of lane marking.
- Type of final-approach marker used.
- Traffic-control plan and march order.
- Up-to-date status of lane marking, maintenance, and so forth.

Full Lane Marking

Expanding breach lanes to full (two-way) lane marking (Figure 10-24, page 10-32) is resource-intensive and is not normally a part of an initial breach operation. A fully matured lane is one that will support uninterrupted, two-way traffic. Expanding a breach lane to a full lane involves expanding the width of the lane to accommodate two-way traffic and modifying the marking pattern to give forward and rearward passing forces the same visual signature. Upgrading to a full lane is normally assigned to follow-on engineer forces, since it is usually beyond the immediate capability of engineers with forward units.

Upgrading intermediate lane marking to full lane marking begins by temporarily closing the lane, rerouting traffic, and expanding the lane width. The initial reduced and proofed lane is always expanded to the left, in relation to the direction of the attack. Engineers reduce and proof the obstacle beginning at the left handrail to give a total lane width of 10 meters (5 meters

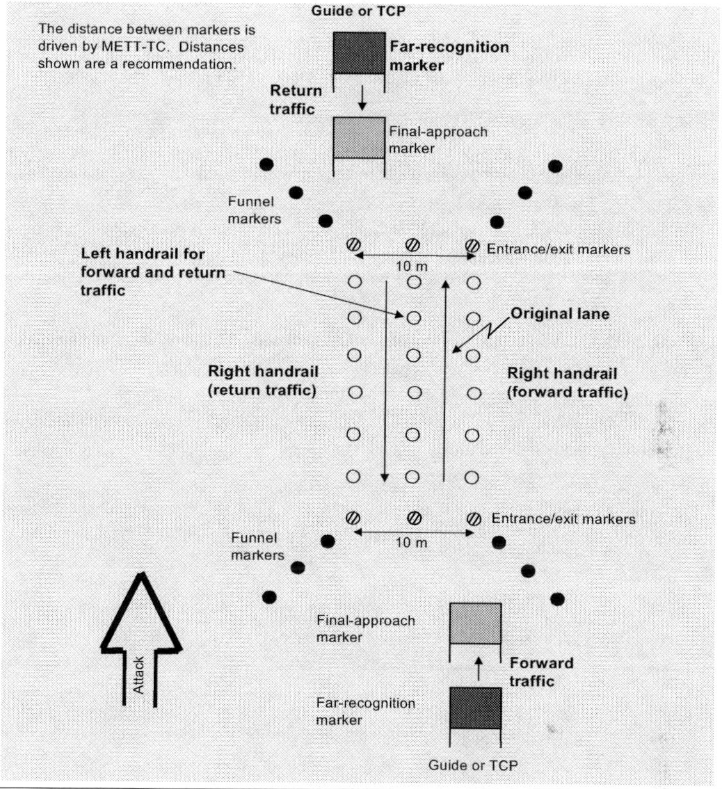

Figure 10-23. Full lane marking

each way). The expansion width requirement is the same for armored and light forces, because both forces must be able to pass mounted sustainment and combat forces during this phase.

Once the engineers expand the lane width to 10 meters, they ensure that entrance, exit, handrail, funnel, and final-approach markers are replaced on the return lane. All markings are the same as described in previous paragraphs.

The full lane-marking pattern has three entrance and three exit markers. They are placed the width of forward and return lanes and are visually different from other markers. Units must be trained to recognize that three entrance markers indicate a two-way traffic lane and that they should always use the rightmost lane.

Entrance and exit funnel markers are placed slightly different from previous marking patterns. They extend out from the entrance and exit markers on the right side only.

Final-approach markers are placed 200 meters from, and centered on, entrances of forward and return lanes. This helps forces clearly identify the entrance points from either direction.

Far recognition markers are placed a maximum of 500 meters from the lane entrance or on the nearest terrain feature from forward and return final-approach markers.

COMMANDER'S GUIDANCE FOR LANE MARKING

Table 10-1 provides a summary of lane-marking levels, guidelines on unit responsibilities, and events that trigger lane upgrade. In the table, *who* refers to the unit responsible for lane upgrade marking and *when* describes events that trigger the need to upgrade.

Table 10-1. Lane-marking levels, unit responsibilities, and trigger events

	Initial	Intermediate	Full (Two Way)
Who	TF breach force	TF breach force	Brigade
When	Obstacle is reduced	Passing battalion- or company-size forces	Passing brigade- or battalion-size forces
	Passing platoon- or company-size assault forces	Passing force which cannot see the lane	Situation requires uninterrupted sustainment traffic
		Passing TF combat trains	
Markers	Entrance	Add right handrail	Expand lane width to 10 meters
	Exit	Add exit funnel	Adjust entrance/exit
	Left handrail	Add farside final approach	Adjust left/right handrails to new width
	Entrance funnel	Add far recognition	Add far recognition
	Final approach	Add guides or TCPs	Add farside guides or TCPs

Lane-Marking Devices

The majority of lane marking in the field is done by using nonstandard marking devices. When adopting a nonstandard marking device, commanders should consider the guidelines summarized in Table 10-2.

Table 10-2. Guidelines for lane-marking devices

Marker	Mounted Forces	Dismounted Forces
Handrail and funnel markers	Visible by TC and driver (buttoned up) from 50 meters	Visible by a dismounted soldier in a prone position from 15 meters
	Quick and easy to emplace, minimizing the need to expose soldiers outside the carrier	Lightweight, quick, and easy to emplace (a dismounted soldier should be able to carry enough markers for the lane and still be able to fire and maneuver)
Entrance and exit markers	Visible by TC buttoned up from 100 meters	Visible by a dismounted soldier from 50 meters
	Visually different from handrail and funnel markers	Visually different from handrail and funnel markers
	Quick and easy to emplace (may require soldiers to dismount to emplace)	Lightweight, quick, and easy to emplace
	Easily man-portable	
Final-approach and far recognition markers	Visible by TC (not buttoned up) from 500 meters	Visible by a dismounted soldier on the march from 100 meters
	Visually different from each other	Visually different from each other
	Visually alterable to facilitate traffic control through multiple lanes	Visually alterable to facilitate traffic control through multiple lanes

Figure 10-25 shows some of the devices that can be utilized for lane marking, and they are easily procured or fabricated. This is not an inclusive listing but is intended to show commanders some of the options.

Some general requirements for lane marking are—

- Markers must be able to withstand the rigors of the terrain, the weather, and the battlefield.
- Markers should be easy to modify, using minimal manpower and equipment, when visibility is limited.
- Lane-marking panels should have thermal and IR reflective marking so that they can be easily identified during limited visibility.
- Enhancements for limited visibility should be a constant source rather than a pulsating strobe. Strobes do not make the marking pattern readily apparent, particularly when approaching from an angle.

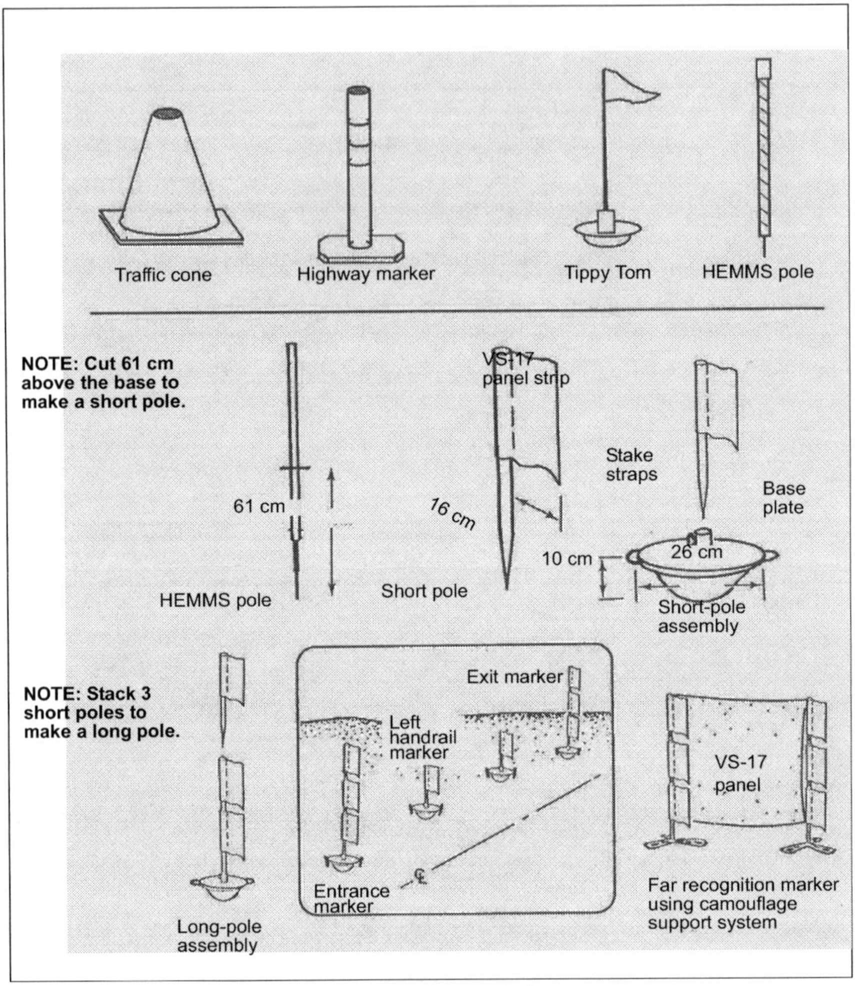

Figure 10-24. Marking devices

The following standard marking sets are available through normal supply channels:

- Minefield marking set number 2, line item number (LIN): M49096, NSN: 9905-00-375-9180.
- HEMMS, LIN: M49483, NSN: 9905-01-019-0140.

Marking Requirements of the North Atlantic Treaty Organization

The following paragraphs paraphrase the lane-marking requirements outlined in STANAGs 2889 and 2036. They also establish the procedures used by US forces to modify intermediate and full lane marking to STANAG standard. With the combined nature of warfare, commanders need to be aware of their responsibilities for marking hazardous areas, particularly breach lanes.

STANAGs 2889 and 2036 state that the type of marking device, pattern, and lighting used to mark breach lanes in forward areas is at the discretion of national authorities or the authorized commander. This gives commanders who are participating in a combined operation the flexibility to mark lanes consistent with their respective Army's standard. It also outlines minimum requirements for the lane-marking pattern before it is used by troops of other nations; however, commanders must plan for converting a lane to NATO standard as early as possible. When converting to NATO standard, the STANAG directs commanders to use lane-marking devices as stated below. Within an offensive operation, marking a lane to NATO standard will not normally occur until after the lane is matured to a full lane.

Marking Pattern and Device

The intermediate lane marking discussed earlier satisfies the minimum lane-marking pattern that must be used before forces from another country are able to pass through a lane. STANAGs 2889 and 2036 state that regardless of the marking device used, the entrance point, exit point, and left and right handrails are the minimum required lane signature. Therefore, once the lane is marked to the intermediate level, allied forces can use the lane without any additional marking.

STANAG 2889 requires that commanders convert marking devices to NATO standard as early as possible. Figure 10-26 shows a NATO standard marker. The marker is placed at right angles to the direction of travel, so that the white portion of the arrow points inward to the lane, indicating the safe side of the lane. The red portion is outward, indicating the lane limit or dangerous side of the lane. STANAG 2889 also requires that markers be large enough to be visible from 50 meters under most daylight conditions and have a field life of 60 days.

Conversion to NATO Standard Marking

To convert intermediate and full lane marking to NATO standard, affix NATO markers to long pickets and replace the existing entrance, exit, funnel, and handrail markers one for one (Figure 10-27).

Two NATO markers are used for entrance and exit markers to make them distinctly different. One NATO marker is affixed to each funnel marker and to

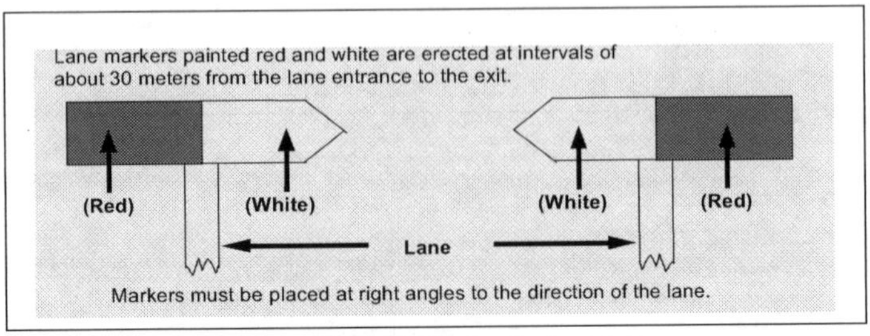

Figure 10-25. NATO standard marker

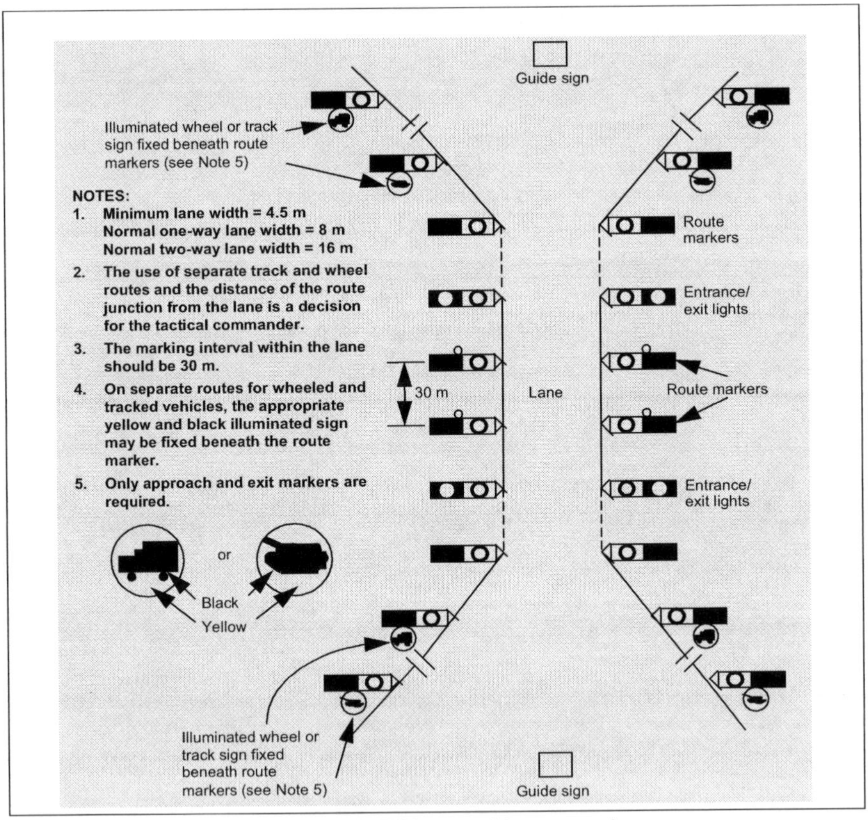

Figure 10-26. NATO lane-marking conversion

each left and right handrail marker. When converting full lane marking, the center handrail is marked with a modified NATO marker. The combination of a modified center handrail marker and directional arrows at each lane entrance provides allied forces with the signature necessary to distinguish two separate lanes. In addition, a barbwire or concertina fence (one strand minimum) is laid 1 meter above the ground to connect funnel markers, entrance markers, handrail markers, and exit pickets.

NATO uses white or green lights to illuminate markers at night (Figure 10-28). Entrance and exit markers are marked with two green or white lights placed horizontally, so that the safe and dangerous markings on them are clearly visible. One white or green light is used on funnel and handrail markers. The commander decides whether the light is placed on top of the NATO marker or placed so that it illuminates the markers. Lights must be visible from a minimum of 50 meters under most conditions and have a continuous life of 12 hours.

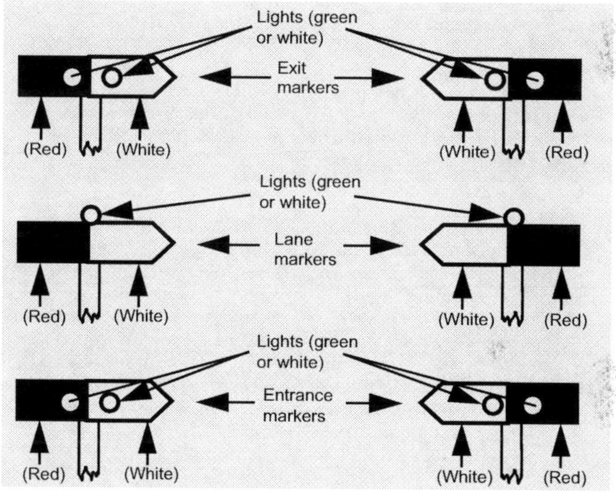

Figure 10-27. NATO standard marking for limited visibility

The mission to convert intermediate or full lane marking to NATO standard is normally assigned to corps-level engineer battalions working in the division rear area. In special cases, divisional engineer battalions may be tasked with NATO marking.

Chapter 11

Route and Area Clearance

The ability to move forces and material to any point in an AO is basic to combat power and often decides the outcome of combat operations. Maneuver relies on the availability of LOC within an AO; and during OOTW, clear LOC is essential to the movement of forces. Units must conduct route and area clearance to ensure that LOC enables safe passage of combat, combat support (CS), and CSS organizations. Clearance operations are best-suited for rear-area and stability support operations.

ROUTE CLEARANCE

Route clearance is a combined arms operation. Units must clear LOC of obstacles and enemy activity that disrupt battlefield circulation.

PLANNING

The principles of breaching operations (Chapter 9) apply to the development and execution of the route-clearance mission. The breaching tenets (intelligence, fundamentals, organization, mass, and synchronization) should be the basis for planning.

Intelligence

Incorporating the IPB and METT-TC factors into route-clearance operations will enable units to predict what the enemy will do and where it will do it. The IPB and the EBA offer ideal methods for establishing a SITEMP. After the S2 and the engineer identify the most probable threat sites, the S2 designates them as NAIs. These NAIs are the focus of the reconnaissance effort. Engineers work in concert with other reconnaissance assets to confirm the presence or absence of ambushes, UXO, and minefields. The information gathered from the IPB and the reconnaissance effort determines the method and the type of route clearance necessary. It also helps the commander determine any outside resources (EOD, SOF) that he may need.

Fundamentals

SOSR may not be executed, but it is planned as it is in breaching operations. Units must be prepared to execute SOSR fundamentals as necessary.

Organization

Task organization for a route clearance is similar to the task organization for a deliberate breach. The clearance company team is organized into breach, support, and assault forces. The breach force conducts clearing operations, the support force isolates the area being cleared, and the assault force performs security functions beyond the clearance site (traffic control points) and assists the breach force in disengagement, as required. Table 11-1 shows a sample task organization for a route clearance.

Table 11-1. Sample task organization for a route clearance

Team	Support Force	Assault Force	Breach Force
Heavy	• Mechanized infantry platoon with dismount capability • Armor platoon	• Mechanized infantry platoon • Engineer squad • Mortar section • Medical team (two ambulances) • PSYOP team • FIST • MP element	• Engineer platoon with organic vehicles • Armor platoon with plows and rollers
Light/Heavy	Two infantry platoons (light)	• Bradley platoon with dismount capability • Engineer squad • 60-mm mortar section • Medical team (two ambulances) • PSYOP team • Forward observer • MP element	• Engineer platoon with organic vehicles • Armor platoon with plows and rollers
Light	Two infantry platoons (light)	• AT/MP section with M60/MK19 mix • 60-mm mortar section • Medical team (two ambulances) • PSYOP team • Forward observer • MP element	• Engineer squad (+) • Infantry platoon (light) • AT/MP section with M60/MK19 mix

Mass

Sufficient maneuver and engineer assets must be allocated to the clearance company team. The length and the width of the route and the type of clearance to be conducted determine the size of the sweep team. Clearing a Class A military road with the deliberate sweep technique requires at least two engineer squads due to the total lane width to be cleared and the requirement for the rotation of mine-detector operators. Depending on the type of sweep operations, the commander can expect a 50 percent loss of sweep assets. Normally, as in breaching, a 50 percent redundancy of engineer assets should be allocated to the sweep team.

Synchronization

All aspects of synchronization should be implemented when planning route clearance. It is especially important that rehearsals be conducted at the combined arms level. Rehearsals should include—

- Reaction to enemy contact.
- Reaction to an ambush.
- Communications exercise.
- Fire support (obscuration smoke, immediate suppression fires, critical friendly zones for counterfire radar, and no-fire area around the clearance site).

- CSS (maneuver, casualty evacuation, marking materials, and demolitions resupply).

PLANNING CONSIDERATIONS

The purpose of breaching is to project combat power to the farside of an obstacle, and breaching usually occurs under enemy fire. Route-clearance operations focus on opening and maintaining LOC to ensure the safe passage of combat, CS, and CSS organizations. Like breaching, route-clearance operations require extensive BOS coordination. The following planning considerations should be used by brigade and battalion TFs when planning route-clearance operations:

Intelligence

- Identify choke points, bridges, tunnels, critical road junctions, and other built-up areas. These are the most suspect areas for obstacle emplacement. However, depending on the enemy's overall mission, it may not always emplace obstacles at these locations. This is especially true if the enemy's goal is to psychologically disrupt our convoys.
- Maintain a situation map with a graphics overlay that reflects the most current intelligence information.
- Maintain an incident map with a graphics overlay to facilitate a pattern analysis.
- Maintain a threat order-of-battle database, such as how the enemy will disrupt unit LOC.
- Develop a detailed R&S plan that incorporates modern battlefield techniques and systems, such as ground sensors, forward-looking airborne radar, and satellite images. As a minimum—
  - — Coordinate for UAV support, if available.
  - — Develop infiltration routes to support recon and security at likely enemy ambush sites.
  - — Develop an estimate of impact to civilians on the battlefield (COBs). COBs include local nationals, nongovernment organizations (NGOs), and private volunteer organizations (PVOs).
  - — Conduct a daily flight over the area to provide up-to-the-minute intelligence. When available, coordinate ASTAMIDS coverage.
  - — Coordinate with the USAF to periodically check the route (for example, using an AC-130 Specter gunship).
- Provide an intelligence update to company team leaders before departure. This should be in the form of a 1:50,000 enemy SITEMP overlay (confirmed and suspected/templated).
- Establish liaison with the host nation, NGOs, and SOF.

Maneuver

- Provide personnel for TCPs.

- Clear and secure flanks (at least 100 meters) and the farside of suspected and known obstacle locations.
- Close the route to US-controlled traffic during route-clearance operations to minimize the target presented to enemy forces.
- Identify and clear potential sniper positions before beginning obstacle reduction or clearance.
- Provide security for the cleared route.
- Give operational control (OPCON) of aviation assets to the route-clearance commander for clearance-support missions.
- Plan the building of static security points along the cleared route to reduce the probability of reseeding.

Fire Support

- Plan smoke for templated locations.
- Position mortars to ensure continuous coverage of the operation (move and set up with the support force).
- Prepare fires within the tactical rules of engagement.
- Ensure that the route-clearance team has a FIST coordinator. The clearance commander should locate the FIST element well forward in the order of march.
- Designate obstacle clearance sites as critical friendly zones for counterfire radar and no-fire areas.

NOTES:

1. Priority targets shift in conjunction with company team movement on the main supply route (MSR).

2. Clearance of fires is the responsibility of the maneuver commander of the sector where the target is located.

3. Adequate Q-36 radar coverage is necessary for deliberate sweep operations.

Mobility/Survivability

- Provide detailed OBSTINTEL on minefields. It must include the—
  - Description of mines or explosive devices most likely encountered.
  - Composition and pattern of obstacle(s).
  - Enemy actions or techniques used during obstacle emplacement.
- Conduct deliberate sweep operations 100 meters past the obstacle or suspected threat.
- Report, clear, and mark mines, obstacles, and explosive devices to facilitate unimpeded movement.
- Ensure that lane marking meets the standards outlined in Chapter 10 and that materials and techniques are standard throughout the route.

- Consider including road repair equipment and material as part of the sweep element (for example, a 5-ton dump truck filled with soil and an ACE to spread the soil).
- Keep all radios, electronic equipment, and aviation assets at a safe distance during reduction operations.
- Block uncleared roads and trails that branch from the route being cleared. This protects units from inadvertently traveling an uncleared route.
- Debrief the chain of command and the TF S2 on the location, the composition, and the orientation of all obstacles cleared and encountered. This assists the S2 and the engineer in IPB/EBA pattern analysis.

Air-Defense Artillery

- Consider the possibility of an air attack.
- Use the following passive air-defense measures:
  — Eliminate glare by using mud, tape, cardboard, or camouflage nets to cover headlights, mirrors, and portions of windshields.
  — Reduce dust clouds by reducing speed.
  — Plan routes that offer natural concealment.
  — Use air guards.
- Increase the distance between vehicles.
- Incorporate Stinger missile teams into the support force.

Combat Service Support

- Ensure that clearance operations are supported by a logistical/combat health support (CHS) package from the brigade support area.
- Plan for air and ground evacuation of casualties. The preferred evacuation method is by air; the routine method is by ground.
  — Conduct an air-mission brief with air ambulance assets, to include pickup zones and markers. Rehearse procedures for evacuation requests.
  — Ensure that the medical team consists of one or two ambulances. Locate the medical team with the support force.
  — Identify the ambulance exchange point along the route to be cleared.
- Ensure that all personnel wear flak vests or IBASIC (Figure 11-1, page 11-6).
- Ensure that all vehicles have tow cables in the front and the rear for extraction purposes.
- Ensure that all vehicles carrying troops have hardening (sandbags on floors and sides).

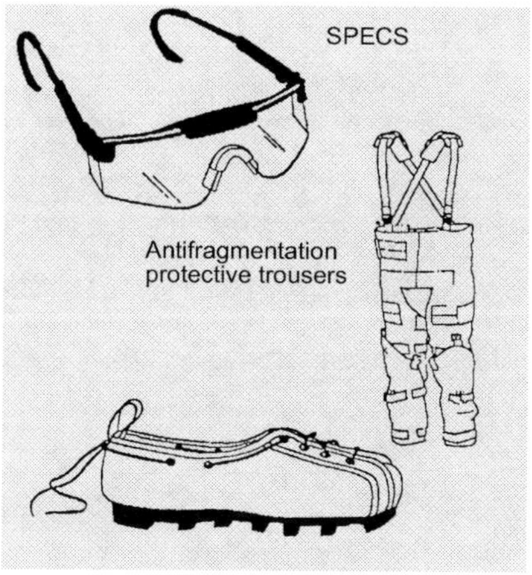

Figure 11-1. IBASIC

- Provide MP and explosive-sniffing dogs to help in clearance and provide security for convoys during and after clearing operations.

Command and Control

NOTE: The company team commander is required to operate on three separate frequencies—battalion command network, company team command network, and fire-support network.

- Designate, recognize, and include minefield indicators (Chapter 10) as part of company team rehearsals.
- Designate a reserve force (at least platoon-size) that is mechanized or air-assault capable.
- Ensure that proper rehearsals are planned and conducted according to FM 3-34.2. As a minimum, the clearance force should rehearse actions on the obstacle, actions on enemy contact, casualty evacuation, and the control of COBs.
- Ensure that the tasked unit has a clear understanding of the mission, intent, and end state. For example, the clearing unit commander should understand that his unit must clear the road width, including the shoulders, and secure the route.
- Assign clearance responsibilities to brigade and battalion assets.
- Ensure that the maneuver commander/TF S3—

— Controls the movement of all personnel and equipment along the route (travel authorization is coordinated through the S4).

— Prepares a mine risk assessment of the mission before issuing the OPORD. (An example of a mine risk assessment is shown in Appendix F.)

— Tracks the status of routes (red, amber, green) in the TF sector, based on the amount of time since the route was cleared and the intelligence and enemy situations.

— Tracks the progress of the clearance operation and integrates it into the maneuver and CSS plans.

— Determines the route length, using clearly definable start and end points.

— Sets priorities for the route-clearance element.

— Coordinates with adjacent units, the host nation, NGOs, PVOs, and SOF.

Special Operations

- Ensure that psychological operations (PSYOP)/civil affairs (CA) support the counterintelligence effort by conducting civilian interviews.
- Direct civilians along the MSR to the displaced-personnel holding areas or along the routes that the brigade has indicated for use.
- Employ PSYOP/CA teams forward to disperse civilians and provide traffic management to isolate the route during clearance operations.

TASK ORGANIZATION

A brigade or battalion TF normally conducts clearance-in-zone operations. To clear a route, the battalion TF focuses a company team as the main effort on the proposed MSR. Table 11-1, page 11-2, shows a sample task organization for a route clearance.

Support Force

This force is comprised of two maneuver platoons and the maneuver company team XO. The support force provides flank security, rear security, and protection to the breach force. It neutralizes hostile forces encountered by the company team. In rugged terrain or highly mined areas, moving the assault force on the flanks would be too risky. Aviation assets can provide flank security while ground forces provide rear security. The assault force also searches for suspected off-route mines.

Assault Force

This force is comprised of a maneuver platoon, an engineer squad, a mortar section, a medical team, a PSYOP team, an EOD team (or one that is on call), and a forward observer. The assault force's mission is the same as in a breaching operation (Chapter 9).

Breach Force

This force is comprised of a maneuver platoon (including the commander) and an engineer platoon (minus). The breach force sweeps the route and reduces mine and explosive threats. It is further task-organized into sweep teams.

A sweep team is a trained detection team that searches for mines and explosive devices. The organization of the sweep team depends on the type of sweep mission and the length, the width, and the surface composition (pavement, gravel, dirt) of the area to be swept. A platoon-size element can normally clear a 4.5-meter-wide path, and a squad-size element can normally clear a 1.5-meter-wide path. If the route is wider or time does not permit multiple passes of the route, additional engineer assets are required. Table 11-2 outlines personnel and equipment requirements for a sweep team.

Table 11-2. Personnel and equipment requirements for a sweep team

Personnel	Support Personnel	Equipment
• NCOIC • Mine-detector operators • Probers/markers • Radio operator • Demolition teams	• Medics • Vehicle operator	• One panel marker • Operational map with required maneuver graphics • Four smoke grenades (minimum) • Six mine detectors (includes three backups) and extra batteries • Two grappling hooks with 60 meters of rope each • One demolition kit or bag for each demolition man • Six probes • Mine marking material • Lane marking tape • Ten stakes

Platoon-Size Sweep Team

The normal configuration for a platoon-size sweep team is twelve soldiers in a modified column (Figure 11-2). The platoon leader supervises the entire operation. This configuration is best suited for sweeping routes in friendly territory that is not under constant surveillance.

When sweeping in areas where 100 percent coverage is required, the team should establish a clear lane from which to operate. This is conducted by the grapple hook being thrown one additional time into the suspected minefield. The cord on the grapple hook is pulled taunt, with the free end secured to a stake.

This establishes a left or right physical boundary marker for the mine-detector operator to use as a guide. The first mine-detector operator will sweep a 1.5-meter lane left or right of this cord, up to the grapple hook. At the hook, a second stake will be emplaced to hold the boundary line and the grapple hook will be thrown again and pulled for trip wires as before. A fourth throw will serve as the boundary marker through the suspected minefield. This procedure will continue until the first mine-detector operator is out of the minefield on the farside.

The boundary marker will then be secured to stakes, using the lane marking tape. A second boundary maker will be placed 1 meter to the left or right of the original boundary marker, depending on which side was swept. This will

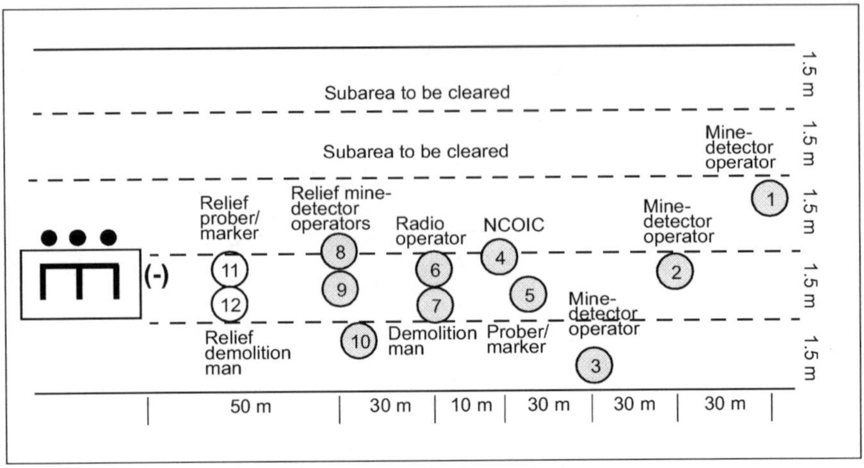

Figure 11-2. Platoon-size sweep team

establish a fixed 1-meter lane. The 0.5-meter area that was not included in the original lane is to ensure that no gaps in coverage were missed on the edge of the first lane.

The unit will then stretch a piece of marking tape out from across the clear footpath, moving it left or right 1.5 meters and staking it down on both ends. This will be repeated as many times as necessary.

Always mark lanes that have not been cleared with a piece of marking tape across the entry to that lane to prevent personnel from using the lane before it has been swept. After multiple lanes are physically marked, the unit can perform a platoon-size sweep. Lanes should be swept and cleared by working out from the original 1-meter footpath. As the cleared area becomes bigger, the marking tape is removed to allow the passage of personnel and equipment though the cleared lane.

As subsequent lanes are being swept, the marking tape will need to be secured on the outside edge of the lane. Always conduct a thorough sweep of the spot where the stake will be placed. Stake intervals will depend on terrain and atmospheric conditions.

Soldiers 1, 2, and 3 (mine-detector operators) lead the sweep team in echelon. Each sweep team covers 1.5 meters of front, and sweep teams are spaced 30 meters apart to prevent fatalities from accidental detonation by other mine-detector operators. If required, a fourth mine-detector operator can be added to the detection column.

Soldiers 4 (NCOIC) and 5 (prober/marker) follow 30 meters behind the last mine-detector operator (Soldier 3) and are centered in the cleared lane. The prober/marker is responsible for marking the cleared lane on both sides.

Soldiers 6 (radio operator) and 7 (demolition man) follow 10 meters behind Soldiers 4 and 5 and are centered in the cleared lane.

Soldiers 8, 9, 10 (relief mine-detector operators), 11 (relief prober/marker), and 12 (reserve demolition man) follow 30 meters behind Soldiers 6 and 7. If a fourth mine-detector operator is added to the column, an additional relief mine-detector operator must also be added.

The remaining platoon members help the support force or act as a reserve force, as required. They should first be integrated into the sweep team as a relief element and then moved forward as needed.

Squad-Size Sweep Team

The normal configuration for a squad-size sweep team is seven soldiers in a modified column (Figure 11-3). The squad leader supervises the entire sweep operation. This configuration is designed for sweeping routes in friendly territory that is not under constant surveillance.

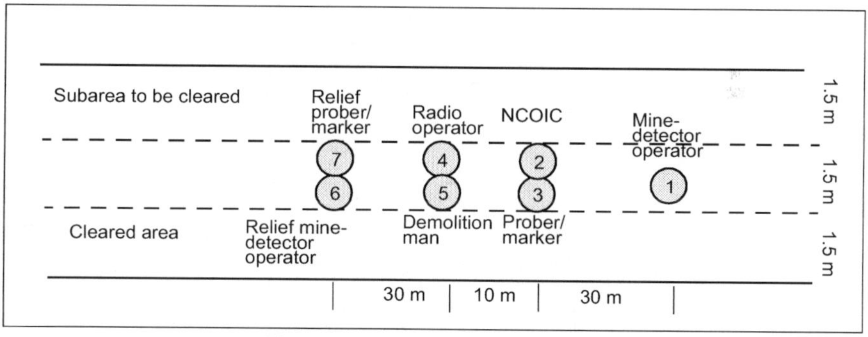

Figure 11-3. Squad-size sweep team

Soldier 1 (mine-detector operator) leads the sweep team and covers a 1.5-meter-wide path.

Soldiers 2 (NCOIC) and 3 (prober/marker) follow 30 meters behind Soldier 1 and are centered in the cleared lane. The prober/marker is responsible for marking the cleared lane on both sides.

Soldiers 4 (radio operator) and 5 (demolition man) follow 10 meters behind Soldiers 2 and 3 and are centered in the cleared lane.

Soldiers 6 (relief mine-detector operator) and 7 (relief prober/marker) follow 30 meters behind Soldiers 4 and 5. If the squad cannot use seven team members, the relief prober/marker position can be eliminated from the formation.

The engineer platoon can configure the platoon into squad-size sweep teams and place them in echelon (Figure 11-4).

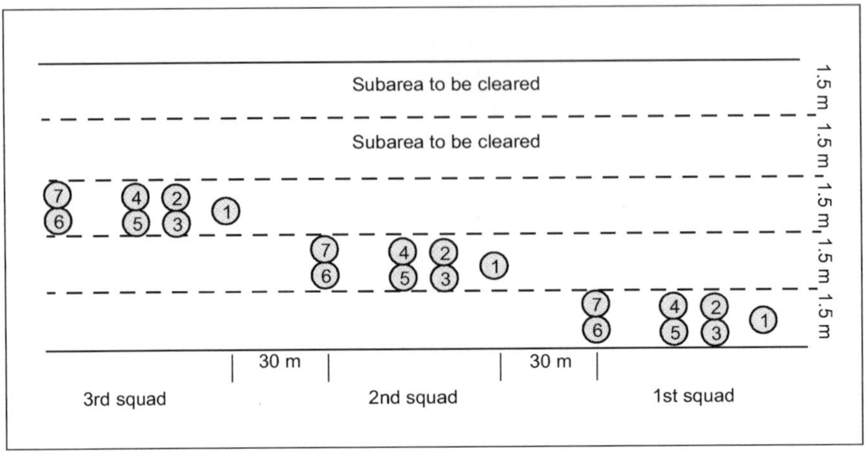

Figure 11-4. Sweep teams in echelon

METHODS AND TYPES

The information gathered from the IPB and the reconnaissance effort determines the method and the type of route clearance to conduct. The determination is based on the situation, the time available, the threat level, and available assets.

During OOTW, it is recommended that former warring faction (FWF) engineer-equivalent clearance teams precede US forces clearance teams within the FWF's AO. Do not assume that FWF clearance teams will be thorough in their clearance operation. Treat the route as unsafe until US or allied force clearance teams have proofed the route to confirm that it is cleared.

Methods

There are three methods of route clearance—linear, combat, and combined. The method employed depends on the situation, the time available, and the clearance assets available. The maneuver force should always establish static security positions at critical locations following the completion of route clearance.

Linear Clearance

In linear clearance (Figure 11-5), sweep and security teams begin route clearance at Point A and complete it at Point B. This method provides the best assurance of route coverage. Although this is an effective method, it is not the most secure method in a high-threat environment. It is also time-intensive and constrains the maneuver commander's flexibility.

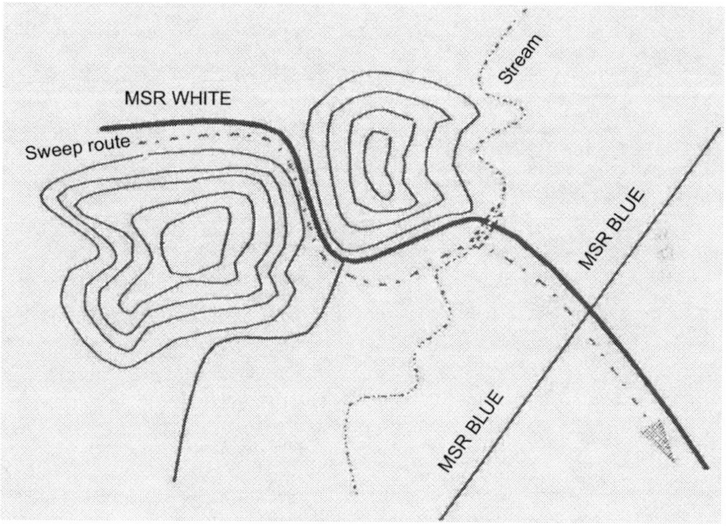

Figure 11-5. Linear clearance method

Combat Clearance

Whereas linear clearance focuses on a specific route, combat clearance (Figure 11-6) focuses on specific points along a route. As mentioned previously, IPB and EBA can identify areas for likely mine and ambush locations. These areas become NAIs or objectives for combat clearance missions. The combat clearance method divides a route into sections according to the number of suspected high-threat areas. Once the sweep element (maneuver and engineer forces) secures and sweeps these areas, the route is considered clear. Combat forces can patrol the route from these objectives to ensure that the route is secure, and if necessary, the sweep element can sweep the surrounding area if a minefield is found. Following the seizure of these objectives, the commander must assume a moderate risk that the S2 and the force engineer have identified all high-threat areas and that the route is clear of mines. Combat clearance is ideal for dismounted (light) forces since it provides the maximum use of surprise and concealment.

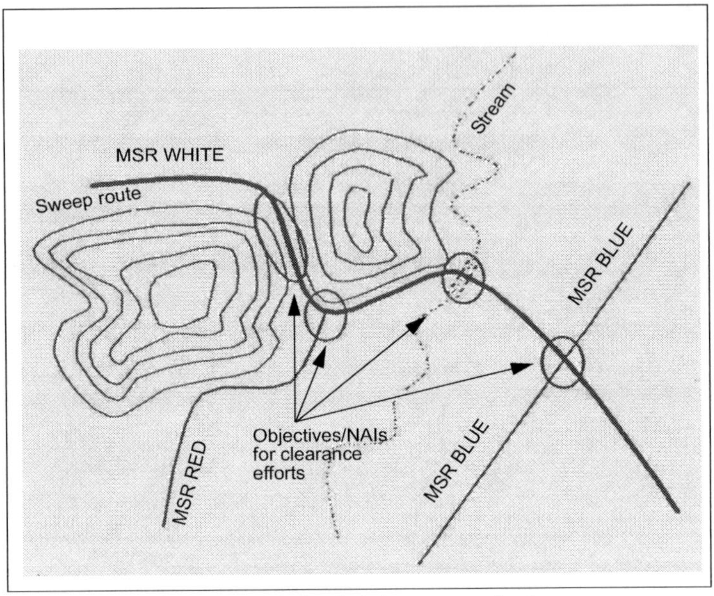

Figure 11-6. Combat clearance method

Combined Clearance

This method combines the complete clearance capabilities of the linear clearance method with the security and surprise elements of the combat clearance method. Combined clearance is a two-phase, force-intensive operation and may require a battalion-size effort, depending on the length of the route. First, identify high-threat areas through IPB/EBA and target them as NAIs and/or objectives to secure. Then, clear obstacles and enemy forces before the movement of sweep elements. The sweep team moves down the road and clears any obstacles missed or not identified during the planning process. The main advantage of this method is that the TF commander has immediately secured MSRs, allowing him to *push out* (expand forces out past the secured area and secure additional areas) and find the enemy with a degree of confidence that follow-on forces will be much safer.

Types

There are two types of sweep operations—deliberate and hasty. Deliberate and hasty clearance operations can be modified to meet the time and equipment limitations of the TF, but the commander assumes greater risk when the clearance type is modified.

Deliberate

A deliberate sweep (Figure 11-7) is very thorough and includes a complete sweep of the entire road (shoulders, culverts, ditches, and bridges). It is the most time-consuming sweep operation and relies on electronic (primary) and visual (secondary) detection systems.

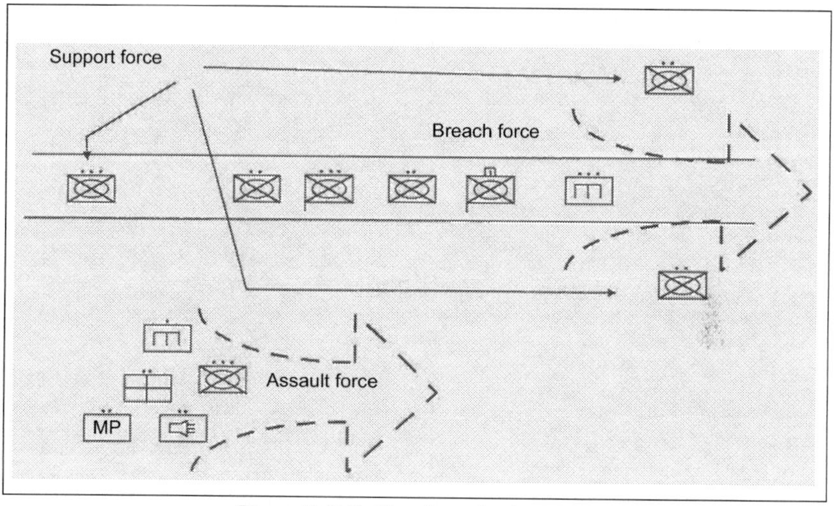

Figure 11-7. Deliberate route clearance

The platoon sweep team (Figure 11-2, page 11-9) is dismounted to focus its attention on the entire length of the route. The support force (company-size) secures at least 100 meters on the flanks and 100 meters forward to clear possible enemy direct-fire systems and overwatching elements in front of the breach force. This not only allows the breach force to focus solely on the route but also clears the area of off-route and command-detonated mines.

If enemy contact is made, the support force fixes the threat while the assault force reacts. The sweep teams withdraw to a location that provides concealment and/or security. Mechanical detection provides a third means of detection and is the method used to proof the route after the sweep team has passed through the area. The deliberate sweep includes a route reconnaissance and looks at all areas of a route, including bypasses. The deliberate sweep focuses on thoroughness rather than speed. This method is very slow and tedious and should only be used when time is not a factor; 80 to 100 meters can be covered per hour.

Hasty

A hasty sweep (Figure 11-8) consists of visual inspection, physical search or probing, and the use of mine detectors. It is the fastest, most risky method and is suited for an armored or mechanized team. It relies primarily upon visual detection (thermal sights or the naked eye) for minefield identification. The breach force looks for mines, wire, and other minefield indicators. The road surface, culverts, ditches, and bridges are inspected and searched. Visual detection is accompanied by a mechanical proofing system. Electronic mine detectors are used by sweep teams to check all suspected areas.

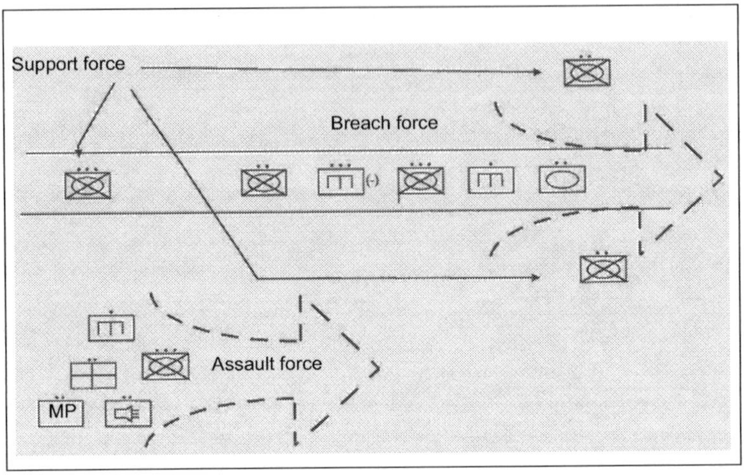

Figure 11-8. Hasty route clearance

The support force includes a maneuver platoon that provides overwatching fire and/or security. Actions upon enemy contact are the same as in a deliberate sweep. The primary objective of this technique is speed, moving approximately 3 to 5 kph. This method is extremely similar to the instride breach method when encountering minefields.

The sweep team focuses on identifying immediate risks to traffic, neutralizing those risks, and continuing on with the mission. A hasty sweep is used during the combat clearance method to validate the areas that were not deliberately cleared by the sweep team. It is also used if the METT-TC analysis does not permit a deliberate sweep or if the need for a road to be opened is urgent. Time and distance factors may be imposed. A light force may not have an MCR system but can conduct the same sweep method with an improvised roller system, or the force can use a sandbagged, 5-ton truck moving backwards as a last-resort method. Using MCRs or their equivalent is absolutely imperative due to the high risk of encountering a minefield. The mine rake or plow is not a satisfactory substitute because it destroys road surfaces.

AREA CLEARANCE

Area clearance is also a combined arms mission. Clearing operations occur when engineers receive a mission to clear an area of mines or to clear a specific minefield in a friendly AO. In most cases, the minefield has been reported and may already be marked on all sides. The engineer unit receiving the mission must base its plans on available information and prepare equipment based on the estimate.

PLANNING

Planning area-clearance operations is very similar to planning breach operations. Commanders and staffs plan and coordinate all the breaching tenets.

Intelligence

Intelligence is particularly important for discovering the types of mines and mine fuses the enemy employs. The engineer uses this information to determine which clearance and neutralization techniques offer the best chance for success and minimize the risk to the sweep teams. Intelligence also helps the commander determine the need for outside resources, such as EOD and SOF elements.

The information needed for area-clearance operations includes—

- Minefield location.
- Minefield orientation.
- Presence of wire as an obstacle.
- Location of gaps and bypasses.
- Minefield composition (buried or surface-laid mines, AT or AP mines, AHDs, trip wires, and minefield depth).
- Types of mines, employment techniques, fuses, and booby-trap configurations.
- Information on whether or not the minefield is marked and, if it is marked, the material used.
- Possibility of hostile forces in the area.

Fundamentals

You must plan to apply the four fundamentals of breaching operations (SOSR), but you may not execute all of them due to the lack of enemy presence.

Organization

Task organization is similar to that used for route-clearance operations. The breach force is the clearance element, and the support force is responsible for all security and maneuver responsibilities. There is no assault force in area-clearance operations. Table 11-3 shows a sample task organization for an area clearance. The size of the force can be tailored, based on the probability of contact.

Mass

Sufficient maneuver and engineer assets must be allocated to the clearance company team. The length and width of the route and the amount of time

Table 11-3. Sample task organization for an area clearance

Team	Support Force	Breach Force
Heavy	• Mechanized infantry platoon • Mortar section • FIST • Armor platoon	• Engineer platoon with organic vehicles • Medical team (two ambulances) • EOD team
Light/heavy	• Bradley platoon with dismount capability • 60-mm mortar section • Forward observer • One infantry platoon (light)	• Engineer platoon with organic vehicles • Medical team (two ambulances) • EOD team
Light	• AT/MP section with M60/MK19 mix • 60-mm mortar section • Forward observer • Two infantry platoons (light)	• Engineer squad (-) • Medical team (two ambulances) • EOD team

available determines the size of the sweep team. A platoon is normally used to clear a 200- by 300-meter minefield; additional assets are required to clear larger and multiple minefields.

Synchronization

All aspects of synchronization should be implemented when planning area clearance. It is especially important that rehearsals be conducted at the combined arms level. Rehearsals should include—

- Reaction to enemy contact.
- Reaction to an ambush.
- Communications exercise.
- Fire support (obscuration smoke, immediate suppression fires, critical friendly zones for counterfire radar, and no-fire area around the clearance site).
- CSS (maneuver, casualty evacuation, marking materials, and demolitions resupply).

PLANNING CONSIDERATIONS

Area-clearance BOS planning considerations are parallel to route-clearance planning considerations. Brigade and battalion TF staffs should use the planning considerations outlined on pages 11-3 through 11-7 plus the ones outlined below:

- Intelligence. Focus on the most probable enemy attack method and AAs.
- Maneuver.
  - — Clear and secure flanks (at least 500 meters) and the farside of the area to be cleared.
  - — Provide security for the cleared area.

- Fire support. Ensure that the area-clearance team has a FIST coordinator. The FIST should be collocated with the support force OIC.
- Mobility/survivability. Establish minefield control points along the area to be cleared.
- CSS.
  - — Ensure that the medical team consists of one or two ambulances and that it is located with the breach force.
  - — Ensure that all personnel wear flak vests or IBASIC (Figure 11-1, page 11-6).
- C2.
  - — Determine the area length, using clearly definable perimeter points.
  - — Coordinate with adjacent units, the host nation, NGOs, PVOs, and SOF.

TASK ORGANIZATION

The battalion TF will focus a company team (minus) as the main effort to conduct area clearance.

Support Force

This force is comprised of two maneuver platoons and an OIC. The support force provides flank security, forward security, and protection for the breach force. It neutralizes hostile forces that are encountered by the company team. The support force secures the area 500 meters beyond the area to be cleared. METT-TC factors will affect the actual distance based on the threat and the weapon systems. The support force OIC establishes static security positions around the area until the clearance operation is complete. He also has control of fires and the responsibility to neutralize any hostile force.

Breach Force

The breach force is comprised of an engineer platoon that is organized into sweep teams, a medical team, and an EOD team (or one that is on call). The sweep team (squad-size) is organized as shown in Figure 11-3, page 11-10. The breach force's mission is to sweep and clear the area of mine and explosive threats.

METHODS AND TYPES

The breach force OIC determines the perimeter of the area to be cleared and ensures that it is marked. The OIC divides the area into sections to be cleared (Figure 11-9). The sections should be no larger than 40 meters wide and 100 meters long. This is an optimal-sized area for a sweep team to clear at one time. The OIC assigns squad-size sweep teams to each section.

The squads clear their assigned sections using the sweeping techniques discussed earlier in this chapter. As the sections are cleared, they are marked for safety and control purposes. This process is continued until the entire area is cleared. Progress is reported to the company team commander as required.

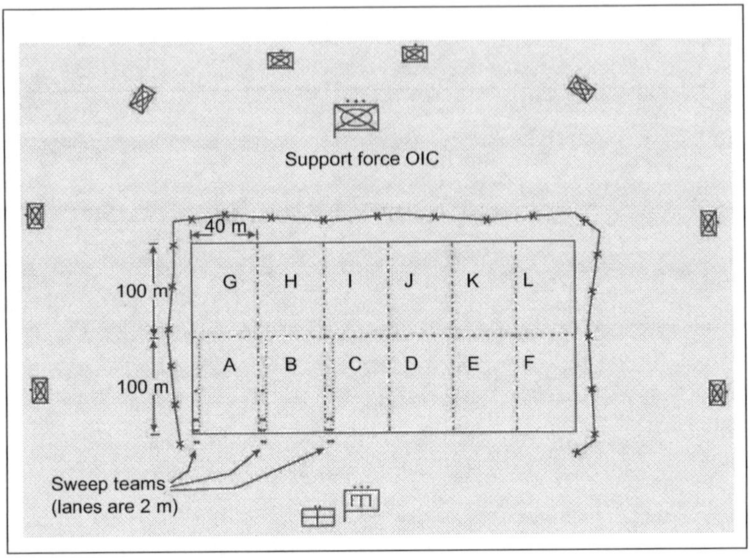

Figure 11-9. Area clearance site layout

IMPROVISED MINE THREAT

Mines are not always employed conventionally by military forces organic to the host nation or its enemies. In many cases, they are also employed by terrorists against allied forces or the host-nation populace. In these cases, the threat increases because of the improvised methods in which the mines were emplaced. In conventional emplacement of mines, a pattern emerges from the emplacing force's doctrine, and the threat can easily be reduced by using this knowledge. There is less pattern in the case of improvised mining methods, and this makes detection and removal very difficult.

Improvised mining has many different employment techniques. In most of the techniques shown below, a UXO can easily be employed in place of a mine:

- Coupling mines. Coupling is done by linking one mine to another, usually with detonating cord. When the initial mine is detonated, it detonates the linked mine. This technique is done to defeat countermine equipment.
- Boosting mines. Buried mines are stacked atop one another, and the farthest mine from the surface is fused. This reduces the probability of detection and increases the force of the blast.
- Sensitizing AT mines. On some nonmetallic AT mines, the pressure plate can be cracked and the spring removed or the mine's explosive can be cut into smaller blocks and employed as powerful AP mines.

The pressure plate can be removed from metallic AT mines and employed in the same manner. Alternatively, a pressure-fused AP mine can be placed on the top of an AT mine.

- Mixing training mines with live mines. Hostile forces can employ training mines at the start of a minefield and emplace live mines toward the end. The sweep element falsely believes that the minefield is phony and becomes complacent in its reduction activities. When this technique is used, live mines are painted to resemble training mines.
- Daisy-chaining mines. Command-detonated AP mines are commonly used in daisy chaining. Hostile forces link the mines with trip wires or detonating cord. When the initial mine is detonated, the other mines will detonate.

MINE LOCATIONS

Hostile forces normally place more than one mine in each mined area. Do not focus the detection effort solely on a horizontal mine threat, such as on the ground or in culverts. The mine threat is also vertical, such as in trees or attached to an overpass. Clearance efforts must accommodate the three-dimensional battlefield. Mines may be placed in—

- Frequently used roadways.
- Brush and other traffic obstructions placed on roadways.
- Bridge bypasses and fording sites.
- Road junctions.
- Obvious turnarounds, bypasses, culverts, ditches, and shoulders.
- Key logistic points (water, fuel, food).
- Debris along a route.

DISPOSITION OF MINES

The following actions should be taken when a suspected mine is found:

- Mark the suspected mine location; do not leave any mine unmarked.
- Search for electric wires and trip wires in the immediate area. Trace the wires in both directions to determine if items are attached to them. If there is nothing attached and the IPB does not state otherwise, cut loose trip wires and electric wires.

DANGER

Never cut taut trip wires. Alert the security element to search for an enemy that may be manning a command-detonated mine. Keep troops away from the mine until all the wires are traced and cut. Be alert for booby traps and ambush. If booby traps are found, use the clearance procedures outlined in Chapter 13.

- Probe the suspected mine location and uncover enough of the object to identify it. Other personnel should stay at least 30 meters away.

- If the object is a mine, the prober withdraws and notifies the OIC. The OIC decides to bypass the mine, destroy it in place, remove it with a grapnel, or notify EOD for hand neutralization.
- If the object is debris, get in a protected position and carefully remove the debris with a grapnel hook. Be alert for booby traps or AHDs wired to the debris.

Mine-Removal Techniques

A mine can be bypassed, detonated in place, pulled out by a rope or a wire, or neutralized and removed by hand. The method used depends on the location of the mine, the type of the mine and the fuse, and the tactical situation. Methods of removal and actions on finding a mine should be addressed in the OPORD and rehearsed prior to executing the mission.

Trip-wire and tilt-rod mines can be detonated by throwing a grapnel, with a rope attached, past the trip wire or tilt rod and pulling the grapnel back to actuate the mine. Grapnels may be improvised from any available material, such as a bent drift pin or scrap material.

A hand-emplaced charge is the standard demolition material used to destroy a mine in place (see FM 5-250). A 1-pound block of explosive placed next to a mine is sufficient to detonate most mines. A charge can be placed next to each mine in a group, then the charges can be connected and fired simultaneously.

Rope or wire can be used to pull a mine out of its installed position. This is a safe method and only detonates mines that are equipped with AHDs. It also reduces noise and cratering. A tripod (Figure 10-21, page 10-23) makes it easier to pull a mine out of a hole on the first attempt. Use the following procedures to remove mines:

- Uncover only enough of the mine to expose a handle or a projection. Attach a 60-meter length of rope or wire to the mine or engage a grapnel. If there is no projection, engage a grapnel on the bottom side of the mine, opposite the direction of pull.

DANGER
Do not move the mine while uncovering it or attaching the rope because movement might detonate an AHD.

- Ensure that the covered area is not mined. Take cover and lie in a prone position at least 50 meters from the mine. Pull the rope to remove the mine from the hole.
- Wait at least 30 seconds before leaving cover and approaching the mine if the mine type is unknown. This guards against the possibility of a delay firing mechanism.
- Dispose of the mine according to the unit directive or SOP.

Hand Neutralization

Appendix A discusses procedures for hand neutralization of US mines. Foreign mines and booby traps should only be neutralized by EOD personnel.

Mines are neutralized by hand, when—

- Units are conducting a covert breach.
- The mine is located on a bridge, building, or other facility required for use by friendly forces.
- Neutralization by other means is not possible.
- The mine can be positively neutralized by hand and is required for reuse.
- The mine type is unknown and recovery must be attempted for intelligence purposes.
- Chemical mines are located in areas where contamination would restrict the use of the area by friendly troops.

SAFETY

The following safety procedures should be observed during route and area clearance:

- Personnel should wear helmets and flak jackets to protect them against fragmentation. Sweep team members should wear IBASIC, if available.
- Vehicle floorboards should be sandbagged.
- Vehicles should be dispersed at 50-meter intervals. This ensures that a mine detonated by one vehicle will not cause casualties in other vehicles.
- One person at a time should be allowed at a suspected mine location.
- Personnel should assume that mines and explosive devices are equipped with AHDs until proven otherwise.
- Personnel should not run and should move only in previously cleared areas.
- Armored vehicles should have their hatches open to vent the pressure pulse from a mine detonation.
- Soldiers should wear ballistic and laser protective spectacles (BLPS) or lightly tinted, protective eyewear to reduce eye fatigue and improve their ability to recognize mine indicators.

REPORTS

Dissemination of information is the key to battlefield management. Units encountering minefields or explosive devices should follow a five-step process—stop, secure, mark, report, and avoid. Units must provide adequate information to their higher headquarters to ensure that follow-on elements are well informed. Information must include the known or suspected minefield location, types of mines (if known), the marking method, the time the minefield was encountered, and any additional information that may be of use to the clearing unit.

Division and maneuver brigades must establish a central control cell for mine clearance information. This cell receives and gathers all mine and explosive threat data within the unit's AO. Mine-contact reports are reported through maneuver command channels with a priority of *flash* or *immediate*. The information is then jointly controlled in the operations cell and the central mine control cell by the engineer staff officer, the G3/S3, and the Assistant Chief of Staff, G2 (Intelligence) (G2)/S2. The mine contact database is jointly maintained by the engineer and the G2/S2, who subsequently conduct pattern analysis and integrate it into intelligence and operational updates.

The central mine control cell performs the following actions:

- Maintains a current situation map and overlay that depict friendly and enemy mines and obstacles.
- Maintains and updates enemy obstacle (Figure 11-10, page 11-24) and route status (Figure 11-11, page 11-25) information.
- Receives and maintains minefield recording forms (US and foreign) within the unit's operational area (this includes host-nation minefield data, if available).
- Maintains a mine-contact database. (This could be a clearinghouse for future operations.)
- Processes, analyzes, and updates information; disseminates the information to subordinate commanders and staff.

SITUATION REPORT

Clearing units submit a situation report to higher headquarters if enemy activity is encountered or if an explosive device is discovered. This information should be tracked in the TOC and the CTCP. Information must be disseminated to subordinate units, especially CSS elements.

PROGRESS REPORT

The clearing unit submits progress and completion reports until the clearance operation is complete. Progress reports must be timely and accurate. Report format and frequency are established in the OPORD before the clearance mission is executed.

ENEMY OBSTACLE REPORT
As of: ____________

Obstacle No	Obstacle Location	Reporting Unit	Info DTG	Obstacle Description (Type of Mines/Wire)	Depth	DTG Marked	DTG Reduced	DTG Cleared	Lane/ Bypass Grids	Remarks

Figure 11-10. Sample enemy obstacle report

ROUTE STATUS REPORT
As of: ____________

Route Name	Responsible Unit	Status (Red, Amber, Green)	As Of	Start Point	End Point	Remarks
		○				
		○				
		○				
		○				
		○				
		○				
		○				
		○				
		○				
		○				
		○				
		○				
		○				

Figure 11-11. Sample route status report

Mine Incident Report

A mine incident includes any unplanned activity involving a mine, UXO, or a booby trap. It also includes near misses that could have resulted in damage or injury. The mine incident report (Figure 11-12) is a technical report that follows a serious incident report (SIR), and it should be submitted as soon as possible (local SOP will indicate time requirements).

MINE INCIDENT REPORT

DATE:
FROM: THRU: TO:

REFERENCE SIR #__________

A. Incident DTG	A1. dd/time/zone/mm/yy	
B. Incident location	B1. Map sheet/UTM/grid reference (8 digit)	Include a site sketch as an attachment.
	B2. Location (road, field, building)	
	B3. Emplacement (buried, surface-laid, off-route)	
C. Effects (to complement information already in the SIR)	C1. Casualties (rank, name, date of awareness training, time in the mission area, protection equipment used)	
	C2. Vehicle damage (number, type, extent of damage)	Include a photo if possible.
	C3. Collateral damage	
D. Device suspected	D1. Type of mine (AT, AP, make, model)	
	D2. Type of booby trap (pull, release, pressure)	
	D3. Type of UXO (dropped, thrown, projected)	
	D4. Unknown (detail, color, shape, size)	
E. Circumstances	E1. Activity at the time of the incident	
	E2. Degree of previous use of the route, area, location	
	E3. Date of previous clearance and proofing by engineers	
	E4. Where the route, area, or location is monitored	
F. Recommendations	F1. Recommendations to prevent reoccurrence	
G. Miscellaneous	G1. Any other pertinent data	

Figure 11-12. Sample mine incident report

PART THREE

Special Mining Operations

Part three provides tactical and technical information on special-mining operations, such as using booby traps and expedient devices. It also discusses mining in rivers, urban terrain, and unique environments. Restrictions and responsibilities are outlined in detail for the employment and the clearance of special mines and devices.

Chapter 12

Mining Operations in Special Environments

Mines are emplaced and encountered in all environments, some of which need special consideration to understand effective employment, detection, and/or removal.

STREAMBED AND RIVER MINING

EMPLOYMENT

Conventional AT mines are much more effective in water than on land because water transmits the shock effect better than air. Vehicle support members, tracks, and wheels are damaged by a mine blast. Small vehicles are overturned and almost completely destroyed. Because water amplifies and transmits shock waves, mines equipped with pressure-actuated fuses are subject to sympathetic detonation at greater distances in water than on land.

M15 and M19 AT mines can be used for streambed and river mining. The M21 AT mine should not be used because it is very difficult to arm and disarm underwater, and it can be easily functioned by drifting debris. To avoid sympathetic detonation, AT mines must be at least 14 meters apart in water that is less than 61 centimeters deep, and at least 25 meters apart in water that is deeper than 61 centimeters. The mined areas are chosen to take advantage of stream and adjacent area characteristics. Water depth within the minefield should not exceed 1 meter because it is difficult to work in deeper water, and pressure-actuated fuses are usually ineffective against waterborne vehicles.

Current velocity must be considered when emplacing mines in a streambed or a river. If the mines are placed deeper than 45 centimeters, they must be recovered by engineer divers:

- A lightweight diver has diving restrictions based on current velocity.
- A scuba diver is restricted to a maximum current velocity of 0.5 meter per second.

- A surface-supplied diver is restricted to a maximum current velocity of 1.3 meters per second.

Seasonal current velocity should also be considered if the minefield is to be in place for an extended period of time. Additional information on diving restrictions can be found in FMs 20-11 and 5-490.

Since sand in inland waters continuously moves downstream, it may be difficult to locate and remove mines planted on sandbars or downstream from sandbars. If the site has a muddy bottom, the mud should not be deeper than 46 centimeters and there must be a hard base underneath it. The enemy is unlikely to choose a fording point where vehicles mire easily. If underwater obstacles (gravel, rock, stumps) are bigger than the mine, the area cannot be easily mined. If such areas must be used, place the mines so that they are exposed to vehicle wheels or tracks. Armored vehicles usually enter and exit streams at points where the incline is less than 45 percent. After entering a stream, vehicles often travel upstream or downstream before exiting. Carefully examine riverbank formations and underwater obstacles to predict the trail a vehicle will use to ford the stream.

EMPLACEMENT

When emplacing mines in streams and rivers, always work in pairs. Prepare the mine on land near the emplacement site. Coat fuse threads and wells with silicone grease (a waterproof lubricant) or a heavy grease to minimize the chances of water leaking into the mine. Waterproof joints between the pressure plate and the mine case with silicone grease. As a rule of thumb, waterproofed mines are reliable up to 3 months when immersed without waterproof coverings. Secure the mine with outriggers to prevent drifting:

- Construct field-improvised outriggers with—
  - Two green limbs that are about 3 centimeters in diameter and 1 meter long. Green limbs are recommended because they are stronger and less likely to float than those which are dried out and dead. (Steel pickets, sign posts, fence rails, or similar items having the proper dimensions may also be used.)
  - Two pieces of clothesline, manila line, or similar material that are about 1 meter long.
- Fasten the limbs to the underside of the mine and secure them with the line (Figure 12-1).
- Approach the emplacement position from the downstream side. To prevent dragging the outrigger or contacting objects in the stream, carry the mine by grasping its sides, not by its carrying handle.
- Place the mine and the outrigger on the stream bottom. Stake down outriggers after they are emplaced to prevent drifting. If staking is impossible, place sandbags or large rocks on the outriggers for better anchorage.
- Arm the fuse.

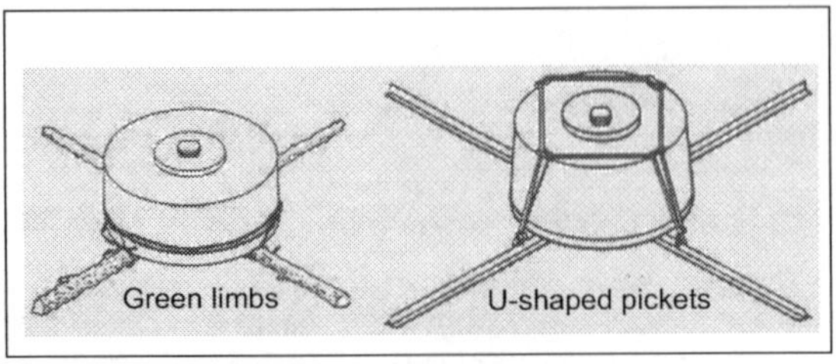

Figure 12-1. Outrigger techniques

RECOVERY

> **WARNING**
> **Mines may have drifted downstream and/or been tampered with by enemy forces. Removal by any method other than explosive breaching (see TM 9-1375-213-12) is extremely hazardous and is not recommended. If the situation demands recovery by hand, proceed with utmost caution.**

A two-person recovery team—

- Slowly proceeds 2 meters downstream from where the mine was emplaced and then carefully probes for the mine. If the mine was placed deeper than 45 centimeters, it must be recovered by engineer divers.
- Removes any foreign material from the top of the mine and disarms it.
- Carries the mine ashore and removes the fuse and the detonator.

Engineer divers normally emplace new mines or mines that have not been submersed in water; but if the mine and the fuse show no evidence of damage or deterioration, the mine can be resused. If the mine is reused, mark it to indicate that it has been immersed in water; for example, place the letter *W* on the pressure plate.

RECORDING

The minefield is recorded on DA Form 1355 (Figures 12-2a and 12-2b, pages 12-4 and 12-5).

SAFETY

In addition to normal safety measures, underwater mining requires evaluation of the tactical situation and application of special safety techniques. The turbidity, the velocity, and the depth of the water and the condition of the bottom require that laying-party personnel be able to swim well. Prolonged immersion of personnel, especially in cold temperatures, must be avoided. Sudden drop-offs, rocks, and other objects that are likely to cause personnel to lose their footing must be considered. Other safety measures include the following:

- Work in pairs.

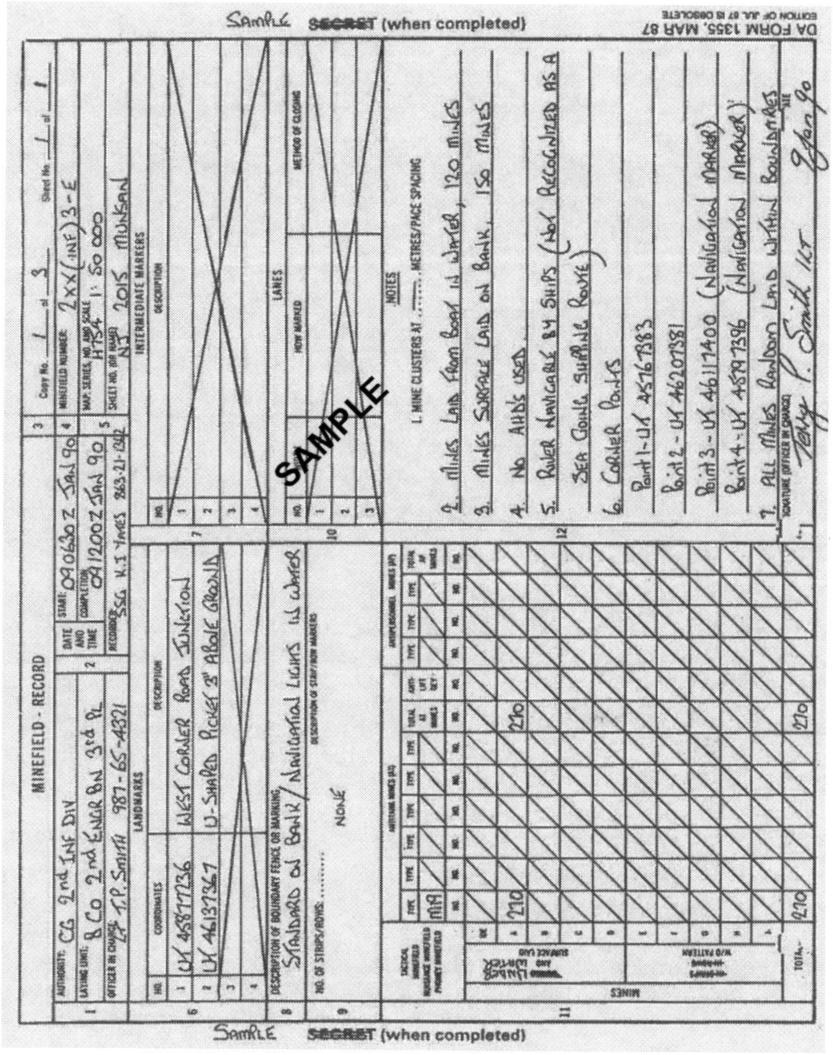

Sample SECRET (when completed)

MINEFIELD - RECORD

3 Copy No. 1 of 3 Sheet No. 1 of 1

1 AUTHORITY: CG 2nd INF DIV
LAYING UNIT: B Co 2nd ENGR BN 3rd PL
OFFICER IN CHARGE: LT T.P. SMITH 987-65-4321

2 DATE AND TIME START: 090630Z JAN 90
COMPLETION: 091200Z JAN 90
RECORDER: SSG K.I YATES 863-21-1342

4 MINEFIELD NUMBER: 2XX(INF)3-E

5 MAP: SERIES, NO. AND SCALE: H754 1:50 000
SHEET NO. (OR NAME): NJ 2015 MUNSAN

6 LANDMARKS

NO.	COORDINATES	DESCRIPTION
1	UK 45877236	WEST CORNER ROAD JUNCTION
2	UK 46137367	U-SHAPED PICKET 3' ABOVE GROUND
3		
4		

7 INTERMEDIATE MARKERS

NO.	DESCRIPTION
1	
2	
3	
4	

8 DESCRIPTION OF BOUNDARY FENCE OR MARKING: STANDARD ON BANK/NAVIGATION LIGHTS IN WATER

9 NO. OF STRIPS/ROWS: ....
DESCRIPTION OF STRIP/ROW MARKERS: NONE

10 LANES

NO.	HOW MARKED	METHOD OF CLOSING
1		
2		
3		

SAMPLE

11 TACTICAL MINEFIELD / NUISANCE MINEFIELD / PHONEY MINEFIELD

MINES		ANTITANK MINES (AT) TYPE M19	TYPE	TYPE	TYPE	TYPE	TYPE	TOTAL AT MINES	ANTI-LIFT DEV	ANTIPERSONNEL MINES (AP) TYPE	TYPE	TYPE	TOTAL AP MINES
		NO.	NO.	NO.	NO.	NO.	NO.	NO.	NO.	NO.	NO.	NO.	NO.
LINEAR AND SURFACE LAID	OE												
	A	270						270					
	B												
	C												
	D												
ROWS/STRIPS W/O PATTERN	E												
	F												
	G												
	H												
	J												
TOTAL		270						270					

12 NOTES

1. MINE CLUSTERS AT .... METRES/PACE SPACING
2. MINES LAID FROM BOAT IN WATER, 120 MINES
3. MINES SURFACE LAID ON BANK, 150 MINES
4. NO AHD'S USED.
5. RIVER NAVIGABLE BY SHIPS (NOT RECOGNIZED AS A SEA GOING SHIPPING ROUTE)
6. CORNER POINTS
Point 1 - UK 45767383
Point 2 - UK 46207381
Point 3 - UK 46117400 (NAVIGATION MARKER)
Point 4 - UK 45797396 (NAVIGATION MARKER)
7. ALL MINES RANDOM LAID WITHIN BOUNDARIES

SIGNATURE (OFFICER IN CHARGE): Terry P. Smith 1LT DATE: 9 Jan 90

DA FORM 1355, MAR 87 EDITION OF JUL 87 IS OBSOLETE

Sample SECRET (when completed)

Figure 12-2a. Sample DA Form 1355 (front side) for river mining

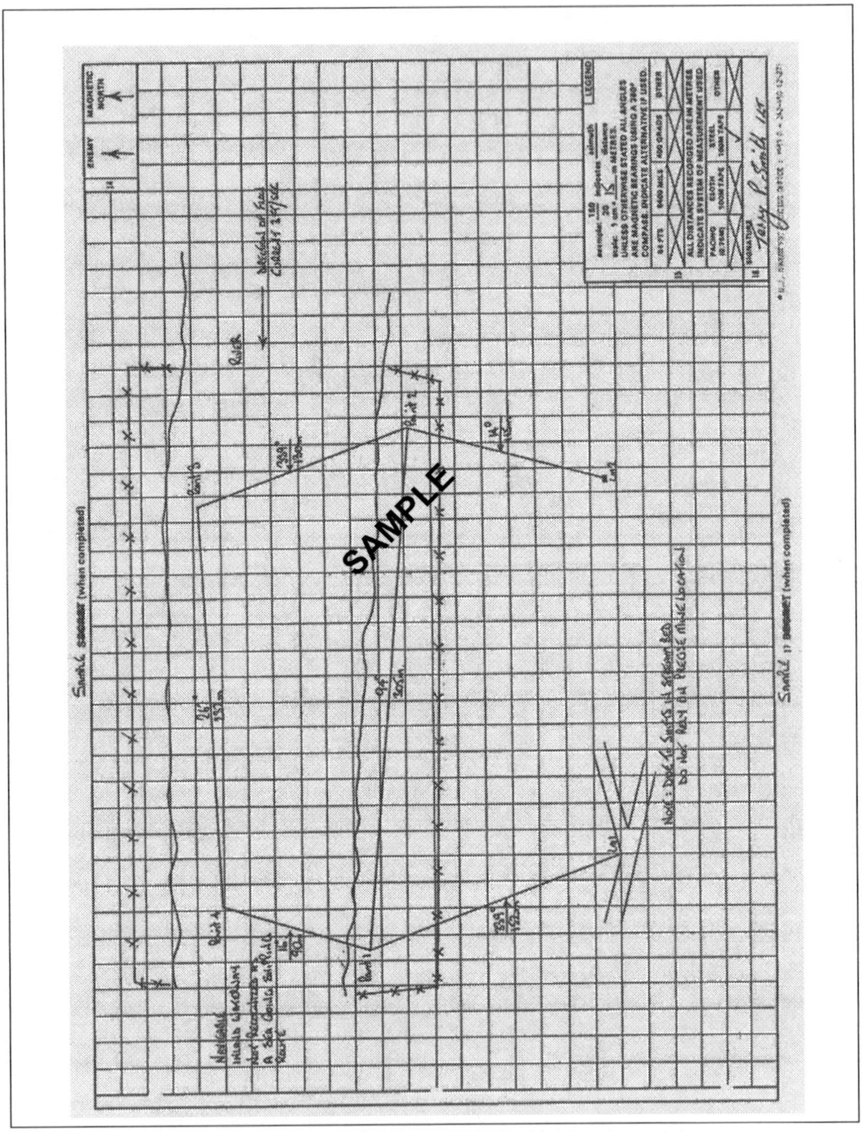

Figure 12-2b. Sample DA Form 1355 (inside) for river mining

- Emplace mines from upstream to downstream to prevent personnel and equipment from being swept into the mined area.
- Stay on the downstream side of the mine when arming the fuse.
- Place the mine as flat as possible on the bottom to prevent drifting. Use green saplings or other nonbuoyant material to construct outriggers.
- Do not arm the mine before it is laid.
- Carry the mine horizontally or edgewise to the current to reduce water resistance on the mine's pressure plate.

URBAN-TERRAIN MINING

Characteristics of urban areas (such as a high proportion of hard-surfaced roads) prohibit a simple transition from open- to urban-area mine employment techniques and doctrine. The advantages of abundant cover and concealment, maneuver restrictions, and observation already possessed by the defender of an urban area can be significantly enhanced by the proper use of mines. Terrain modified through the process of urbanization provides a unique battle environment.

The major characteristics of urban terrain which are likely to impact on mine warfare include the following:

- Multistoried buildings add a vertical dimension to the battle. Basements and floors become part of the battle scene. The vulnerability of combat vehicles increases because attack from above or below is likely.
- Fighting is done at close range, often face-to-face, and seldom exceeds 50 meters. Some weapons, particularly large-caliber weapons, are unsuitable at a short range.
- Sewers, subways, and tunnels provide covered and concealed passageways for the movement of troops on both sides. Detailed knowledge of the location and the status of these tunnels is needed to successfully wage an urban battle.
- Streets and parking lots are modified to withstand continuous use by vehicles. Major routes and lots are paved. A high density and complex pattern of streets provide numerous avenues of advance. Burying mines is extremely difficult. Most mines are surface-laid and camouflaged with rubble and debris to avoid detection.
- Movement by vehicle is difficult. Streets are littered by rubble and cratered if the city has been bombed or subjected to artillery attack. Bridges and overpasses are likely to be destroyed or blocked. Traffic flow is highly channelized.
- Extensive map and chart data are needed by the commander. For example, the commander should know the locations of telephone, electric, gas, water, and sewer connections; substations; and generating and pumping stations.

Antipersonnel Mines

US policy prohibits the use of non-self-destructing AP mines for all US forces except those on the Korean Peninsula. However, US forces can expect to encounter AP mines that are emplaced by other countries in support of MOBA. They are employed to block infantry approaches through or over underground passageways; open spaces; street, roof, and building obstacles; and dead spaces.

When AP mines are encountered **(Korea Only: or used)** in MOBA, mine locations are recorded on DA Form 1355 as shown in Figure 12-3)

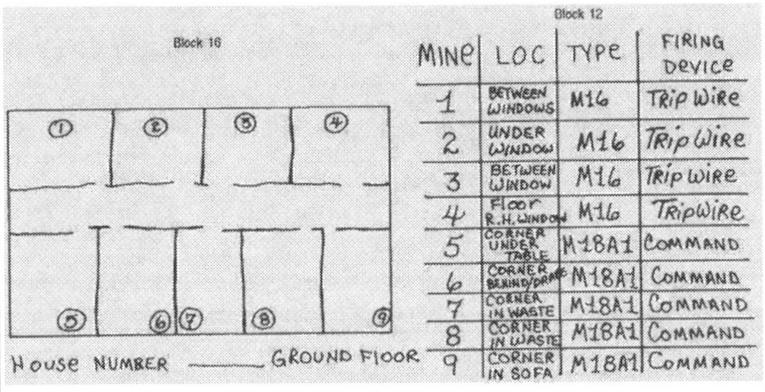

Figure 12-3. Building sketch and mine plan (DA Form 1355)

Underground Passageways

Subways, sewers, cellars, and utility tunnels provide protected movement routes for troops. In large cities where underground systems are numerous and complex, limited manpower may dictate that forces employ obstacles to block key passageways with wire and AP mines. (See Figure 12-4, page 12-8.)

Open Spaces

Open spaces include gaps between buildings, courtyards, residential yards, gardens, parks, and parking lots. In some cases, mines can be concealed in rubble or buried. However, the characteristics of most terrain surfaces, coupled with limited time and resources, dictate that mines be surface-laid. (See Figure 12-5, page 12-8.)

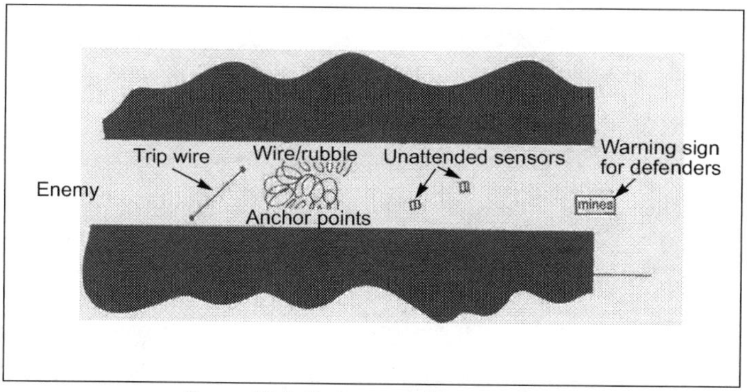

Figure 12-4. Underground passageway

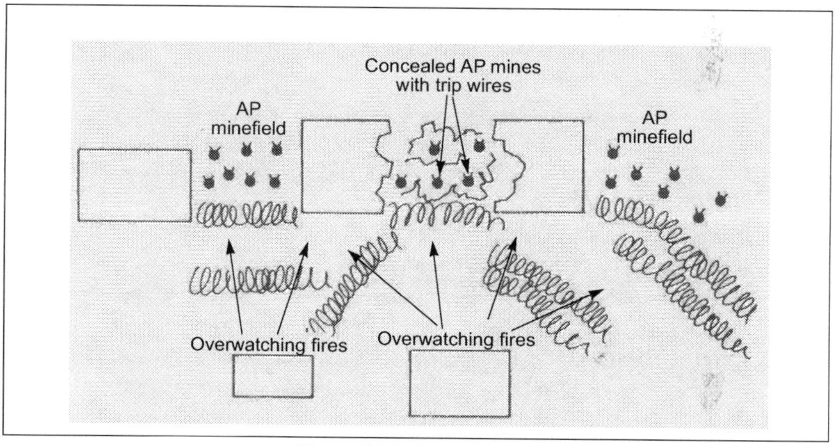

Figure 12-5. Open spaces

Street Obstacles

Hand-emplaced AP mines can be emplaced on street surfaces, on railroad lines, and in areas along shallow waterways. (See Figure 12-6.)

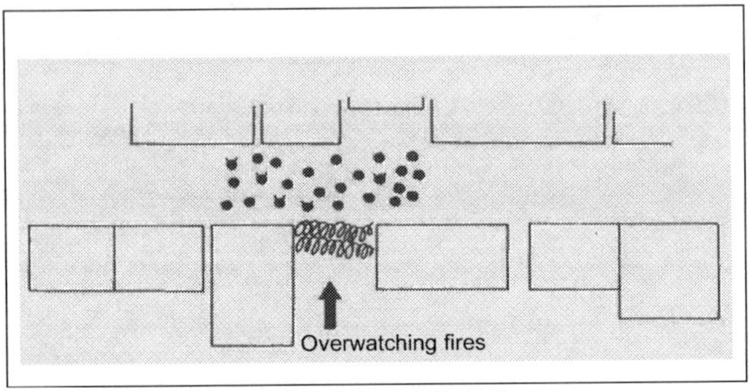

Figure 12-6. Street obstacles

Roof Obstacles

Mines and booby traps supplement wire obstacles to deny operations that require air assault onto rooftops. They also prevent occupation on roofs that afford good observation points and fields of fire. (See Figure 12-7.)

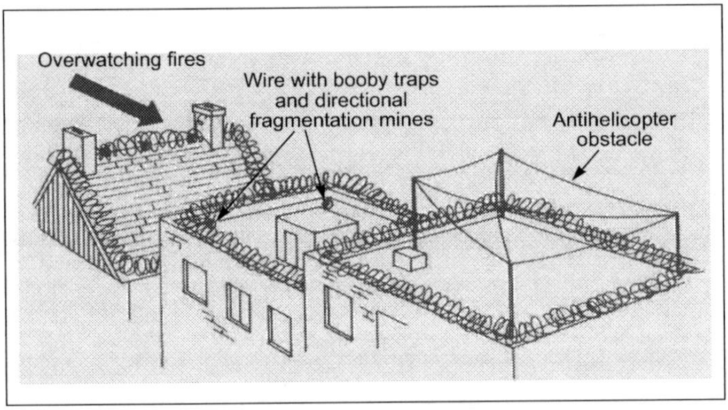

Figure 12-7. Roof obstacles

Building Obstacles

Building obstacles include areas within and adjacent to buildings. Forces can lay mines in conjunction with wire obstacles to deny infantry access to covered routes and weapon positions (Figure 12-8).

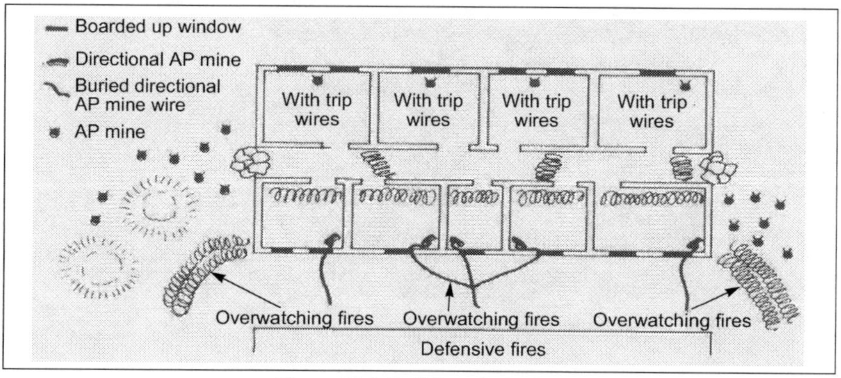

Figure 12-8. Building obstacles

Dead Spaces

Obstacles and mines can be emplaced to restrict infantry movement in areas that cannot be observed and in areas that are protected from direct fire.

Employment

The following AP mines are effective in urban terrain:

- **M14 (used by US forces in Korea only). Its small size makes it ideal for obscure places, such as stairs and cellars. It can be used in conjunction with metallic AP and AT mines to confuse and hinder breaching attempts. (See Figure 12-9.)**
- **M16 (used by US forces in Korea only). With trip-wire actuation, its lethal radius covers large areas such as rooftops, backyards, and cellars. An added advantage can be gained by attaching twine or wire to the release-pin ring to expediently rig the mine for command detonation. (See Figure 12-10.)**
- M18A1 (claymore). Numerous innovative applications of claymore munition deployment can be found for defensive warfare in urban areas (Figure 12-11, page 12-12). With remote firing, a series of claymore mines along a street establishes a highly effective ambush zone. Mines can also be employed on the sides of buildings, in abandoned vehicles, or in any other sturdy structure. Numerous opportunities exist for effectively sited, well-concealed mine employment above the terrain surface. Claymore munitions can be used to fill the dead space in the FPF of automatic weapons. They present a hazard when used in confined, built-up areas. Exercise caution when using them close to friendly forces because there is a danger of backblast.

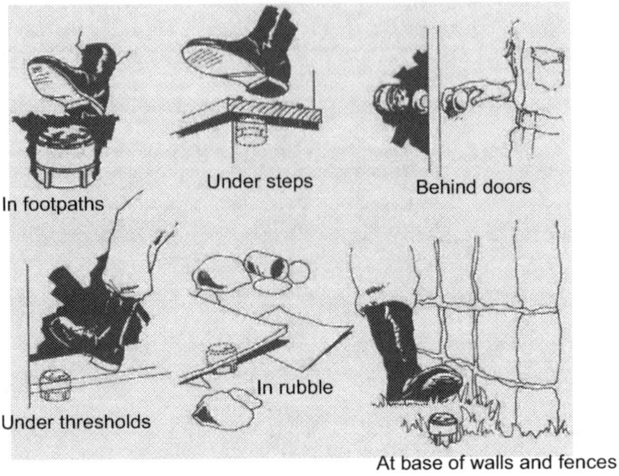

Figure 12-9. Probable M14 AP mine emplacement

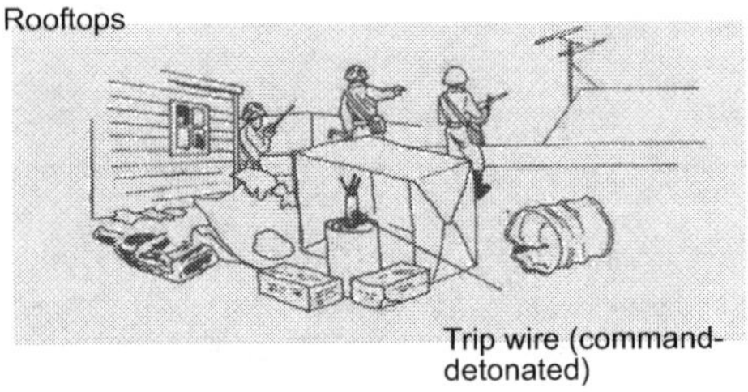

Figure 12-10. Probable M16 AP mine emplacement

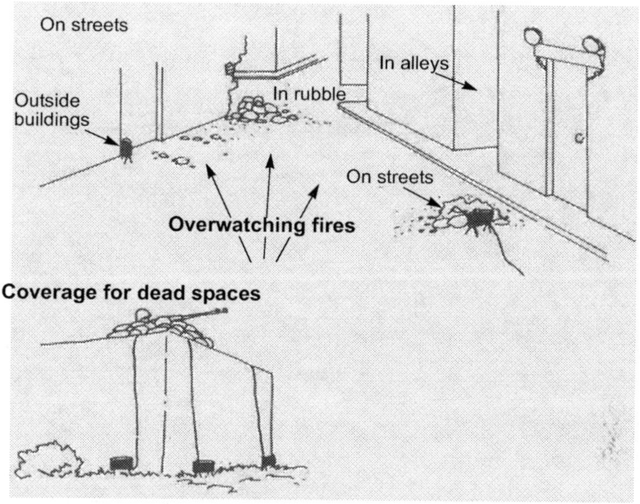

Figure 12-11. Probable M18A1 munition emplacement

CONVENTIONAL ANTITANK MINES

Enemy tanks, infantry fighting vehicles (IFVs), and direct-fire support weapons are restricted to streets, railroad lines, and, in some instances, waterways. (See Figure 12-12.) M15, M19, and M21 mines are used primarily in tactical and nuisance minefields; but they are occasionally used in protective minefields. They should be employed with other obstacles and covered by fire. Conventional AT mines emplaced in streets or alleys block routes of advance in narrow defiles. Concealment of large AT mines is accomplished by placing them in and around rubble and other obstacles. Extensive labor requirements generally prohibit burying mines in difficult terrain types.

In dispersed residential areas, obstacles are required to reduce the enemy's infantry mobility through and between houses and in open areas. They also prevent armored vehicles from moving between houses and along streets. AT minefield patterns should extend outward from the streets, incorporating open areas between buildings and streets to prevent easy bypass of street obstacles.

Significant labor and mine materials are required to deploy conventional mines between widely spaced buildings, in high-rise construction, and in industrial and transportation areas. Therefore, SCATMINEs should be seriously considered as viable alternatives. Some situations, such as the one shown in Figure 12-13, provide opportunities for the effective employment of mines in tactical and nuisance minefields.

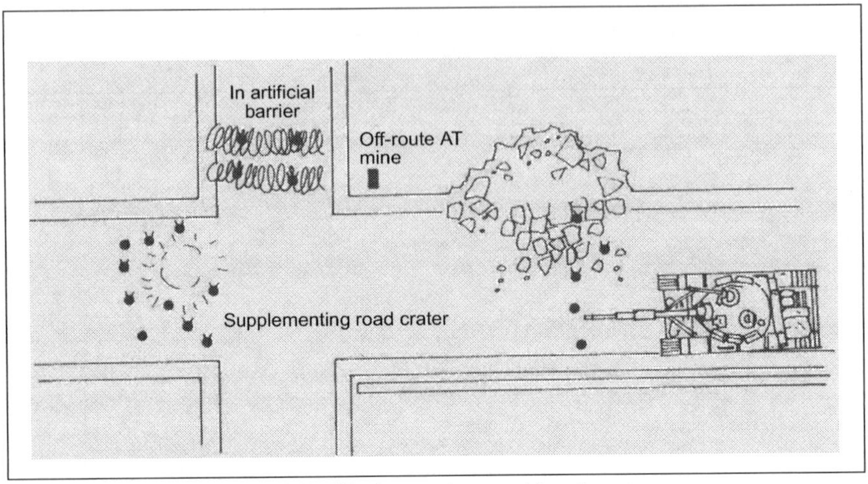

Figure 12-12. AT mine emplacement in urban areas

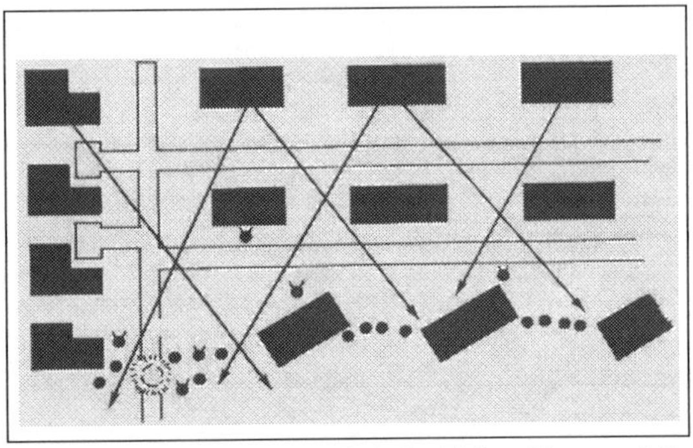

Figure 12-13. AT mine emplacement in industrial and transportation areas

SCATTERABLE MINES

Area-Denial Artillery Munitions and Remote Antiarmor Mines

In addition to the advantages (such as reducing required resources and emplacement time) applicable to all SCATMINE systems, ADAMs and RAAMs have two specific advantages. They are the most rapidly deployed SCATMINE systems, and preplanning artillery-delivered minefields increases the rate at which nuisance minefields can be emplaced. Secondly, these mines can be delivered under enemy fire. Employment of ADAMs and RAAMs is most effective when the enemy's intentions are known and their forces are committed to an avenue of advance. (See Figure 12-14, page 12-14.)

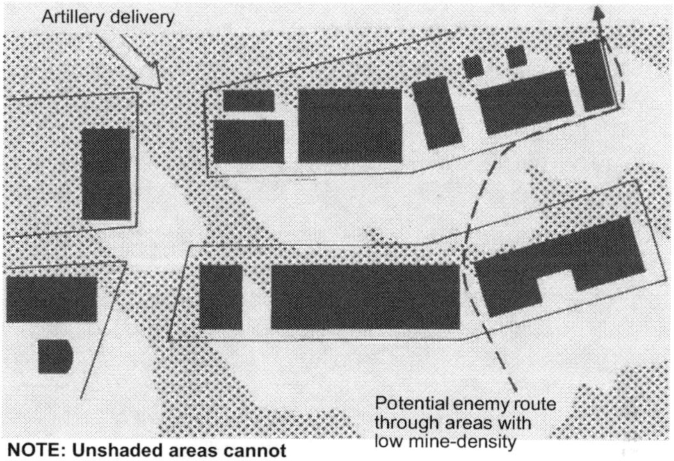

Figure 12-14. ADAM/RAAM employment

Using ADAMs/RAAMs in urban terrain involves five specific problem areas:

- Difficulty in precise minefield siting. Accurate siting is extremely critical due to the typically restrictive avenues of advance and may be futile due to the difficulty in adjusting artillery rounds in an environment that obscures observation. Further, buildings tend to create unmined shadow zones.
- Uncertainty of ADAM/RAAM survivability upon impact with a building or ground surfaces that are characteristic in urban areas.
- Likely availability of artillery firing units. ADAM/RAAM emplacement may not be a priority of the maneuver commander, because his supporting units may not have enough ADAM/RAAM munitions on hand. Assuming the availability of artillery assets for an ADAM/RAAM mission could prove disastrous for defending forces.
- High detectability of these mines on bare and lightly covered surfaces. This permits the enemy to seek out unmined passageways or pick through lightly seeded areas. If you use the doctrinal guidelines for emplacing artillery-delivered mines on top of the advancing enemy or immediately in front of them, the desired obstacle intent (disrupt, turn, fix, block) and enhanced fires are achieved.
- Difficulty in achieving a good, random pattern. Hard-surfaced areas cause mines to bounce and roll. Some mines (especially AT mines) will land on top of buildings and are ineffective.

Air Volcano

The primary advantage of the air Volcano system is its capability to site and emplace minefields accurately. This depends on the helicopter's maneuverability over the selected minefield terrain and the proper coordination between ground forces and aviation support. Disadvantages include vulnerability and the high replacement cost of the helicopter. However, in view of the system's operational concept, employment in urban terrain (which provides little exposure of the helicopter) actually increases the practicality of employing this system in urban areas. Mine survival rate on impact with a hard surface is another potential problem.

Ground Volcano

Three aspects of the ground Volcano distinguish it from other SCATMINE systems:

- The dispenser is organic to supporting combat engineers, making it readily available to support the maneuver commander's defensive plan.
- Delivery siting is accurately pinpointed to the ground.
- Better opportunities exist to record the presence of a minefield. In contrast to artillery-delivered and air Volcano systems, the ground Volcano is delivered by engineers who are normally located with and report directly to the maneuver commander.

Some primary factors may degrade ground Volcano deployment in urban terrain. The requirement to emplace minefields before an actual attack in order to reduce system vulnerability is the most significant factor. This makes the minefield detectable and provides more reaction time for the enemy to alter their scheme of maneuver. The delivery of mines depends on terrain trafficability. The prime mover and the launch vehicle must negotiate the terrain over which mines are to be dispensed.

Modular Pack Mine System

The MOPMS is ideally suited for employment in urban terrain (Figure 12-15, page 12-16). The module can be hidden from enemy view, and the mines can be dispensed after attackers are committed to a route of advance. Additionally, mines can be emplaced rapidly under enemy fire. In contrast to other SCATMINE systems, the commander controls when and where mines are dispensed and how they are detonated, regardless of the enemy situation.

Gator

When considered for employment in urban terrain, Gators encompass the same problems as artillery-delivered and air Volcano mine systems.

DECEPTION MEASURES

Phony minefields can be established rapidly with negligible effort and cost. They have the distinct advantage of blocking the enemy but not friendly forces. Although it is difficult to fake a surface-laid minefield, expedients such as soup pans, seat cushions, and cardboard boxes have historically proven effective in delaying and channelizing attacking forces. These objects, as well

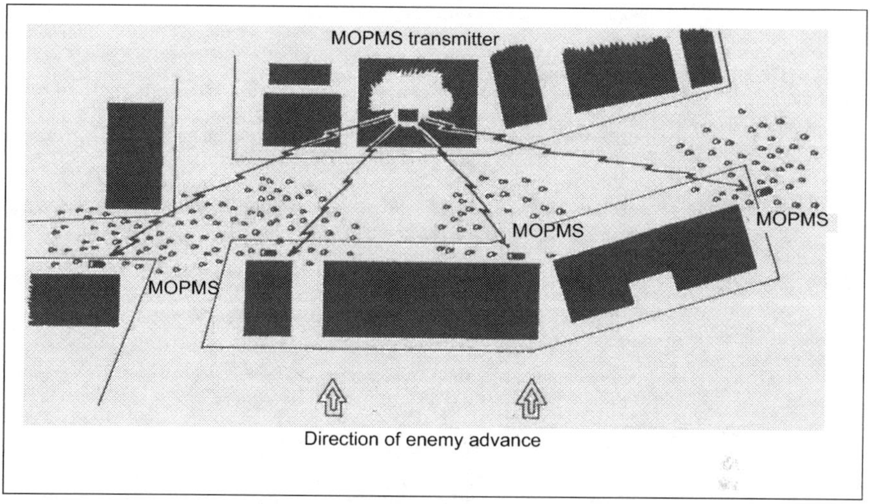

Figure 12-15. MOPMS employment

as other ones readily available in urban areas, can be used as phony minefields or used to cover real mines. A more realistic phony minefield could be created with inert or training mines.

Inadequate minefield camouflage in urban terrain is viewed as a critical constraint in deploying conventional mines and SCATMINEs. Smoke can be deployed from various dispensers, but it must be dense and accurately employed and released.

SPECIAL ENVIRONMENTS

COLD REGIONS

Mine employment in cold regions poses special problems—the principal one being emplacement. Mine burial is extremely difficult in frozen ground. The freezing water in soil causes it to have high strength and penetration resistance, so digging times are greatly increased if not impractical. However, there are several means to overcome this problem. In some cases, the minefield can be laid out before the soil freezes. To do this, dig holes for each individual mine and insert a plug into the hole to protect its shape and prevent it from being filled in. A wide variety of material can be used for plugs. Ideally, the plug should be economical, easy to remove, and rigid enough to maintain the depth and shape of the hole. Sandbags, plastic bags filled with sand or sawdust, or logs make excellent plugs. If the minefield cannot be prechambered, mechanical means can be used to dig holes. When available, civilian construction equipment (particularly large earth augers) can be used to drill holes for mine emplacement.

To assure detonation of buried, pressure-actuated mines, they should be placed in a hole that is shallow enough for the pressure plate to be above ground. Covering spoil should be a maximum of 1 centimeter deep.

When burial is impossible, mines are placed on the surface. Heavy snow cover may reduce the effectiveness of both buried and surface-laid mines by causing them to be bridged. Mines laid in deep snow should be placed as close as possible to the surface and supported by boards or compacted snow. Waterproof mines before emplacement in cold regions. Mines can also be placed in plastic bags before burial. In some cases, a layer of ice may form over the top of the pressure plate. Although the load required to break the ice is slightly higher than that required to activate the fuse, thin layers of external ice will have little effect on mine functioning. When possible, tilt-rod mines should be used in cold regions because they are less susceptible to ice and snow. Magnetic mines are not significantly affected by snow, although cold weather decreases battery life.

Camouflaging a minefield in a cold region is difficult. Mines should be painted white when snow is expected to remain on the ground for extended periods of time. Minefield signature tracks should be swept away, or deliberate tracks should be made to give the impression of a safe area.

Korea Only: When trip-wire mines are employed in snow, the wire should be about 10 meters long, with a slight amount of slack left in the wire. The trip wires should be supported approximately 46 centimeters above the ground to avoid degradation by snowfall.

JUNGLES

Fuses and explosive components deteriorate very rapidly in jungle climates. As a result, mines and mine material require more frequent and extensive maintenance and inspection. Waterproof mines employed in humid climates. The rapid growth of vegetation hinders maintenance recovery and removal. Dense vegetation may cause mines to become inoperable or windblown foliage can detonate them. FM 90-5 provides detailed information on jungle operations.

DESERTS

In desert climates, fuses and explosive components deteriorate slowly. The terrain and the situation determine how mines will be emplaced. Mine boards will normally be required to provide support in soft, shifting sand. Mines emplaced in the desert have a tendency to shift position, and the spacing between mines and rows should be increased to prevent sympathetic detonation. Blowing sand may expose buried mines or cover surface-laid mines. Sand may also cause mines to malfunction. It is important to realize the difficulty of accurately recording minefield locations in vast, open, desert areas void of recognizable terrain features. More mines are required for desert operations. Typically, desert minefields are much larger and have a lower density than those used in Europe or Korea. FM 90-3 provides detailed information on desert operations.

Chapter 13

Booby Traps and Expedient Devices

During war and OOTW, booby traps can be found anywhere at anytime. They can kill or incapacitate their unsuspecting victims. This chapter provides information on booby-trap employment concepts, detection techniques, marking and recording procedures, and removal guidelines.

This chapter also provides an overview of expedient devices and their employment considerations.

SECTION I. SETTING BOOBY TRAPS

US policy restricts the use of booby traps by US personnel. This does not preclude their use by other countries, so US forces may encounter them during operations.

The use of booby traps is limited only by the imagination of the force employing them. They—

- Are usually explosive in nature.
- Are actuated when an unsuspecting person disturbs an apparently harmless object or performs a presumably safe act.
- Are designed to kill or incapacitate.
- Cause unexpected, random casualties and damage.
- Create an attitude of uncertainty and suspicion in the enemy's mind, thereby, lowering his morale and inducing a degree of caution that restricts or slows his movement.

Many booby traps are constructed using military equipment and ammunition. Improvised traps are used during counterinsurgency missions in low-intensity conflicts.

The corps commander is the employment authority for booby traps. He can delegate this authority to the division commander. If authority is given to set booby traps, US personnel will adhere to the rules for international law applicable to armed conflict. There are several uses of booby traps that are prohibited. Remember, these restrictions are not observed by all countries; US personnel must still be cautious when approaching objects in areas where booby traps are supposedly prohibited.

International law prohibits the use of booby traps as follows:

- Booby traps and other devices are prohibited if they are attached to or associated with—

  - Internationally recognized protective emblems, signs, or signals.
  - Sick, wounded, or dead personnel.
  - Burial or cremation sites or graves.
  - Medical facilities, equipment, or supplies.
  - Children's toys or other portable objects or products that are designed for their feeding, health, hygiene, clothing, or education.
  - Food or drink.
  - Kitchen utensils or appliances except in military establishments, military locations, or supply depots.
  - Objects that are clearly religious in nature.
  - Historic monuments, works of art, or places of worship.
  - Animals or their carcasses.
- Booby traps are prohibited in cities, villages, and other areas that contain civilians if combat between ground forces is not taking place or does not appear to be imminent, unless—
  - Booby traps are placed on or in the close vicinity of a military objective.
  - Measures (guards, warning, or fences) are taken to protect civilians from booby-trap effects.

TACTICS

Booby traps are psychological weapons. They make the enemy cautious and slow it down. These actions, in turn, cause enemy casualties. Do not waste time attempting to set elaborate traps that are undetectable or impossible to disarm. Also, do not waste time developing difficult sites, because simple traps usually have the same chance of catching the enemy. Even if booby traps are detected and cleared, their aim is achieved.

The principles governing the use of booby traps and nuisance mines are identical, so consider using them in conjunction with one another. They have characteristics that make them suitable for use in different situations:

- Nuisance mines are quicker to lay and safer to use than booby traps, and they are normally used in outside locations where they can be buried.
- Booby traps are normally used in urban areas, structures, and places where mines are easily detected.

Booby traps and nuisance mines are particularly suited for defensive operations. They are used to—

- Slow the enemy's advance.
- Deny the enemy use of facilities and material.
- Warn of enemy approach.

- Deter the enemy from using ground not covered by direct fire.
- Plan defensive operations.

In offensive operations, booby traps and nuisance mines are employed on an opportunity basis during raids and patrols. Formal instruction is not usually issued by the staff.

Exercise caution when using bobby traps in offensive operations because they may hinder the operation. In advance and pursuit operations, they are primarily used by patrols and raiding parties. They slow down enemy follow-up actions and hinder the enemy's repair and maintenance teams after raids.

The following considerations pertain to defensive operations but may be relevant to offensive operations and must be considered when briefing troops:

- Booby trapping is rarely given a high priority and is usually peripheral to other engineer tasks.
- Nuisance mines are more cost-effective than booby traps, unless booby traps are used in situations that allow their full potential to be exploited. If it is easier, use nuisance mines instead of booby traps.

To maximize the effect of booby traps and nuisance mines, the staff provides engineer commanders with the following information:

- Purpose. Booby traps are time-consuming and dangerous to set. Do not waste time and effort setting traps that are unlikely to be actuated or that are not specifically designed to achieve the required aim. For example, if booby traps are being used against troops, small, simple traps designed to incapacitate will achieve this result just as well as complicated ones with large charges. If the aim is to destroy vehicles, use mines.
- Location. The precise location for booby traps and nuisance mines can only be determined by the setting unit. Areas must be delineated and recorded so that there is no threat to friendly forces in the event of reoccupation.
- Time setting starts and time available for setting. The time setting starts affects other engineer tasks, and the length of time available for setting governs the number of men required.
- Number of safe routes required. Safe routes are important during general withdrawals where authority has been given to booby-trap positions as they are evacuated. They also provide safe areas for the covering force to launch counterattacks.
- Likelihood of reoccupation. Even if the enemy has not detonated booby traps, they might have interfered with them. Therefore, do not set booby traps when areas are to be vacated to meet short-term tactical requirements or when reoccupation is expected soon.

Intelligence personnel provide information to assist the setting unit in maximizing the effect of booby traps. The nature and the type of traps required depend on the enemy unit. For example, while paying particular attention to dead space and defilade positions, use mines or widely dispersed traps (with large charges) against a mechanized enemy. Conversely, use small traps and AP mines (in places that afford cover) against an infantry enemy.

SITING

If the first obstacle or installation the enemy strikes is booby-trapped or nuisance-mined, he is delayed while he clears it. The enemy is further delayed by an increased degree of caution. His troops know that additional traps and mines can be encountered. Booby traps and nuisance mines are generally located—

- In and around buildings, installations, and field defenses.
- In and around road craters or any obstacle that must be cleared.
- In natural, covered resting places along routes.
- In likely assembly areas.
- In the vicinity of stocks of fuel, supplies, or materials.
- At focal points and bottlenecks in the road or rail systems (particularly the ones that cannot be bypassed).

The setting-party commander is responsible for the detailed siting and design of booby traps. Consider all the information about the enemy soldier and his operating procedures when selecting places and objects to trap. Also, consider the traps from the enemy's point of view and assess the courses open to the enemy when he encounters them. This can expose weaknesses in your initial plan and bring about changes to the proposed layout, or it can result in a different location being selected. In addition, determine the effort required by the enemy to bypass the traps. This shows whether the imposed delay justifies the effort required to set the booby traps in the selected location.

TYPES OF TRAPS

Booby traps are designed to—

- Be actuated by persons carrying out their normal duties.
- Take advantage of human nature.

The following booby traps can often be detected because they are designed to make the person do something:

- Bait. Usually consists of objects that arouse someone's interest, such as attractive or interesting items that have apparently been left behind or discarded during a rapid evacuation.
- Decoy. The most common decoy consists of two traps—one designed to be detected, the other designed to actuate when personnel deal with the first one. The first trap can be a dummy. A classic form of a decoy is to place booby traps or nuisance mines in locations from which the decoy mine can be removed.
- Bluff. A bluff is a hoax and usually consists of a dummy trap.

- Double bluff. A double bluff only appears to be a bluff. Personnel believe the trap is safe or can be disarmed. For example, a number of traps can be set that are disarmed when the detonating cord is removed from the charge. The double bluff is achieved by setting another trap that appears to be the same, but it actually explodes when the detonating cord is removed from the charge. Double bluffs rely on a reduced awareness and alertness caused by repetition.

COMPONENTS AND PRINCIPLES

There are two initiation methods for explosive booby traps—electric and nonelectric. Both methods can be constructed using many different types of FDs. FDs can be secured to the charge (direct connection) or located away from it (remote connection). They are actuated by one or more methods. It is impossible to describe every type of trap that may be encountered; however, most are constructed and operated by using components similar to those listed below:

- FD.
- Power source (battery, for example).
- Connection (usually detonating cord or electric wires).
- Blasting cap.
- Main charge.

Figure 13-1, page 13-6, shows how typical electric and nonelectric traps can be made.

ACTUATION METHODS

Many sophisticated booby-trap devices are now being manufactured that operate on vibration, sound, temperature change, and other methods. Current intelligence on the booby trap being used in the AO should be gathered so that countermeasures can be developed and practiced. Most FDs found in the combat zone are simple mechanisms designed to be actuated by pull, pressure, pressure release, or tension release (Figure 13-2, page 13-7).

METHODS OF CONNECTION

Procedures can be varied when it is safe to do so. For example, instead of connecting the FD to a charge already in position, preconnect trap components and then position the trap.

Small charges (up to 1 kilogram) are sufficient for AP traps, but larger quantities can be used to increase their effect. Shrapnel can be produced by packing stones, scrap metal, nails, or other material around the charge. AT traps require large charges (up to 6.75 kilograms for wheeled vehicles and 11.25 kilograms or more for tracked vehicles).

REMOTE

Follow the procedures listed below when assembling a remotely connected trap using an M142 FD (similar to the illustration in Figure 13-3, page 13-7):

- Design the trap and collect necessary materials.

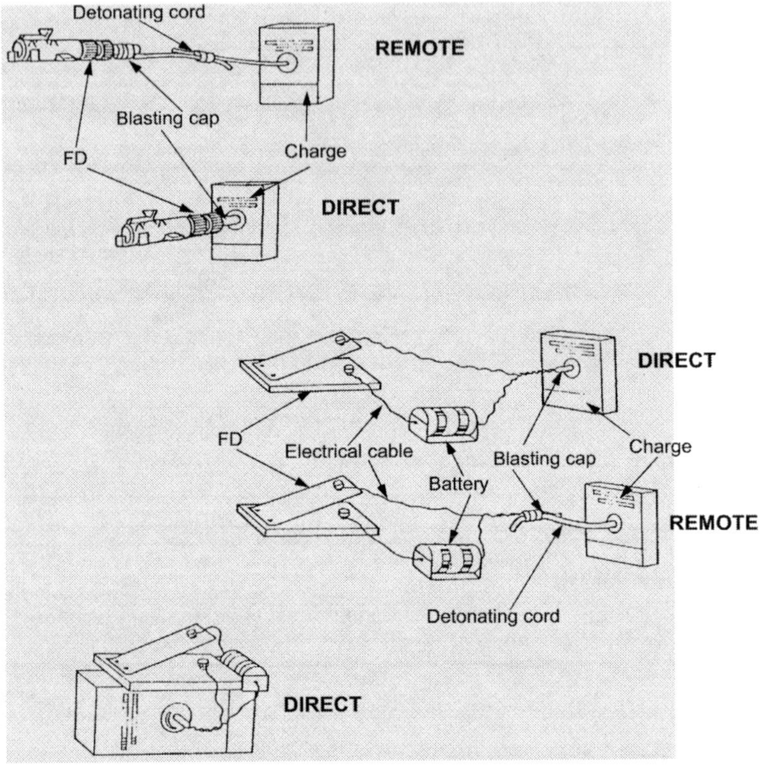

Figure 13-1. Typical electric and nonelectric booby traps

- Test the M142 FD.
- Lay the detonating cord from the charge location to the FD location.
- Position the charge.
- Connect the detonating cord to the charge.
- Prepare the coupler.
- Tape a length (46 centimeters, minimum) of detonating cord to the coupler's blasting end.
- Prepare and position the M142, set it to operate in the desired manner, and remove the round- or square-headed pin.

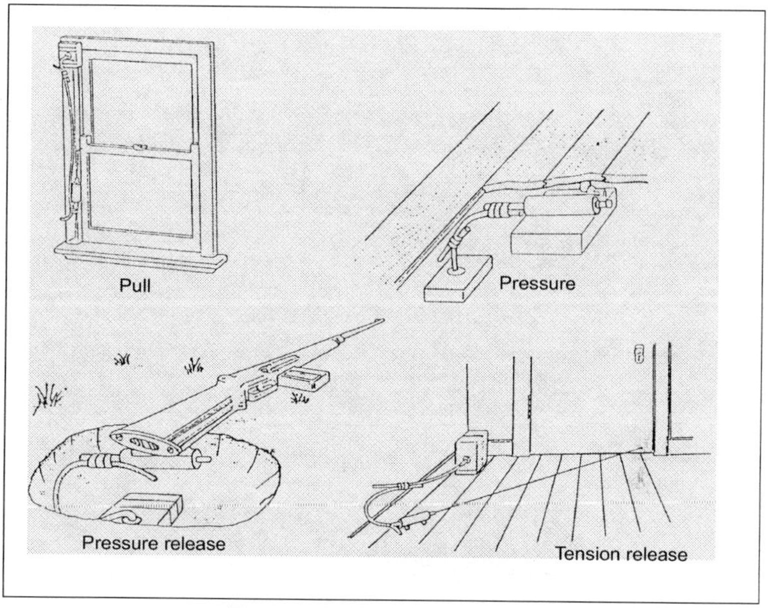

Figure 13-2. Methods of actuation

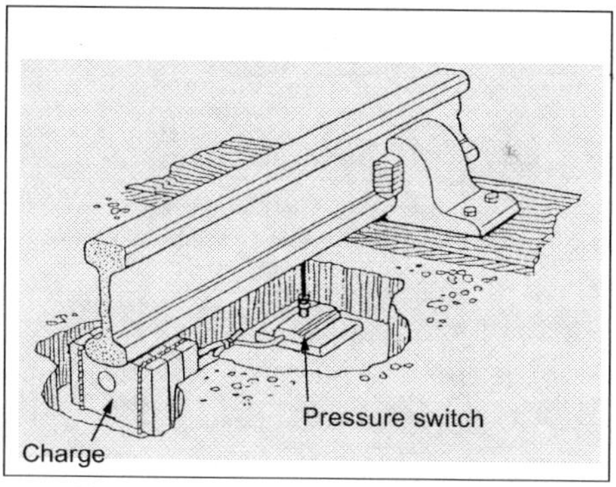

Figure 13-3. Remotely connected traps

- Ensure that the two detonating cords are not touching but can be conveniently connected when necessary.
- Remove any materials and other signs of laying, and fully camouflage the area.

- Arm the FD by removing the positive safety pin.
- Arm the trap by connecting the two detonating cords.

DIRECT

Follow the procedures listed below when assembling a nonelectric, directly connected trap using an M142 FD (similar to the illustration in Figure 13-1, page 13-6):

- Design the trap and collect necessary materials.
- Test the M142 FD.
- Prepare and position the charge for coupling to the FD, and prepare the explosive to receive a blasting cap.
- Prepare the coupler and insert it into the charge.
- Prepare the M142, set it to operate in the desired manner, and remove the round- or square-headed pin.
- Remove any materials and other signs of laying, and fully camouflage the area.
- Arm the trap by removing the positive safety pin.

DANGER
Do not attempt to further camouflage the area after the trap is armed.

PLANNING, SETTING, AND RECORDING

TIMELINESS

The setting-party commander must be given all the information and sufficient time to carry out a reconnaissance. This allows his plan and the soldier's briefing to be as complete as possible.

ORDERS AND BRIEFING

Once the commander has finished the reconnaissance, he makes a detailed plan, prepares the order, and briefs soldiers. The following points are covered:

- Enemy intelligence.
- Number and types of traps to be set.
- Location and design of traps (precise or general).
- Tasking and allocation of areas.
- Recording, marking, and arming procedures.
- Control measures.
- Time requirements.
- Material and equipment availability.
- Rehearsal and equipment preparation.

REHEARSAL

Whenever possible, rehearse laying procedures. Thoroughly exercise control and safety measures so that any flaws in the system will be discovered. Lay practice traps and carry out arming and recording procedures. If you cannot rehearse at the trap location, hold the rehearsal in a similar area.

ORGANIZATION AND PROCEDURE

Setting booby traps is a dangerous task and it must be carefully controlled. Exact drills cannot be prescribed, but the following procedures should be followed as closely as possible:

- Control point. The commander must establish a control point upon arrival at the trap area. The control point forms a headquarters and a material holding area. Also, safe routes start at the control point. If the rehearsal is conducted at another area, assemble all the troops at the control point before setting starts. At this time, relate the ground to the plan, confirm control measures, and check FDs and equipment for serviceability and adequacy.
- Control of parties. Each setting party consists of one or two men. Clearly define the area where each setting party will work (use tape, if necessary). Position traps so that accidental detonation will not cause friendly casualties. Strictly control the entry to areas where traps are being set. In most situations, only the commander, setting parties, and a recorder enter areas. Troops who are not assigned tasks should remain at the control point unless otherwise authorized by the commander.
- Marking. As with nuisance minefields, mark booby-trapped areas until they are evacuated. When possible, use standard booby-trap signs (Figure 13-4). When signs are not available, use temporary markings (any barrier or fence) to clearly indicate a booby-trapped area. Remove the temporary marking after the area is evacuated and before the enemy reconnoiters the area.

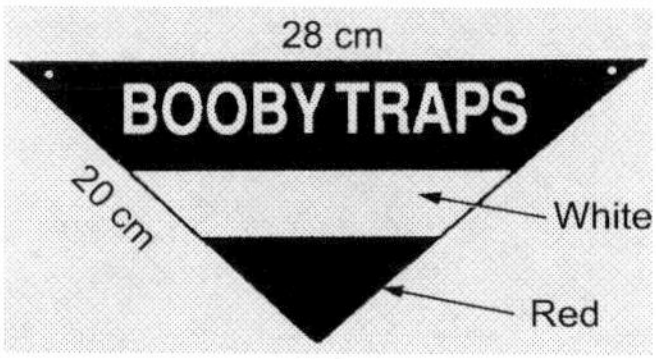

Figure 13-4. Standard booby-trap sign

- **Setting.** Setting parties transport required explosives and material to the site. They do not arm FDs or traps until ordered to do so. If the setting party consists of one man, he remains with the trap until it is armed or until the commander directs him to leave. In a two-man party, one man returns to the control point and reports the completion of the task, and the other man remains with the trap. A setting party enters and leaves its area by the same route. If the route is not clearly apparent, personnel lay tapes to guide the commander and the recorder to the area.

- Inspection and arming. The setting-party commander inspects traps for safety and camouflage before giving the order to arm them. He devises an arming plan that enables the setting parties to withdraw to the control point in a safe and logical manner. For example, in open areas, start arming at the furthest mine from the control point; in buildings, arm from the top down.

Reporting and Recording

Booby traps do not distinguish between friend and foe. Correct reporting and recording procedures must be observed to avoid friendly casualties. The procedures for recording booby traps are the same as for nuisance mines (see Chapter 8). When traps are set in areas occupied by the enemy, they are not recorded. In this case, it is sufficient to note their location and type in after-action reports.

Tactical Reports

Information about booby-trapped areas must be provided in stages from the time they are established until they are cleared. This procedure allows up-to-date information to be annotated on operational maps at all concerned units and headquarters (see Table 13-1).

Booby-Trap Records

The setting-party commander compiles records for all booby-trapped areas on DA Form 1355 (Figures 13-5a and 13-5b, pages 13-12 and 13-13). Strictly observe the instructions printed on the back of the form. Booby-trap records provide detailed information about the composition of an area, and they allow the area to be cleared quickly, with minimum casualty risk. They contain complete information on the number and types of traps set, as well as accurate and precise details on the location and design of individual traps. If a single record is not large enough to record all the traps set in a definable area, clearly note this fact and reference serial numbers of adjacent records. Submit the completed DA Form 1355, in triplicate, to the next higher headquarters.

Table 13-1. Tactical reports

Report	By Whom and When Initiated	To Whom Forwarded	Details to be Included
Intention to set (required)	Division commander, on deciding to set booby traps	• Corps HQ • Controlling HQ	• Purpose • General area to be booby-trapped (by grid coordinates) • Estimated number and types of traps • Estimated starting date and time • Estimated completion date and time
Siting (optional)	• Setting unit, on completion of reconnaissance • Division ordering setting, only if materially different from intention-to-set report	• Division ordering setting • Next higher HQ	• Boundaries of booby-trapped area (by grid coordinates) • Number and approximate positions of safe routes, if any • Estimated number and types of traps
Start of setting (required)	• Setting unit, when setting starts • Division ordering setting	• Division HQ ordering setting • Next higher HQ	• Time setting started • Estimated time of completion • Number and types of traps being set • Any change to the detail given in the siting report, and whether or not it was a result of orders
Completion (required)	• Setting unit, when setting is completed • Division HQ ordering setting	• Division HQ ordering setting • Next higher HQ	• A trace on the largest scale map available, showing the boundaries of the trapped area • Full details of the area and buildings trapped (with sketch maps) • Full details of traps set, including design and location • Full details of all safe routes • Total number and types of traps • Time and date of completion
Changes (required)	Division HQ ordering change	• Next higher HQ • Any other unit holding information on existing traps	Full details of change

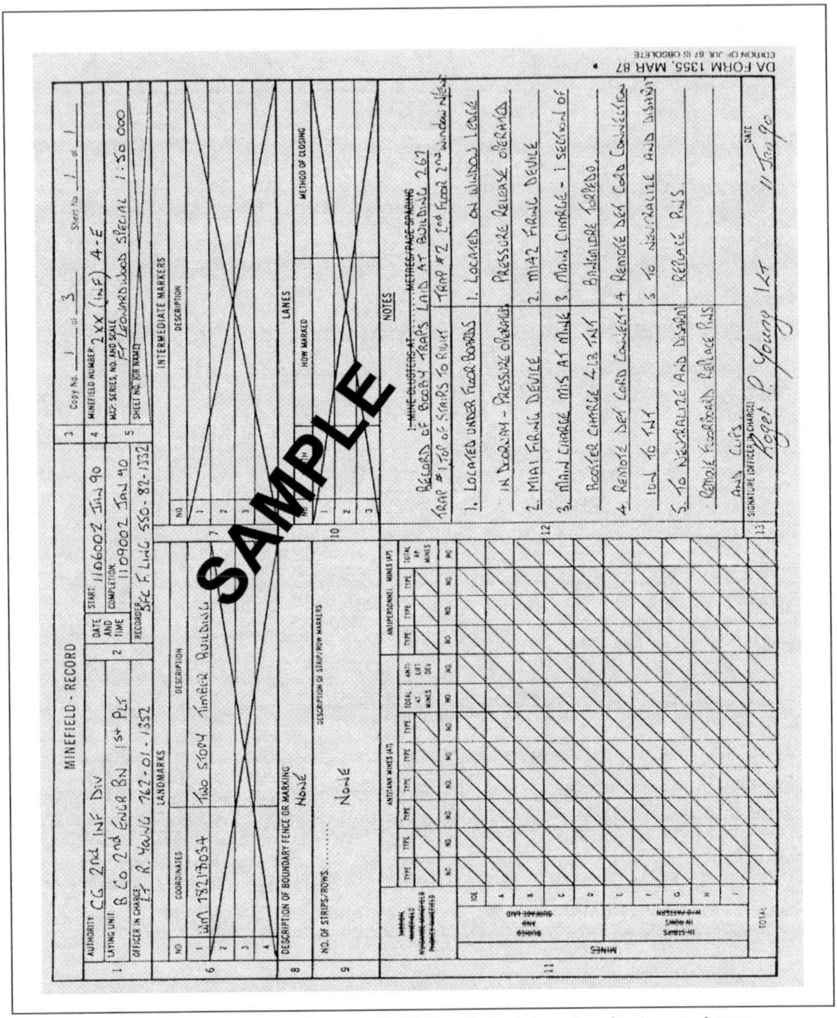
MINEFIELD - RECORD

1 AUTHORITY: CG 2nd INF DIV
LAYING UNIT: B Co 2nd ENGR BN 1st PLT
OFFICER IN CHARGE: LT R. YOUNG 262-01-1352

2 DATE AND TIME: START: 1106002 JAN 90; COMPLETION: 1109002 JAN 90
RECORDER: SFC F. LING 550-82-1332

3 Copy No. 1 of 3 Sheet No. 1 of 1
4 MINEFIELD NUMBER: 2XX (INF) A-E
MAP: SERIES, NO. AND SCALE: FT LEONARD WOOD SPECIAL 1:50 000
5 SHEET NO. (OR NAME)

6 LANDMARKS

NO	COORDINATES	DESCRIPTION
1	WM 15217034	TWO STORY TIMBER BUILDING
2		
3		
4		

7 INTERMEDIATE MARKERS

NO	DESCRIPTION
1	
2	
3	

8 DESCRIPTION OF BOUNDARY FENCE OR MARKING: NONE

9 NO. OF STRIPS/ROWS .... DESCRIPTION OF STRIP/ROW MARKERS: NONE

10 LANES

NO	WIDTH	HOW MARKED	METHOD OF CLOSING
1			
2			
3			

11 MINES: ANTITANK MINES (AT) / ANTIPERSONNEL MINES (AP)

12 NOTES

~~MINE CLUSTERS AT~~ .... METRES/PACE SPACING
RECORD OF BOOBY TRAPS LAID AT BUILDING 261

TRAP #1 TOP OF STAIRS TO RIGHT
1. LOCATED UNDER FLOOR BOARDS
IN DOORWAY - PRESSURE OPERATED
2. M1A1 FIRING DEVICE
3. MAIN CHARGE M15 AT MINE
BOOSTER CHARGE 1/2 lb TNT
4. REMOTE DET CORD CONNECT 1 lb TO TNT
5. TO NEUTRALIZE AND DISARM REMOVE FLOORBOARDS. REPLACE PINS AND CLIPS.

TRAP #2 2nd FLOOR 2nd WINDOW N/E
1. LOCATED ON WINDOW LEDGE
PRESSURE RELEASE OPERATED
2. M142 FIRING DEVICE
3. MAIN CHARGE - 1 SECTION OF BANGALORE TORPEDO.
4. REMOTE DET CORD CONNECTION
5. TO NEUTRALIZE AND DISARM REPLACE PINS.

13 SIGNATURE (OFFICER IN CHARGE): Roger P. Young 1LT DATE: 11 JAN 90

DA FORM 1355, MAR 87
EDITION OF JUL 67 IS OBSOLETE

SAMPLE

Figure 13-5a. Sample DA Form 1355 (front side) for a booby-trapped area

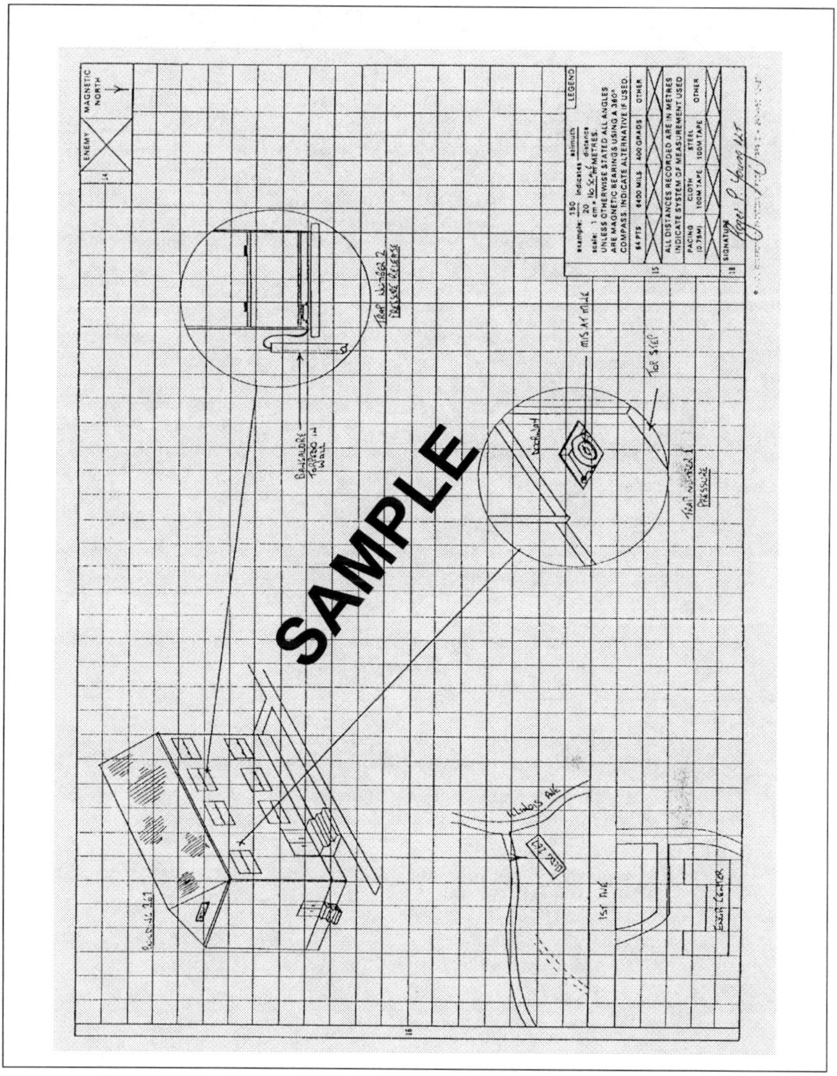

Figure 13-5b. Sample DA Form 1355 (inside) for a booby-trapped area

SECTION II. CLEARING BOOBY TRAPS

Soldiers must be aware of the threat that is presented by booby traps. They must also receive sufficient training to recognize locations and items that lend themselves to booby-trapping. Avoid overemphasis and strike a balance between what is possible and what is probable. For example, in medium-level operations, overemphasis of the booby-trap threat can be counterproductive by slowing momentum. This causes casualties that might otherwise have been avoided. In low-intensity conflicts where the use of booby traps and improvised explosive devices are probably widespread, training must be given high priority and emphasized at all levels.

SITES

Although many booby-trapped sites are similar, the items selected, the reason for their use, and the scale of the threat are quite different. In medium-level conflicts, booby traps are mainly used in recently contested areas. The items selected and the reasons for using traps are the same as those taught to our troops. Therefore, by anticipating the presence of traps, it is possible to isolate and bypass trapped areas. If this is not practicable, you can plan countermeasures, such as avoiding convenient and covered resting places along routes where mines can be located. At bridge or ferry sites that cannot be avoided, you must ensure that they are free of traps.

In low-intensity conflicts where booby traps are used to cause casualties, delays, or disruptions, no items or areas can be considered safe. Quality collective training in booby-trap awareness is necessary for all units. Rapidly disseminate booby-trap incident reports to all levels. This allows personnel to develop an understanding of the enemy's method of operation and a feel for what might or might not be targets.

SAFETY

The following rules and safety procedures can save your life and the lives of others—learn and remember them:

- Suspect any object that appears to be out of place or artificial in its surroundings. Remember, what you see may well be what the enemy wants you to see. If you did not put it there, do not pick it up.
- Examine mines and booby traps from all angles, and check for alternative means of firing before approaching them.
- Ensure that only one man works on a booby trap.
- Do not use force. Stop if force becomes necessary.
- Do not touch a trip wire until both ends have been investigated and all devices are disarmed or neutralized.
- Trace trip wires and check for additional traps located along and beneath them.
- Treat all parts of a trap with suspicion, because each part may be set to actuate the trap.

- Wait at least 30 seconds after pulling a booby trap or a mine. There might be a delay fuse.
- Mark all traps until they are cleared.
- Expect constant change in enemy techniques.
- Never attempt to clear booby traps by hand if pulling them or destroying them in place is possible and acceptable.

INDICATIONS

Successful detection depends on two things—being aware of what might be trapped and why, and being able to recognize the evidence of setting. The first requirement demands a well-developed sense of intuition; the second, a keen eye. Intuition, like mine sense, is gained through experience and an understanding of the enemy's techniques and habits. A keen eye is the result of training and practice in the recognition of things that might indicate the presence of a trap. The presence of booby traps or nuisance mines is indicated by—

- Disturbance of ground surface or scattered, loose soil.
- Wrappers, seals, loose shell caps, safety pins, nails, and pieces of wire or cord.
- Improvised methods of marking traps, such as piles of stones or marks on walls or trees.
- Evidence of camouflage, such as withered vegetation or signs of cutting.
- Breaks in the continuity of dust, paint, or vegetation.
- Trampled earth or vegetation; foot marks.
- Lumps or bulges under carpet or in furniture.

DETECTION

Detection methods depend on the nature of the environment. In open areas, methods used to detect mines can usually detect booby traps. Look for trip wires and other signs suggesting the presence of an actuating mechanism. In urban areas, mine detectors are probably of little use. You have to rely on manual search techniques and, if available, special equipment.

CLEARING METHODS

The method used to neutralize or disarm a trap depends on many things, including time constraints, personnel assets, and the type of trap. Remember, a trap cannot be considered safe until the blasting cap or the detonating cord has been removed from the charge. This is your first objective and is particularly important for electric traps, which may contain a collapsing circuit.

Use the safest method to neutralize a trap. For example, if the FD and the detonating cord are accessible, it is usually safer to cut the detonating cord. This method does not actuate the trap, but inserting pins in the FD might.

COMBAT CLEARANCE

Clearing booby traps and nuisance mines in AOs is done primarily by engineers. Therefore, engineer advice is important during the planning stages of any operation where booby traps are likely to be encountered. Intelligence regarding the possible presence and types of traps must be provided to engineer units as soon as possible. This allows the unit to take necessary action and provide relevant training. Clearance of booby traps cannot be undertaken as a secondary task, because engineer clearing teams might require protection that necessitates combined arms training. Before engineer planning can start, the staff provides commanders with the latest intelligence information and, if possible, the following information:

- Amount of clearance required.
- Acceptable damage.
- Time requirements.
- Availability of special equipment.
- Security requirements.

Intelligence information regarding the nature, type, and location of traps has a direct bearing on the number of clearing parties necessary and the degree of protection required. For example, in built-up areas where traps have to be cleared in buildings that offer protection from enemy fire, direct protection is usually provided by the normal combat situation. On the other hand, in open areas where clearing parties may be required to clear traps covered by direct enemy fire, protection arrangements must be more specific.

Engineer commanders must be aware of the time needed to clear various types of traps in differing terrain situations. Remember, increasing the number of clearance parties may not necessarily reduce the time required to clear traps. This is particularly true when traps are set close together or set deep along a narrow front that is the only available route.

Initially, clear areas of immediate tactical importance and traps that present a specific threat. For example, clear only the portion of a building required for observation and those traps presenting an immediate hazard. This enables clearing parties to concentrate on other areas of tactical importance.

Clearing traps by hand is the only way that damage can be avoided and security guaranteed. When it is vital to avoid equipment or structure damage, consider using available EOD assets. It is often necessary to balance the requirement to remain silent and avoid damage with the requirement to maintain momentum.

When traps are being cleared in direct-support combat situations, they are normally dealt with by using unit resources and locally manufactured or acquired aids. Specified equipment is rarely available. Equipment varies with the situation but usually consists of selected items from Table 13-2. In areas with a high incidence of booby traps, assemble and reserve special clearing kits.

Table 13-2. Clearing equipment

Item	Remarks
Unit Equipment	
Mine detector	
Mine probe/knife/bayonet	
Eye protection	
Body armor	Flak jacket, Kevlar helmet
Booby-trap signs	
Mine marking tape	
Mine markers	Chalk, felt-tipped pens
Electrician's adhesive tape	
Flashlight	
Tape measure	
Mine marking cones	
Trademan's tools	Saws, pliers, hammers, screwdrivers
Self-developing camera and film	
Locally Manufactured/Acquired	
Trip-wire feeler	
Grapnel and rope	
Hand mirrors	
Pins, wire, nails	For use as safety pins
Meat hooks	For use as block and tackle

CLEARANCE IN SECURE AREAS

POLICY AND PLANNING

Formal clearing procedures must be followed in secure areas. This type of clearance can be done by engineers or EOD personnel. It is subject to time constraints when traps or suspected traps are located in urgently required installations or facilities, such as supply dumps or telecommunication centers.

Commanders reconnoiter their areas of responsibility, make detailed plans, prepare orders, and brief their men. The following points are covered:

- Intelligence.
- How the operation is to be conducted (include acceptable damage and methods of clearing).
- Action on finding traps.
- Marking.
- Disarming procedures.
- Tasking and allocation of areas.
- Equipment available.
- Control measures.

- Time requirements.
- Rehearsal and equipment preparation.

Control Point

The operation commander establishes a control point upon arrival at the area to be cleared. The control point functions as the headquarters and the material holding area. It is also the point from which all clearance starts. Its suitability might not have escaped the enemy's notice, so it must be cleared before it can be used.

Control and Size of Parties

The size of clearing parties varies depending on the location being cleared. The following rules apply:

- Each party is controlled by an NCO.
- Only one party works in a particular subarea.
- The distance between parties is sufficient so that detonation in one area does not endanger persons in other areas.
- The NCOIC of each party is in visual, radio, or voice contact with every person in his party.
- Only one person works on a booby trap.

Marking

Booby traps and nuisance mines can be placed in diverse locations. It is impossible to standardize a method for marking areas, individual traps, or safe routes. Any form of prominent, permanent marking can be used.

Uncleared Areas

The perimeter marking of uncleared areas can take any clearly recognized form. Standard minefield or booby-trap signs suspended from a single-strand fence are recommended. The spacing of the signs is the same as standard minefield marking (see Chapter 2). As the area is cleared, the fence should be progressively moved.

Individual Traps

Because booby traps can have more than one means of actuation, do not attempt to place a cone or other marker over any part of a trap. Use sufficient signs to ensure that the trap can be detected and accurately located. In buildings, clearly mark rooms containing traps and, where possible, indicate the exact location of traps.

Internal Marking

The internal marking system depends on the area being cleared. One good system is to divide the total area into subareas, clear and mark safe lanes between the subareas, and then use the lanes as safe routes.

Clearing of Open Areas

Roads, road shoulders, bridges, obstacles, and other structures must be cleared in open areas. The main threat comes from nuisance mines, so regard each potential site as a nuisance minefield and use established minefield clearing procedures. After the site is cleared, adapt the drill to suit the situation.

CLEARING OF BUILDINGS

Buildings are excellent locations for booby traps. Booby traps are concentrated inside the building, so carefully organize search and clearing procedures. In most cases, buildings are required for use, and excessive damage must be avoided.

A two-man clearing party can clear an average-size structure. No one else should enter the building until it is cleared. In very large buildings, two or more parties can be employed, if they work as far apart as possible and have clearly defined boundaries. As a general rule, two walls or two floors is the minimum distance between parties.

EXTERIOR RECONNAISSANCE AND ENTRY

Before approaching a building, check the surrounding area for booby traps and nuisance mines. Carry out a reconnaissance to determine the point of entry and clear the way to it. When selecting the point of entry, consider the following:

- Doorways. Never consider doorways to be safe unless the door is fully open and the entrance is clear. If a house is built on a concrete slab, it is not likely to have a pressure FD in the floor.
- Windows. Windows are excellent locations for booby traps. Pay particular attention to the ground outside and the floor inside because they are classic sites for pressure FDs. Use the following procedures if access must be gained through a window:
  - — Pull the window if it is unlocked and can be moved. If it is locked, use a small charge or a heavy object to break the glass.
  - — Select a stationary window that cannot be opened versus a window that can be opened when both types are available, because the window that can be opened is more likely to contain a booby trap.
  - — Deal with blinds and curtains in a manner similar to procedures used for windows.
  - — Use a mirror to examine inside the window frame before entering the building.
- Mouseholes. If you decide not to enter the structure through a door or a window, use explosives to make a mousehole in the wall, roof, or floor. This offers a remote, safe method of creating an access point, but it can also detonate nearby traps. Exercise caution if entering through the roof of a two-story building, because it is far more difficult to clear booby traps when going down steps than it is when going up them.

SEARCH TECHNIQUES

It is impossible to establish a search drill that is suitable for use in all buildings. It is essential, however, for each working party to develop its own drill and follow it. Domestic dwellings should be searched in the following order:

1. Floors and furniture.
2. Walls (including doors, windows, fireplaces, and cupboards).
3. Fittings (including light switches and pictures).
4. Ceilings.

Mark each area or item *SAFE* as it is cleared. This can be conveniently done by using chalk or a felt-tipped pen.

Use the following techniques and precautions when searching buildings. They can save your life—learn and remember them!

- Check both sides of a door before opening it. Do this by drilling a hole through the door and using a mirror to check the other side. You can also check or open doors by pulling or blowing the lock and hinges with a small charge.
- Examine floor coverings for signs of disturbance. The presence of FDs is often indicated by loose floorboards, bulges or tears in carpets, or loose tiles.
- Use a pulling cable to move furniture and to open cupboard doors and drawers.
- Check upholstered furniture and beds by remotely dropping a heavy object onto them.
- Treat every switch with suspicion, since electrical wiring provides a ready-made circuit for booby traps. To explode all the traps connected to the normal power supply, disconnect the power at the fuse board, turn all the switches on, and then reconnect the power. Repeat the procedure with the switch turned off in case the switch has been reversed. Remember, this procedure will not disclose traps that use a battery. Exercise caution when using switches, even if the power is disconnected.
- Leave doors, windows, cupboards, and drawers open after they have been cleared.
- Mark all routes, areas, and items that have been cleared.
- Check plumbing by remotely turning on all water taps and allowing the water to run for at least one minute.
- Check toilet tanks before flushing.
- Beware of light-sensitive devices in dark places (closets, cabinets, basements, attics, chimneys).
- Allow at least 30 seconds for an explosion after pulling anything, because there might be a delay fuse.

Clearing Installations and Facilities

Clearing by hand is necessary in installations and facilities (fuel dumps, ammunition dumps, electric substations) where an explosion could result in the loss of resources. In other situations, the item's importance or the resulting damage might not be obvious. For example, a small charge placed against the control valves of a dam or against the main cable entering a telephone exchange results in unforeseen damage that can take days to repair. Therefore, you should seek a specialist's advice, if possible, when clearing booby traps in industrial areas and unfamiliar locations.

Clearing Obstacles

If an enemy has time to create obstacles, he also has time to set booby traps and lay nuisance mines. The obstacle itself is usually clear of traps to encourage a false sense of security and lead troops into more dangerous areas. Therefore, regard all obstacles as booby-trapped until proven otherwise. The simplest, safest way to deal with movable obstacles is to pull them. Before an obstacle can be pulled, you must first clear the area from which the pull will be made.

Clearing Secure Areas

When clearing secure areas and time is not a major factor, use specialized clearance equipment as much as possible. The following equipment might be available for use:

- Cameras. Cameras have a wide range of applications. They can be used with different types of film, such as infrared and ultraviolet, to disclose evidence that is indiscernible to the naked eye. For example, infrared photography reveals differences in the heat emitted by objects and can often disclose recent digging and buried or concealed objects.
- Explosive detector dogs (EDDs). Although EDDs can detect minute quantities of explosives and the presence of trip wires, they are trained to detect the charge and not the FD. This extremely limits their usefulness in detecting booby traps. They also tend to become confused if the area contains explosive odors other than those emitting from booby traps.
- Electronic countermeasures. Electronic countermeasures can be used to explode electric booby traps and to prevent remotely controlled, improvised explosive devices from being detonated by radio.
- Robots. In their simplest form, robots can be used to detonate or neutralize booby traps. More sophisticated models can be remotely controlled to carry out simple tasks, such as videotaping or cutting wires.
- Body armor.
- Electric meters.
- X-ray equipment.

CLEARANCE METHODS

- Pulling. This method uses a grapnel and a rope to pull the trap. It is used when the resulting damage is acceptable. It is the safest method and is particularly applicable to traps set in open areas. Do not disturb any part of a booby trap when placing the grapnel and pulling the cable. Carefully select the site from where the pull is to be made because it might be mined or trapped. When a booby trap is pulled and does not explode, wait at least 30 seconds before approaching it in case delay devices have been used. Disposal of unexploded traps depends on their condition when inspected. The procedure for pulling booby traps is similar to that for pulling mines (see Chapter 11).
- Trip wires. Check the area for AP devices before proceeding. Place a grapnel hook as close as possible to the trip wire. Do not touch the trip wire until the pulling party is in a covered area.
- Pull and release. Pull away objects that conceal and operate pull and release mechanisms.
- Pressure mechanism. Pull pressure mechanisms from under objects that conceal and operate them. If this is impossible, blow them in place. In many cases, it might be easier to pull the charge rather than the FD. Take extreme care when attempting this, because additional mechanisms are often concealed in or under the main charge.
- Destroying in place. When destroying booby traps in place, explode a small charge near the booby trap's charge. Again, use this method only if damage from the explosion is acceptable. When it is impossible to place the explosive close enough to ensure detonation of the main charge, carefully place it alongside the mechanism. Do not assume the main charge is safe to handle just because the mechanism has been destroyed. Actuate pressure mechanisms by suspending one-half pound of explosive above the pressure plate.
- Clearing by hand. This method involves neutralizing, disarming, removing, and disposing of traps without causing damage. It is extremely hazardous and should only be used when pulling or destructing traps in place is impossible or unacceptable. Clearance should only be conducted by EOD personnel or experienced engineers. Carefully examine all aspects of the trap before deciding how to clear it.
- Explosive line charge. Using this device produces quick results when only a narrow path is required through a booby-trapped area. It gives clearance for the same distance to either side, only where it is in contact with the ground.
- Armor. This method is used where traps with small charges (designed as AP devices) are located in open areas. Armored vehicles track back and forth over the area. This shortens the clearing time with little risk of casualties.

- Fire. If traps are set in grass or dense vegetation, fire can be used to burn away camouflage material and expose traps. In most cases, sufficient heat is generated to burn or explode the traps. Unexploded traps are considered extremely sensitive and must not be cleared by hand.

IMPROVISED TRAPS

In low-intensity conflicts, there is a high incidence of improvised mines and booby traps. It is impossible to provide a complete list of FDs and improvised demolition material that can be used. However, most improvised traps are variations of those described below.

An electric FD requires a current to be passed between two contacts. The ways in which this can be achieved are limited only by the imagination. Examples of simple electric FDs that can be manufactured using household items or appliances are shown in Figure 13-6, page 13-24. Examples of improvised mechanical FDs are shown in Figures 13-7 and 13-8, page 13-25.

The simplest and most accurate method of incorporating delay is to use some form of clock or timer. However, any mechanism or chemical reaction that takes a measurable time to complete its function can be used. Examples of simple improvised delay devices are shown in Figures 13-9 and 13-10, page 13-26.

NONEXPLOSIVE TRAPS

Nonexplosive traps are typically used in tropical or rain-forest regions. Ideal construction materials abound, and concealment in surrounding vegetation is relatively easy. No prescribed procedures exist for clearing nonexplosive traps. Each trap must be cleared according to its nature.

PUNJI

The punji (Figure 13-11, page 13-27) is one of the simplest traps. It is normally made from locally available material, such as sharpened stakes, nails, or heavy-gauge wire. It is placed in concealed pits or in places that give cover from fire. The enemy normally smears the spikes or cutting edges with excrement or poison. After clearing a punji, sterilize cuts and abrasions immediately and see a medical officer as soon as possible.

CLOSING TRAP

The side-closing trap (Figure 13-12, page 13-27) consists of two wooden slats that are studded with spikes. The slats slide along a pair of guide rods controlled by heavy bands. When the prop holding the slats apart is dislodged, the slats spring together and implant the spikes into the portion of the body passing between them.

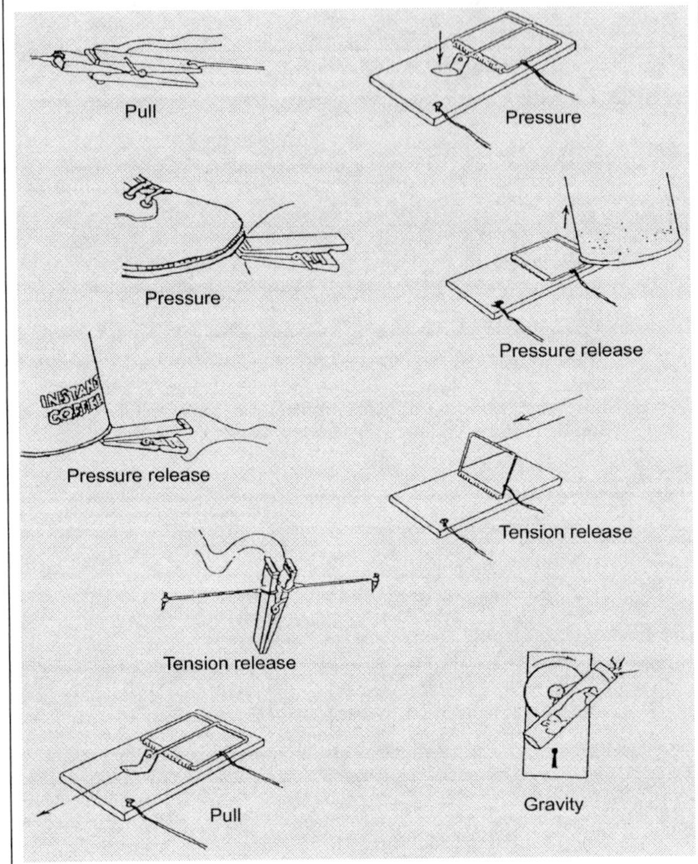

Figure 13-6. Improvised electrical FDs

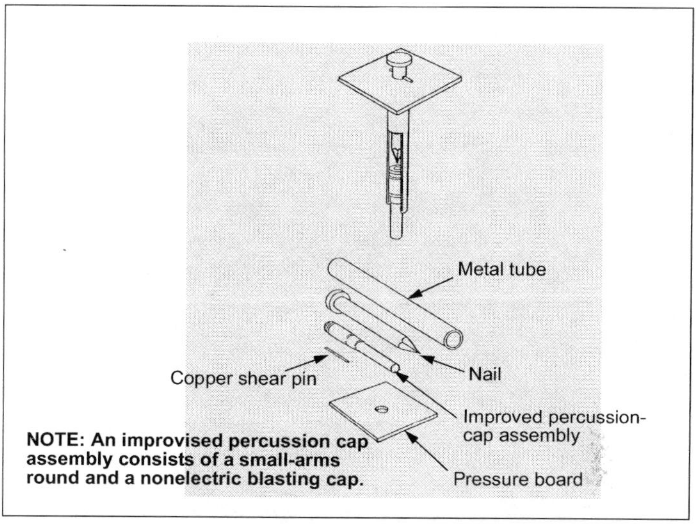

Figure 13-7. Improvised nonelectric FDs (shear-pin operated)

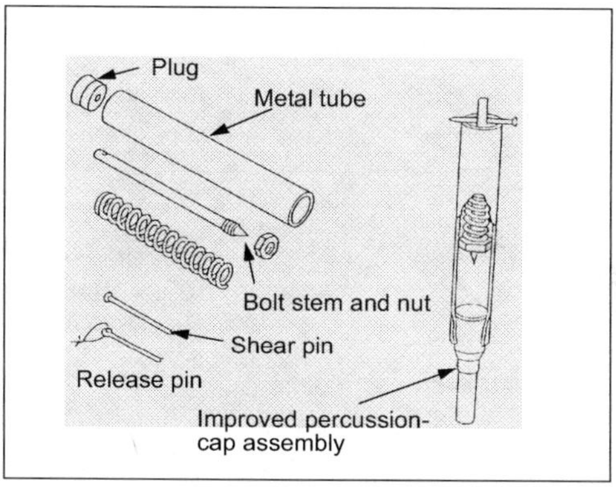

Figure 13-8. Improvised nonelectric FDs (spring-operated)

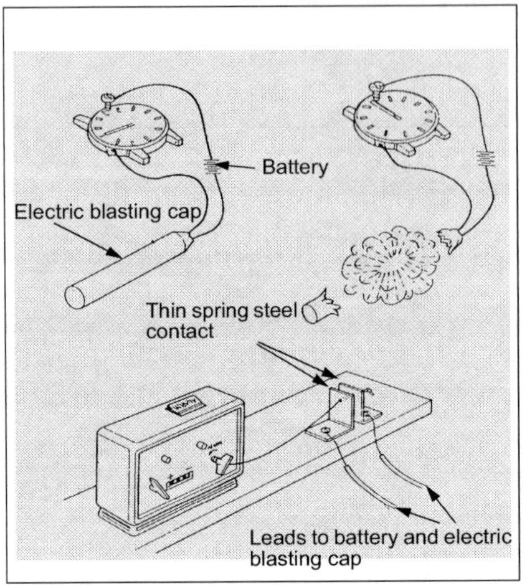

Figure 13-9. Improvised, electric delay devices

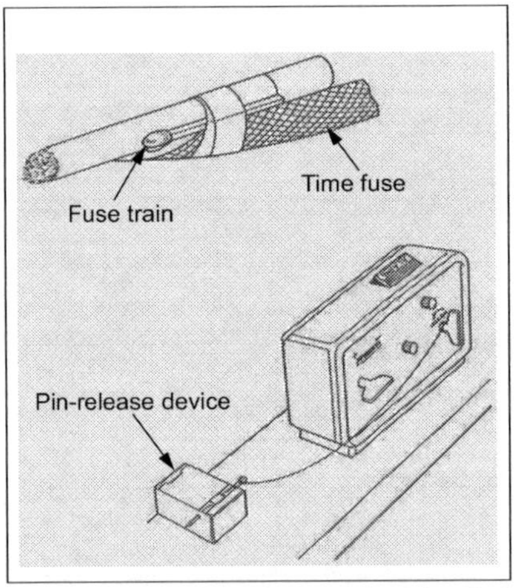

Figure 13-10. Improvised, nonelectric delay devices

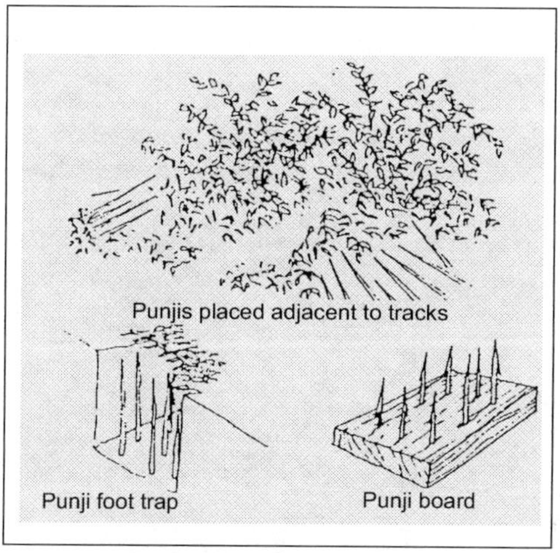

Figure 13-11. Typical punjis

Figure 13-12. Side-closing trap

Spike Board

The spike board (Figure 13-13) is used in a pit and consists of a treadle board with one end spiked. When a man steps on the board, the spiked end flies up and strikes him in the face or chest.

Figure 13-13. Spike board

Venus Flytrap

The Venus flytrap consists of a rectangular framework with overlapping barbs emplaced in a pit. The one shown in Figure 13-14 is made from a metal container that is sunk into the ground until the top is flush. It is then covered with grass and/or leaves. The barbs inflict injury, especially when the victim attempts to withdraw his leg from the trap. These traps are typically located on tracks and along road edges.

Figure 13-14. Venus fly trap

SECTION III. EXPEDIENT DEVICES

Expedient devices are constructed in the field with locally available material. They are employed against vehicles or personnel in the same manner as other mine systems. Expedient devices—

- Supplement a unit's low supply of conventional mines.
- Hinder reconnaissance, clearance, and neutralization of minefields.
- Create enemy attitudes of uncertainty and suspicion to lower morale and slow movement).

AUTHORIZATION

Because expedient devices have nonstandard design and functioning, take special precautions to protect friendly forces. Consider neutralization, disarming requirements, and adequate marking procedures. The use of expedient devices is restricted under the Convention of Conventional Warfare. Expedient devices have the same international restrictions as booby traps. The corps commander is the employment authority for expedient devices. He can delegate this authority to the division commander. If authority is given to use expedient devices, US personnel will adhere to the rules for international law that are applicable to armed conflict.

EMPLOYMENT AND CONSTRUCTION TECHNIQUES

If issued mines are not readily available on the battlefield, expedient devices can be manufactured in the field. Construction varies based on available materials and the ingenuity of the personnel who are fabricating the devices. Expedient devices pose a potential safety hazard to friendly forces—those who are constructing them and those who may later encounter them. Construction should be performed by personnel who are familiar with the materials being used. Innovative designs should be checked and tested before arming and emplacing the devices.

As a minimum, test the fusing mechanism separately to ensure that it functions as designed. Improper fuse operation is the most common cause of malfunction. Also, test the fuse and the firing chain (base charge, blasting cap, and detonating cord) without the main charge to ensure proper operation. Emplace the device after satisfactory performance of the firing mechanism. First, emplace heavy items (such as artillery shells) that are used as the main charge, and then add the firing mechanism. Take care when moving or emplacing expedient devices because their nonstandard manufacture and potentially faulty construction make them highly sensitive to jars and shocks. Construct devices at the emplacement site whenever possible.

Expedient devices are prepared in the field using standard US FDs, detonators, and demolition materials. All devices discussed in this chapter can be made to function electrically or nonelectrically using modernized demolition initiators (MDIs). AP devices must be command-detonated.

High-Explosive, Artillery-Shell Device

The HE, artillery-shell device (Figure 13-15) can be readily adapted to expedient mining. Remove the artillery fuse and replace it with a standard FD and a length of detonating cord or with an MDI blasting cap. If properly assembled, a destructor may also be used. If a destructor is not available, firmly pack the fuse well with composition C4 explosive and insert a length of knotted detonating cord or a blasting cap.

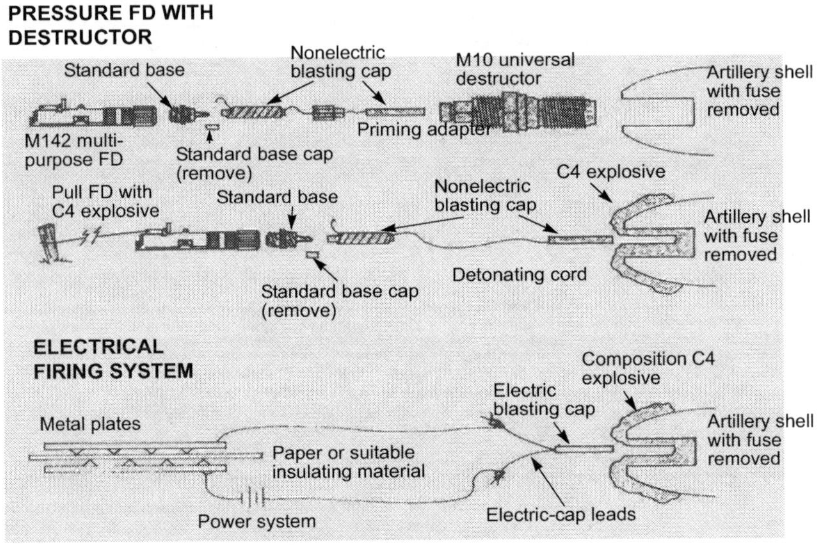

Figure 13-15. HE, artillery-shell device

The device can be activated by a variety of methods depending on the type of FD used. When MDI blasting caps are used, the device is command-detonated. The device can also be adapted to function electrically by adding an electric cap and a power source.

This device can be used as an AT or an AP device. When used as an AP device, it must be command-detonated.

NOTE: Use only serviceable US ammunition that has remained in the possession of US forces. Never use captured ammunition or UXO found on the battlefield. It may be armed, booby-trapped, or deteriorated.

PLATTER CHARGE

The platter charge (Figure 13-16) consists of a suitable container that is filled with uniformly packed explosive and placed behind a platter. The platter is metal (preferably round, but square is satisfactory) and weighs 1 to 3 kilograms. The explosive required is equal to the weight of the platter. The container may not be necessary if the explosive can be held firmly against the platter (tape can be used). The charge should be primed from the exact rear center, and the blasting cap should be secured with a small amount of C4 to ensure detonation.

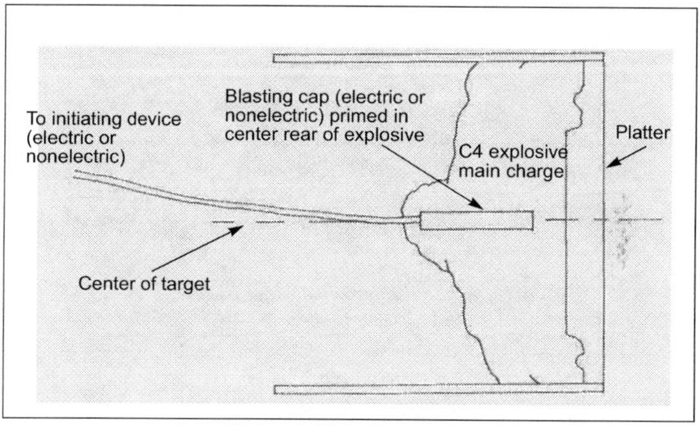

Figure 13-16. Platter charge

The charge should be aimed at the direct center of the target. The effective range (primarily a matter of aim) is approximately 35 meters for a small target. With practice, experienced personnel can hit a 55-gallon drum (a relatively small target) at 25 meters with about 90 percent accuracy.

The platter charge can be used as an AT or an AP device. When used as an AP device, it must be command-detonated.

IMPROVISED CLAYMORE

For the improvised claymore device (Figure 13-17, page 13-32), a layer of plastic explosive is attached to the convex side of a suitably dense, curved base (such as wood or metal). A hole must be made in the exact rear of the base. A blasting cap is placed in the hole to prime the device. Shrapnel is fixed to the explosive with a suitable retainer (cloth, tape, mesh screen).

The device must be command-detonated. Command detonation is best achieved with electrical priming or an MDI. A blasting device is attached to the electric cap via firing wires laid at least 50 meters from the device. Ensure that personnel have adequate cover when detonating the improvised claymore.

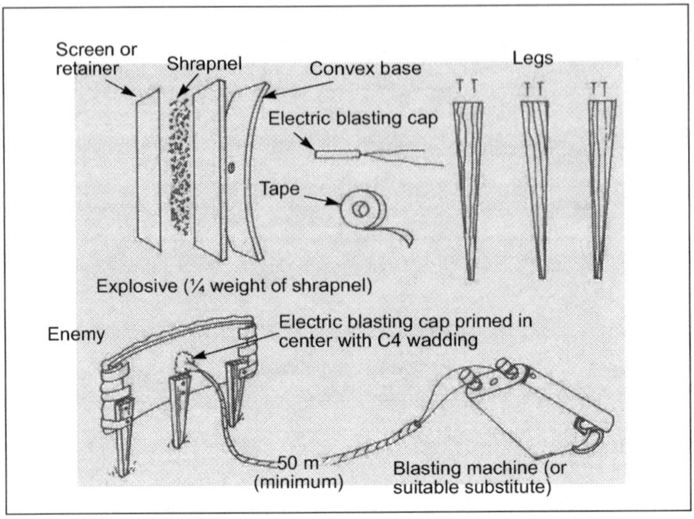

Figure 13-17. Improvised claymore device

Grapeshot Antipersonnel Device

Place shrapnel in the bottom of a cylindrical container to make a grapeshot AP device (Figure 13-18). The shrapnel is tamped and held in place with a suitable separator (wadding). Explosive (approximately one-quarter the weight of the shrapnel) is packed to a uniform density behind the wadding. The device is primed in the center of the explosive with an electric cap or an MDI.

NOTE: The United Nations Convention of Certain Conventional Weapons mandates that all fragment munitions produce fragments that are visible by X ray (such as metal or rock).

This device must be command-detonated. The explosive propels the shrapnel outward from the container. The grapeshot is very effective against personnel targets.

Barbwire Antipersonnel Device

The barbwire AP device (Figure 13-19) can be made directional by placing the wire against an embankment or a fixed object. This causes the force of the explosion to expel the barbwire fragments in the desired direction. One roll of standard barbwire is placed into position, and one block of C4 is placed in the center of the roll and primed. This device must be command-detonated.

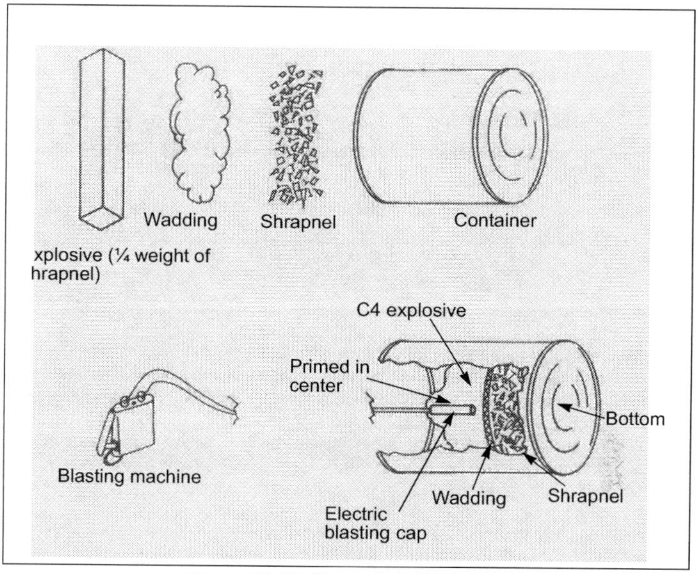

Figure 13-18. Grapeshot AP device

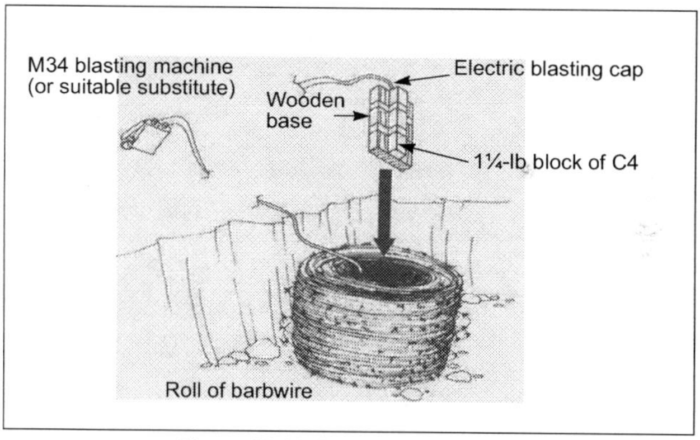

Figure 13-19. Barbwire AP device

Appendix A

Installation and Removal of US Mines and Firing Devices

This appendix provides installation and removal procedures for AT mines, AP mines, and FDs. The scope of this appendix is limited to US hand-emplaced mines that require manual arming.

WARNINGS

1. If there is a problem when performing any installation or removal step, notify the NCOIC.
2. If you hear a click when removing the safety clip or if the pressure plate snaps downward so it is level with the body of the mine, notify the NCOIC. DO NOT use the mine.
3. If the safety clip cannot be reinserted, notify the NCOIC.
4. DO NOT apply pressure to the pressure plate, tilt rod, or fuse at any time.
5. Before attempting to disarm and remove the mine, check for AHDs, damage, and malfunctions. If any of these conditions exist, notify the NCOIC. DO NOT attempt to disarm the mine.
6. If you feel a jar or hear a metallic click when removing the locking safety pin, stop and notify the NCOIC. The firing pin has gone forward and is resting on the positive safety pin. DO NOT remove the positive safety pin.
7. After removing the positive safety pin, proceed with extreme caution. The slider pin can detonate the mine if it is accidentally pushed in.
8. When attaching trip wires to the release-pin ring on the fuse, leave a little slack in the wires. This prevents pull on the release-pin ring, which could set off the mine when the safety pins are removed.
9. Ensure that the extension rod is vertical and is not tilted in any direction. A 20-degree tilt of the extension rod will detonate the mine.
10. Ensure that the safety fork moves freely. If there is pressure on the fork, DO NOT remove it.
11. DO NOT apply pressure to the pressure plate of the fuse when inserting it into the fuse well.
12. If the setting knob is difficult to turn, DO NOT force it; notify the NCOIC.
13. DO NOT adjust the setting knob while the detonator is in the detonator well.
14. If any cracks are noted in the plastic collar, slowly and carefully reassemble the stop and safety pin on the fuse. Carefully remove the extension rod and the fuse from the mine. Give the fuse to the NCOIC and replace it with a new fuse.
15. Before cutting loose trip wires, look at each end to ensure that there are no electric-producing devices that might initiate another system.
16. Arm and disarm all mines while in the prone position.

SECTION I. ANTIPERSONNEL MINES

AP mines can kill or incapacitate their victims. They can be fused by pressure, wire, or command detonation and contain a blast, bounding-fragmentation, or direct-fragmentation warhead.

M14

The M14 mine (Figure A-1) is a low-metallic, blast AP mine. It has a plastic body and an integral plastic fuse with a steel firing pin.

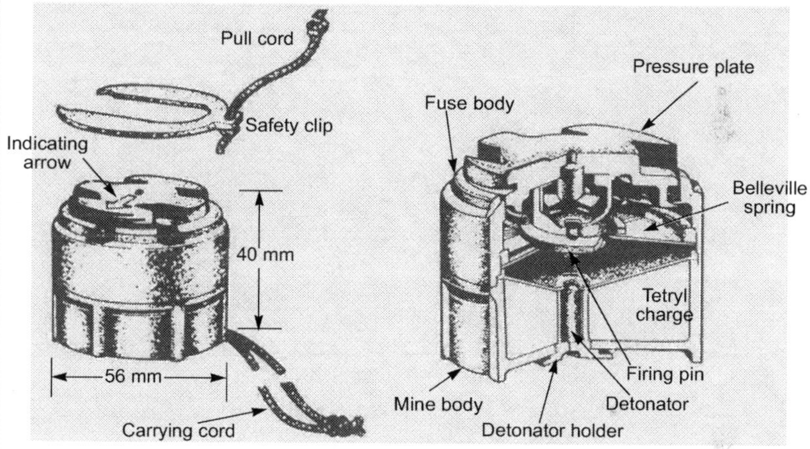

Figure A-1. M14 AP mine

CHARACTERISTICS

Main Charge	Diameter	Height	Weight	No Mines per Box	Weight per Box
Tetryl, 28.4 g	56 mm	40 mm	99.4 g (fused)	90	19.8 kg

- **Korea Only: The M14 is employed in tactical and nuisance minefields.**
- The M14's size allows for employment in large numbers, and rapid concealment is possible.
- The M14 is buried to prevent the target from knocking it over.

- The M14 is not designed to kill but to penetrate a boot or a foot.
- The M14 is difficult to detect because of plastic construction.
- The M14 requires an operating force of 11.5 to 13.5 kilograms to activate.

INSTALLATION

> **WARNING**
> **Emplace and remove the mine while in the prone position.**

- Inspect the mine.
  - Do not use the mine if it is dented, cracked, or damaged.
  - Use the M22 wrench (Figure A-2) to remove the shipping plug from the detonator well.

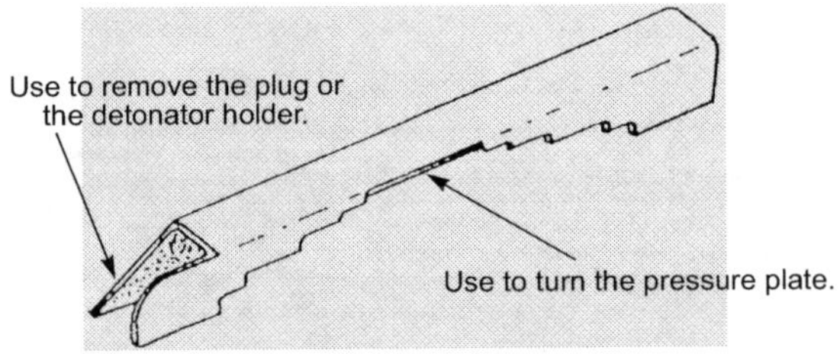

Figure A-2. M22 wrench

  - Inspect the position of the firing pin. Do not use the mine if the firing pin extends into the detonator well.
  - Inspect the detonator well for foreign material. If foreign material is present, remove it by carefully tapping the mine against the palm of your hand. If you cannot remove the debris, replace the shipping plug and do not use the mine.
- Test the pressure plate.
  - Ensure that the arrow is in the *SAFE* position.
  - Use the M22 wrench to turn the pressure plate from the *SAFE* position to the *ARMED* position (Figure A-3, page A-4).
  - Grasp the mine in one hand and remove the safety clip with the other hand. Listen for a click, indicating that the firing pin has dropped. Recheck the fuse well (Figure A-4, page A-4).
  - Replace the safety clip.

— Use the M22 wrench to turn the pressure plate back to the *SAFE* position.

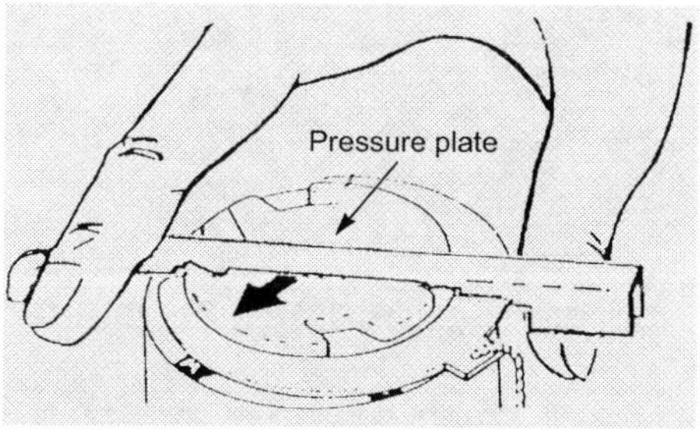

Figure A-3. M14 mine in *ARMED* position

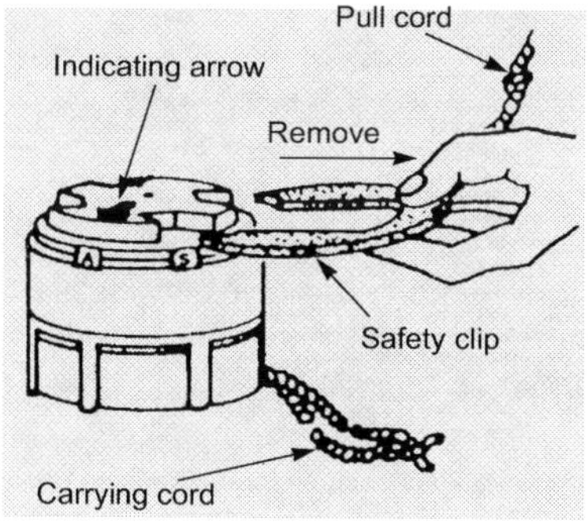

Figure A-4. Removal of safety clip

- Ensure that a metallic washer is attached (glued) to the bottom of the mine. If a washer is not attached, notify the NCOIC and do **not** use the mine. Apply silicone on the outside of the washer to prevent water damage, and screw the detonator into the well on the bottom of the mine (Figure A-5). **NOTE: Ensure that the gasket is tightly wedged between the detonator and the washer and that the washer is properly seated between the gasket and the body of the mine so that water cannot enter the mine.**
- Dig a hole to fit the mine.

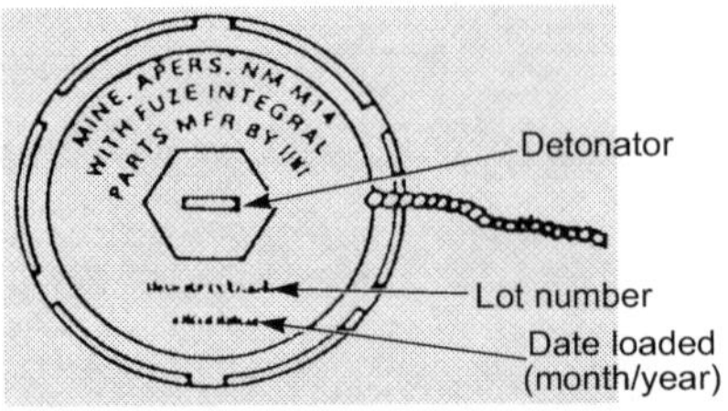

Figure A-5. Bottom view of M14 mine

- Dig a hole approximately 10 centimeters in diameter and deep enough (approximately 3.5 centimeters) so that the pressure plate extends above the ground.
- Inspect the ground at the bottom of the hole.
  > Ensure that the ground is hard enough to support the mine when pressure is applied to the pressure plate.

  > Place a flat object in the bottom of the hole for the mine to rest on if the ground is too soft. Allow additional depth for the objects.

- Arm the mine with an M22 wrench by turning the pressure plate to the *ARMED* position.
- Place the mine in the hole.
- Remove the safety clip carefully, while holding the mine body firmly in the hole.
- Camouflage the mine.

REMOVAL

- Disarm the mine.
  - Clear the soil away from the mine carefully.
  - Grasp the body of the mine firmly with one hand, and insert the safety clip with the other hand.
  - Use the M22 wrench to turn the pressure plate to the *SAFE* position.
- Remove the mine from the hole.
  - Turn the mine over, and carefully remove the detonator from the detonator well.
  - Screw the shipping plug into the detonator well.
  - Give the detonator to the NCOIC.

M16

The M16-series mine (Figures A-6, page A-6) is a bounding-fragmentation AP mine. Once actuated, the mine is propelled out of the ground (to a height of approximately 1.8 meters) and explodes. The mine consists of an M605 fuse, a propelling charge, and a projectile that are contained in a sheet-steel case. The fuse is screwed into the top of the case and extends through the center of the projectile to the bottom of the case, where the propelling charge is located. The remaining space inside the case is occupied by the projectile. Earlier versions of the mine are also available for issue. The principal difference between the old and new versions is the construction of the detonators and boosters.

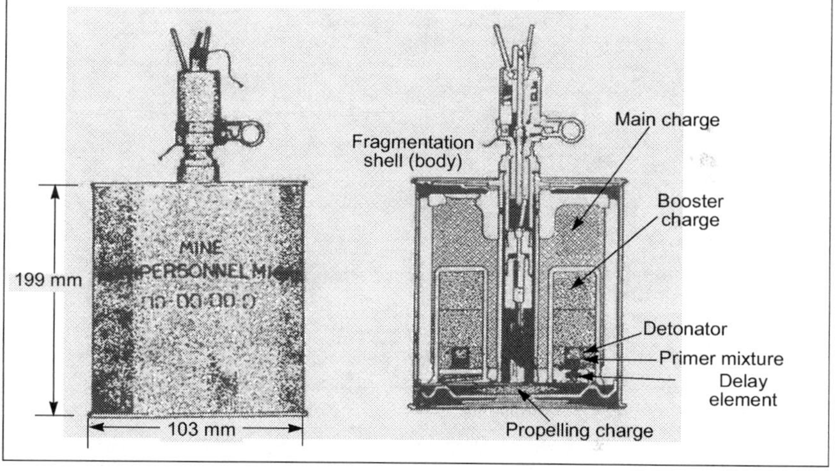

Figure A-6. M16A1 AP mine

CHARACTERISTICS

Main Charge	Diameter	Height	Weight	No Mines per Box	Weight per Box
TNT	103 mm	199 mm	3.5 kg	4	20.25 kg

- **Korea Only: The M16 is employed in protective, tactical, and nuisance minefields.**
- The M16 is used to defeat dismounted assaults and breaching operations.
- The M16 is pressure-actuated (3.6 to 9 kilograms) or pull-actuated (1.4 to 4.5 kilograms).

- The M16 and M16A1 have a casualty radius of 27 meters; the M16A2 has a casualty radius of 30 meters. The danger radius for friendly forces is 183 meters for all M16 mines.

INSTALLATION

> **WARNING**
> **Emplace and remove the mine while in the prone position.**

- Inspect the mine.
  - — Do not use the mine if it is dented, cracked, or damaged.
  - — Use the closed end of an M25 wrench to remove the shipping plug from the fuse well (Figure A-7).

Figure A-7. M16A1 mine and M25 wrench

  - — Inspect the fuse well for foreign material. If foreign material is present, turn the mine upside down and gently tap the bottom with your hand to dislodge the material. If you cannot remove the debris, replace the shipping plug and do not use the mine.
  - — Examine the fuse assembly carefully for evidence of damage or missing safety pins. Ensure that the safety pins move freely in their holes and that the rubber gasket is around the fuse base (Figure A-8, page A-8).

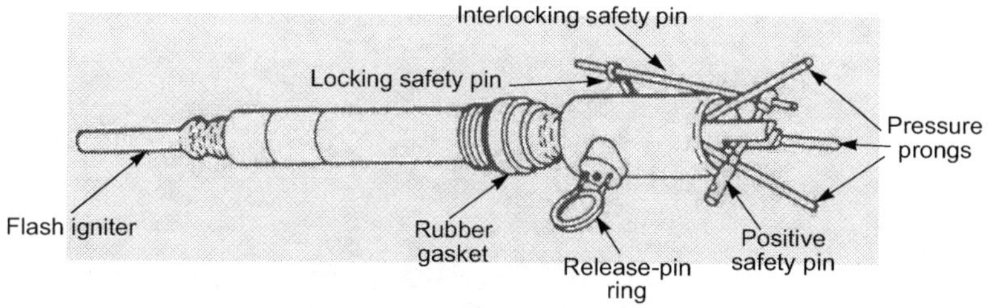

Figure A-8. M605 fuse

> **WARNING**
> **DO NOT use any fuse that was manufactured before 1957.**

- Fuse the mine.
  - Use the open end of the M25 wrench to tighten the bushing adapter on the fuse well.
  - Screw the fuse assembly into the fuse well by hand and tighten the assembly with the M25 wrench. Ensure that the rubber gasket is between the fuse body and the bushing adapter. **NOTE: For long-term use, smear a thin layer of silicone grease or similar lubricant on the fuse and threads.**
- Dig a hole to fit the mine.
  - Pressure installation. Dig the hole so that only the pressure-prong tips are above ground level.
  - Trip-wire installation. Dig the hole so that the release-pin ring is above ground level.
- Emplace the mine.
  - Place the mine in the hole and ensure that the safety pins remain in place.
  - Cover the mine with soil to the bottom of the release-pin ring.
  - Press the soil firmly around the sides of the mine.

Pressure Installation

- Remove the locking safety pin (Figure A-9). The interlocking safety pin will come free.

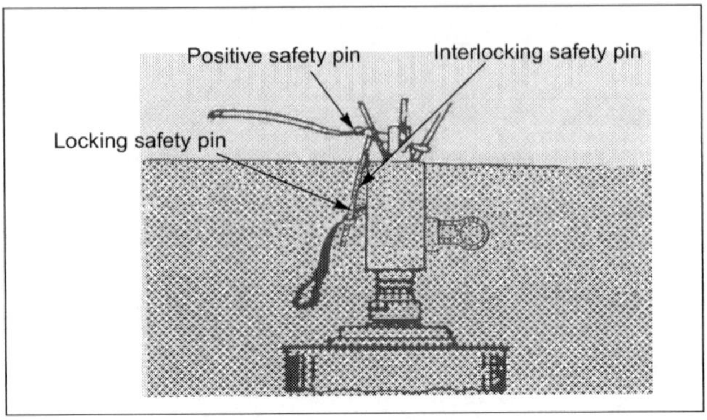

Figure A-9. Safety pins

- Arrange the pull cord on the positive safety pin so that it withdraws easily. Remove the metal collar.
- Finish covering the mine with soil until only the pressure prongs are above ground level.
- Camouflage the mine, place excess soil in sandbags, and remove sandbags from the area.
- Arm the mine by removing the positive safety pin (Figure A-9).
- Give the safety pins and the shipping plug to the NCOIC.

Trip-Wire Installation

- Install the trip wire (Figure A-10).

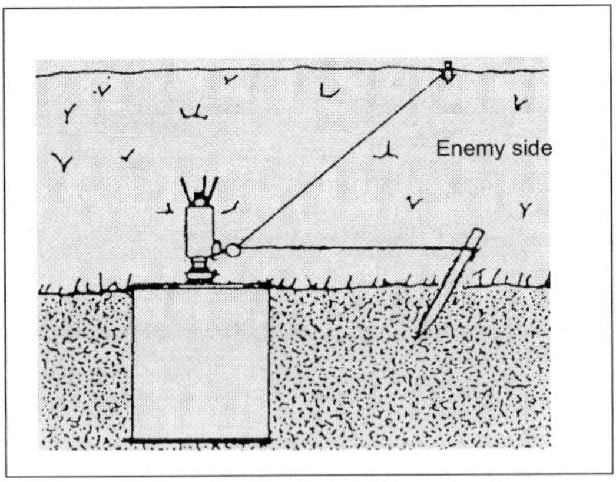

Figure A-10. Buried mine with a trip wire

- Cover the mine with soil and press it firmly around the sides of the mine. Leave the release-pin ring and the pressure prongs exposed.
- Tie off trip wires, approximately 10 meters from the mine. The wires should form a wide V, with the opening toward the enemy.
- Attach trip wires to the release-pin ring on the fuse.
- Remove the locking safety pin. The interlocking safety pin will come free.
- Arrange the pull cord on the positive safety pin so that it withdraws easily.
- Camouflage the mine, place excess soil in sandbags, and remove sandbags from the area.
- Arm the mine by removing the positive safety pin.
- Give the safety pins and the shipping plug to the NCOIC.

REMOVAL

> **WARNING**
> **Before attempting to disarm and remove the mine, ensure that the metal collar (Figure A-11) over the top of the striker on the M605 fuse is in place.**

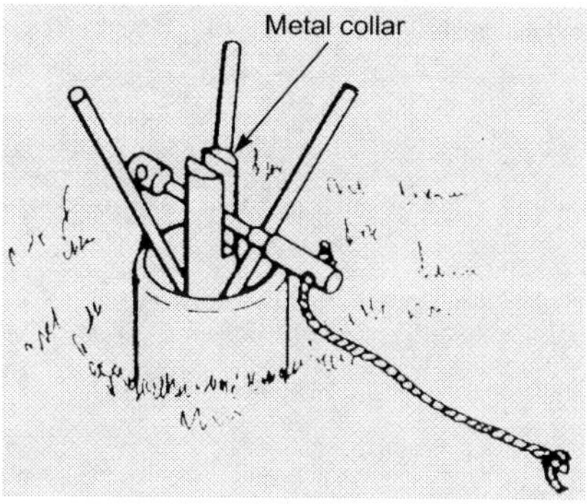

Figure A-11. Metal collar on an M605 fuse

- Disarm the mine.
  - Clear the soil carefully from the top of the fuse to the positive safety-pin hole. When using the M605 fuse, clear away all the soil from the fuse area.
  - Insert the positive safety pin through the positive safety-pin hole.
  - Insert the locking safety pin through the locking safety-pin hole.
  - Cut the slack trip wires that are attached to the release-pin ring.
- Check for AHDs.
  - Hold the mine body firmly in place with one hand.
  - Feel for AHDs with the other hand by digging around the sides of and underneath the mine.
- Remove the mine.
  - Remove the mine from the hole. Ensure that the safety pins remain in place.
  - Remove the M605 fuse with the M25 wrench.
  - Replace the shipping plug in the fuse well.

SECTION II. ANTITANK MINES

AT mines are designed to immobilize or destroy tanks and vehicles and their crews. They perform this function by producing an M-Kill or a K-Kill. An M-Kill is achieved by destroying one or more of the vehicle's vital drive components (usually breaking the track on a tank), causing the target to be immobilized. The weapon system and the crew are not destroyed in an M-Kill; the weapon system is immobile but continues to function. A K-Kill results when the weapon system or the crew is destroyed.

Conventional AT mines are distinguished by their effects and their fusing systems. Blast AT mines, such as the M15 and M19, derive their effectiveness through the blast generated by their detonation. These usually produce an M-Kill, but a K-Kill may result. Mines such as the M21 use a shaped charge or an SFF designed to penetrate the underside of a vehicle's armor. A K-Kill normally results unless the mine detonates under the vehicle's track.

M15

The M15 (Figure A-12, page A-12) is a blast AT mine that is contained in a round sheet-steel casing. The primary fuse well is located in the top center of the mine. There are two secondary fuse wells—one on the side and one on the bottom. The primary fuse well accepts the M603 pressure-actuated fuse. Standard FDs can be used in the secondary fuse wells with the M1 activator. The M624 tilt-rod-actuated fuse can also be used with this mine.

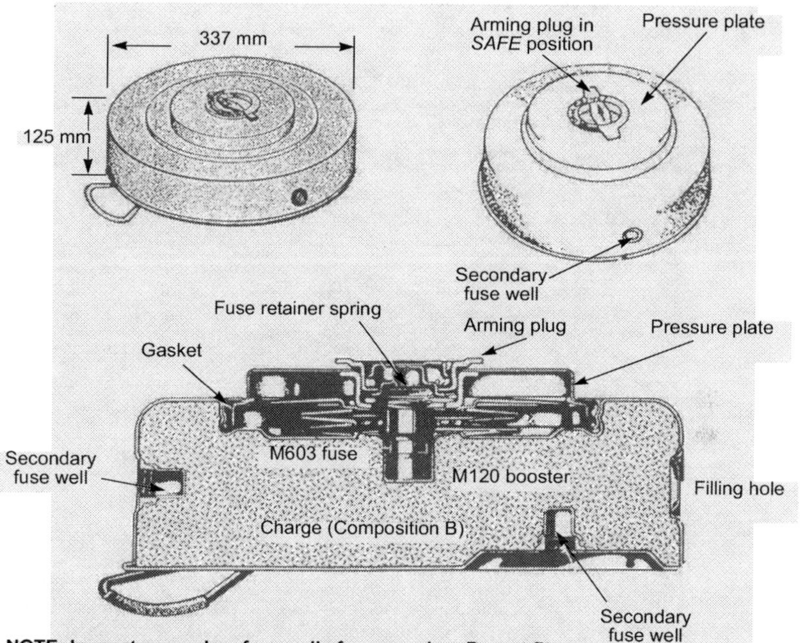

NOTE: Inspect secondary fuze wells for corrosion. Do not fit an M1 activator into a corroded fuze well. In training, return any mine with a corroded fuze well to the ASP as unserviceable.

Figure A-12. M15 AT mine

CHARACTERISTICS

Main Charge	Diameter	Height	Weight	No Mines per Box	Weight per Box
Comp B, 9.9 kg	337 mm	125 mm	13.5 kg	1	18 kg

- The M15 is employed in protective, tactical, and nuisance minefields.
- The M15 is surface-laid or buried.
- The M15 requires a force of 158 to 338 kilograms to detonate the M603 fuse and a force of 1.7 kilograms to deflect the tilt rod and detonate the M624 fuse.
- The M15 is designed to defeat heavy tanks.
- The M15 produces an M-Kill upon contact.

INSTALLATION USING THE M624 FUSE

> **WARNING**
> **Emplace and remove the mine while in the prone position.**

- Inspect the mine.
  - — Do not use the mine if it is dented, cracked, or damaged.
  - — Use the M20 wrench (Figure A-13) to remove the arming plug.

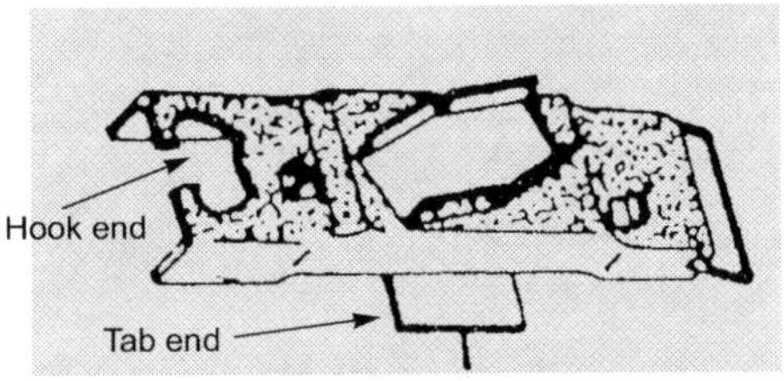

Figure A-13. M20 wrench

  - — Inspect the fuse well for foreign material. If foreign material is present, turn the mine upside down and gently tap the bottom with your hand to dislodge the material. If you cannot remove the debris, replace the arming plug and do not use the mine.
  - — Ensure that the booster retainer ring is seated in the fuse well. If the retainer ring is missing, replace the mine.
- Inspect the fuse.
  - — Remove the M624 fuse from the metal shipping container and inspect it for serviceability.
  - — Inspect the plastic collar of the fuse by looking down through the top of the pressure ring. If the safety pin is missing or improperly assembled, do not use the fuse (Figure A-14, page A-14).
  - — Do not use the fuse if the plastic collar appears to be cracked.
- Fuse the mine.
  - — Remove the M624 fuse from its fiber sleeve.
  - — Remove the end closure on the M624 fuse. **NOTE: For long-term emplacement, coat the fuse threads and gasket with silicone grease before removing the end closure (Figure A-15, page A-14).**

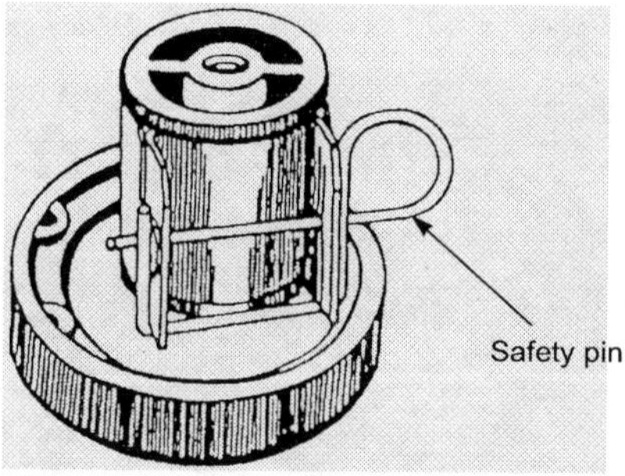

Figure A-14. Correct safety-pin configuration

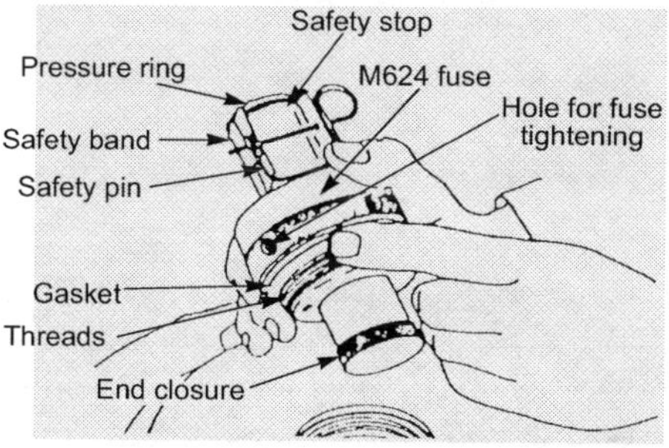

Figure A-15. Greasing the M624 fuse

- Screw the fuse hand-tight into the fuse well.
- Remove the extension rod from its packaging.
- Tighten the fuse by inserting the unthreaded end of one extension rod piece into the hole on the side of the fuse. Turn the fuse a quarter turn (Figure A-16).
- Remove the extension rod for further use after the fuse is secure.

NOTE: The M15 AT mine (with the M624 fuse) can be buried or surface-laid. If surface-laid, it must be staked in place.

- Dig a hole to fit the mine.
  - Dig a hole deep enough so that the top of the pressure plate will be at ground level.

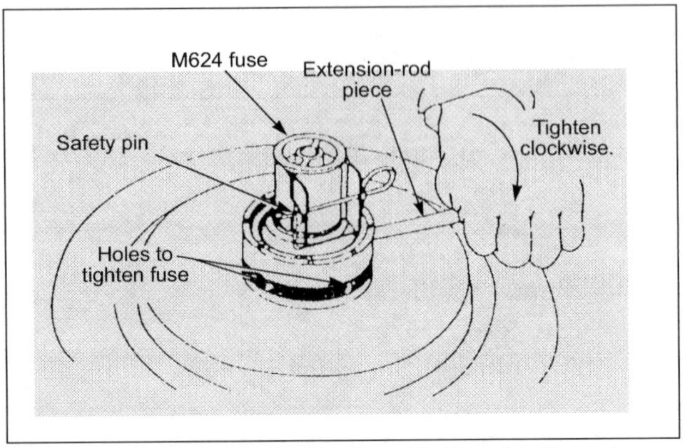

Figure A-16. Tightening the fuse with the extension rod

  - Dig the sides of the hole at a 45-degree angle to prevent vehicles from bridging the mine.
- Emplace the mine.
  - Place the mine in the hole.
  - Cover the mine with 2 centimeters of soil (Figure A-17).

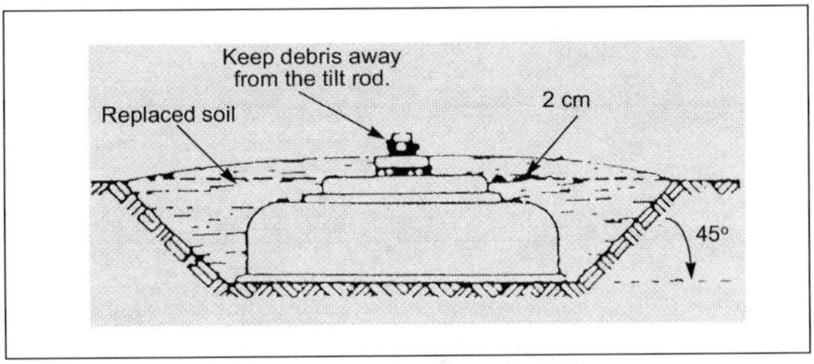

Figure A-17. M15 mine in the hole

NOTE: The M15 AT mine (with the M624 fuse) can be used in the tilt-rod or pressure role. In the tilt-rod role only, assemble all three pieces of the extension rod (Figure A-18, page A-16) and thread the extension rod into the threaded pressure ring of the fuse (Figure A-19, page A-16).

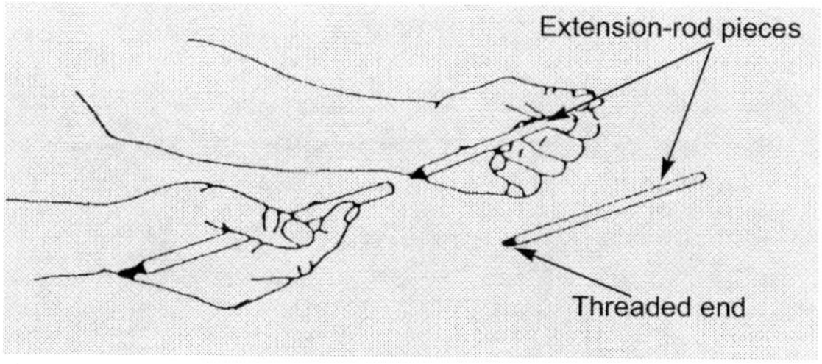

Figure A-18. Extension-rod assembly

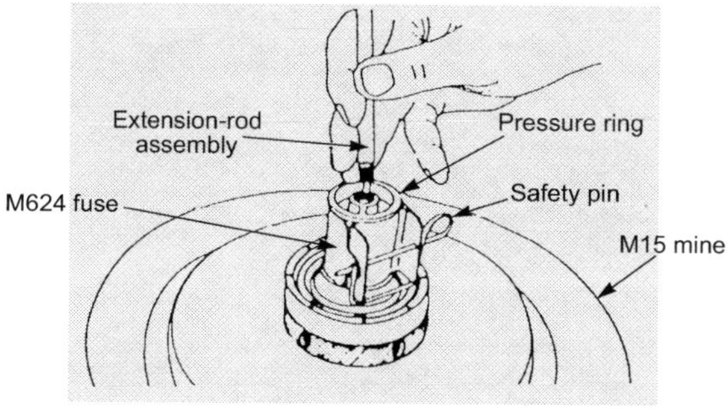

Figure A-19. Assembly of the extension rod into the fuse ring

- Arm the mine.
  - Use your right hand to raise the safety pin to the horizontal position, and grasp the safety band and safety stop with your left hand. Note the position of the thumb in Figure A-20.
  - Remove the safety pin with your right index finger, pulling it to the right.
  - Remove the safety stop carefully while holding the safety band in place.
  - Remove the safety band.
- Camouflage the mine.
  - Camouflage the mine with twigs, grass, or other material in the area. Place mines with extension rods in tall grass, if possible. Ensure that no pressure is applied to the tilt rod or the fuse.

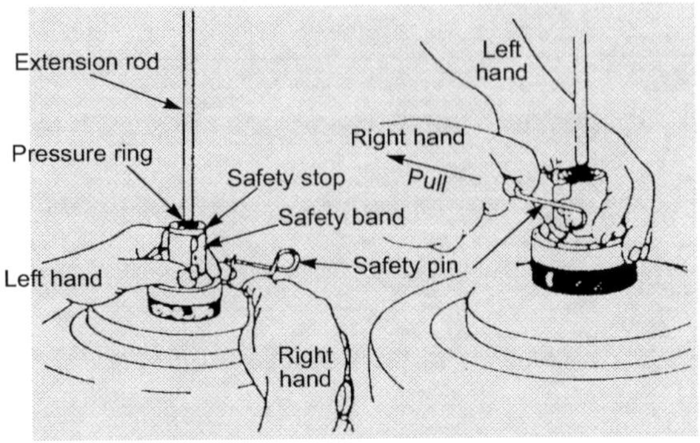

Figure A-20. Removal of safety pin

- Place excess soil in sandbags and remove them from the area.
- Give the band, the stop, the pull-ring assembly, the arming plug, and the end closure to the NCOIC.

Removal Using the M624 Fuse

- Disarm the mine.
  - Clear camouflage away from the mine carefully.
  - Assemble the band, the stop, and the safety-pin assembly on the fuse so that the pressure ring is immobilized.
  - Remove the extension rod.
- Check for AHDs.
  - Hold the mine firmly in place with one hand, without putting pressure on the fuse.
  - Feel for AHDs with the other hand by digging around the sides of and underneath the mine.
- Remove the mine.
  - Remove the mine from the hole.
  - Remove the fuse from the mine; use the extension rod if necessary.
  - Replace the end closure on the fuse.
  - Install the arming plug into the fuse well.

Installation Using the M603 Fuse

- Inspect the mine.
  - Do not use the mine if it is dented, cracked, or damaged.

— Use the M20 wrench to remove the arming plug from the mine.

— Inspect the fuse well for foreign material. If foreign material is present, turn the mine upside down and gently tap the bottom with your hand to dislodge the material. If you cannot remove the debris, replace the arming plug and do not use the mine.

— Ensure that the booster retainer ring is seated in the fuse well. If the retainer ring is missing, replace the mine.

- Perform a function check with the arming plug.

— Turn the setting knob to the *ARMED* position. Ensure that the shutter bar moves across the bottom of the arming plug (Figure A-21)

— Turn the setting knob to the *SAFE* position. Ensure that the shutter bar moves back across the bottom of the arming plug (Figure A-22). **NOTE: If the shutter bar does not go into the *SAFE* or *ARMED* position, notify the NCOIC.**

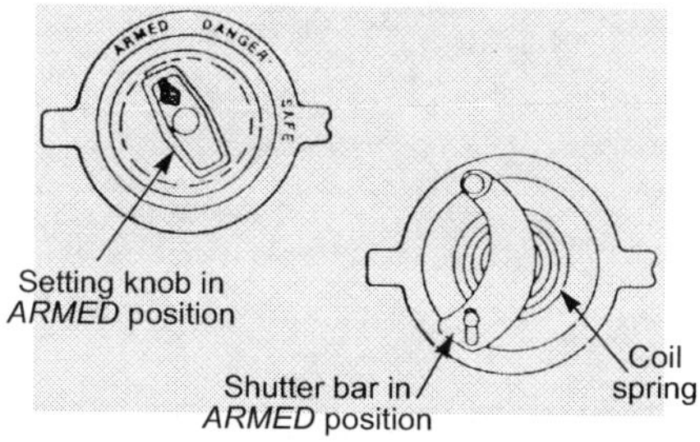

NOTE: A coil spring may not be present in older models.

Figure A-21. *ARMED* position

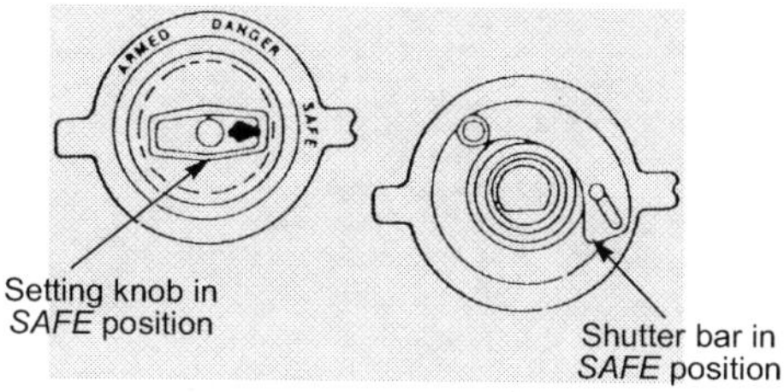

Figure A-22. *SAFE* position

- Fuse the mine.
  - Remove the M603 fuse from the metal shipping container and inspect it for serviceability. The green end of the detonator must show in the bottom of the fuse.
  - Remove the safety fork; use the hooked end of an M20 wrench if necessary (Figure A-23).

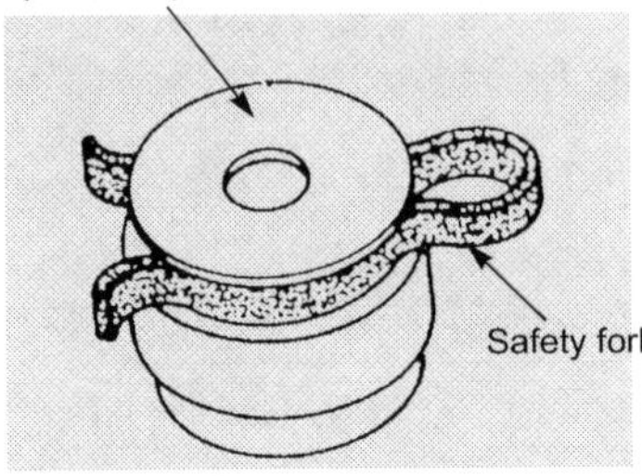

Figure A-23. Safety fork

  - Insert the fuse into the fuse well carefully until it seats securely on top of the booster retaining ring.
  - Perform a clearance test using the tab end of the M20 wrench (Figure A-24, page A-20).

WARNING
If the fuse pressure plate interferes with the tab end of the M20 wrench, investigate the cause and notify the NCOIC. DO NOT arm the mine.

NOTE: For long-term emplacement, smear a thin layer of silicone grease or similar lubricant on the arming plug, the threads, and the gasket.

  - Ensure that the setting knob is in the *SAFE* position.
  - Screw the arming plug into the mine by hand. Ensure a watertight seal by tightening the arming plug with the M20 wrench.
- Dig a hole to fit the mine.
  - Dig a hole deep enough so that the top of the pressure plate is about 3 centimeters below ground level.
  - Dig the sides of the hole at a 45-degree angle to prevent vehicles from bridging the mine (Figure A-25, page A-20).
- Emplace the mine.

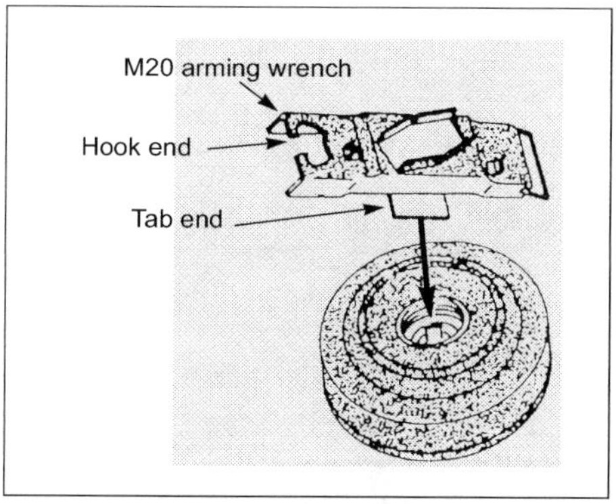

Figure A-24. Clearance test

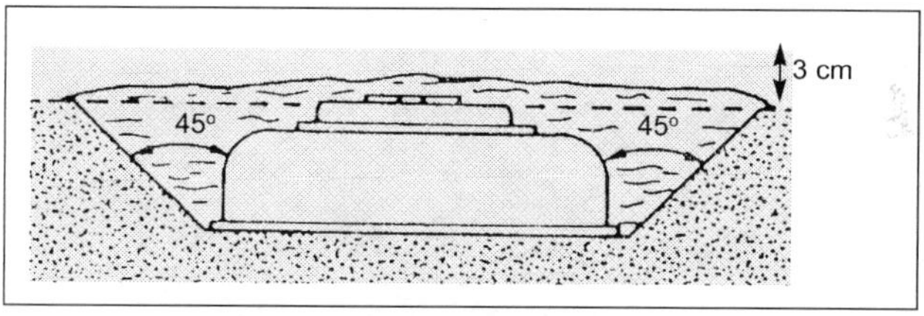

Figure A-25. M15 mine in the hole

- Place the mine in the hole.
- Cover the mine with soil until it is level with the top of the pressure plate.

- Use the M20 wrench to arm the mine by turning the setting knob from the *SAFE* position to the *ARMED* position.
- Camouflage the mine.
  - Cover the mine with 3 to 5 centimeters of soil.
  - Camouflage the mine, place excess soil in sandbags, and remove sandbags from the area.
  - Give the safety clip to the NCOIC.

Removal Using the M603 Fuse

- Disarm the mine.
  - Clear the soil from the top of the mine carefully.
  - Hold the mine firmly in place with one hand, without putting pressure on the pressure plate.
  - Feel for AHDs with the other hand by digging around the sides of and underneath the mine.

  — Use the M20 wrench to turn the setting knob to the *SAFE* position.

- Remove the mine.
  — Remove the mine from the hole.
  — Use the M20 wrench to turn the arming plug counterclockwise, and remove the arming plug.
  — Remove the M603 fuse from the fuse well and replace the safety fork.
  — Install the arming plug.

M19

The M19 AT mine (Figure A-26) is housed in a square, plastic case and holds 9.45 kilograms of Composition B (HE charge). It consists of an M606 integral pressure fuse and two secondary fuse wells—one in the side and one in the bottom. The fuse body contains a pressure plate, a Belleville spring, a setting knob, a step plate, a firing-pin assembly, and a detonator.

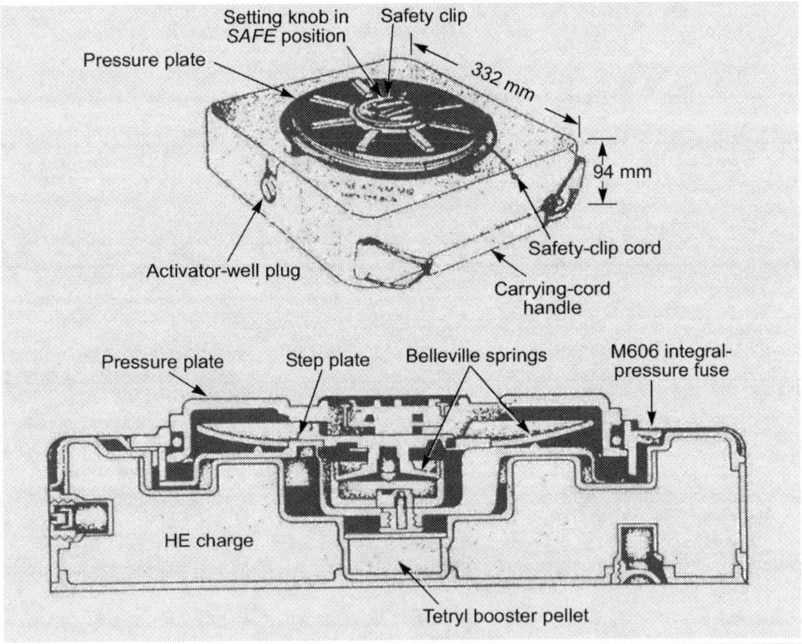

Figure A-26. M19 AT mine

CHARACTERISTICS

Main Charge	Diameter	Height	Weight	No Mines per Box	Weight per Box
Comp B, 9.45 kg	332 mm	94 mm	12.6 kg	2	32.5 kg

- The M19 is employed in tactical and nuisance minefields.
- The M19 is buried or surface-laid.
- The M19 requires a force of 157 to 225 kilograms to detonate.
- Standard FDs may be used with the M2 activator in any of the secondary fuse wells of the M19.

INSTALLATION

> **WARNING**
> **Emplace and remove the mine while in the prone position.**

- Inspect the mine.
  - Do not use the mine if it is dented, cracked, or damaged.
  - Remove the M606 fuse from the fuse well by turning it counterclockwise a quarter turn (Figure A-27).
  - Ensure that the rubber gasket is on the M606 fuse.

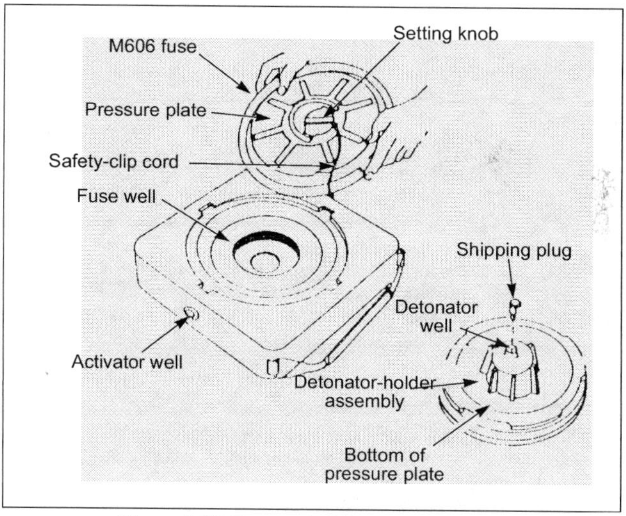

Figure A-27. Removal of the pressure plate

— Remove any foreign material from the fuse well.

— Ensure that the setting knob is in the *SAFE* position and that the safety clip is in place.

— Use the M22 wrench to remove the shipping plug from the detonator well.

— Inspect the detonator well for foreign material. If foreign material is present, gently tap the pressure plate with your hand to dislodge the material.

- Test the position of the firing pin (Figure A-28).

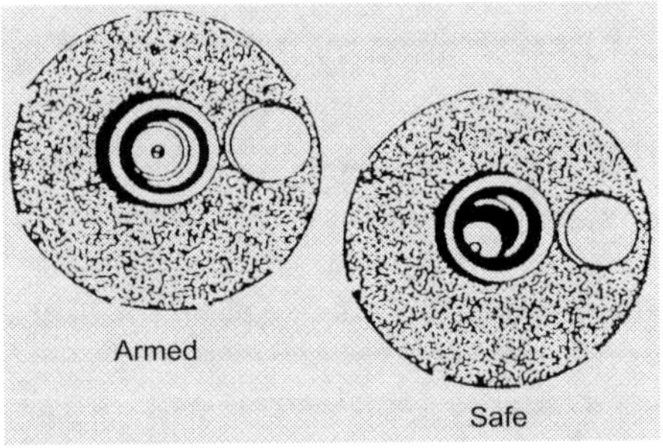

Figure A-28. Firing pin

— Ensure that the firing pin is at the edge of the well when the setting knob is in the *SAFE* position.

— Remove the safety clip.

— Use the M22 wrench to turn the setting knob to the *ARMED* position. Ensure that the firing pin is in the center of the well.

— Use the M22 wrench to turn the setting knob back to the *SAFE* position. Ensure that the firing pin moves back to the side of the well. **NOTE: If the firing pin is not in the correct position when the setting knob is in the *ARMED* or *SAFE* position, notify the NCOIC.**

— Replace the safety clip.

- Use the M22 wrench to screw the M50 detonator into the detonator well.
- Use the M22 wrench to tighten the M606 fuse into the fuse well.
- Dig a hole to fit the mine.

— Dig a hole deep enough so that the top of the pressure plate will be even with or slightly below ground level.

- Dig the sides of the hole at a 45-degree angle to prevent vehicles from bridging the mine.

- Emplace the mine.
  - Place the mine in the hole.
  - Cover the mine with soil until it is level with the top of the pressure plate.
- Arm the mine.
  - Remove the safety clip.
  - Use the M22 wrench to turn the setting knob from the *SAFE* position to the *ARMED* position.
- Camouflage the mine.
  - Cover the mine with 3 centimeters of soil.
  - Camouflage the mine, place excess soil in sandbags, and remove sandbags from the area.
  - Give the safety clip and the shipping plug to the NCOIC.

REMOVAL

- Disarm the mine.
  - Clear the soil from the top of the mine carefully.
  - Hold the mine firmly in place with one hand, without putting pressure on the pressure plate.
  - Feel for AHDs with the other hand by digging around the sides of and underneath the mine.
  - Use the M22 wrench to turn the setting knob to the *SAFE* position.
  - Replace the safety clip on the M606 fuse.
- Remove the mine.
  - Remove the mine from the hole.
  - Use the M22 wrench to remove the M606 fuse by turning it counterclockwise and lifting it out of the fuse well.
  - Use the M22 wrench to remove the detonator from the detonator well.
  - Replace the shipping plug in the detonator well.
  - Replace the pressure plate in the mine.

M21

The M21 AT mine (Figure A-29) utilizes a direct-energy warhead that is designed to produce a K-Kill. It is used in conjunction with the M607 fuse. The M21 produces a K-Kill against heavy tanks, unless the mine is activated under the track. It can be buried with a tilt rod, or it can be surface-laid with or without a tilt rod. If the M21 is surface-laid with a tilt rod, it must be staked to prevent it from being knocked over and causing the warhead to be directed away from the target. The M21 is not compatible with any mechanical mine-dispensing system.

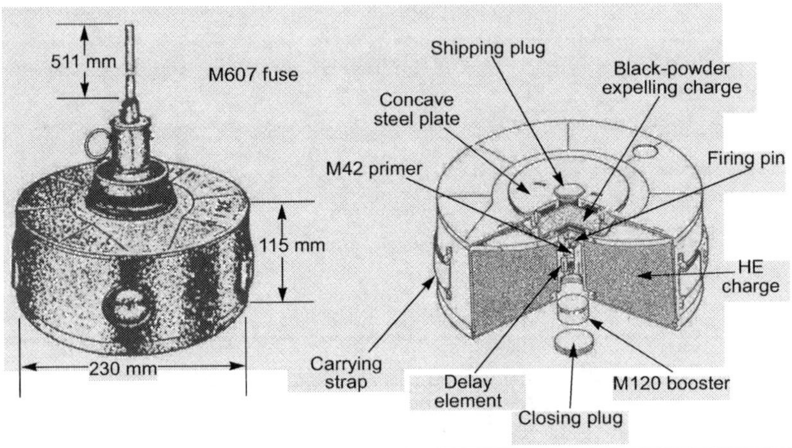

Figure A-29. M21 AT mine

CHARACTERISTICS

Main Charge	Diameter	Height	Weight	No Mines per Box	Weight per Box
Comp H6, 4.95 kg	230 mm	115 mm	7.8 kg	4	41 kg

- The M21 requires a minimum of 130.5 kilograms of pressure to detonate.
- The M21 can be used with a tilt-rod assembly, requiring 1.7 kilograms of pressure on the extension rod to cause a 20-degree deflection.
- The M21 is the only conventional US AT mine with a direct-energy warhead.

INSTALLATION

> **WARNING**
> **Emplace and remove the mine while in the prone position.**

- Inspect the mine.
  - — Do not use the mine if it is dented, cracked, or damaged.
  - — Ensure that the cotter pins on the fuse pull-ring assembly and the fuse-closure assembly are securely in place (Figure A-30).
  - — Inspect the fuse to ensure that the neck portion behind the tile collar is not cracked.

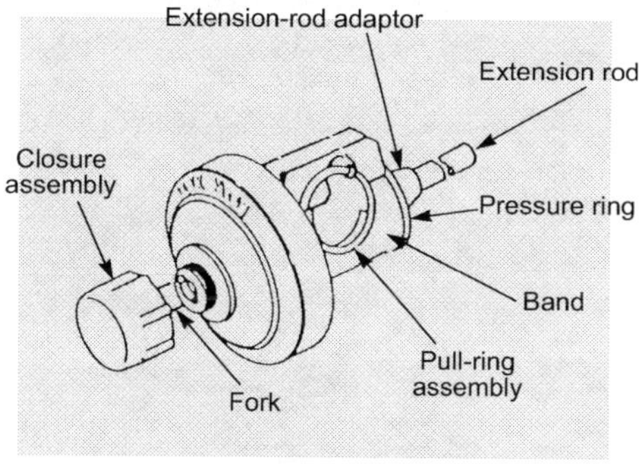

Figure A-30. M607 fuse

- Insert the booster.
  - — Use the screwdriver end of the M26 wrench (Figure A-31) to remove the closing plug from the bottom of the mine.

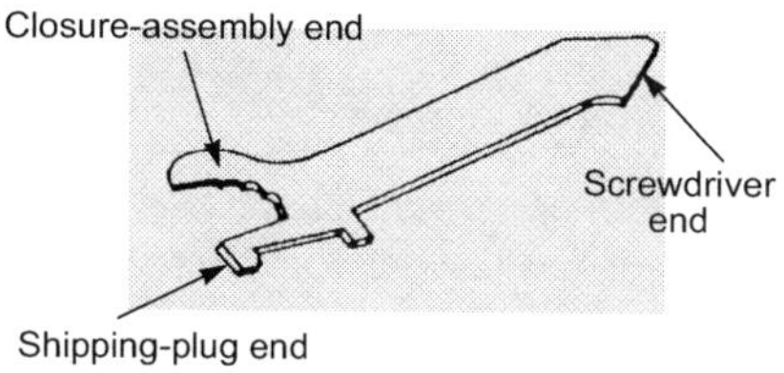

Figure A-31. M26 wrench

- — Inspect the booster well for foreign material. If foreign material is present, gently tap the top of the mine with your hand to dislodge the material. If you cannot remove the debris, replace the closing plug and do not use the mine.
- — Insert the M120 booster (with the washer side toward the fuse) into the booster well.
- — Replace the closing plug with the M26 wrench.

- Fuse the mine.
  - — Remove the shipping plug from the fuse well with the M26 wrench.
  - — Inspect the fuse well for foreign material. If foreign material is present, gently shake the mine to dislodge the material. If black powder falls out of the fuse well or you cannot remove the foreign material, do not use the mine.
  - — Remove the closure assembly from the M607 fuse with the M26 wrench. Ensure that the gasket remains in place on the fuse.
  - — Screw the fuse hand-tight into the fuse well.
- Dig a hole to fit the mine.
  - — Dig a hole deep enough so that the top of the mine will be at ground level (Figure A-32).

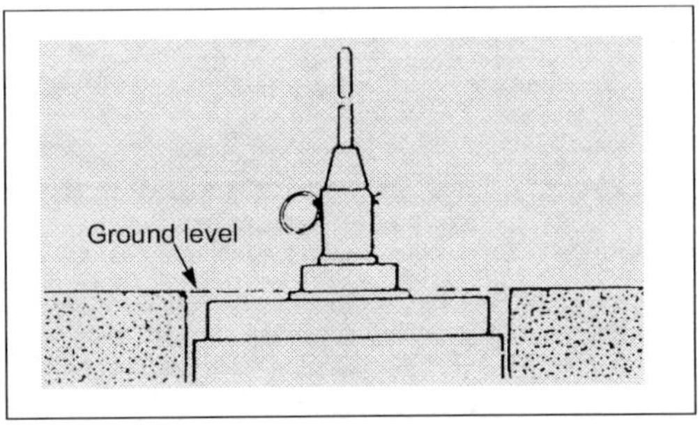

Figure A-32. Buried M21 mine

  - — Check the bottom of the hole to ensure that the ground is solid enough to support the mine. If necessary, place a flat object under the mine to provide a firm foundation. Allow additional depth for the object.
- Emplace the mine.
  - — Place the mine in the hole.
  - — Cover the mine with soil until it is level with the top of the mine. Ensure that no soil falls around or under the plastic collar.

- Press the soil firmly around the sides of the mine.

• Assemble the extension rod. **NOTE: For pressure operation, do not use the extension-rod assembly.**

- Screw the extension rod onto the M607 fuse.
- Ensure that the extension rod is pointing straight up.

• Arm the mine.

- Squeeze the end of the cotter pin together on the pull ring.
- Remove the cotter pin by holding the fuse firmly in one hand and pulling the pull ring with the other hand.
- Remove the band and the stop slowly and carefully from the neck of the fuse (Figure A-33).

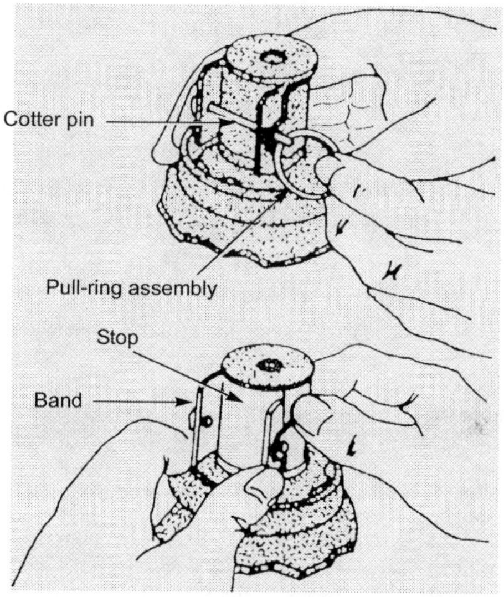

Figure A-33. Removing the band and the stop

• Camouflage the mine.

- Camouflage the mine with twigs, grass, or other material in the area. Place mines with extension rods in tall grass, if possible. Ensure that no pressure is applied to the tilt rod or the fuse.
- Place the excess soil in sandbags and remove them from the area.

- Give the band, the stop, the pull ring, the shipping plugs, and the closure assembly to the NCOIC.

REMOVAL

- Disarm the mine.
  - Clear the camouflage away from the mine carefully.
  - Attach the band and the stop to the fuse.
  - Insert the cotter pin into the band and the stop. Spread the ends of the cotter pin.
  - Remove the extension rod.
- Check for AHDs.
  - Hold the mine firmly in place with one hand, without putting pressure on the fuse.
  - Feel for AHDs with the other hand by digging around the sides of and underneath the mine.
- Remove the mine.
  - Remove the mine from the hole.
  - Remove the fuse from the mine.
  - Install the closure assembly on the fuse.
  - Install the shipping plug into the fuse well of the mine.
  - Remove the closing plug from the bottom of the mine.
  - Remove the booster from the mine.
  - Install the closing plug into the booster well.

SECTION III. FIRING DEVICES AND ACTIVATORS

An FD performs the function of a mine fuse by providing an alternative means to detonate the mine. It is normally used in conjunction with a standard fuse so that a mine will have two separate explosive chains. The purpose of the second firing chain is to prevent the enemy from disarming or removing mines after emplacement. When used for this purpose, the FD is called an AHD and it is designed to function by detonating the attached mine or another explosive charge nearby if unauthorized personnel attempt to remove or tamper with the mine. **NOTE: US forces will not employ AHDs on AP mines.** Both the M19 and the M15 have two secondary fuse wells for attaching an FD and an activator.

There are two standard US FDs—M5 pressure release and M142 multipurpose. They utilize a spring-loaded striker and a standard base and are designed to function in one or more of the following modes:

- Pressure.
- Pressure release.
- Tension.
- Tension release.

M5 PRESSURE-RELEASE FIRING DEVICE (MOUSETRAP)

The M5 FD (Figure A-34) is activated by the release of pressure. Lifting or removing a restraining weight releases the striker to fire the cap.

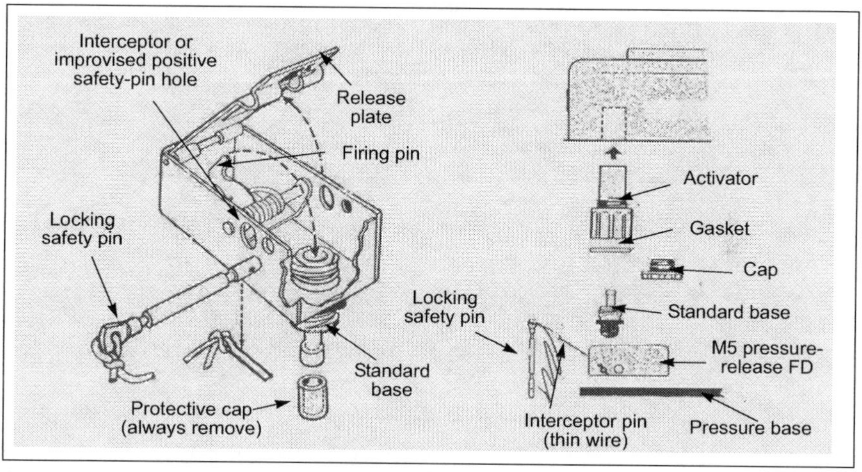

Figure A-34. M5 FD

CHARACTERISTICS

- Case: Metal.
- Color: Olive-drab.
- Length: 445 millimeters.
- Width: 239 millimeters.
- Height: 175 millimeters.
- Internal action: Mechanical with hinged striker release.
- Initiating action: Removal of restraining weight, 2.25 kilograms or more.
- Accessories: Pressure board.
- Safeties: Safety pin and hole for interceptor pin.
- Packaging: Four complete FDs and four plywood pressure boards are packaged in a paper carton, five cartons are packaged in a fiberboard box, and 10 fiberboard boxes are shipped in a wooden box.

INSTALLATION

- Dig a hole deep enough to bury the mine on a firm foundation. The pressure plate should be slightly above ground level.

> **WARNING**
> **Ensure that the mine and the FD are resting on a firm foundation before removing the pins.**

- Remove the protective cap from the standard base. Assemble the FD.
- Use a coat hanger as a positive safety pin. Place the mine in the hole. Leave enough room to remove the pins.
- Bury and camouflage the mine.
- Arm the mine.
- Remove the locking safety pin (Figure A-35).

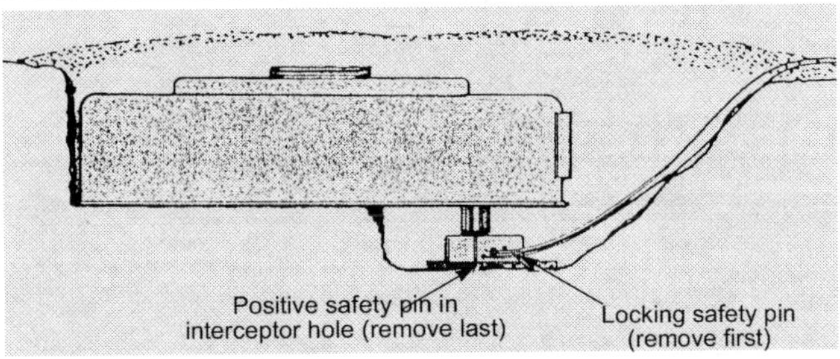

Figure A-35. Arming the M15

- Remove the positive safety pin (interceptor pin). **NOTE: If the positive safety pin is difficult to remove or if you hear a click when removing the locking safety pin, carefully remove the mine and replace the FD.**

REMOVAL

- Uncover the mine carefully and inspect it for tampering. Locate and carefully uncover the FD.

> **WARNING**
> **DO NOT release the pressure being applied to the device.**

- Insert the positive safety pin into the interceptor hole, and then insert the locking safety pin into the safety-pin hole.

- Disarm and remove the mine. Recover the FD.

M142 MULTIPURPOSE FIRING DEVICE

The M142 FD (Figure A-36) can be designed to function in the following modes:

- Pressure (11 kilograms or more).
- Pressure release (between 2 and 67 kilograms).
- Tension (3 kilograms or more).
- Tension release.

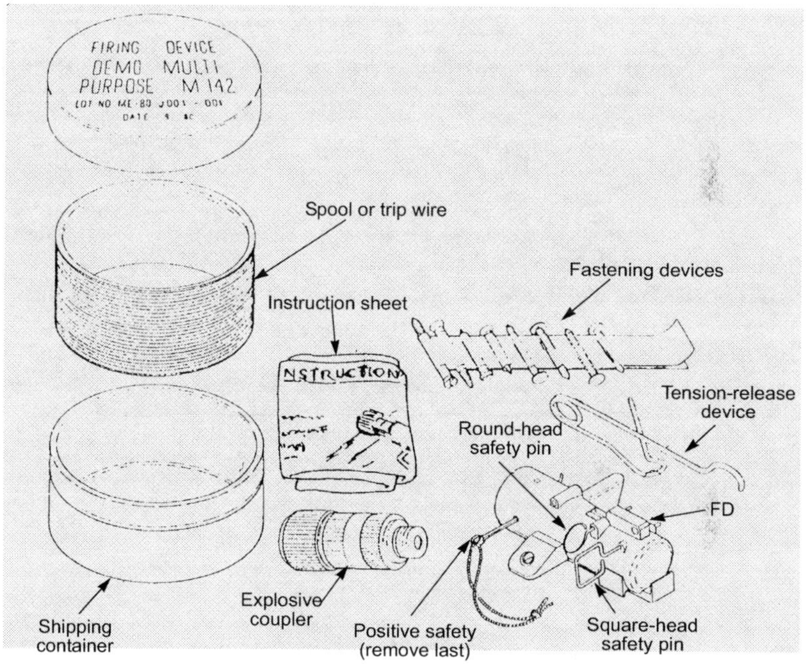

Figure A-36. M142 FD

Although primarily intended for booby-trap applications, the M142 can be readily adapted as an AHD. The M142 comes with a coupling device and a primer that accepts a standard nonelectric blasting cap. The initiating action sets off an explosive chain that passes from the FD and the primer to the blasting cap, and then via the detonating cord to the main charge. However, the coupling device with the primer will not initiate the detonating cord alone without a blasting cap attached, and it is not adaptable to any activator or secondary fuse well. When the M142 is used as an AHD, the coupling device is removed and an M1 or M2 standard base is used.

CHARACTERISTICS

- Case: Plastic.
- Color: Olive-drab.
- Diameter: 190.5 millimeters.
- Length: 571.5 millimeters.
- Internal action: Spring-driver striker.
- Safeties: Positive safety pin, square-head pivot pin, round-head pivot pin, and alternative safety-pin hole.
- Accessories: Nail and screw fasteners, coupling assembly, tension-release attachment, 15-meter spool of trip wire, and vinyl instruction sheet.
- Packaging: Round, metal can containing FD with accessories.

ARMING AND DISARMING

Arming and disarming procedures vary based on the activation mode. Detailed instructions are printed on a weatherproof, vinyl sheet included in each FD package.

M1 AND M2 ACTIVATORS

When FDs are employed with M15 and M19 AT mines, they require the use of an M1 or M2 activator.

Activators are essentially detonator boosters that are designed to magnify the explosive force generated by an FD with a standard base and transfer the force to the main charge. Activators may be used with either type of FD to supply an AT mine with a secondary fuse for antihandling purposes. The M1 activator is used with the M15 AT mine, and the M2 activator is used with the M19 AT mine. The activator also performs the function of an adapter for attaching the FD to the mine. One end of the activator is threaded externally for insertion in the secondary well of the mine; the other end is threaded internally to receive the standard base coupling of the FD.

The M1 activator (Figure A-37, page A-34) is 54 millimeters long (with cap), is made of olive-drab plastic, contains a detonator, and has a threaded closing plug and a gasket. It has a cylindrical, unthreaded cap that is cemented to the opposite end of the body and contains a tetryl booster charge. The threaded end, which screws into the mine, is 25 millimeters in diameter.

The M2 activator is similar to the M1 except that it contains an HE pellet, and its overall length, with cap, is 53 millimeters.

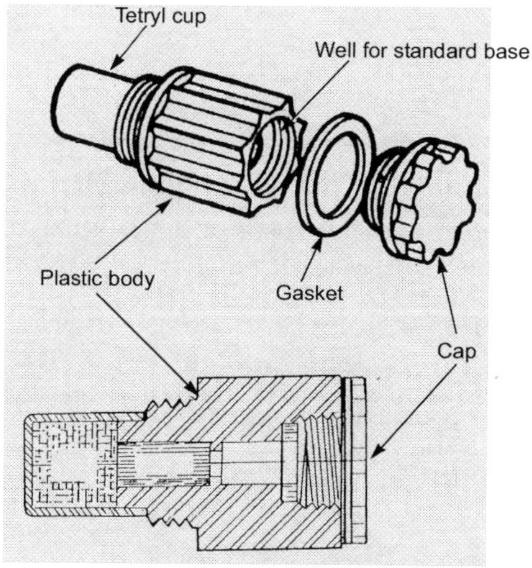

Figure A-37. M1 activator

Appendix B

Controls and Components of Special-Purpose Munitions

This appendix provides characteristics and detailed descriptions of US special-purpose munitions. The use of these munitions is outlined in Chapter 4.

SELECTABLE LIGHTWEIGHT ATTACK MUNITION

The SLAM is a multipurpose munition with antidisturbance and antitamper features. There are two models of the SLAM—one is self-neutralizing (M2) and the other is self-destructing (M4). The M2 is solid green and has no labels, brands, or other distinguishing marks. The M4 is green with a black warhead (EFP) face.

Employment methods for the SLAM are outlined in Chapter 4.

Figure B-1 describes and illustrates the major components of the SLAM.

Component	Description
Mounting holes (1)	The mounting holes are used to secure the carrying strap or the mounting wire to the SLAM when attaching the SLAM to trees and so forth.
Bore sights (2)	Two bore sights and an omega sight are located on the top of the SLAM and are used to aim the SLAM at targets.
Selector switch (3)	The selector switch is used to select operating modes and times. It has eight detent positions. The switch is against a stop (in the shipping position), which is the only switch position that allows the SLAM to fit in the reusable environmental protective pack. Turning clockwise, there are three positions for selecting the operating time (4, 10, and 24 hours). Setting any of these positions will select an internal sensor mode of operation, which is a magnetic sensor for mine mode and a passive infrared sensor for side-attack mode. These three positions will cause the SLAM to self-destruct (M4) or self-neutralize (M2) at the end of the selected operating time. Continuing clockwise, the last four positions select an internal timer, which sets the minutes until demolition. These positions are 15, 30, 45, and 60 minutes.
Activation-lever shear pin (4)	There is a shear pin mounted across the SLAM's lever slot. If the shear pin is sheared, thereby breaking the seal, the lever may have been pulled and the SLAM may be an electronic dud. If the shear pin is broken, it should only be used in the command-detonation mode.
Safety pin (7)	The safety pin slides from the body and starts the SLAM's timing. It is pried from its latch with the tip of the lever. Once the safety pin is pulled, it cannot be reinserted.

Figure B-1. SLAM components

Component	Description
Passive infrared sensor (8) and cover (9)	The SLAM is equipped with a passive infrared sensor that detects trucks and light armored vehicles by sensing the change in background temperature as vehicles cross in front of the SLAM. The sensor is directional and is aligned with the EFP. The sensor is active when the SLAM is operating with the selector switch set to 4, 10, or 24 hours and the sensor cover is removed to expose the infrared sensor (such as, during the side-attack mode). The SLAM will self-destruct (M4) or self-neutralize (M2) if the selected time expires before it is detonated by vehicle passage.
Blasting-cap well and plug (10)	The threaded plug seals the blasting-cap well. It is removed to mount a standard military blasting cap with a priming adapter.
Warhead (11)	The warhead is an EFP that is designed to defeat light armored vehicles. The EFP forms within the first 5 inches of flight and has an effective range of 25 feet.
Housing assembly (12)	The housing assembly contains the fusing, electronics, and S&A components. It also provides a structural interface for the warhead, the sights, the activation lever, the passive infrared sensor, the selector switch, and the safety pin.

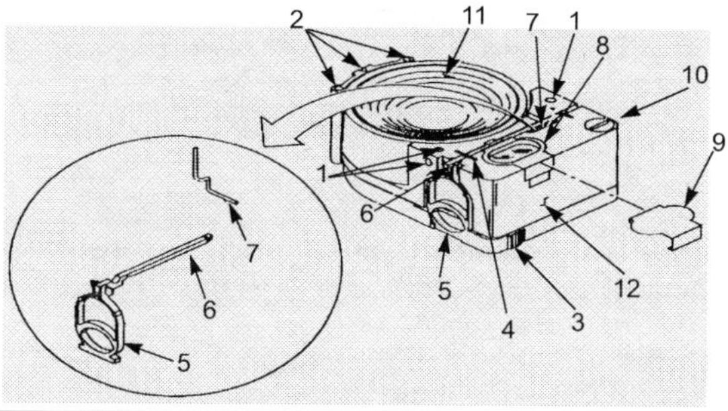

Figure B-1. SLAM components (continued)

M93 HORNET

The M93 Hornet is a lightweight (35 pounds) AT/antivehicular munition that one person can carry and employ. It is a one-time use, nonrecoverable munition that is capable of destroying vehicles using sound and motion as detection methods. The Hornet will automatically search, detect, recognize, and engage moving targets, using top attack at a maximum standoff distance of 100 meters. It is employed by units equipped with an M71 RCU. The RCU is a hand-held encoding unit that interfaces with the Hornet when the remote mode is selected at the time of employment. After encoding, the RCU can be used to arm the Hornet, reset SD times, and destruct the Hornet. Employment methods of the Hornet are outlined in Chapter 4.

Figure B-2 describes and illustrates the major components of the Hornet. Figure B-3, page B-4, describes and illustrates the controls and indicators of the Hornet.

Component	Description
Support legs (1)	Support legs are used to stabilize the Hornet when it is deployed.
Active battery-pack cover (2)	The active battery-pack cover provides a seal to protect and secure the active battery pack. The latch is lifted up to remove the cover, the active battery pack is installed, and the cover is then reinstalled and latched down. A line secures the battery-pack cover to the control panel of the munition.
SD switch (3)	The SD switch is a six-position rotary switch that is used to select the SD time and unlock the arm control switch. The SD switch is also used to unlock the arming lever. This is done by rotating the switch to the setting "U." A red lock element is extended 1/8 inch from the side of the munition when the SD switch is in the unlock position. The SD time is preset to Setting 1 when the Hornet is shipped. SD times are as follows: Setting — Time 1 — 4 hours 2 — 48 hours 3 — 5 days 4 — 15 days 5 — 30 days
Arm control switch (4)	The arm control switch consists of an arming lever interlocked with the SD switch and the S&H band assembly to prevent inadvertent actuation. Until the S&H band assembly is removed and the SD switch is placed in the unlock position, the arming lever cannot be moved to the arm position. An internal lock secures the arming lever in the arm position.
Microphones (5)	When the geophone seismic sensor detects a potential target, usually at ranges up to 600 meters, it alerts the munition to start listening with the three microphones that extend from the munition body. They track the two loudest noise sources that are heard.
Antenna (6)	The antenna provides a means for the Hornet to receive M71 RCU commands.
Capture screws (7)	These are four flat-head screws that secure the bottom plate to the munition body. They are removed along with the bottom plate to access the battery compartment.
Bottom plate (8)	The bottom plate provides a seal to protect and secure the battery compartment and connect the batteries once they are installed.
D-cell batteries (9)	The battery compartment houses four D-cell batteries. A drawing on the inside of each battery tube shows battery orientation.
Dowel pin (10)	The dowel pin ensures that the bottom plate is in the correct orientation to properly connect the batteries.

Figure B-2. Hornet components

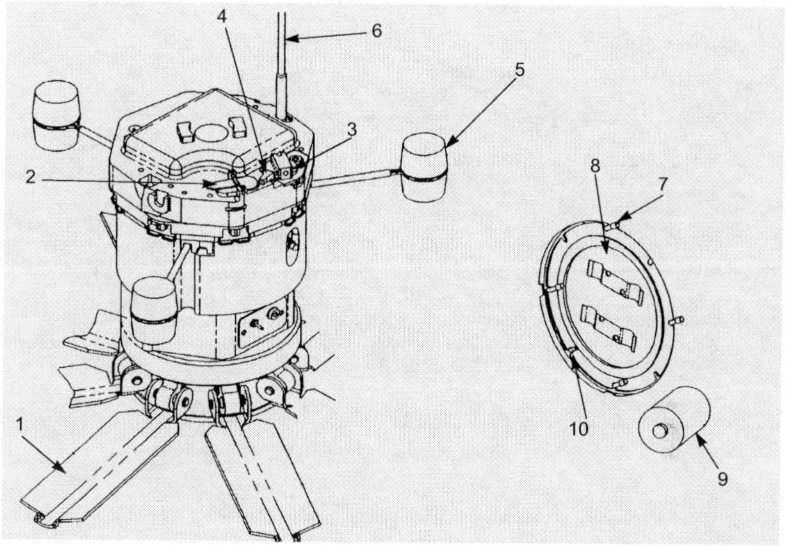

Figure B-2. Hornet components (continued)

Component	Description
Magnetic coupling device (MCD) (1)	This device is used as part of the RCU interface. The RCU interface consists of the MCD and keyed tabs. In the remote arming mode, the RCU is placed on top of the MCD and minefield code data is transferred to the munition. Upon successful encoding, the status light begins to flash.
Target switch (2)	The target switch is a toggle switch used to select the type of target engagement. This gives the operator the choice between detecting and destroying only heavy armored vehicles or all vehicles.
Manual select switch (3)	The manual select switch is a push-button switch, protected by a plastic cover that must be removed to access the switch. Successful activation of the switch will cause the status light to flash. This switch is used to allow the operator to employ the Hornet without the RCU.
Status light (4)	The status light is a visual indicator for the operator during the munition setup. It is a green light-emitting diode (LED) that indicates a self-test was successfully performed or an operating-mode selection was successfully selected.
SD switch (5)	See Figure B-2, page B-3.
Arming lever (6)	See Figure B-2.
Active battery-pack cover (7)	See Figure B-2.

Figure B-3. Hornet controls and indicators

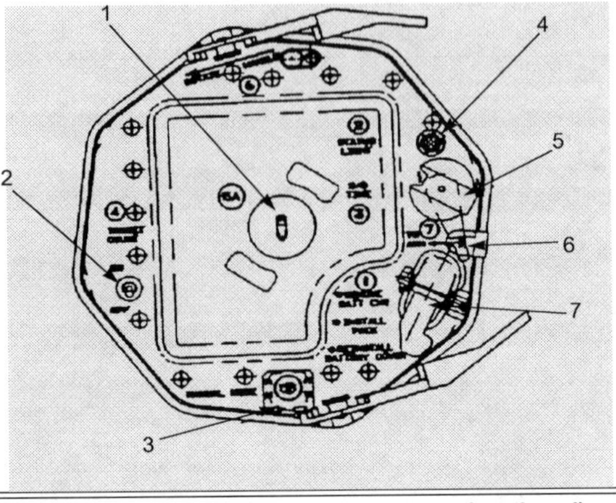

Figure B-3. Hornet controls and indicators (continued)

Appendix C

Threat Mine/Countermine Operations

This appendix is intended to complement the information presented in other manuals on threat obstacle tactics. It applies to most threat armies and their surrogates. Commanders should use this information to give added realism to unclassified training, although obstacle employment norms can change with METT-TC factors for a given AO. Therefore, preoperational training on templating, intelligence, reconnaissance, and reduction procedures must be based on the best information available before deployment.

Appendix G contains a compilation of countermine data.

MINE OPERATIONS

Threat formations contain considerable organic minefield emplacement capability. Threat rapid-mining capability presents a serious challenge to friendly maneuver.

To lay mines and place obstacles rapidly during offensive operations, threat armies form a special team from regimental and divisional assets. This team is called a *mobile obstacle detachment* (MOD). The MOD places AT mines on the most likely avenues for armored attacks or counterattacks. MODs are positioned on the flanks of a march formation for rapid deployment and are normally close to AT reserves. During the march, MODs reconnoiter avenues into the flanks and identify the most likely avenues for tank movement. At secured objectives, MODs reinforce existing obstacles and place new obstacles to assist in the defeat of counterattacks.

The combined arms commander orders the organization of MODs and determines their composition based on the combat situation and available troops. Engineer elements in a division MOD come from the divisional engineer battalion and normally consist of three armored tracked mine layers known as *GMZs* (Figure C-1, page C-2). This platoon-sized element has two or three trucks that carry mines for immediate resupply. For the regimental MOD, the regimental engineer company normally provides a platoon-sized unit equipped with two or three GMZs. The platoon travels in BTR-50/60s and has 600 AT mines.

The GMZ dispenses mines at a predetermined spacing of 5.5 meters. Mine-laying helicopters also support the MOD. The HIP and HIND-D helicopters carry two or three dispenser pods of AP or AT mines. Artillery-fired SCATMINEs can also support the MOD. Three GMZs can lay a 1,200-meter, three-row minefield, containing 624 mines, in 26 minutes. Doctrinally, this minefield would be broken into several minefields, each 200 to 300 meters long.

Threat armies use obstacles extensively throughout the depth of their defense, and their tactics are chosen well. Shallow obstacles are reduced quickly and easily. For example, a shallow, one-row minefield is essentially reduced by blowing one or two mines in the row. A threat rapidly emplaced minefield

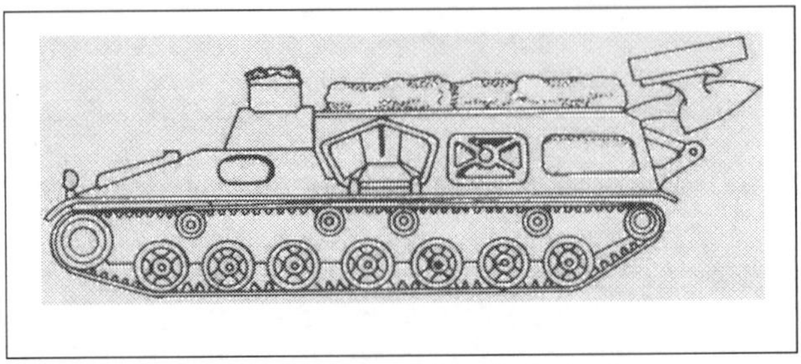

Figure C-1. GMZ armored tracked mine layer

consists of three or four 200- to 300-meter rows, spaced 20 to 40 meters apart, with mines spaced 4 to 6 meters apart. As a rule, the minefield covers the depth of a football field.

Table C-1 provides detailed information on standard threat AT and AP minefields. Terrain and tactical situations dictate the actual dimensions and distances of minefields.

Table C-1. Normal parameters for threat-style minefields

AT Minefields	
Front (situation-dependent)	200 to 300 meters
Depth	40 to 120 meters
Number of rows	3 or 4
Distance between rows	20 to 40 meters
Distance between mines	4 to 6 meters for antitrack mines; 9 to 12 meters for anithull mines
Outlay, normal	550 to 750 antitrack mines per kilometer; 300 to 400 antihull mines per kilometer
Outlay, increased effect	1,000+ antitrack mines per kilometer; 500+ antihull mines per kilometer
Probability of destruction	57% for antitrack mines (750 per kilometer); 85% for antihull mines (400 per kilometer)
AP Minefields	
Front (situation-dependent)	30 to 300 meters
Depth	10 to 150 meters
Number of rows	3 or 4
Distance between rows	5+ meters for blast mines; 25 to 50 meters for fragmentation mines
Distance between mines	1 meter for blast mines; 50 meters (or twice the lethal radius of fragmentation) for fragmentation mines
Outlay, normal	2,000 to 3,000 HE/blast mines per kilometer; 100 to 300 fragmentation mines per kilometer
Outlay, increased effect	2 to 3 times the normal outlay
Probability of destruction	15 to 20% for HE/blast mines (2,000 per kilometer); 10 to 15% for fragmentation mines (100 per kilometer)

Figure C-2 shows a standard rapidly emplaced minefield. The threat army typically uses such a minefield when they are in a hasty defense (offense is temporarily stalled).

Figures C-3 shows a standard antitrack minefield.

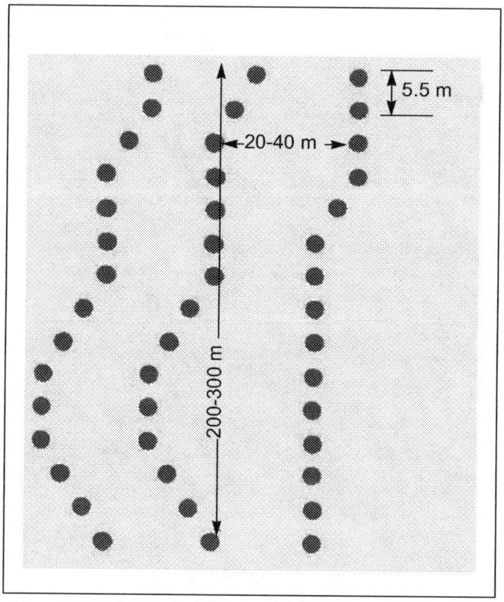

Figure C-2. Threat-style rapidly emplaced minefield

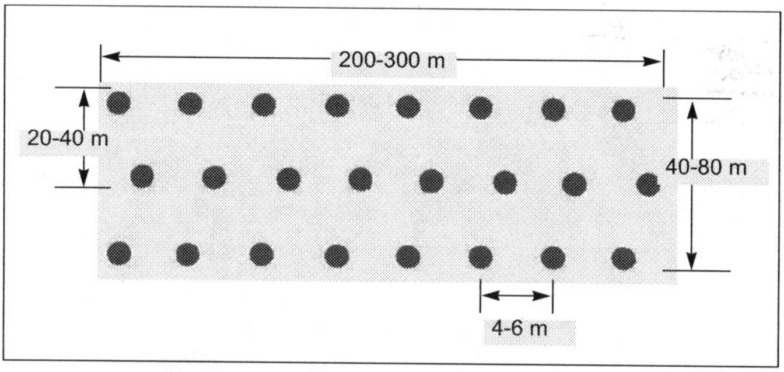

Figure C-3. Threat-style antitrack minefield

Figure C-4 shows a standard antihull minefield.

Figure C-5 shows a standard AP minefield.

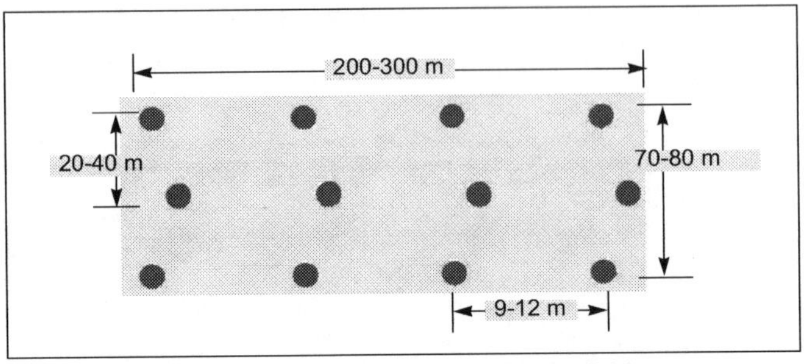

Figure C-4. Threat-style antihull minefield

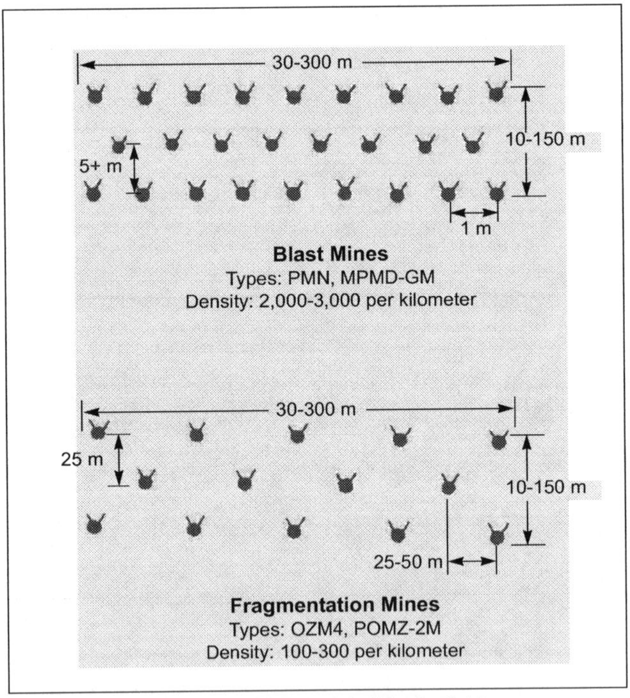

Figure C-5. Threat-style AP minefield

Threat armies also emplace mixed minefields. They are not the same as US mixed minefields. Threat armies normally emplace three rows of AT mines, then several rows of AP mines. AT and AP mines **are not** mixed in the same row.

Threat engineers use two fundamental drills to emplace mines:

- When emplacing armed mines, the drill uses a crew of five sappers. The first crew member (the senior man and operator) is in the mine-layer's seat and monitors the operation of the mine layer and the motion of the mines in the guide chute. He also sets the mine spacing and controls the actions of the GMZ. The second and third members take mines out of containers and place them in the intake chute at intervals between the guide tray's drive chain. The GMZ driver steers the vehicle along the indicated route at the established speed.
- When emplacing unarmed mines, two or three additional sappers are assigned to arm the mines. After emplacing the mines, one sapper trails the mine layer, marks emplaced mines with pennants, and partially camouflages the mines. The remaining sapper(s) then arm the mines.

Special precautions are taken when emplacing AP minefields. Threat doctrine only allows PMN mines to be surface-laid from mine layers. POMZ-2M mines are emplaced with the truck-and-tray technique. Extra effort is required to assemble, emplace, and deploy the trip wire and to camouflage the POMZ-2M mine.

Using three GMZs, a threat MOD can emplace 1,200 meters of a three-row AT, surface-laid minefield, containing 624 AT mines, in 26 minutes. This does not include the 12- to 15-minute reload and travel times. Travel and reload times increase during limited visibility.

Threat forces can also have ground-emplaced SCATMINE capability. One such system is the UMZ SCATMINE system (Figure C-6, page C-6). There are three UMZ truck-mounted SCATMINE systems in each combat regiment. The UMZ consists of six firing modules mounted on the back of a Zil-131 truck. Each module has 30 firing tubes, for a total of 180 firing tubes per system. Depending on the type of minefield desired, the UMZ can lay 180 to 11,520 mines without reloading. The UMZ can launch an AT or AP minefield 30 to 60 meters from the vehicle while the truck is driving 10 to 40 kph. It takes two men 1½ to 2 hours to reload the UMZ. One UMZ can lay a three-row minefield, 150 to 1,500 meters long, depending on the type of mine that is used.

UMZ vehicles are usually deployed together as a mobile obstacle/mine-laying detachment. The UMZ is used to lay minefields that protect subunit positions and flanks and the boundaries between subunits. UMZ-laid minefields also cover firing lines and gaps in combat formations. The UMZ can quickly close gaps in existing minefields and increase the density of mines on armor avenues of approach.

For hand-emplaced SCATMINEs, there is a man-portable SCATMINE dispenser. The PKM weighs 2.63 kilograms (without the mine canister) and consists of a single launch tube with a base mount, a blasting machine, and a reel of electric ignition wire. The operator loads a propelling charge and a mine canister into the launch tube and mounts the tube on the edge of a trench or firing parapet. He then aims the tube, connects the ignition wire to

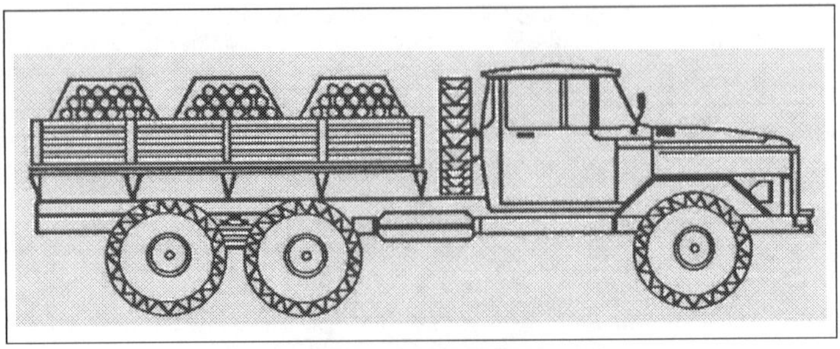

Figure C-6. UMZ SCATMINE system

the tube, and moves to a safe distance. At an initiating point, the operator connects the ignition wire to the blasting machine and initiates the system. The PKM propels the canister 30 to 100 meters, depending on the type of mine. It lays an AP minefield that is 10 by 20 meters (POM-1S mine canister), 10 by 40 meters (POM-2S mine canister), or 20 by 10 meters (PFM-1S mine canister). It takes a trained operator 5 minutes to set up the PKM and create a minefield. The PKM can also be used to launch the PTM-1S and PTM-3 AT mine canisters.

Threat forces use the PKM to lay minefields that protect subunit positions and flanks and the boundaries between subunits. PKM-laid minefields also cover firing lines and gaps in combat formations. The PKM can quickly close breaches in existing minefields and increase the density of mines on armor avenues of approach.

The type and complexity of an obstacle depends on the installing unit. Maneuver and artillery soldiers usually install simple single-system minefields that are protective in nature. Engineer soldiers install complex obstacles that can include AHDs. Engineer obstacle placement is usually equipment-intensive. Threat engineer effort generally concentrates on tactical obstacles unless maneuver soldiers are unable to employ the necessary protective obstacles. Threat units continue to improve the obstacles, supporting their positions by marking the friendly side of the obstacles, burying mines, and adding AHDs.

CHEMICAL MINES

Chemical land mines are AP mines with command- or target-detonated fuses, and they are filled with a persistent chemical (nerve or blister) agent. US policy prohibits their use by US personnel. However, this does not preclude their use by other countries, and US forces may encounter them during operations. When used, they are normally used in defense and retrograde operations. They are mixed with HE mines to form a HE chemical minefield. Chemical mines are normally encountered in tactical or nuisance minefields, and some countries use them in protective minefields. When an integrated HE chemical minefield is laid, it serves the following purposes:

- Chemical mines discourage the use of explosive, rapid mine-clearing devices because they create a chemical hazard in the area.

- HE mines reduce the speed of enemy forces crossing the minefield. Speed is further reduced by forcing the enemy to use protective clothing and masks.

Chemical mines will usually be added to existing HE minefields by laying additional strips of chemical mines in a random pattern or by adding HE chemical strips or rows to the front or rear of existing fields (Figure C-7).

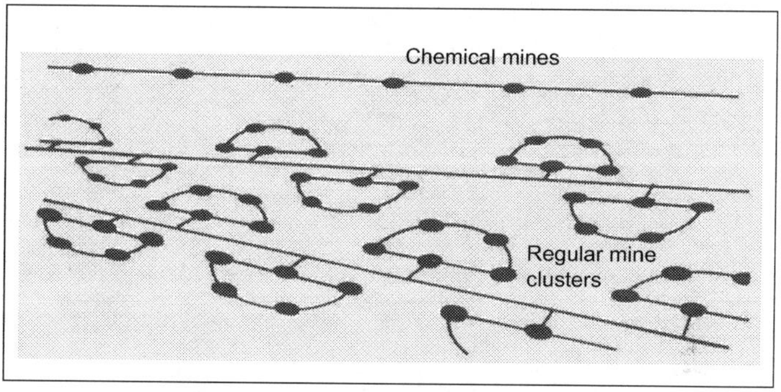

Figure C-7. Chemical-mine employment

No particular branch is responsible for clearing chemical mines. Planning chemical countermine operations is a brigade-level responsibility. When reducing chemical mines, consider prevailing and expected wind conditions. Commanders must ensure that friendly troops are protected when chemical agents are released. The release of chemical agents occurs as a result of enemy fire or friendly breaching attempts. Contact-actuated chemical mines are not likely to create a major downwind hazard because only single mines or small groups may be set off at one time.

COUNTERMINE OPERATIONS

In offensive operations, threat engineers clear lanes through obstacles when they cannot be bypassed. Although clearing obstacles applies to the march and the defense, the most critical performance of this task occurs during the attack. Engineers can be required to clear mines delivered by air, artillery, and rockets well ahead of NATO's forward edge. They must breach obstacles contained within NATO strongpoints. Threat forces must also clear their own minefields when making the transition from defense to offense. In the offense, threat forces breach or bypass remotely delivered minefields in their form-up areas or routes of movement to the attack line. They also breach obstacles along the forward edge of the battle area and deep within NATO defenses.

Although clearing passages through obstacles is a primary task for threat engineers, any maneuver element may encounter mines. Engineers may not be able to respond to every encounter, so maneuver troops are also required to breach through remotely emplaced obstacles.

ORGANIZATION

A movement support detachment (MSD) supports the movement of maneuver forces. It is task-organized from divisional or regimental engineer assets and can be platoon- to company-size. The MSD is equipped with route- and mine-clearing vehicles and devices. Depending on the mission (which comes directly from the combined arms commander or the chief of engineer services), an MSD is capable of filling craters, clearing minefields, preparing bypasses around major obstructions, and identifying NBC-contaminated areas.

The divisional engineer battalion can form two or three MSDs. During marches, MSDs travel in advance of the main body and clear obstructions reported by division reconnaissance elements. When they are deployed on main routes, they are under the protection of an advance guard or forward security element. When deployed on other routes, the leading regiments provide MSDs from organic engineer assets. An MSD at this level might consist of an engineer platoon, with one or two dozers and up to three tanks fitted with dozer blades. MSDs can be protected by a platoon of infantry or tanks and are usually accompanied by chemical-reconnaissance personnel. They can detect, mark, and breach hasty minefields that are not properly covered by fire. If MSDs encounter properly defended minefields, their clearing capabilities are limited.

Each battalion forms an obstacle-clearing group to create gaps in explosive and nonexplosive obstacles. Normally a part of a battalion-level MSD, the group follows first-echelon companies in APCs and creates gaps for those forces. These units may possess BAT-M vehicles with BTU bulldozer blades (Figure C-8) or KMT-series mine plows (Figure C-9).

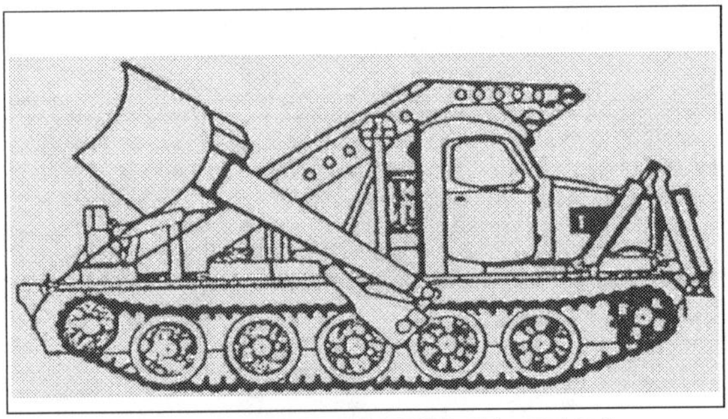

Figure C-8. BAT-M with BTU bulldozer blade

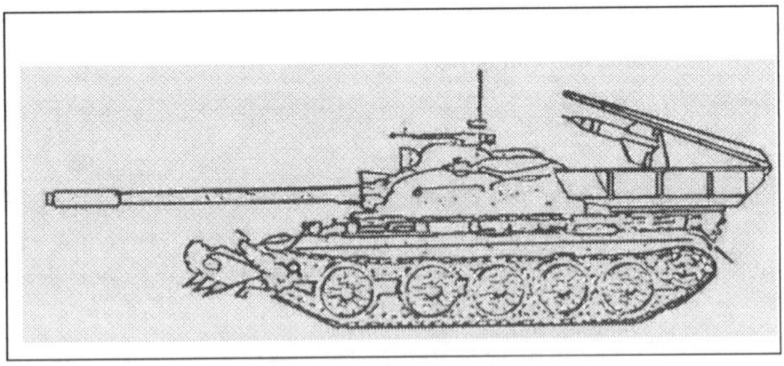

Figure C-9. KMT-4 plow

An obstacle-clearing detachment is created when more resources are needed to clear obstacles and debris. This usually occurs in urban environments and under conditions of massive destruction. An obstacle-clearing detachment is similar to an MSD, but its sole mission is to clear debris. Like an MSD, its composition depends on the mission scope, the mission objective, and the tempo of the offensive.

The divisional engineer battalion of the motorized rifle or tank division has a sapper company to clear obstacles. The company commander receives a mission to clear minefields. He then determines the exact location of the obstacle, ascertains the assets to devote to the task, and plans the methodology for success. Teams may be created to manually breach lanes using probes, IMP portable mine detectors (Figure C-10), and shovels. Larger tasks may necessitate the use of vehicle-mounted DIM mine detectors (Figure C-11), armored vehicle mine plows and/or rollers (Figure C-12, page C-10), and explosive line charges. When necessary or more practical, mines are explosively destroyed in place.

Figure C-10. IMP portable mine detector

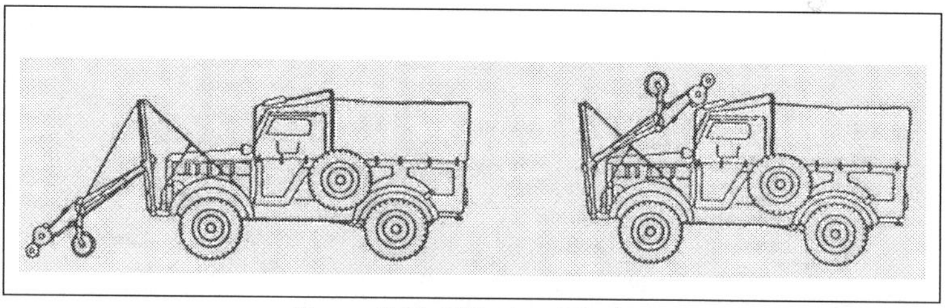

Figure C-11. DIM mine detector

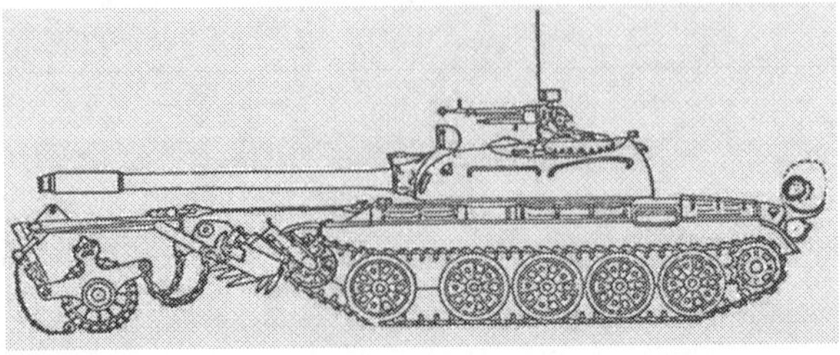

Figure C-12. KMT-5 plow-roller combination

The engineer company of the motorized rifle or tank regiment has breaching equipment such as KMT-series mine plows and rollers and BTU bulldozer blades located in its technical platoon. Because of limited assets in the technical platoon, coupled with the responsibility of forming its own MSD, the regiment can receive a sapper section from the divisional sapper company. An additional IMR armored engineer tractor (Figure C-13), BTR-50/60, and M1979 armored mine clearer (Figure C-14) and manual breaching equipment come with the sapper section.

Figure C-13. IMR armored engineer tractor

Maneuver units usually breach remotely emplaced obstacles by using attached, built-in breaching equipment (BTUs and KMTs). In order to carry out this task successfully, all subunit commanders must organize constant reconnaissance, notify subordinates about mined areas in a timely manner, train personnel on the means and methods for handling remotely emplaced mines, and clear terrain in a timely manner. They must also train their own teams for independent actions when removing combat equipment from mined areas. Plows in the threat army are considered maneuver-force assets, and one plow is assigned to each tank platoon. The BMP has recently been equipped with track-width mine plows, but the allocation has not been determined.

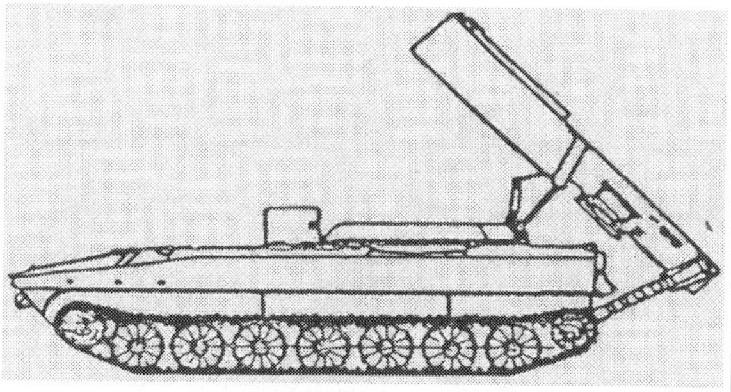

Figure C-14. M1979 armored mine clearer

EQUIPMENT

Several pieces of equipment are used by threat armies to detect and clear mines.

BAT-M Dozer

The BAT-M dozer (Figure C-8, page C-8) is a modified artillery tractor with a hydraulically operated bulldozer blade and crane. It is sometimes called a *roader* by Russians. The BAT-M dozer clears obstacles, fills craters, prepares bridge approaches, and performs other heavy pioneer tasks. It can also be configured for snowplowing.

The second generation BAT-M is the BAT-2. The BAT-2 is able to carry an 8-man engineer squad and operate in an NBC environment. It is replacing the BAT-M.

KMT-Series Plows and Rollers

KMT-4

The KMT-4 mine-clearing plow (Figure C-9, page C-8) was developed in the 1960s to fit on a T-545/55 tank. It actually consists of two plows (one mounted in front of each track), and each plow has five attached teeth. When the plow is lowered, the teeth dig into the ground and remove mines from the path of the tank. A plow is lighter than a roller and permits tanks to retain their cross-country mobility. The estimated clearing speed is 10 kph, and the depth of clearance is 10 centimeters.

Three plows are issued per tank company (one per platoon). However, these assets are normally held in the engineer company of a tank or MRR.

KMT-5

The KMT-5 mine-clearing plow-roller combination (Figure C-12, page C-10) consists of two plows and two rollers attached to the front of a tank hull. The plows or the rollers can be used, depending on terrain features, the type of soil, and the mine fuse. Plows and rollers cannot be used simultaneously. The rollers function against pressure-fused mines. The system can survive 5 to 6 kilograms of explosives, five or six times. The KMT-5 also includes a luminous lane-marking device for night operations.

KMT-6

The KMT-6 mine-clearing plow was introduced with the T-64 and T-72 tanks in the early 1970s. It has operating characteristics similar to those of the KMT-4.

KMT-10

The KMT-10 mine-clearing plow is fitted to the BMP-2 infantry combat vehicle.

IMP Portable Mine Detector

The IMP portable mine detector (Figure C-10, page C-9) weighs 7 kilograms and can detect mines buried to a depth of 45 centimeters. It has a tubular search head (one transmitting and two receiving coils encased in plastic) and a four-section handle. Power is furnished by four flashlight batteries that permit 20 hours of continuous operation. Two tuning controls are mounted on the handle. The coils in the search head compromise an induction bridge and are initially balanced for zero coupling. When the head passes over a metallic object, the induction bridge becomes unbalanced and produces an audible signal in the headset.

DIM Vehicle-Mounted Mine Detector

The DIM vehicle-mounted mine detector (Figure C-11, page C-9) is primarily used to clear roads during convoys and road marches. It sweeps at a speed of 10 kph with a 2.2-meter width. It can detect metallic mines at a depth of 25 centimeters. The brakes on the DIM automatically engage when a mine is detected. Cross-country use of the DIM is limited.

IMR Armored Engineer Tractor

The IMR armored engineer tractor (Figure C-13, page C-10) is mounted on a modified T-54/55 chassis. The turret is removed and a hydraulic crane, which can be fitted with either a grab or an excavator bucket, is emplaced. An adjustable, hydraulically operated blade is mounted on the front. The crane operator is provided with an armored cupola. The IMR can operate in an NBC environment.

M1979 Armored Mine Clearer

The M1979 armored mine clearer (Figure C-14, page C-11) is mounted on the chassis of an amphibious 122-millimeter, 2S1 self-propelled howitzer. It has a turret-like superstructure that contains three rockets on launch ramps. These, along with the upper part of the superstructure, are hydraulically elevated for firing. The rocket range is estimated at 200 to 400 meters. Each rocket is connected to 170 meters of mine-clearing hose via a towing line. The hose is folded and stowed in the uncovered base of the turret and connected to the vehicle with a cable. The cable allows the vehicle crew to reposition the hose after launching.

This chapter implements STANAG 2990.

Appendix D

Air Volcano

The air Volcano system provides a three-dimensional capability that allows units to emplace minefields in deep, close, and rear operations. It provides US forces with the capability to employ minefields rapidly under varied conditions. The air Volcano can be used to emplace tactical minefields; reinforce existing obstacles; close lanes, gaps, and defiles; protect flanks; and deny the enemy use of potential air-defense sites. Volcano minefields are ideal for flank protection of advancing forces and for operating in concert with air and ground cavalry units on flank-guard or screen missions.

COMPONENTS

The air Volcano system (Figure D-1) consists of an M87-series mine canister, an M139 dispenser, and vehicle-specific mounting hardware (UH-60 Blackhawks require a jettison kit).

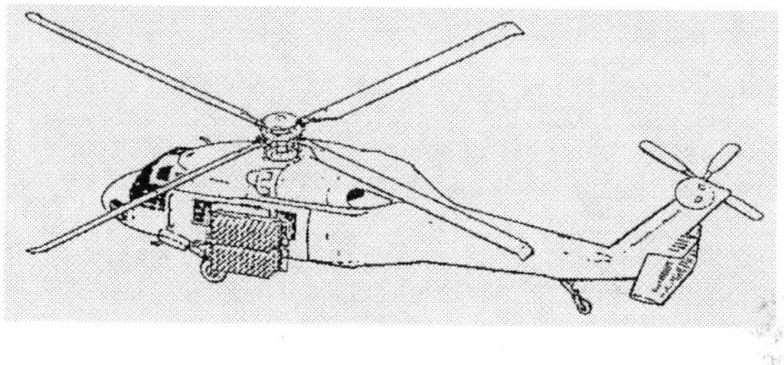

Figure D-1. Air Volcano system

M87-Series Mine Canister

The M87-series mine canister is the same canister used for the ground Volcano system:

- M87. Prepackaged with five AT mines, one AP mine, and a propulsion device inside a tube housing.
- M87A1. Prepackaged with six AT mines and a propulsion device inside a tube housing.

The mixture of mines is fixed and cannot be altered in the field. The mines in each canister are electrically connected with a web that functions as a lateral dispersion device as the mines exit the canister. Spring fingers mounted on each mine prevent the mine from coming to rest on its edge. AT mines have a delay-arm time of 2 minutes 30 seconds; AP mines have a delay-arm time of 4 minutes. All canisters are capable of dispensing mines with 4-hour, 48-hour, or 15-day SD times. SD times are selected prior to dispensing and do not require a change or modification in the base M87-series mine canister.

M139 Dispenser

The M139 dispenser consists of an electronic DCU and four launcher racks; each rack holds 40 M87-series mine canisters. The racks provide the structural strength and the mechanical support required for launch and provide the electrical interface between the mine canisters and the DCU. Mounting hardware for the UH-60A Blackhawk includes a jettison subassembly to propel the Volcano racks and canisters away from the aircraft in the event of an emergency.

The operator uses the DCU to electrically control the dispensing operation from within the aircraft. The DCU provides controls for the arming sequence and the delivery speed selection, and it sets mine SD times. The DCU allows the operator to start and stop mine dispensing at anytime. A counter on the DCU indicates the number of canisters remaining on each side of the aircraft.

Mines are dispensed from their canisters by an explosive propelling charge, 35 to 70 meters from the aircraft's line of flight. The aircraft flies at a minimum altitude of 1.5 meters, at speeds of 20 to 120 knots. It can deliver up to 960 mines per sortie.

LIMITATIONS

The total weight of the air Volcano system is 2,886 kilograms. An aircraft will be close to its maximum gross weight when it contains the Volcano system and a full crew. Based on weather and environmental conditions, the aircraft may be required to execute the mission without a full fuel load, thus reducing enroute time.

The flight crew cannot operate the M60D machine gun with the air Volcano system installed, and it takes 3 to 4 hours to install the air Volcano system on a UH-60A Blackhawk.

EMPLOYMENT

The air Volcano is the fastest method for emplacing large tactical minefields. When employed by combat aviation elements in support of maneuver units, close coordination between aviation and ground units ensures that mines are emplaced accurately and quickly. Although placement is not as precise as it is with ground systems, air Volcano minefields can be placed accurately enough to avoid the dangers inherent in minefields emplaced by artillery or jet aircraft.

Air Volcano minefields can be emplaced in friendly and enemy territories. They should not be planned in areas of enemy observation and fire, because the helicopter is extremely vulnerable while flying at the steady altitude, the speed, and the path required to emplace the minefield. The air Volcano is the best form of a situational obstacle because of its short emplacement time. Its employment varies depending on the type of operation.

DEEP OPERATIONS

Employment

The air Volcano is employed in deep operations to—

- Disrupt enemy formations along key AAs and choke points.
- Fix enemy formations in EAs to enhance target acquisition by attack helicopters, CAS, artillery, or a combination of weapon platforms.
- Provide area denial of possible enemy artillery and ADA sites that will affect future friendly schemes of maneuver.

AH-64 security is essential to air Volcano missions because the target area must not be in the enemy formation's direct line of sight. It may require two suppression of enemy air defense (SEAD) missions to get the air Volcano across the FLOT to the target location. This removes the security aircraft from the battle. All air Volcano missions require planning, designation, and control during the execution of air corridors (routes).

Aviation Configuration

The air Volcano can be used in deep operations, but it has distinct limitations. The aircraft can maintain a speed of 80 to 90 knots (UH-60A) for 90 to 120 kilometers, depending on the wind and the temperature (see Table D-1, page D-4).

The aircraft cannot employ the door guns for self-defense with the Volcano system mounted. The attack aircraft averages 100 to 120 knots while targeting enemy formations that are 150 to 200 kilometers deep. In order for the air Volcano to go deep, the attack aircraft must slow down or special security/escort aircraft must be assigned for protection. The security/escort can be AH-64s or OH-58D Kiowa Warriors (KWs). KWs have many advantages—they are not a primary killing platform, and they are very good at quickly securing an area. The air Volcano requires one or two security aircraft for protection, and there should also be a backup aircraft for the mission.

Fire-Support Coordination

The division main (light force) or brigade main (mechanized force) FSE is responsible for coordinating and executing fires in support of air Volcano missions. Upon completion of the target meeting, the FSE, the assistant division/brigade engineer, and the G3/S3 air representative will coordinate to ensure that the air-coordination/tasking order will support the mission and the planned SEAD fires.

If the mission is a deep aviation attack, the aviation element FSE is responsible for coordinating through the forward command post to the division/brigade main FSE. If the forward command post has *jumped forward* or has not deployed, the aviation element FSE will coordinate directly with the division/brigade main FSE.

Table D-1. Air Volcano capabilities and limitations

Pressure Altitude (ft)	Temperature (°C)	Maximum Weight (lb)[1]	Cruise Speed (knots/kph)	Fuel Burn Rate (lb per hr)	Endurance (hr + min)[2]	Maximum Range (nm/km)[3]	Planning Combat Radius (nm/km)[4]
0	+20	22,000	80/148	965	1 + 57	154/285	60/111
			90/166	1,040	1 + 48	160/296	
			100/182	1,145	1 + 35	158/292	
	+30	22,000	80/148	975	1 + 56	156/289	59/109
			90/166	1,045	1 + 48	160/296	
			100/182	1,160	1 + 34	156/289	
	+40	21,600	80/148	995	1 + 53	150/277	58/107
			90/166	1,055	1 + 45	156/289	
			100/182[7]	1,175	1 + 33	154/285	
2,000	+20	22,000	80/148	960	1 + 58	157/290	60/111
			90/166	1,060	1 + 45	157/290	
			100/182	1,155	1 + 35	158/292	
	+30	21,000[5]	80/148	1,000	1 + 28	117/217	45/83
			90/166	1,045	1 + 23	124/229	
			100/182[7]	1,175	1 + 12	120/222	
	+40	20,000[6]	80/148	1,005	1 + 21	28/52	7/13
			90/166[7]	1,075	1 + 19	29/53	
			100/182[7]	1,150	1 + 16	27/49	

NOTE: Based on UH-60A with an aircraft torque factor of 97.5.

[1]Maximum weight for deep operations only; no interdiction capability.

[2]Endurance includes 20-minute fuel reserve.

[3]Does not compensate for winds. This is the maximum range line-of-site distance.

[4]The general planning figures used for time-on-target missions; includes compensation for 10-knot winds, makeup for lost time, and a 5-minute on-station time.

[5]1,800 pounds of fuel at takeoff

[6]700 pounds of fuel at takeoff

[7]Operation at this air speed may be limited to 30 minutes due to target limits.

LIMITATIONS:

- There are no modifications to mount the M60D machine gun on the air Volcano aircraft.
- Crews cannot use the M60D with the system mounted on an aircraft.
- Crews cannot fly the aircraft with full tanks above +30°C and 2,000-ft pressure altitude.
- Aircraft may require a rolling takeoff and landing, depending on ambient weather conditions.

CLOSE OPERATIONS

Employment

The air Volcano is employed in close operations to fix enemy formations in EAs, turn advancing formations into desirable terrain that supports the friendly scheme of maneuver, disrupt formations enough to slow the enemy down, and block key AAs. Multiple missions may be required, depending on the intended effect. The air Volcano can be used to reseed existing minefields or to close lanes and gaps. The target area must be clear of friendly forces before an air Volcano mission is executed.

Use of the air Volcano in close operations should be a primary planning consideration. It can quickly reach the outer edge of the forward operating base where AAs need a minefield obstacle. The threat level will be lower, and the station time will increase.

Aviation Configuration

Two air Volcano aircraft should be used (one primary, one backup). The requirement for security aircraft depends on METT-TC factors, but security should be used whenever possible.

Fire-Support Coordination

The forward command post FSE coordinates and executes fires in support of air Volcano missions. The FSE, the engineer liaison officer, and the G3/S3 representative coordinate to ensure that the air coordination/tasking order supports the mission and the planned SEAD fires. The division/brigade main will be available to support the forward command post as necessary.

The brigade/TF FSE is responsible for coordinating through the forward command post to the division/brigade main FSE. If the forward command post has jumped, the brigade/TF FSE coordinates directly with the division/brigade main FSE.

REAR OPERATIONS

Employment

The primary purposes of the air Volcano in rear areas is to protect key terrain from possible airborne/air-assault forces and to fix/disrupt enemy forces long enough to allow the tactical combat force or ready-reserve force time to react and meet the changing enemy situation.

The least preferred employment method is to deliver tactical minefields to brigade and corps support areas. This employment tactic is normally used when all other available assets have been exhausted. The flexibility of the air Volcano system makes it ideal for employment against a mounted Level III threat in the rear. The target area should be out of the direct view/fire of the threat and on a choke point that allows cover for the reacting forces.

Aviation Configuration

The air Volcano aircraft could be employed individually or with security/escort aircraft. The use of OH-58D KWs as security aircraft allows units to develop the situation and helps place minefields in the proper location to assist inbound attack aircraft or fires. If the air Volcano aircraft is not provided security aircraft, it is recommended that ground forces provide covering fires.

Fire-Support Coordination

The division/brigade rear FSE coordinates and executes fires in support of air Volcano missions. The FSE, the engineer liaison officer, and the G3/S3 representative coordinate to ensure that the air coordination/tasking order supports the mission and the planned SEAD fires. The division/brigade main will be available to support the division/brigade rear as necessary.

The headquarters element that controls the rear area coordinates with the division/brigade rear FSE. The division/brigade rear FSE coordinates with the division/brigade FSE for fire support and air assets.

MINEFIELD EFFECTS

Turn

A turn minefield manipulates enemy maneuver in a desired direction. It forces or entices enemy formations to move in a different direction rather than breach the obstacle. This means the bypass must be easily identified. Turn minefields are extremely lethal, with approximately 80 percent probability of mine encounter. The typical width is 557 by 320 meters for air Volcano. Figure D-2 shows two turn minefields combined to create a turn-effect obstacle group. It takes 160 canisters (800 AT/160 AP mines) to emplace one turn minefield. One air Volcano aircraft can lay one turn minefield (see Table D-2).

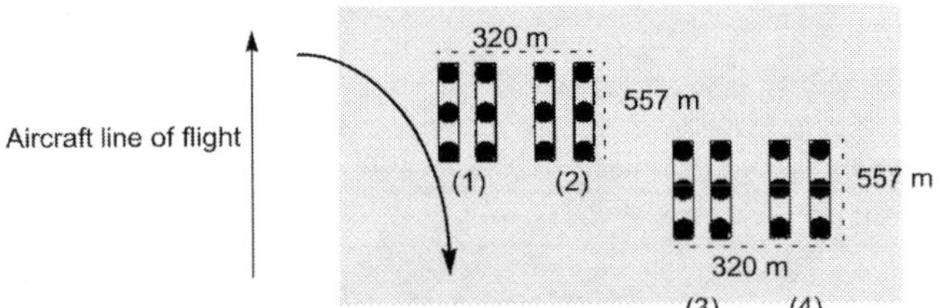

NOTE: Numbers correspond to the aircraft pass.

Figure D-2. Turn obstacle

Table D-2. Air Volcano minefield data

Type of Minefield	Depth (m)	Frontage of Minefield (m)	Number of Strips	Canisters per Strip	Total Canisters	Minefields per Aircraft
Disrupt	120	278	1	40 (20 each side)	40	4
Fix	120	278	1	40 (20 each side)	40	4
Turn	320	557	1	80 (40 each side)	160	1
Block	320	557	1	80 (40 each side)	160	1

Block

A block minefield (Figure D-3) is designed to stop an enemy advance along a specific AA or allow it to advance at an extremely high cost. Block minefields are obstacles with intensive integrated fires. They should be employed in a complex obstacle scheme with road craters or bridge demolitions enhancing the effectiveness of the minefield. One air Volcano aircraft can lay a 557- by 320-meter block minefield, using all 160 canisters (800 AT/160 AP mines). The probability of mine encounter is more than 80 percent. One Volcano aircraft can lay one block minefield, making two passes side by side (see Table D-2).

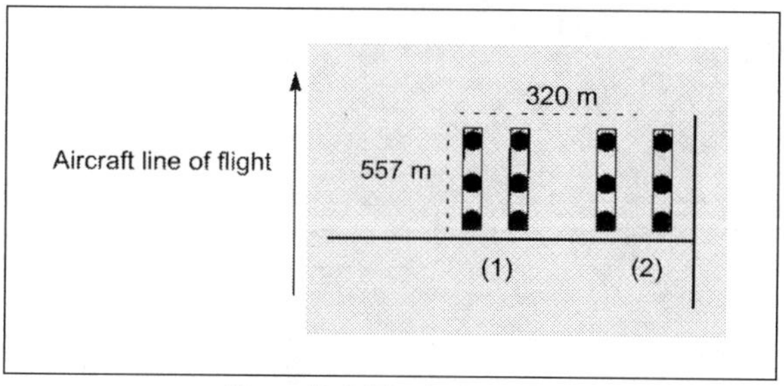

Figure D-3. Block obstacle

Disrupt

A disrupt minefield (Figure D-4) fractures and breaks up enemy formations. It causes premature commitment of reduction assets, interrupts C^2, and alters timing. A disrupt minefield is not resource- or time-intensive. The probability of mine encounter is approximately 50 percent, and the typical width is 278 by 120 meters. It takes 40 canisters (200 AT/40 AP mines) to emplace one disrupt minefield. One air Volcano aircraft can lay four disrupt minefields (see Table D-2).

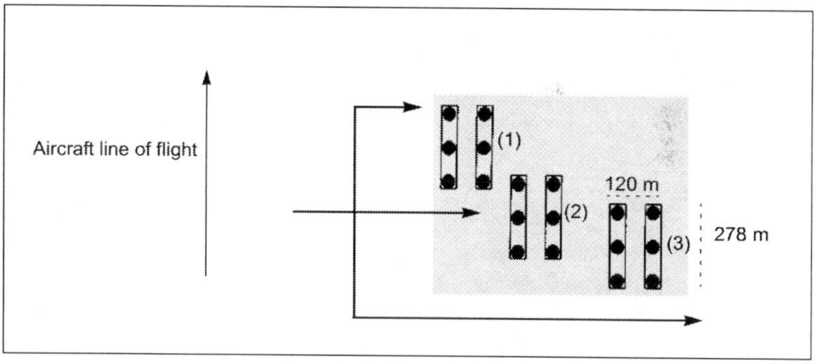

Figure D-4. Disrupt obstacle

Fix

A fix minefield (Figure D-5) slows enemy formations within a specified EA. It gives friendly forces time to acquire, target, and destroy enemy formations. A fix minefield is employed in depth and causes enemy formations to react and breach repeatedly. The air Volcano fix minefield does not look impenetrable to enemy formations. The probability of mine encounter is 50 percent, and the typical size is 278 by 120 meters. It takes 40 canisters (200 AT/40 AP mines) to emplace one fix minefield. One air Volcano aircraft can lay four fix minefields (see Table D-2, page D-6).

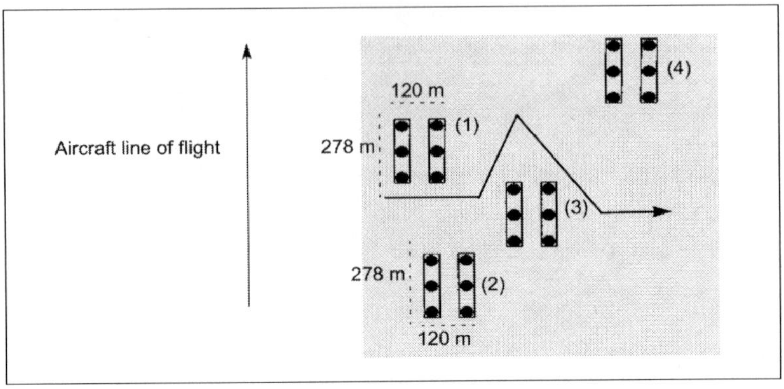

Figure D-5. Fix obstacle

Linear

If the threat situation allows, the aircrew makes a pass to confirm the minefield end points and the suitability of the terrain. In a high-threat situation, the aircrew emplaces the minefield on the first pass. This hasty minefield is linear in configuration and is 1,115 by 75 meters. All 160 canisters are fired.

PLANNING

Responsibilities

Division Commander

The division commander approves air Volcano employment and integration into deep, close, and rear operations. He is also the authority for SCATMINE employment.

Maneuver Brigade Commander

When authority is delegated by the division commander, the maneuver brigade commander is responsible for employing air Volcano in close operations and in supporting follow-on missions. He is responsible for approving target nominations to be submitted to the division. The maneuver brigade commander receives the air Volcano aircraft and its crew in OPCON status.

Aviation Brigade Commander

The aviation brigade commander is responsible for integrating the air Volcano into deep aviation attacks and for shaping EAs in the division AO. He executes the air Volcano missions in deep, close, and rear operations. The aviation brigade commander submits target nominations to the division targeting cell through the aviation brigade engineer and provides support for transporting and loading Volcano mines.

Division Engineer

The division engineer is responsible for target nominations that support division missions or objectives in deep operations. He—

- Submits the nominations to the division targeting cell.
- Determines minefield characteristics.
- Incorporates the Volcano minefield, DTG of SD times, and safety zones into the division obstacle plan.
- Estimates the requirements for mine canisters and Class IV/V supplies.
- Determines the intent of the Volcano minefield as it is integrated into the division obstacle plan.
- Disseminates the SCATMINWARN to adjacent and subordinate units before the minefield is laid and one hour before the SD sequence of the minefield is initiated.

Brigade Engineer

The brigade engineer is responsible for target nominations that support objectives within his respective fight. He—

- Submits target nominations to the assistant division engineer.
- Synchronizes air Volcano missions into brigade operations.
- Assists the brigade FSE in planning SEAD, CAS, and Apache escort.
- Provides logistics estimates to the brigade S4 for coordination of ammunition requirements.
- Posts the operations map with the minefield's location, aviation graphics, the safety zone, and the DTG of SD times.
- Disseminates the SCATMINWARN to adjacent and subordinate units before the minefield is laid and one hour before SD sequence of the minefield is initiated.

Deputy Fire-Support Coordinator

The deputy FSCOORD recommends commanding general/chief of staff approval for target nominations developed by the division targeting cell. He is responsible for submitting division-approved air Volcano target nominations to the G3 air for inclusion in the air coordination order. He coordinates SEAD, CAS, and intelligence and electronic warfare for division-directed air Volcano missions. The deputy FSCOORD is also responsible for—

- Targeting (intelligence and asset coordination).
- The air tasking order.
- The G3 air coordination order.
- G2 collection and assessment.

G3 Air

The G3 air synchronizes coordination and deconfliction of division air space for air missions, SEAD, and CAS. When necessary, he submits the air coordination order to higher headquarters with the division-approved target list.

Emplacing Unit/UH-60 Company

The emplacing unit or the designated UH-60 company sets up and loads the air Volcano system in conjunction with the forward-area refuel point. The Volcano system is loaded on the UH-60 at the designated point. The emplacing unit/UH-60 company is also responsible for—

- Maintaining the unit basic load at the forward-area refuel point or ATP.
- Preparing the scatterable minefield report and record, forwarding it to the authorizing commander or the aviation brigade engineer, and verifying that the assistant division/brigade engineer receives the entire report.
- Coordinating air routes and corridors.
- Requesting SEAD and security aircraft.
- Planning the air-mission coordination meeting to refine or develop the aviation scheme of maneuver.
- Ensuring that pilots attend the unit's rehearsal.
- Posting aviation graphics on the current operations overlay.

Process

All target nominations are submitted through the division targeting cell. In the brigade, all target nominations go to the assistant division engineer for submission to the board. Nominations should be submitted 96 hours prior to the execution time (see Table D-3).

- The targeting cell validates the targets based on the SITEMP and recommends approval through the commanding general/chief of staff.
- Upon approval, the deputy FSCOORD turns the target numbers and the air requirements over to the G3 air, who adds the targets to the air coordination order.
- The deputy FSCOORD begins to plan SEAD/CAS requirements for division missions.
- The G3 issues the division WO so that units can be prepared to execute specific air Volcano missions.

Table D-3. Planning process (H-hour sequence)

H-96+00. Submit target nominations/receive approval. Submit CAS requests for enemy ADA sites along ingress and egress routes.

H-72+00. Receive mission. Division WO issued.

- Conduct air-mission coordination meeting.
- Conduct S2 update on enemy situation and pass down to executing unit.
- Submit additions or changes to the CAS request to FSE for enemy ADA sites.
- Post the air Volcano mission in the air coordination order.
- Disseminate the SCATMINWARN.

H-48+00. Mount the system. The system takes 3 to 4 hours with a crew of three on a hard surface. The time is doubled in a field environment and requires three or four personnel to assist in loading.

H-24+00. Conduct detailed mission analysis and planning. Issue WO to attack assets, including—

- NLT time for mission planning and upload completion.
- Time and location of air-mission brief.
- Units attending.
- Minefield location and type.
- Establishment of C^2 relationship.
- Face-to-face coordination with attack assets.
- Liaison officers planning multiple routes to minefield.
- Liaison officers coordinating with engineers, ADA, and FSEs for support. G4/S4 must be notified of reload plans and any other Class III/V requirements.
- Intelligence update.

H-20+00. Volcano integrated into deep attack or air-assault matrix by the aviation brigade.

H-8+00. Conditions-check matrix approved by the commanding general or designated authority, including—

- Weather.
- Threat suppression along route.
- Attack support.
- FLOT crossing coordination completion.
- FLOT initial crossing point.
- SEAD fire coordination.
- Mines available.
- Integration into execution matrix.
- Crews updated on ADA threat.
- Backup system and aircraft available.
- CAS coordination.
- Rehearsal completion with attack assets.
- Intelligence update.
- Close operation mission, including—
  - — Location of friendly forces.
  - — Type of marking and when it will be in place.

H-6+00. Conduct air-mission brief, including—

- S2 intelligence update.
- ADA threat update.
- Rehearsal held after brief.
- Brigade engineer and brigade fire-support officer being present.

H-4+00. Latest time for link-up of aviation assets. Fly to forward-area refuel point or tactical assembly area for mine load-up. Remember that—

- Loading canisters takes 1 to 3 hours with a crew of four to eight. Time varies greatly based on whether or not the canisters are in shipping containers.
- The loading time in a field environment is decreased if one pad is dedicated to air Volcano aircraft loading and arming.

H-0+30. Cross FLOT (estimated; exact time is based on route), to include—

- SEAD fired.
- CAS sorties.

H-hour. Minefield deployed, to include—

- Aircraft reports to the supporting brigade engineer if an infantry brigade mission.
- Aircraft reports to the aviation brigade engineer if a division mission.
- Engineer disseminates the scatterable minefield report and record to appropriate units.

- The division issues a fragmentary order (FRAGO) to order the execution of the air Volcano mission.
- The planning staff gives the aviation brigade a date and a time for the air Volcano system to be uploaded and prepared for employment.
- Air Volcano minefields are integrated into the scheme of maneuver as a directed, situational, or reserve obstacle as stated in FM 90-7. This includes integration into the COA that is synchronized during war gaming and included in the OPORD as part of the rehearsal.

Logistics

Air Volcano munitions are transported and handled in the same manner as Class V mines and explosives. The only significant difference in handling is whether air Volcano operations are conducted from the tactical assembly area or the forward operating base.

Echelon-above-corps transportation assets will transfer the air Volcano to corps storage areas (ammunition). Based on forecasts by the division ammunition officer, the corps support area will push air Volcanos to ASPs or ATPs in the division area (tactical assembly area). ASPs and ATPs in or near the tactical assembly area are normally operated by a direct-support ammunition company attached to the corps support group (forward). ATPs in brigade support areas are normally operated by forward support battalions. Mines are moved from the corps support area to ASPs/ATPs by transportation units from the corps support group (rear). Further throughput of mines from ASPs to ATPs is supported by transportation assets from the corps support group (forward). Based on the division ammunition officer's forecast and the availability of transportation, the corps support area will attempt to transfer bulk mines to the ASP/ATP that is best positioned to support requirements. ATPs issue mines to using units while ASPs handle bulk ammunition. ASPs can do emergency issue to using units, but this is usually done only for units using large quantities of bulk ammunition and having their own ammunition transportation support available (division or corps artillery).

Preparation and Coordination

If an air Volcano mission has been approved for a unit, an aviation brigade liaison officer coordinates with the S3 air, the engineer, and the air defense officer to outline air Volcano requirements.

Logistical Requirements

The engineer planner calculates Class IV/V supplies and requests them from the Assistant Chief of Staff, G4 (Logistics)(G4)/S4. The engineer and the assistant division aviation officer coordinate the location (forward-area refuel point, tactical assembly area, forward operating base) of the ATP where the UH-60 will be loaded and fueled. The ALO must provide the amount of air time available and the fuel required, and he must be prepared to discuss emplacement times based on Volcano locations. The air-mission brief will facilitate dissemination of this information.

Concept of the Operation

The scheme of maneuver, fires, and engineer operation must be outlined. The scheme of engineer operations outlines exact grid coordinates, the obstacle intent, and the minefield composition and size. The air Volcano can be emplaced under enemy contact, but additional control measures must be used to protect the aircraft and the crew. If the air Volcano is triggered by enemy action, the DST must be briefed during the air-mission brief. The DST must outline NAIs, including the type of sensors and triggers (long-range surveillance device [LRSD], SOF, UAV), TAIs, decision points, and execution criteria.

Control Points and Markers

The initial point, the approach marker, and minefield markers are designated. Approach and minefield markers must be visible from the air and be distinctly different from one another:

- The initial point is an easily identifiable terrain feature used for coordinating the entry point of the UH-60 into the sector.
- The approach marker allows the aircraft to set the altitude, the speed, and the final orientation to the minefield. The approach marker can also be a terrain feature.
- Minefield markers establish the limits of the desired minefield. Deep area minefields do not require minefield markers.

Terrain Analysis

A combined (aviation, maneuver, fire-support, and engineer planners) terrain analysis should be conducted using Terrabase or a similar product. Terrabase enables planners to analyze the effects of the terrain in a three-dimensional format. AAs, terrain references, TAIs, NAIs, line-of-sight profiles, and minefield locations can be confirmed during this analysis.

Air-Mission Brief

The air-mission brief is the most critical planning and coordination meeting at the execution level, and it must occur no later than H-6 in the planning process. During the air-mission brief—

- C^2 is established.
- Updates and changes to the situation are exchanged between the brigade or TF engineer, the fire-support officer, and the air-mission commander.
- The engineer and the air-mission commander use four control techniques (discussed later in this appendix) to ensure mission success. The primary technique for emplacement and the responsibilities for each control technique are outlined.
- Radio frequencies; points of contact; code words; identification, friend or foe, modes; and challenge and passwords are exchanged and disseminated.

- The forward-area refuel point location, the security aircraft, flight routes, and lethal/nonlethal SEAD are identified (if already approved).

Mounting

The system must be mounted no later than H-48 in the planning process.

Execution

Siting

The key to proper emplacement is the location of the minefield in relation to existing terrain features. If appropriate, the minefield should tie into an existing terrain feature to prevent easy bypass or fording. Using Terrabase enhances siting and emplacement procedures in deep operations. Reconnaissance must be conducted to verify the location.

Movement

Loading. The launcher rack functions as the carrier and launcher platform for 40 mine canisters. The rack has 40 keyholes for mine canisters, a green latch that latches the mine canister to the rack, and a red latch that arms the mine canisters. The rack has two electrical receptacles—one for the connector and one for the launcher rack cable from the DCU. While looking at the canister side of the rack, rows are 1 through 4 from bottom to top and columns are 1 through 10 from left to right.

Arming.

- Due to the weight of the Volcano mine system, a large, open area that is clear of obstacles must be selected. The site should have a hard surface if possible. If a hard surface is unavailable, inspect the ground to ensure that it is firm enough to support the weight of the aircraft. Perforated steel planking or two pieces of 1-inch plywood (4 by 4 feet) may be used as a field-expedient surface in soft areas.
- Concentrations of nonessential personnel or frequently traveled vehicular routes should not be within 1,000 meters of the site. This distance is based on the total weight of explosives and the safe fragmentation distance found in FM 5-250. When using the M88 training canisters, the minimum distance is 30 meters.
- Two Underwriters Laboratories, Incorporated (UL)-listed 10BC fire extinguishers and a grounding rod (minimum safety equipment) must be available at the arming point. This equipment is provided by personnel who deliver the mines.
- The number of personnel allowed access to the site should be held to a minimum. All personnel involved in the arming will receive a safety brief that includes—
  - — Ammunition handling and inspection procedures.
  - — Loading procedures.
  - — Emergency procedures and rendezvous points.
- Emergency procedures.

- Fire. In the event of a fire away from the mines, attempt to contain or extinguish the fire by any available means. If the fire is near the mines or in them, clear the area to a minimum distance of 1,000 meters and notify fire-fighting personnel immediately. When training with M88 canisters, clear the area to a minimum distance of 30 meters.
- Accidental discharge. Immediately clear the area to a distance of 640 meters and notify EOD. The mines arm approximately 2½ minutes after firing. When training with M88 canisters, terminate arming until the problem can be identified and corrected.
- Failure to fire. Remove the canister from the aircraft and place it in the dud pit. Notify EOD immediately. When training with M88 canisters, remove the canister from the aircraft, separate it from the other canisters, repack it, and return it to the ASP.

• Site layout (Figure D-6).

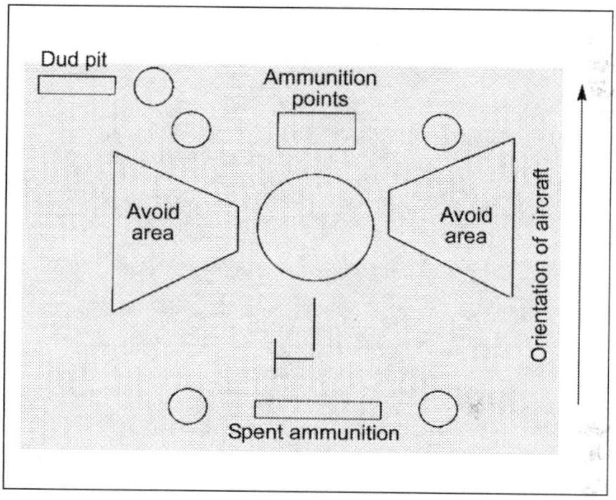

Figure D-6. Site layout

- Berming of the site is not required for a tactical arming point.
- The following rules apply when the site is located next to a refuel point:
  > A minimum of 1,000 meters must exist between arming points and refuel points when the total quantity of explosives is less than 600 kilograms. For quantities greater than 600 kilograms, refer to FM 5-250.

NOTE: Each M87 canister contains 3.4 kilograms of explosives; a full load (160 canisters) contains 550 kilograms of explosives.

    - > The refuel point for armed aircraft must be located at least 375 meters from other aircraft refueling points.
    - > Parked, armed aircraft must be at least 36 meters from other armed aircraft to prevent the detonation of explosives on adjacent aircraft. This distance will not prevent damage to adjacent aircraft; a 130-meter distance is required to prevent damage by fragments and to ensure that the aircraft remains operational.
  - — A dud pit (bermed when possible) for damaged or misfired ammunition should be established beyond the ammunition points.
  - — Arming points should be laid out as shown in Figure D-6.

Dearming. After the mission is complete, the aircraft returns to the arming point for dearming. Spent canisters should be discarded at least 30 meters from the aircraft, at the 4- and 8-o'clock positions. Live canisters should be returned to ASPs for future use or repackaging. Canisters that misfire should be placed in the dud pit.

Flight Planning and Preflight.

- The flight crew analyzes the mission using METT-TC factors and determines the flight profile to be used during mine emplacement. It will select (or have designated) one or more of the following control measures to be used during mine emplacement:
  - — Visual identification (start and stop markers on the ground).
  - — Time-lapse (tables to determine the minefield length).
  - — Number of canisters fired.
  - — Doppler/GPS (start and stop coordinates).
- The crew member(s) will ensure that the air Volcano is installed properly, that all installation checks are completed, and that mine canister pallets are loaded as directed by the pilot or the SOP.
- The flight crew conducts ground checks according to the checklist in TM 1-1520-237-10 to confirm proper operation of the air Volcano prior to takeoff.

Before Arrival at the Target Area.

- During the equipment check, the crew chief turns on the DCU power-control switch, verifies that no malfunctions were indicated during the initial built-in test, and turns off the DCU power-control switch.
- After completion of *run-up* with the aircraft at flight idle, the crew chief turns on the DCU power-control switch.
- Before arrival at the release point, the pilot will make the following checks (listed on the Volcano card [a sample is shown in Figure D-7]):
  - — Verify that the DCU is on.
  - — Verify that the mine SD time is properly set.

VOLCANO CARD

Emplacement Date:

↑ N	**Unit**	**Engr**	**ASP**
	Call Sign		
	Frequency		
	Location		
	Remarks		

Not to scale

	1st pass	**2nd pass**	**3rd pass**	**4th pass**
Initial Point	A	D	G	J
SD Time				
Ground Speed	knots kph	knots kph	knots kph	knots kph
No Tubes				
Altitude				
Track				
Start Time	B	E	H	K
Stop Time	C	F	I	L

Figure D-7. Sample Volcano card

— Announce the ground speed in knots or kph, as required. (The pilot and the crew chief will acknowledge.)

— Announce the number of canisters the crew chief will count down. If deployment is 40 canisters per run (20 per side) and there are two runs, the crew chief will count down from 80 to 60 on the first run and 60 to 40 on the second run. (The crew chief will acknowledge.)

— Announce the altitude for employment. (The pilot will acknowledge.)

— Announce the course for the delivery track. (The pilot will acknowledge.)

- Announce the delivery time based on the setting of the ground speed. (The crew chief will acknowledge.)

- Before arrival at the initial point, the crew chief will—
  - Ensure that the DCU fire-circuit switch's safety pin and streamer are removed.
  - Ensure that the DCU fire-circuit switch is enabled.
  - Place the interface control-panel arming switch to the *ARM* position. He will verify that the jettison advisory light indicates *armed* and that no fault codes are displayed on the DCU.

At the Target Area.

- The pilot simultaneously announces "mark," presses the go-around switch, and starts timing the run when he is over the minefield start point.
- The pilot maintains a ground speed of 5 kph/3 knots and an altitude of 3 meters during the mine-dispensing pass. The pilot is responsible for flying the aircraft within the prescribed limitations.
- The crew chief announces the mine-canister count, as the canisters are dispensed, by counting down in 10s. He then announces the last three canisters. For example, the pilot announces a canister count of 60, the crew chief calls out "80, 70, 3, 2, 1, mark."
- The pilot terminates mine dispensing when the grid location is reached.

After Mission Completion.

The crew chief will—

- Place the interface control-panel arming switch to the *SAFE* position and verify that the armed advisory capsule is extinguished.
- Place the DCU fire-circuit switch to the *OFF* position. Install the safety pin and the streamer.
- Prepare and submit a SCATMINWARN.

EMPLACEMENT

The air Volcano is fast and flexible, but it is difficult to accurately dispense mines within the confines of the minefield marking. The desired obstacle-effect norms for the air Volcano require extensive planning, preparation, coordination, and positive control during emplacement. The critical aspect of the air Volcano is getting the right amount of mines in the specified location and in the desired density.

The detailed coordination focuses on positive control. Positive control of an air Volcano mission requires a redundancy of control techniques to minimize errors in minefield size and location. These control techniques must compensate for poor visibility, wind speed and direction, and navigational errors.

The following control techniques are used by the engineer and the air-mission commander to ensure that Volcano minefields match specific obstacle-effect norms. Units rely on these techniques to accomplish the mission, and they are part of the Volcano air-mission brief:

- Visual identification. Focuses on the visual identification of minefield emplacement. As part of the preparation for a Volcano minefield, an engineer element erects airfield panel markers to mark start and end points. This provides a visual signal for the engineer and the air-mission commander to start and stop firing Volcano canisters. The pilot depresses the launch switch over the first marker to start firing and depresses it again over the second marker to stop firing. This control technique is good for open terrain with adequate visibility and little canopy coverage.
- Time lapse. Focuses on when to stop firing Volcano canisters. The UH-60's air speed and the type of minefield being laid determines the amount of time it takes to lay a minefield. The air Volcano has six air speed settings—20, 30, 40, 55, 80, and 120 knots. Table D-4 shows the time required to lay minefields and the full load time.

Table D-4. Air Volcano dispensing times based on air speed

Knots	Disrupt and Fix Minefields	Turn and Block Minefields	160 Canisters per Load
20	27 seconds	54 seconds	108 seconds
30	18 seconds	36 seconds	72 seconds
40[1]	13 seconds	27 seconds	54 seconds
55	9 seconds[2]	18 seconds	39 seconds
80	6 seconds[2]	13 seconds[2]	27 seconds
120	4 seconds[2]	9 seconds[2]	18 seconds
Width of minefield (meters)	278.8	557.5	1,115
No passes per minefield	1	2[3]	1
No canisters per pass	40	80	160

[1]Recommended air speed

[2]Recommended only if absolutely necessary

[3]Blackhawks in pairs can lay turn and block minefields in one pass, firing 80 canisters each.

The following example is provided to show how Table D-4 is used:

Example: The mission is to install an air Volcano disrupt minefield. The UH-60 is traveling at 40 knots (this is entered on the DCU), and the pilot initiates (depresses the launch switch) at the identification of the Volcano start marker or the grid location on the ground. The pilot depresses the launch switch a second time after 13 seconds have elapsed.

- Number of canisters fired. Focuses on when to stop firing Volcano canisters. The number of Volcano canisters dispensed also determines when firing is terminated. There is a digital readout on the DCU (for the left and right side) that shows the number of canisters remaining. The pilot stops firing when the required number of canisters have been fired (see Table D-4, page D-19).

Using the disrupt minefield example above, the UH-60 starts the mission with a full load (80 canisters on each side of the aircraft). The pilot initiates (depresses the launch switch) at the identification of the Volcano start marker or the grid location on the ground. The pilot depresses the launch switch a second time after 20 canisters have been expended on each side. The DCU counts down from the total number of canisters. When the DCU reads 60 right/60 left, the pilot depresses the switch to end the firing process. Ideally, the timing of delivery and the number of canisters fired are done simultaneously. As the crew chief counts down the timer, the pilot and the crew chief monitor the number of canisters remaining on the DCU digital readout.

- Doppler/GPS. Focuses on when to start and stop firing Volcano canisters using the UH-60's Doppler/GPS guidance and navigation set. This set provides the present position or destination in latitude and longitude (degrees and minutes) or grid coordinates. As part of the preparation for the Volcano minefield, exact grid coordinates are needed to determine the approach points and the limits of the minefield. These coordinates are provided to the air-mission commander during the air-mission brief. The pilot enters the grid coordinates into the Doppler/GPS on the primary and the backup aircraft. During execution, the air-mission commander monitors the Doppler/GPS and determines the time to target, when to initiate firing, and when to terminate firing.

Outside Friendly Territory

Reconnaissance of the proposed site for the air Volcano minefield will be conducted before mines are emplaced. This could include—

- LRSD.
- Apache gun tapes.
- UAV overflights.
- Imagery.

Key terrain or landmarks are used to identify start and end points for aviation assets. It is unlikely that military marking will be employed based on the proximity to enemy forces and the probability of early detection if man-made markers are present. Fencing the minefield is not required until the area has been secured by friendly forces.

Within Friendly Territory

Within friendly territory, air Volcano minefields should be fenced and marked with NATO identification signs to protect friendly forces.

Fencing is installed before the air Volcano minefield is delivered, and it is located 100 meters from the centerline of the minefield and 100 meters from the start and end points (Figure D-8).

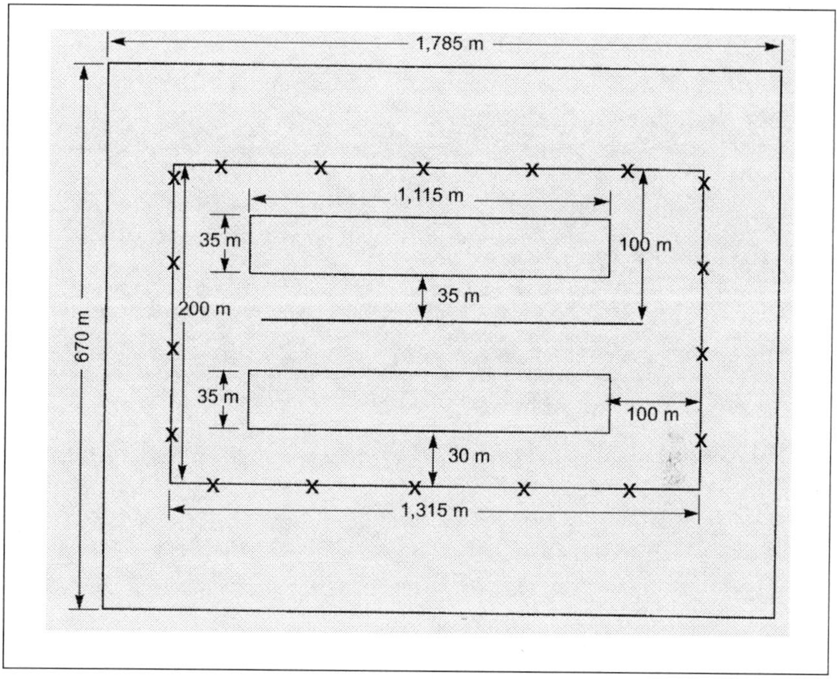

Figure D-8. Fencing for an air Volcano minefield

Start and end points should be marked with man-made devices such as VS-17 panels. During limited visibility, start and end points should be marked with infrared or heat-producing sources. Key terrain features and landmarks should still be used to identify start and end points.

Fencing the minefield is not viable when the minefield duration is short or civilians on the battlefield are an issue. In this case, CA, public affairs, and PSYOP personnel should be involved in letting friendly personnel know the minefield location. This could include—

- Leaflet drops.
- CA teams disseminating information.
- Host-nation support.
- PSYOP.

REPORTING

SCATTERABLE MINEFIELD WARNING

The emplacing unit is responsible for issuing the SCATMINWARN (Figure 8-7, page 8-23) to adjacent units and higher headquarters. The brigade engineers and the assistant division engineer assist in this process. They disseminate the warning based on whether or not it is a brigade (brigade engineer) or division (aviation brigade engineer) mission. To ensure that all units are informed, the assistant division engineer forwards the SCATMINWARN to the G3 for dissemination through operational channels.

SCATTERABLE MINEFIELD REPORT AND RECORD

The aircraft emplacing the minefield reports initiation and completion times to the engineer of the emplacing unit. The engineer prepares the scatterable minefield report and record (Figure 8-8, page 8-23) and forwards it through his unit to the assistant division engineer. The assistant division engineer forwards the report to the G3 who provides the information to higher and subordinate units through operational channels.

Appendix E

Safety and Training

Mine training is inherently dangerous, in part, because several different types of mines and fuse systems are used throughout the world. Detailed safety instructions for each type of mine are provided throughout this manual. This appendix merely points out the safety aspects of live-mine training that are common to all types of mines.

Conduct mine training as if the mines were live. This is the only way soldiers form a habit of correctly and safely handling mines and gain a true appreciation of the requirements and the time it takes to perform an actual mine-warfare mission. Live-mine training gives soldiers the confidence they need to handle mines and their components. Accidents can usually be traced to ignorance, negligence, deliberate mishandling, overconfidence, mechanical failure, or fright. The first four can be overcome by training and proper supervision. Mechanical failure rarely happens; but if it does, it can be controlled by training and proper supervision. The last item, fright, is mastered through well-controlled, live-mine training.

STORAGE

There are three types of mines used in mine training:

- Inert. Does not contain explosives.
- Practice. Contains an LE charge or a smoke-producing component to simulate detonation.
- HE. Involves actual mines used in combat

Conventional mines are painted to enhance concealment, retard rusting of exposed metal parts, and help identify the type of mine and filler (HE, LE, or chemical agent). Older manufactured mines are painted according to the Five-Element Marking System; newer mines use the Standard Ammunition Color-Coding System (see Table E-1, page E-2).

NOTE: Mines that are color-coded and marked according to the old system have been on hand for several years. Ensure that all ammunition, whether color-coded according to the old or new system, is properly and fully identified.

Always handle mines with care. The explosive elements in fuses, primers, detonators, and boosters are particularly sensitive to mechanical shock, friction, static electricity, and high temperatures. Boxes and crates containing mines should not be dropped, dragged, tumbled, walked on, or struck. Do not smoke within 50 meters of a mine or its components.

Table E-1. Mine color-coding system

Type of Ammunition	Five-Element Marking System (Old)	Standard Ammunition Color-Coding System (New)*
Persistent casualty chemical agent	Gray with green markings and two green bands	Gray with green markings and two 12-mm green bands
Nerve agents	Gray with green markings and two or three green bands	Gray with green markings and three 12-mm green bands
Incendiary	Gray with violet markings and one violet band	Light red with black markings and one yellow band
HE	Olive drab with yellow markings	Olive drab with yellow markings
Practice mines	Blue with white markings	Blue with white markings
Inert mines	Black with the word *INERT* in white	Blue with the word *INERT* in white
*Chemical ammunition containing an HE has one 6-mm yellow band in addition to the other markings.		

When it is necessary to leave mines in the open—

- Set them on dunnage at least 5 centimeters above the ground.
- Place a waterproof cover (such as canvas) over them, and leave enough space for air circulation.
- Dig drainage trenches around stacks of mines to prevent water from collecting under them.
- Protect mines and their components against moisture by waterproofing them with grease coatings, tar paper, or tarpaulins.

Additional maintenance procedures are as follows:

- Do not open mine boxes in a magazine, at an ammunition dump, or within 30 meters of an explosive store. Use copper or wooden safety tools, if available, to unpack and repack mines.
- Do not fuse mines within 30 meters of an explosive or ammunition holding area. Mines can be fused at the mine dump.
- Use specifics authorized by the US Army Materiel Command and applicable TMs to disassemble mines and their components.
- Remove safety pins, safety forks (clips), and other safety devices as the last step when arming the mine; and replace them before the mine is moved again. These devices prevent accidental initiation of the mine while it is being handled.
- Place tape over open fuse cavities and secondary fuse wells. Ensure that they are clear of obstruction and free of foreign matter before attempting to install the fuse, the detonator, or the FD.
- Take steps to prevent moisture or water from accumulating around the mine and subsequently freezing if the temperature fluctuates around freezing. Mines usually function satisfactorily at temperatures between 40 and 160°F. Most mines are not appreciably affected by temperature changes, but mines can become neutralized by ice formations (see Chapter 12).

- Observe proper procedures when recovering mines. Ensure that components do not show evidence of damage or deterioration.
- Ensure that practice or inert mines or their components are not present when live mines or their components are being used.
- Do not mix inert mines with live mines.
- Do not display live mines or their components in museums, demonstrations, models, or similar layouts. Only inert equipment can be used for displays.
- Handle explosive materials with appropriate care. The explosive elements in primers, blasting caps, and fuses are particularly sensitive to shock and high temperatures.
- Assemble activators, standard bases, and FDs before installing them. Do not carry them in the pockets of your clothing.
- Do not point FDs at anyone.
- Camouflage the mine before removing the positive safety pin when possible.

NOTE: Additional storage and safety precautions are outlined in TM 9-1300-206.

LIVE-MINE TRAINING

NOTE: No live-mine training is authorized with M14 mines. Units outside Korea will not use live M16A1 mines in tactical or protective minefield training.

Live-mine training is conducted by preparing, laying, arming, neutralizing, and disarming live mines (with live fuses and components) in a training environment.

Supervisors must adhere to the following safety considerations when conducting live-mine training:

- Only personnel who are qualified and certified according to the local range SOP are allowed to supervise activities or training in which live mines or their components are used.
- Minimum personnel requirements to conduct live-mine training are—
  - Range officer (OIC).
  - Range safety officer (RSO).
  - One NCO supervisor for each arming bay.
  - Mine-explosive breakdown NCO.
  - One medic per four arming bays.
  - Guards, as required by the range SOP.
- Sound organization is a must before live-mine training can begin. The OIC and supervising NCOs conduct a demonstration/briefing to ensure that the practice runs smoothly.

- The training officer must foresee hazards that can occur through personnel nervousness or material failure. The commander should conduct a risk assessment according to AR 385-10.
- The OIC takes his place at the control point or post. Once he is satisfied that all safety regulations have been observed, he orders the first detail to start training.
- Soldiers are trained on inert and practice mines before arming live mines, according to the guidelines established by the Standards in Training Commission.
- Fuses are not inserted into mines until ordered by the OIC.
- An NCO supervisor must be present when soldiers arm live mines. He ensures that soldiers adhere to the proper procedures and regulations.
- Only one soldier arms a mine at any given time.
- Personnel disarm one mine before arming the next one.
- Personnel never arm an M16 AP mine in the trip-wire mode during live-mine training.
- Personnel never remove the positive safety pin from the M16 AP mine during live-mine training.
- Instructors inspect fuses and mines for serviceability before starting practice.
- Instructors inspect mines and their components for damage and excessive wear after each student has gone through the station. Replace the mine and the fuse if damage or wear is found.
- All personnel wear a helmet (with the serviceable chin strap fastened) and body armor when arming and disarming mines.
- Ear protection is not permitted in the arming bays. The student must be able to hear the supervisor and certain distinct noises (such as a firing pin dropping).
- Instructors post guards at all entrances to the range. The guards communicate with the RSO by radio, wire, voice, or signal. No one enters the range without permission from the RSO.
- Instructors keep mine records and inventory sheets. They maintain accountability of all mines and fuses, before and after each exercise.
- Instructors draw and return supplies; check equipment for issue; and ensure that live mines are safe, serviceable, and unarmed. They ensure that the requirements contained in AR 385-63, range regulations, and SOPs are observed and that no one does anything to prejudice safety.
- Instructors clearly mark the word *LIVE* on all live mines and their components that are used for live-mine training. Live mines are maintained separately from practice and inert mines.

- Live AHDs are not used with live mines during training, but they can be used with practice and inert mines.
- Arming and disarming are conducted in the prone position.
- Waiting personnel are located in a bunker, behind a suitable barricade, or at a safe distance from live-mine training.
- Supervisors ensure that live-mine training is not rushed. There are no shortcuts. Supervisors must allow soldiers ample time to arm and disarm mines. Most soldiers are already in a high state of stress from dealing with live munitions, and rushing them only serves to heighten their stress level.

LIVE-MINE DEMONSTRATIONS

Live-mine demonstrations show mine characteristics and capabilities using M16 and M18 AP mines and M15, M19, and M21 AT mines. The appropriate authority must authorize the demonstration, and firing personnel must be fully conversant with all safety and technical aspects pertaining to live-mine firing.

An OIC and an RSO are appointed for each activity involving live-mine firing. The amount of explosive contained in the mine cannot exceed the maximum amount allowed for the range, and one mine is fired at a time.

Upon arriving at the range, the instructor and his assistants establish areas according to the following rules (signs are posted for large demonstrations):

- Firing point. Sited outside the danger area and near the OIC to facilitate coordination, commentaries, and firing.
- Spectator area. Sited outside the danger area and within earshot of the commentator. It is large enough to provide a good view of the explosion.
- Supply area. Any suitable area away from spectators.
- Explosive area. Sited away from supplies and spectators.
- Mine area. Mines are set out in full view of the OIC and spectators. Individual mines are at least 25 meters apart.
- Target area. Targets are positioned and inspected by spectators before the blasting cap is inserted into the mine.

M16 Antipersonnel Mine

- Safety distance. 300 meters.
- Firing procedures.
  - Roll out 300 meters of firing cable and attach it to a stake or picket in the ground (leave at least 1 meter of free end). Test the firing cable for continuity.
  - Place the mine in the ground (dig in level with the surface). Remove the shipping plug.
  - Test a blasting cap (under a sandbag) with the demolition test set.

— Attach the ends of the blasting-cap leads to the ends of the electric cable and insulate the joints with tape. Place the blasting cap into the fuse well (see Figure E-1, page E-6).

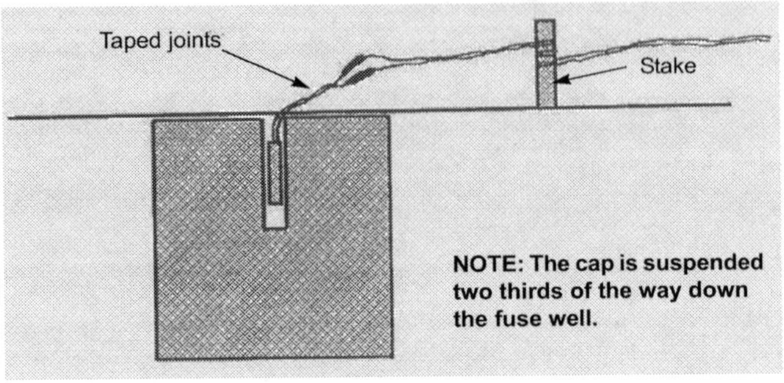

Figure E-1. M16 AP mine

- Suggested target. A circle of tar paper, 6 meters in diameter, supported by 1.8-meter pickets. Spectators can later view shrapnel effects.

NOTE: The procedure detailed here dispenses with the M605 igniter. The mine cannot be detonated by pull or pressure. The expulsion charge and millisecond delay fuses are still operated, and the mine bounds out of its casing (which remains in the ground) before exploding in the air. Although the normal firing delay is removed, it does not detract from the demonstration. The blasting cap is suspended two-thirds of the way down the fuse well to initiate the expelling charge and delay elements.

- Misfires. In the event of a misfire, the RSO disposes of the mine by placing a block of C4 as close to the mine as possible, without touching it. He destroys the mine by normal nonelectric means.

M18A1 ANTIPERSONNEL MUNITION

- Safety distance. 300 meters.
- Firing procedures.
  - — Roll out 300 meters of firing cable and attach it to a stake or picket in the ground (leave at least 1 meter of free end). Test the firing cable for continuity.
  - — Place the mine on the ground and ensure that the front of the mine faces away from the firing point. Remove the shipping plug.
  - — Test an electric blasting cap (under a sandbag) with the demolition test set.

— Attach the ends of the blasting cap leads to the ends of the electric cable and insulate the joints with tape. Place the blasting cap into the detonator well (see Figure E-2).

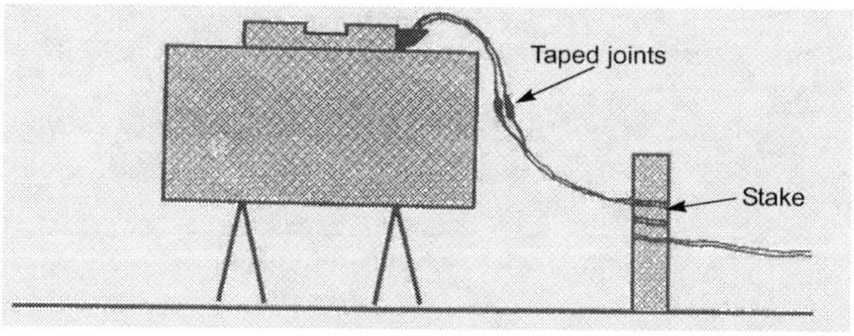

Figure E-2. M18A1 AP mine

- Suggested target. Several E-type silhouette targets, 15 to 100 meters from the mine.

NOTE: The procedure detailed here applies only to demonstration firings. Standard accessories are used on all other occasions. The mine explodes instantaneously and clearly illustrates the sound of an M18A1 explosion.

- Misfires. In the event of a misfire, the RSO disposes of the mine by placing a block of C4 as close to the mine as possible, without touching it. He destroys the mine by normal nonelectric means.

M15, M19, AND M21 ANTITANK MINES

- Safety distance. 1,000 meters.
- Firing procedures.
  - — Roll out 1,000 meters of firing cable and attach it to a stake or picket in the ground (leave at least 1 meter of free end). Test the firing cable for continuity.
  - — Place the mine in the ground and leave the top exposed. A target is used only when the mine can be placed without disturbing the target. A derelict vehicle is a suitable target.
  - — Place a block of C4 on top of M15s and M19s (see Figure E-3, page E-8).
  - — Remove the shipping plug from the booster well of the M21 and pack the well with C4. Insert an electric blasting cap into the C4 (see Figure E-4, page E-8).

NOTE: Do not remove safety devices. Keep arming dials in the *SAFE* position. The mine explodes instantaneously and clearly demonstrates the blast/shaped-charge effect.

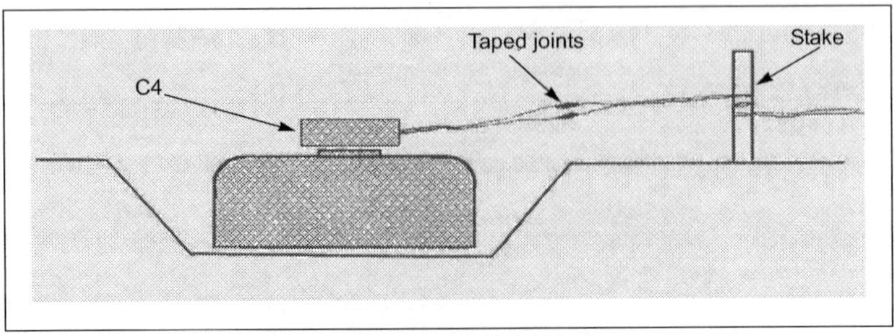

Figure E-3. M15 and M19 AT mines

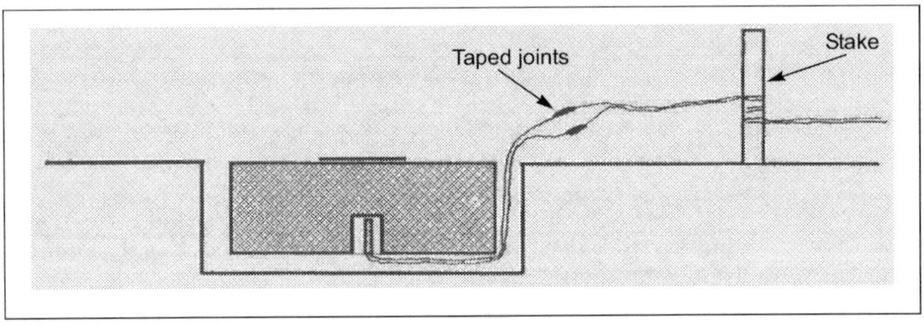

Figure E-4. M21 AT mine

- Misfires. In the event of a misfire, the RSO disposes of the mine by placing a block of C4 as close to the mine as possible, without touching it. He destroys the mine by normal nonelectric means.

RISK ASSESSMENT FOR LIVE-MINE DEMONSTRATIONS

The following risk assessment is provided as a guideline for live-mine demonstrations using an M16 AP mine. It must be carefully reviewed before conducting a demonstration. Live-mine demonstrations can be conducted in a safe manner. The risk of injury to personnel is significantly minimized if you adhere to established procedures.

During the demonstration, mines are not armed with standard fuses. They are activated by electric blasting caps placed inside the fuse wells.

A demonstration shows the effectiveness of an M16 AP mine. Spectators do not handle the mines or explosives. To show the effectiveness of an M16 mine, a sheet of paper is placed in a semicircle around the mine. Spectators remain in bunkers or at a safe distance while mines are primed with electric blasting caps and detonated. After the mines have been detonated and the RSO has cleared the area, spectators are allowed to view the results of the detonated mine. Misfires are handled by the RSO.

Figure E-5 is a risk assessment prepared by the Department of Transportation.

QUALITATIVE RISK ASSESSMENT

Qualitative risk-assessment techniques are used to place a value on the level of risks created by hazards in an operation. The principal qualitative technique is the risk-assessment code (RAC) described in MILSTD-882C. This method was established as a common way to set priorities for Department of Defense (DOD)-wide hazard-abatement programs and uses an RAC matrix format to combine the concepts of frequency and severity into a single, numerical code. It is very useful in comparing different risks (such as those from different programs) or even differences (such as health versus safety risks).

RACs are implemented for the Army in Army Regulation (AR) 385-10, which describes the two qualities of hazard severity and hazard probability as follows:

- Category I—CATASTROPHIC. May cause death or loss of a facility. In this case, loss does not mean a period of interrupted service; it means destruction of the facility or operation.
- Category II—CRITICAL. May cause severe injury, severe occupational illness, or major property damage.
- Category III—MARGINAL. May cause minor injury, minor occupational illness, or minor property damage.
- Category IV—NEGLIGIBLE. Probably would not affect personnel safety or health, but is in violation of specific standards.

Mishap probability is the probability that a hazard will result in a mishap, based on an assessment of factors such as location, exposure in terms of cycles or hours of operation, and affected population. The expression combines the idea of the probability of an event and the exposure to the event. These probabilities are expressed as letters conforming to the following system:

- Subcategory A—Likely to occur immediately.
- Subcategory B—Probably will occur in time.
- Subcategory C—May occur in time.
- Subcategory D—Remote.
- Subcategory E—Improbable.

The two qualities are combined to yield an RAC by using the following table:

Risk-Assessment Code Table

Hazard Severity	Mishap Probability				
	A	B	C	D	E
I	1	1	2	3	4
II	1	2	3	4	5
III	2	3	4	5	5
IV	—	—	—	—	—

In using the RAC system, it is important to note that the IA and IIA classifications are termed "imminent danger." Though their RAC codes of 1 are the same as that of the IB entry, their immediacy makes them more critical. The codes are useful in assessing an operation as it begins, but they must be updated as the operation continues, the facility ages, and so forth to account for degrading condition or performance.

For risk managers, there are some important organizational/management considerations to RAC codes. AR 40-10 also contains an RAC system, but due to a difference in definitions, the health RAC code may be a lower number, indicating a higher degree of risk. This is important to managers who are comparing health risks to others risks under AR 385-10; the health issue would always win if no compensation or consideration was factored into the codes. From the managerial standpoint, it must be remembered that RACs are judgmental and not necessarily held to be the same by different managers or evaluators. When differences in perception occur, the differences are likely based on either the understanding of the operation's behavior or the criteria for selecting the probability and severity. It is wise to listen to the basis of others' RAC choices and attempt to develop a common understanding.

Figure E-5. Excerpt from Risk-Assessment Techniques Manual, prepared by the Department of Transportation's Transportation Safety Institute, August 1986

Qualitative risk-assessment techniques are used to prepare estimates of risk levels using performance data, when available, to improve the accuracy of risk estimates used in risk-acceptance decision making. These assessments are numeric values representing the safety risk of an Army activity, system operation, or comparable endeavor, based on actuarial or derived numeric data. Though RACs are numerical, they are derived from judgments and are not demonstrable in records of performance. If it is desirable that performance be measured, it is necessary that quantified estimates of risk levels be established, that risk levels be predictive so that future performance has a base of comparison, and that risk levels be assigned numeric values.

Figure E-5. Excerpt from Risk-Assessment Techniques Manual, prepared by the Department of Transportation's Transportation Safety Institute, August 1986 (continued)

RISK ASSESSMENT FOR LIVE-MINE TRAINING

The United States Army Engineer Center, Department of Instruction, obtained information for the following risk assessment from the Collective Training Branch, Department of Training and Doctrine, and from the Engineering Branch for Engineer Officer Basic Course demolitions training. Hazards are identified and analyzed on preliminary hazard-analysis work sheets (see Figures E-6 through E-15, pages E-11 through E-21). Risk-assessment codes are assigned to each hazard based on the severity and the probability of occurrence.

References used in the risk-analysis process include this manual and the following publications:

- DA Pamphlet (Pam) 350-38.
- AR 385-10.
- AR 385-16.
- AR 385-63.
- Soldier Training Publication (STP) 5-12B1-SM.
- Training Circular (TC) 25-8.
- TM 9-1345-203-12.
- TM 43-0001-36.

Operation Name: Arming the M15 AT Mine
Date Prepared: 11 October 1990
Sheet Number: 1 of 1

Hazards	System Effects	Causal Factors	RAC	Actions	Controlled RAC	Standards/Comments
Damaged mine used	None	Lack of training/improper training Improper supervision Incomplete or no inspection	IV B (-)	Do not use mine. Contact QASAS to investigate. Turn mine in for investigation/ disposal.	IV D (-)	Proper training is conducted. Soldiers are proficient on inert mines first. One instructor is assigned to one soldier. All training is conducted in the prone position. Mine training is done after identified by a unit's METL and only with select personnel. Training is conducted in the proper environment. Soldiers are in the proper protective gear. No trip wires or AHDs are used with an M15 mine. Using the cap to perform the fuse depth check is not authorized. Use the wrench.
Too much force on extension rod/ pressure ring (M624 fuse)	Mine detonates	Soldier bends/pushes on rod or fuse	I B (1)	Do not remove the safety stop and band if the extension rod is used. **OR** Remove the extension rod before removing safety stop and band.	I D (3)	
Too much pressure on pressure plate (M603 fuse)	Mine detonates	Soldier applies excessive pressure	I D (3)	See comments.	I D (3)	
Fuse improperly seated in the fuse well (M603 fuse)	Mine detonates	Excessive debris/corrosion in well Lack of training/improper training Improper supervision Incomplete or no inspection	I C (2)	Perform depth check with wrench.	I D (3)	
Improper handling of fuse (M603 fuse)	Fuse detonates	Pressure applied to fuse with safety fork removed	II D (4)	Handle fuse properly.	II D (4)	

Figure E-6. Preliminary hazard-analysis work sheet (arming M15)

Operation Name: Disarming the M15 AT Mine
Date Prepared: 11 October 1990
Sheet Number: 1 of 1

Hazards	System Effects	Causal Factors	RAC	Actions	Controlled RAC	Standards/Comments
Camouflage removed	Mine detonates	Too much pressure on pressure plate Damage to the mine	I D (3)	Do not apply pressure to pressure plate. See comments.	I D (3)	See comments for arming an M15 AT mine (Figure E-8, page E-13). See risk assessment for command-detonating a mine in place with explosives (Figure E-16, page E-21).
Unable to turn the mine to SAFE or reinstall safety ring, safety stop, or safety pin	Mine detonates	Too much resistance (M603 fuse) is applied to turning indicator. Safety band (M624), safety stop, or safety pin has become damaged. Foreign material has entered the fuse well.	I C (2)	Perform controlled detonation in place.	None	
Fuse removed/ safety fork installed	Fuse detonates	Pressure is placed on the pressure plate of fuse. Fuse is not removed.	II D (4)	Follow proper procedures.	II D (4)	

Figure E-7. Preliminary hazard-analysis work sheet (disarming M15)

Operation Name: Arming the M16-Series AP Mine
Date Prepared: 11 October 1990
Sheet Number: 1 of 1

Hazards	System Effects	Causal Factors	RAC	Actions	Controlled RAC	Standards/Comments
Damaged mine used	None	Lack of training/improper training Improper supervision Incomplete or no inspection	IV C (-)	Do not use mine. Contact QASAS to investigate. Turn in the mine for investigation/ disposal.	None	Proper training is conducted. Soldiers are proficient on inert mines first. One instructor is assigned to one soldier. All training is conducted in the prone position. Mine training is done after the need is identified by a unit's METL and then only with select personnel. Training is conducted in the proper environment. Soldiers are in the proper protective gear. No trip wires or live AHDs are used with an M16 mine. In the event of damage or loss of any safety pins, stop training with this particular mine and destroy it.
Faulty fuse used (M605)	Mine detonates	Safety pins missing Prongs bent Fuse head does not turn freely Pins installed incorrectly	I B (1)	Do not use fuse. Contact QASAS to investigate. Turn mine in for investigation/ disposal. Conduct quality control check before training.	I D (3)	
Pressure applied to prongs after the positive safety pin is removed	Mine detonates	Pressure is applied inadvertently	I B (1)	Do not remove the positive safety pin at any time.	I D (3)	
Fuse improperly armed	Mine detonates	Individual pushes in on release pin while or after positive safety pin is removed Wrong sequence for removing pins Locking safety pin is removed, click is heard, soldier continues	I B (1)	Do not remove the positive safety pin at any time.	I D (3)	

Figure E-8. Preliminary hazard-analysis work sheet (arming M16)

Operation Name: Disarming the M16-Series AP Mine
Date Prepared: 11 October 1990
Sheet Number: 1 of 1

Hazards	System Effects	Causal Factors	RAC	Actions	Controlled RAC	Standards/Comments
Camouflage removed	Mine detonates	Too much pressure on M605 fuse pressure prongs	I C (2)	Provide close supervision. Do not rush the procedure. Do not camouflage mine.	I D (3)	Do not fully arm this mine; leave the positive safety pin in the fuse at all times. See comments for arming an M16 AP mine (Figure E-10, page E-15). See risk assessment for command-detonating a mine in place with explosives (Figure E-16, page E-21).
Replace safety pins	Mine detonates	Pins not replaced Pull on release-pin ring Pressure on fuse pressure prongs Pins replaced in wrong order	I C (2)	Provide close supervision. Do not rush the procedure.	I D (3)	
M605 fuse removed	Mine detonates	Safety pins not properly in place, thus allowing pressure to be applied to the fuse, pressure prongs, or release pin	I C (2)	Provide close supervision. Assure that pins are secure and replaced correctly.	I D (3)	

Figure E-9. Preliminary hazard-analysis work sheet (disarming M16)

Operation Name: Arming the M19 AT Mine
Date Prepared: 11 October 1990
Sheet Number: 1 of 1

Hazards	System Effects	Causal Factors	RAC	Actions	Controlled RAC	Standards/Comments
Damaged mine used	None	Lack of training/improper training Improper supervision Incomplete or no inspection	IV B (-)	Do not use mine. Contact QASAS to investigate.	IV B (-)	Proper training is conducted. Soldiers are proficient on inert mines first. One instructor is assigned to one soldier. All training is conducted in the prone position. Mine training is done after identified by a unit's METL and only with select personnel. Training is conducted in the proper environment. Soldiers are in the proper protective gear. No trip wires or AHDs are used with a live M19 mine.
Improper handling of detonator	Detonator detonates	Soldier handles detonator improperly Lack of training/improper training Improper supervision Incomplete or no inspection	I D (4)	Handle detonator properly.	III D (5)	
Too much pressure on pressure plate	Mine detonates	Soldier applies too much pressure	I D (3)	See comments.	I D (3)	
Weakened Belleville spring	Mine detonates at pressure lower than designed pressure	Belleville spring weakening	I C (2)	Inform soldiers of less pressure required to detonate.	I D (3)	

Figure E-10. Preliminary hazard-analysis work sheet (arming M19)

Operation Name: Disarming the M19 AT Mine
Date Prepared: 11 October 1990
Sheet Number: 1 of 1

Hazards	System Effects	Causal Factors	RAC	Actions	Controlled RAC	Standards/Comments
Camouflage removed	Mine detonates	Too much pressure on pressure plate Damaged or malfunctions	I D (3)	Do not camouflage mine.	I D (3)	See comments for arming an M19 AT mine (Figure E-12, page E-17). See risk assessment for command-detonating a mine in place with explosives (Figure E-16, page E-21).
Unable to turn the mine to SAFE	Mine detonates	Too much resistance turning the dial indicator Foreign material in the fuse well	I C (2)	Perform controlled detonation in place.	I D (3)	
Fuse assembly removed and detonator replaced with shipping plug	Detonator detonates	Dropped fuse assembly with detonator Dropped the detonator	II C (3)	Handle fuse assembly and detonator properly.	II D (4)	

Figure E-11. Preliminary hazard-analysis work sheet (disarming M19)

Operation Name: Arming the M21 AT Mine
Date Prepared: 11 October 1990
Sheet Number: 1 of 1

Hazards	System Effects	Causal Factors	RAC	Actions	Controlled RAC	Standards/Comments
Damaged mine used	None	Lack of training/improper training Improper supervision Incomplete or no inspection	IV B (-)	Do not use mine. Contact QASAS to investigate. Turn mine in for investigation/ disposal.	IV B (-)	Proper training is conducted. Soldiers are proficient on inert mines first. One instructor is assigned to one soldier. All training is conducted in the prone position. Mine training is done after identified by a unit's METL and only with select personnel. Training is conducted in the proper environment. Soldiers are in the proper protective gear. No trip wires or AHDs are used with a live M21 mine.
Too much force on extension rod/ pressure ring (M607 fuse)	Mine detonates	Soldier bends/pushes on rod or fuse	I B (1)	Do not remove the safety stop and band if the extension rod is used. **OR** Remove the extension rod before removing the safety stop and band.	I D (3)	
Fuse improperly seated in the fuse well	Mine detonates	Excessive debris/corrosion in well Lack of training/improper training Improper supervision Incomplete or no inspection	I C (2)	See comments.	I D (3)	
Improper handling of fuse	Fuse detonates	Pressure applied to fuse while mine is armed	II D (4)	Handle fuse properly.	None	

Figure E-12. Preliminary hazard-analysis work sheet (arming M21)

Operation Name: Disarming the M21 AT Mine
Date Prepared: 11 October 1990
Sheet Number: 1 of 1

Hazards	System Effects	Causal Factors	RAC	Actions	Controlled RAC	Standards/Comments
Camouflage removed	Mine detonates	Too much pressure Tilted extension rod Damaged or malfunctions	I B (1)	Do not use extension rod. Ensure close supervision.	I D (3)	See comment for arming an M21 AT mine (Figure E-14, page E-19).
Band, stop, and cotter pin replaced	Mine detonates	Too much pressure Tilted extension rod Damaged stop, band, or cotter pin	I B (1)	Do not use extension rod. Ensure close supervision.	I C (2)	Stop training with this particular mine and destroy it in the event of damage or loss of stop, band, or cotter pin.
Extension rod and/or fuse removed	Mine detonates	Too much pressure Tilted extension rod	I C (2)	Do not use extension rod. Ensure proper placement of band, stop, and cotter pin.	I D (3)	See risk assessment for command-detonating a mine in place with explosives (Figure E-16, page E-21).
Booster removed	Booster detonates	Soldier improperly handles booster	II D (4)	Handle booster properly.	None	

Figure E-13. Preliminary hazard-analysis work sheet (disarming M21)

Operation Name: Preparing charges and priming explosives to be used for mine demolition (detonated in place)
Date Prepared: 11 October 1990
Sheet Number: 1 of 1

Hazards	System Effects	Causal Factors	RAC	Actions	Controlled RAC	Standards/Comments
Defective wire or faulty test of wire	No detonation	Wire will not carry electrical charge to detonate explosive	IV D (-)	Do not use wire (replace it).	None	After testing, ensure that wire ends are twisted together. Only one demolitions person and one safety person will be at the mine when the blasting cap is inserted into the demolition. RSO is responsible for clearing misfires. RSO keeps a *misfire kit* under his control.
Electric blasting cap check done incorrectly or not at all	Blasting cap detonates	Cap improperly handled Cap not put under sandbag	III C (4)	Handle cap properly. Place cap under sandbag.	IV D (-)	
Blasting cap attached incorrectly to firing wire	Blasting cap detonates	Cap/firing wire improperly handled End of firing wire not shunted	III C (4)	Ensure that firing wire is shunted. Ensure proper handling of blasting cap.	IV D (-)	
Blasting cap inserted incorrectly/too forcefully into 1-pound demolition charge	Blasting cap and/or demolition charge detonates	Too much pressure on blasting cap Cap improperly handled	I D (3)	Ensure proper handling of blasting cap.	I D (3)	
Charge does not detonate	No detonation	Faulty blasting cap Faulty blasting machine Faulty firing wire	IV C (-)	OIC declares a misfire. Follow RSO procedures for a misfire.	IV C (-)	

Figure E-14. Preliminary hazard-analysis work sheet (command detonation)

Operation Name: Mine arming/disarming peripheral factors (not all-inclusive)
Date Prepared: 11 October 1990
Sheet Number: 1 of 2

Hazards	System Effects	Causal Factors	RAC	Actions	Controlled RAC	Standards/Comments
Good weather	No direct	Temperature between 45 and 70°F Clear or partly cloudy Wind speed less than 5 mph	IV A (-)	None	None	This list is not conclusive and will depend on the particulars of the unit, the training, and the range facilities. Additional peripheral factors a unit commander may wish to consider are— • Level of proficiency. • Time of event (day or night). • Availability and extent of emergency-response assets. • Train up (rehearsals and dry runs). • Terrain. • Location of instructor in relation to soldier during the training.
Minimal weather	No direct	Temperature between 32 and 44°F or 71 and 80°F Drizzle Wind speed between 5 and 15 mph	III A (2)	Ensure close supervision. Do not rush training to get in shelter. Postpone or cancel training.	III C (4)	
Bad weather	Wet, miserable soldiers Wet munitions	Temperature less than 32°F or greater than 80°F Moderate to heavy rain Snow or ice Wind speed greater than 15 mph	I C (2)	Postpone or cancel training.	I D (3)	
Lack of references	The sequential arming or disarming not done correctly, causing the fuse and/or mine to detonate	Current doctrine not available Questions or confirmation of techniques not quantified	I B (1)	Stop training if unsure of a task. Have proper references available.	I D (3)	

Figure E-15. Preliminary hazard-analysis work sheet (peripheral factors)

Operation Name: Mine arming/disarming peripheral factors (not all-inclusive)
Date Prepared: 11 October 1990
Sheet Number: 2 of 2

Hazards	System Effects	Causal Factors	RAC	Actions	Controlled RAC	Standards/Comments
Lack of support materials or components	Improper procedures or practices introduced, causing the fuse and/or the mine to detonate	Not enough or incorrect fuses, wrenches, and/or washers used	I C (2)	Ensure that proper support material is available. Stop training if the proper material is not available.	I D (3)	
Condition of soldiers	Improper procedures or practices introduced, causing the fuse and/or the mine to detonate	Long amount of training with little or no sleep Strenuous training conducted before or during live-mine training	I D (3)	Ensure that soldiers have adequate sleep. Allow for breaks during training. Stop training if soldiers appear to be heavily fatigued.	None	
Protective clothing	Improper or lacking protective clothing, increasing the severity of an accident if it does occur	Improper or lacking protective clothing	I C (2)	All soldiers participating in mine training will have Kevlar helmets, flak vests, and boots.	None	

Figure E-15. Preliminary hazard-analysis work sheet (peripheral factors) (continued)

Appendix F

Mine Awareness

Mine awareness should actually be entitled mine/UXO awareness. If only mines are emphasized, ordnance (bomblets, submunitions) may be overlooked, and it has equal if not greater killing potential. The main objective of mine awareness is to save lives, so it is important to ensure that soldiers are well-informed and thoroughly trained. This appendix outlines the tasks needed for soldiers and units to survive in a mined/UXO environment.

Mine awareness should be emphasized at all levels of command, and it involves soldier and leader skills. Soldier skills are a mix of individual and collective tasks that are required for an element to maintain its combat effectiveness in and around a mined environment. It is important to note that a soldier's basic mine-awareness skills are critical to his and the unit's survival. Leader skills involve planning missions, assessing situations, and tracking/disseminating mine information. A unit must be proficient in all mine-awareness skills to effectively operate in a mined environment.

SOLDIER

Soldier skills involve individual and collective tasks that are required for basic survival in a mined/UXO environment. They include minefield indicators, probing techniques, mine-detector operation, extraction drills, survival rules, casualty treatment, and evacuation drills.

VISUAL INDICATORS

Mine/UXO indicators are part of all combat operations. Understanding and recognizing mine indicators could determine whether or not a soldier becomes a casualty. The following may indicate the presence of mines/UXO:

- Trip wires.
- Signs of road repair (such as new fill or paving, road patches, ditching, culvert work).
- Signs placed on trees, posts, or stakes. Threat forces mark their minefields to protect their own forces.
- Dead animals.
- Damaged vehicles.
- Disturbances in previous tire tracks or tracks that stop unexplainably.
- Wires leading away from the side of the road. They may be firing wires that are partially buried.

- Odd features in the ground or patterns that are not present in nature. Plant growth may wilt or change color, rain may wash away some of the cover, the cover may sink or crack around the edges, or the material covering the mines may look like mounds of dirt.
- Civilians. They may know where mines or booby traps are located in the residential area. Civilians staying away from certain places or out of certain buildings are good indications of the presence of mines or booby traps. Question civilians to determine the exact locations.
- Pieces of wood or other debris on a road. They may be indicative of pressure or pressure-release FDs. These devices may be on the surface or partially buried.
- Patterns of objects that could be used as a sighting line. The enemy can use mines that are fired by command, so road shoulders and areas close to the objects should be searched.

PROBING

Probing is very time-consuming and is used primarily for clearing operations, self-extraction, and covert breaching operations. Detection of mines by visual or electronic methods should be confirmed by probing. Use the following procedures and techniques when probing for mines:

- Roll up your sleeves and remove your jewelry to increase sensitivity. Wear a Kevlar helmet, with the chin strap buckled, and a protective fragmentation vest.
- Stay close to the ground and move in a prone position to reduce the effects of an accidental blast. When moving into a prone position, the prober should—
  - Squat down without touching his knees to the ground.
  - Scan forward up to 2 meters and to the sides up to 3 meters for mine indicators.
  - Probe the area around his feet and as far forward as possible.
  - Kneel on the ground after the area is found to be clear and continue probing forward until he is in a prone position.
- Use sight and touch to detect trip wires, fuses, and pressure prongs.
- Use a slender, nonmetallic object as a probe.
- Probe every 5 centimeters across a 1-meter front.
- Gently push the probe into the ground at an angle that is less than 45 degrees.

DANGER

Use extreme caution when probing. If the probe is pushed straight down, its tip may detonate a pressure fuse.

- Apply just enough pressure on the probe to sink it slowly into the ground.

- If the probe encounters resistance and does not go into the ground freely, carefully pick the soil away with the tip of the probe and remove the loose dirt by hand. Care must be taken to prevent functioning the mine.
- When a solid object is touched, stop probing and use two fingers from each hand to carefully remove the surrounding soil and identify the object.
- If the object is a mine, remove enough soil to show the mine type and mark its location. Do not attempt to remove or disarm the mine. Use explosives to destroy detected mines in place or use a grappling hook and rope to cause mines to self-detonate. Metal grappling hooks should not be used on magnetic-fused mines.

Probing is extremely stressful and tedious. The senior leader must set a limit to the time a prober is actually probing in the minefield. To determine a reasonable time, the leader must consider METT-TC factors, weather conditions, the threat level, the unit's stress level, and the prober's fatigue level and state of mind. As a rule, 20 to 30 minutes is the maximum amount of time that an individual can probe effectively.

AN/PSS-12 Metallic Mine Detector

The AN/PSS-12 mine detector (Figure F-1, page F-4) is a man-portable metallic mine-detection system that is used to detect AT and AP land mines. Its search head contains two concentric coils—the transmitting (outer) coil and the receiving (inner) coil. During operation, the transmitting coil emits a magnetic field symmetrically above and below the search head. When this field encounters metal objects, it induces currents in these objects. When the induced currents reach the receiving coil, an auditory signal is provided to the operator. The detector is able to detect large metal mines from a distance of several feet, but can detect low-metal mines from only a few inches.

WARNING
It is important to understand that the mine detector is only effective when there is a sufficient amount of metal in the mine.

Unpacking

The system is stored and transported in a single carrying case.

- Open the pressure-relief valve in the carrying case.
- Release the latches on the carrying case and open the top.
- Remove the bag that contains system components.
- Unzip the bag and ensure that all components are present (Figure F-2, page F-4).
- Remove the following items from the bag carefully:
  - — Telescopic pole and search-head assembly with cable and plug.
  - — Electronic unit.

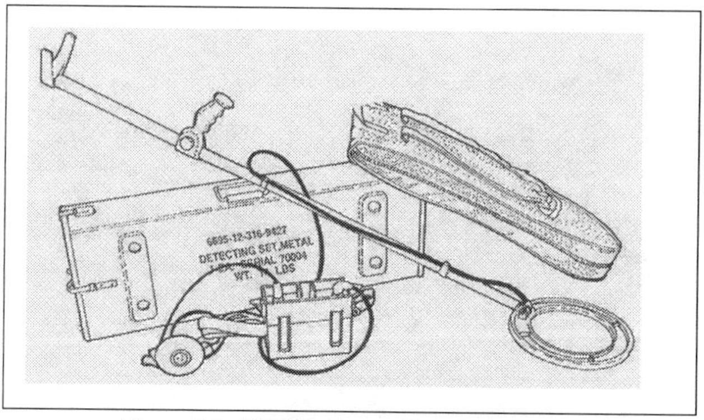

Figure F-1. AN/PSS-12 metallic mine detector

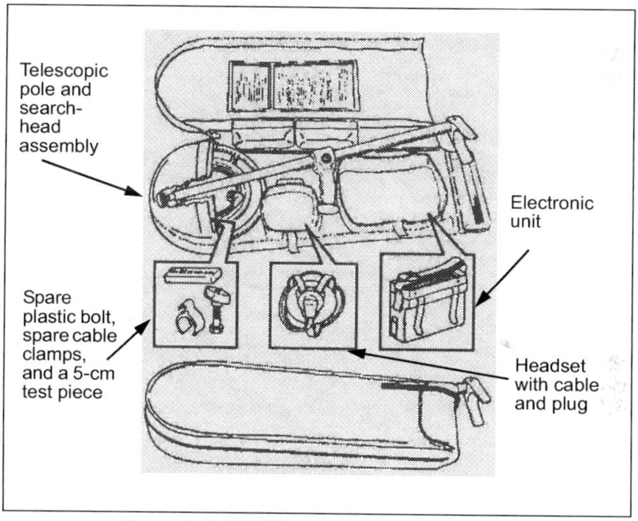

Figure F-2. AN/PSS-12 packed components

  — Headset with cable and plug.
- Ensure that the bag contains the following spare parts and test items:
  — Spare plastic bolt.
  — Spare cable clamps.
  — 5-centimeter test piece.
- Inspect the search head for cracks or damage.

- Inspect cable connectors for damage and check for bent pins.
- Inspect hook and pile material on headphones for serviceability.
- Inspect cables to ensure that they are not cut, broken, or frayed.
- Inspect the electronic unit for cracks, damage, and completeness. This includes ensuring that all switches and knobs are present and functional.
- Ensure that telescopic pole is not bent, dented, or damaged and can be extended and retracted.

Electronic-Unit Setup

- Ensure that the power switch on the electronic unit is in the *OFF* position (Figure F-3).

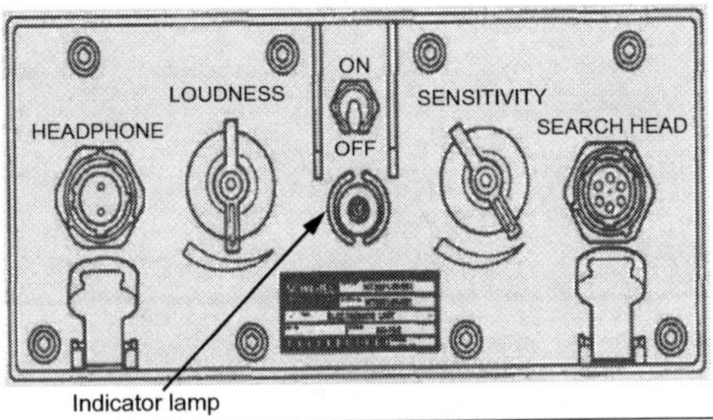

Figure F-3. Electronic unit

- Release the latches on the battery-compartment cover and remove the cover (Figure F-4).

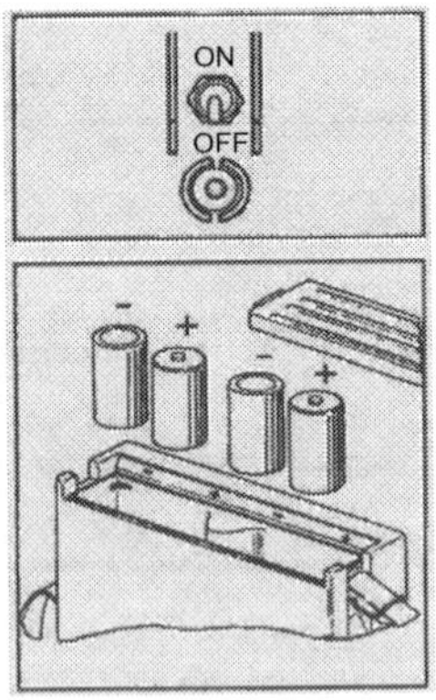

Figure F-4. Battery installation

- Insert batteries according to markings.

> **WARNING**
> **Ensure that the battery cover is completely closed and the latches are in the proper position. This prevents the inadvertent opening of the battery compartment during operation. Failure to do this could result in injury, damage to the equipment, and/or improper operation.**

- Reinstall and latch the battery cover. Ensure that the latches are in the proper closed position.
- Press in the catch (located just below the arm support) to unlock the telescopic pole from the transport position. Turn the outer tube until the catch snaps into the guide groove. Pull the telescopic pole out to a suitable length. Ensure that the telescopic pole is locked in one of three fixed positions.
- Connect the headphones and the detector-head cables to the electronic unit. Fit the protective caps of the cable connectors to the corresponding rubber caps on the electronic unit.
- Put on the headphones. Place the straps of the headphones over the forehead and on top of the head. The hook and pile material goes behind the head at the nape of the neck.
- Connect the detector-head cable to the lower (plastic) portion of the telescopic pole.
  - — Attach the lower cable clamp 5 centimeters above the wing nut.
  - — Attach the second cable clamp 5 centimeters down from the junction of the lower and upper (aluminum) portion of the telescopic pole.
  - — Attach the third cable clamp between the other two cable clamps. Do not attach the cable to this middle clamp. This clamp will become the sensitivity marker clamp.

> **WARNING**
> **The cable should not be attached to the aluminum portion of the pole, because it causes the sensitivity of the mine detector to fluctuate.**

- Adjust the position of the handle by loosening the knurl nut. Retighten the nut when the handle is properly positioned.
- Adjust the position of the search head so that it will be parallel to the ground.
- Attach the electronic unit to the operator's load-bearing equipment (LBE) belt using the belt clips. Right-handed soldiers should put the electronic unit on the left hip. Left-handed soldiers should put the electronic unit on the right hip. Put the safety strap over the shoulder.

Electronic-Unit Controls and Indicators

The AN/PSS-12 electronic unit has the following controls and indicators (Figure F-3, page F-5):

- On/off switch. This switch is the power control for the AN/PSS-12. It must be in the ON position when detection is being conducted.
- Loudness control. This is the volume control for the audible alarm. It should be adjusted to a comfortable listening level during use.
- Sensitivity control. The sensitivity control is used to adjust the detection characteristics of the mine-detector head. Details on how to set the sensitivity are provided below.
- Indicator lamp. This lamp indicates low battery voltage or a system malfunction. It provides a short flash when the unit is turned on and flashes continuously when the batteries are low or there is a malfunction.
- Audible signals.
  - The system provides an audible signal through the headset when the search head is over or very near a metal object. The signal is a continuous tone that at first resembles a growl. The tone increases to a higher pitch as the search head is moved closer to a metal source (such as a mine) and decreases in pitch when moved away.
  - The AN/PSS-l2 has a check tone that is provided every 1 to 2 seconds. The tone resembles a clicking sound, and its purpose is to continuously inform the operator that the system is functioning satisfactorily. If the check tone disappears or its frequency decreases, discontinue searching and determine the malfunction.
  - A flashing indicator lamp and a continuous tone (the same tone heard with a target return, except it does not abate when moved away) on the headphone indicates low battery voltage. If the unit's indicator lamp flashes, change the batteries and readjust the unit.

Operation

Once the system is assembled, it is ready for adjustments prior to operation. Remove rings, watches, and other jewelry before adjusting or using the system.

Power Up/Sensitivity Adjustments

- Turn the sensitivity and loudness knobs completely counterclockwise.
- Look at the indicator lamp on the electronic unit and turn the on/off switch to the ON position. The lamp should give a short flash. If it does not, ensure that the batteries are correctly installed or insert new batteries. If the lamp flashes continuously, the battery voltage may be low.
- Hold the search head approximately 0.6 meter above the ground, and turn the sensitivity knob clockwise until a continuous tone is heard.

While this is being done, adjust the loudness control to a comfortable listening level.

- Turn the sensitivity knob slightly counterclockwise until the tone ceases. The check tone (a clicking sound) should still be heard every 1 to 2 seconds. Readjust the loudness control if necessary. If the check tone disappears or its frequency decreases, discontinue searching and determine the malfunction.
- Set the sensitivity in either of the following two ways:
  - The preferred method is to use the most difficult to detect type of mine (made safe without changing the metal content) that is expected to be encountered.
  - In the absence an actual mine, use the 5-centimeter test piece that comes with the AN/PSS-12 packed components.
- Bury the mine (if you have a mine) at the deepest depth that threat mines are expected to be found. If you do not have a mine, bury the 5-centimeter test piece vertically so that the metal portion is at the deepest depth which threat mines are expected to be found. In the absence of more specific information, bury the test piece at a depth of 5 centimeters). This places the top of the test piece flush with the ground's surface.
- Place the detector head lightly on the ground directly above the mine/test piece and adjust the sensitivity knob so that the mine is detected and you have an easily heard signal from the headset. Now, move the detector away from the mine/test piece until the signal ceases to be heard.
- Float the detector head lightly on the ground at a 0.3-meter-per-second movement rate across the mine/test piece. With the detector head in motion, check to see if the signal can again be easily heard. If not, increase the sensitivity until the signal can be easily heard.

NOTE: The higher the sensitivity is set (between just being able to detect the mine/test piece and where the signal from the soil first becomes audible) the less the chance of missing a mine, but the more the chance of a false alarm rate.

WARNING

In some conditions, you may not be able to detect a target mine/test piece at the deepest depth the threat mine is expected to be encountered. Do not use the AN/PSS-12 to detect mines in those conditions.

NOTE: Adjusting the sensitivity with the mine/test piece buried in soil that is similar to the soil where the detector will be used is very important. Otherwise, the sensitivity setting is simply a guess, because different soil types and moisture content influence the sensitivity of the detector.

Sensitivity Level

The AN/PSS-12's sensitivity setting drifts over time. During use, soldiers should check the detector's sensitivity every 1 to 2 meters of forward advance in the mine lane.

- With the sensitivity properly set using a buried mine/test piece, the detector is moved to an adjacent place on the ground over a clear area.
- To check the detector, a metal object (bayonet or other metal object) is slid down the lower (plastic) portion of the telescopic pole until the operator hears the same auditory signal that was emitted when the search head was placed over the buried mine/test piece.
- At this point on the lower plastic shaft, the middle plastic cable clamp (the middle clamp previously attached without the cable) is positioned. This clamp now becomes the sensitivity marker clamp.
- At approximately every 1 to 2 meters of forward advance in the mine lane (or more often if desired), the same metal object (bayonet) is moved down the shaft to the sensitivity marker clamp and the operator listens to determine if the same auditory signal is emitted. When the same auditory signal is not heard, the sensitivity knob is adjusted up or down until the same auditory signal is replicated and maintained.

NOTE: It is important that the operator maintain a consistent body posture and angle between the telescopic pole and the search head each time a sensitivity check is made.

DANGER

Low batteries may reduce detector performance well before the indicator light comes on. If you need to frequently adjust the sensitivity knob to maintain a constant sensitivity setting or if the indicator light comes on, discard all batteries and replace them with new ones. Failure to do this could result in injury or death.

DANGER

Discontinue searching and readjust the sensitivity if the check tone disappears or its frequency decreases. Failure to do this could result in injury or death.

Search Methods

- Sweep across the lane, keeping the entire search head in light contact with the ground. Sweeping speed should be approximately 0.3 meter per second. Each sweep must be a minimum of 1.5 meters and must overlap the edge of the lane by at least one-half the width of the search head. Each sweep should advance forward at no more than 6 inches per sweep. The sweep and advance must be maintained as described, otherwise gaps will be left between sweep paths or at the edge of the lane and a low-metal mine can be missed.

- Perform a sensitivity check at approximately every 1 to 2 meters of forward advance in the mine lane (or more often if desired) and make adjustments as necessary. It is important that the operator maintain a consistent body posture and angle between the telescopic pole and the search head each time a sensitivity check is made.
- Pat the search head lightly on the ground (each pat advancing no more than one-half the width of the search head) in low vegetation or rocky or uneven ground where smooth sweeping cannot be performed. Ensure that the wing nut is tight, so that the position between the head and telescopic pole is fixed.

NOTE: Many conditions of vegetation may not allow getting the search head on the ground. Do not push through vegetation to get the search head on the ground. Different environments will require slightly different techniques. If you cannot get the search head on or very near the surface of the ground, the detector will not detect low-metal mines.

NOTE: The AN-PSS/12 detector's performance is reduced when the cable between the electronic unit and the top cable clamp is permitted to hang unrestrained during sweeps. The soldier can control this problem by holding the cable in his free hand.

WARNING

Always be aware of the potential for booby traps. The search head should never be swept where its path cannot be visually cleared. If trip wires are a threat, other techniques to detect trip wires should be used before sweeping.

Alarm Investigation/Mine Identification

When an auditory signal is heard, the soldier must investigate further to determine if the signal is a reliable indication of a mine.

- End the sweep procedure and begin the investigation procedure at the first auditory indication from the detector of metal in the ground. The purpose of the investigation is to determine if the signal is repeatable and, therefore, a likely indication of a mine. If so, the investigation continues with the purpose of gaining more information concerning the size, type (high-metal or low-metal mine), and specific location of the signal source.

NOTE: In some cases, the detector may provide a weak signal that cannot be repeated during the initial investigation process. Do not give up on the reliability of the signal until the sensitivity of the detector has been checked to ensure that it has not drifted to the low side.

NOTE: A footprint is defined as the entire area on the ground where a mine or metallic source causes the detector to generate an auditory signal. Small footprints, often as small as 10 to 15 centimeters in diameter, will indicate low-metal mines. Footprints as large as 60 to 122 centimeters in diameter may indicate high-metal mines.

- Note the location on the ground (at the center of the search head) where the signal is first heard. Then attempt to develop a set of points on the ground that identifies the source footprint.
  - Develop this footprint by first moving the search head away from the signal source until no signal is heard and then sliding the search head toward the signal from several clock positions.
  - Note the specific location of the center of the search head at each location where the auditory signal begins. Once the signal is detected, the detector is no longer slid toward the potential source to avoid coming closer than necessary to a potential mine.
  - Repeat this process and mark or note the locations on the ground until the size and shape of the footprint can be understood. Typically, the pattern will resemble a semicircle with the 6 o'clock position nearest the operator.

WARNING
The large footprints of high-metal mines may mask signals from low-metal mines within the footprint. Always assume that there are low-metal mines within the large footprint area.

- Identify the center of very large footprints (if important). The airborne technique may be quicker than the method previously discussed to identify the perimeter.
  - The search head must be adjusted so that it can be held above the ground with the head parallel to the surface.
  - The detector can then be manipulated above the source until the signal can be heard at only a single point. Move the search head in a crossing pattern, which should produce a smaller and smaller signal area as the pattern is repeated at gradually increasing heights. Near the surface, the signal can be heard over a broad lateral area, but as the search head is raised higher off the ground, this area becomes progressively smaller. As the search head is raised higher off the ground, the area where the signal can still be heard is reduced to a point.
  - The center of the mine or metal source should be directly below this point. Do not ignore the potential for low-metal mines to be hidden in this large footprint on the ground.

WARNING
Always be aware of the potential for booby traps and UXO. The search head should never be swept where its path cannot be visually cleared. If trip wires are a threat, other techniques to detect trip wires should be used before sweeping.

Other Operational Actions

- Keep mine detectors at least 2 meters apart during setting, adjustment, and operational phases to prevent interference.
- Change the batteries immediately if the indicator lamp flashes, a constant audible tone is heard, or the check tone (clicking sound) stops and readjust the sensitivity. The search sensitivity is affected long before the lamp begins flashing, and the unit will be unusable until fresh batteries are installed.
- Ensure that only the lower part of the telescopic pole is used when the equipment is operated by a soldier in the prone position.
- Turn the unit off after completing the search operations.

Disassembly and Packing

- Ensure that the on/off switch on the electronic unit is in the *OFF* position.
- Disconnect the headphones from the electronic unit.
- Detach the cable connection on the electronic unit for the magnetic search head, and replace the protective caps on the plug and socket.
- Release the electronic unit's battery-cover latches, and remove the battery cover.
- Remove the batteries, and ensure that none of the battery cases have ruptured; if they have, notify your supervisor. Reinstall the battery cover and latch it.
- Remove the cable clips from the telescopic pole.
- Collapse the telescopic pole to its travel length, and turn its outer tube until it is locked by the catch. Loosen the plastic restraining bolt, and fold in the magnetic search head.
- Pack the components in the carry bag as shown in Figure F-2, page F-4. For long-term storage, do not put batteries in the carry bag. Close and zip the carry bag.
- Place the carry bag in the metal transport case, and latch the case. Close the pressure-relief valve.

As in probing, the senior leader must set a limit to the time an individual can use the mine detector. The time limit is determined by METT-TC factors, weather conditions, the threat level, the unit's stress level, and the individual's fatigue level and state of mind. As a rule, 20 to 30 minutes is the maximum amount of time an individual can use the detector effectively.

EVACUATION DRILLS

A well-developed, well-rehearsed evacuation drill is necessary to extract an individual or a unit from a mined area. Units must develop evacuation drills for dismounted and mounted operations. Each type of operation should include two drills—one using a mine detector (mounted extraction) and one without using a mine detector (dismounted extraction).

Mounted Extraction

- The convoy commander halts the convoy and reports to higher headquarters.
- No vehicles move and no troops dismount unless directed to do so.
- Elements provide 360-degree security from vehicles.
- Troops thrown from vehicles should not move. Personnel are extracted by using dismounted evacuation procedures if electronic detectors are not available.
- The senior leader, if engineers are not available, assesses the situation and directs vehicles to back up along the entry-route tracks. If an immediate threat exists, occupants of damaged vehicles evacuate out the rear of the vehicle and along the vehicle-entry tracks. If no immediate threat exists, occupants of damaged vehicles remain in the vehicle until it is extracted.
- Engineers, if available, sweep the area and provide a cleared path for movement. They—
  - Clear a lane that is wide enough for the towing vehicle.
  - Use all available tow cables to increase the distance before towing if an M88 is unavailable. Remember, an M88 has a wider track base than other tracked vehicles.
  - Ensure that all towing-shackle sets are complete and mounted.
  - Ensure that the towing vehicle has tow cables on the front and the rear if possible.
  - Ensure that rear cables are attached to the lower mounts. This allows the crew to recover the vehicle without touching the ground.
  - Pull the vehicle out at least two-vehicle lengths before switching to a tow bar. When towing a vehicle after a mine strike, the chance of fire is greater because of possible damage to the vehicle.
  - Provide first aid and conduct casualty evacuation or have medics provide treatment and medical evacuation.
  - Guide vehicles through the safe area.
  - Mark, record, and report the threat.
- Continue the mission.

Dismounted Extraction

- All personnel freeze and crouch into a low-silhouette position. Be cautious when making this movement to ensure that you do not detonate another mine. If a protective mask is worn on your hip, do not allow it to come in contact with the ground because contact may detonate a mine. Do not help casualties because you could also become a casualty.
- The leader designates a security element and a soldier to assist in casualty evacuation.
- Soldiers extract along the path they entered. If possible, they step in the same places as before; if impossible, they probe their way out.
- The security element, consisting of individuals who are not in the minefield, sets up security for the unit.
- The soldier extracting the casualty—
  - — Probes a 1-meter-wide path to the casualty and marks the cleared path with foot powder or marking tape as it is probed.
  - — Probes around the casualty to clear the area.
  - — Performs first aid.
  - — Carries the casualty out of the minefield along the cleared path. (Stretcher parties do not enter the area unless a 2-meter-wide path has been cleared to the casualty.)
- The unit marks the threat and assembles back at the rally point.
- The unit reports the incident when it is 50 to 100 meters away from the minefield. If soldiers are in the minefield and radio transmission is required, move the transmitter at least 300 meters from the minefield. This prevents accidental mine detonation from the radio signal.
- The unit provides first aid and conducts casualty evacuation or has medics provide treatment and medical evacuation.
- The unit continues the mission.

Extraction from Scatterable Mines

- The individual who discovers the mine initiates the alarm according to the unit SOP.
- Unit personnel at the command post receive the alarm and alert others.
- The unit TOC requests counterbattery fire if the mines are artillery-delivered.
- Vehicle commanders check the immediate area. They do not dismount. Inspect the vehicle for mines and/or trip wires. Note and record the location of all mines found on or around the vehicle.
- Personnel leave any vehicle that is touching or is blocked by AT mines in place until the remainder of the unit is out of the minefield.

- Unit leaders identify unmovable vehicles and designate one or more lanes for remaining personnel and vehicles to use when exiting the minefield.
- Leaders identify a clearance team to extract remaining vehicles and personnel. The clearance team—
  - Uses visual means to locate mines and marks vehicle lanes at least 4 meters wide. Personnel mark lanes according to the tactical situation and the threat in the area; however, they mark them so that team personnel can reenter the minefield and recover equipment and vehicles.
  - Destroys or removes all mines in lanes, using a grapnel hook or other means as necessary, in the sequence directed by the team leader. Personnel detonate unmovable mines to reduce personnel injury and equipment damage.
- Vehicle commanders direct personnel to ground-guide vehicles out of the minefield. Ground guides—
  - Ensure that individual elements move only when directed by the chain of command.
  - Place equipment that is not in contact with a mine or a trip wire onto vehicles.
  - Direct vehicles to the designated exit lane or, if safer, allow vehicles to exit the minefield on their own.
- Clearance team personnel, aided by unit personnel, remove equipment and vehicles remaining after initial extraction from the minefield. They—
  - Reenter the minefield using the same exit routes.
  - Detonate the minimum number of mines necessary to remove vehicles and equipment from the minefield.
  - Avoid touching mines. Personnel should take all possible precautions to ensure that mines are not jarred.
  - Place sandbags near mines, if possible, to minimize damage to vehicles and equipment.
  - Remove mines from equipment by using a line or other remote means. Ensure that the entire team is far enough from the mine to avoid casualties if it explodes.
  - Place explosive charges to minimize vehicle damage when detonating mines on the ground.
- Clearance team personnel—
  - Clear sufficient mines to allow for mission accomplishment if the position cannot be evacuated.
  - Clear and mark communication lanes between positions.
- Continue the mission.

Survival Rules

Many of our allies have extensive experience in mine operations. Among them is Canada. They have produced several manuals and videotapes on mine awareness and have developed the following *survival rules*. They are very practical and can be applied to our soldiers as well.

- *If you did not drop it, do not pick it up.*
- *All terrain and structures are potentially mined or booby-trapped.*
- *Beware of areas associated with basic human needs. They could be mined or booby-trapped.*
- *Immediately report all confirmed or suspected mines.*
- *Leave mine disposal to the EOD personnel and combat engineers.*
- *Avoid touching or removing foreign objects, no matter how attractive. They could be mined or booby-trapped.*
- *Avoid verges because they could be mined. Stay on the traveled road.*
- *Mark and avoid UXOs if possible. Consider them unstable.*
- *Develop and rehearse effective evacuation drills.*

A convoy provides better protection against mine and UXO threat than a single vehicle. In convoy movement, some rules of thumb should be followed:

- The lead vehicle should be one of the heaviest (2½-ton, 5-ton) vehicles in the unit and be hardened against a mine threat.
- A high-mobility, multipurpose, wheeled vehicle (HMMWV) or a HEMMT should not lead a convoy unless absolutely necessary. These vehicles are extremely vulnerable to mine and UXO threat and are difficult to harden without commercial products.

Casualty Treatment and Evacuation

Casualty treatment and evacuation should be a part of every mission order. Incorporate the following elements into the unit SOP, and ensure that soldiers understand them. Emphasize that rushing to help a mine victim can lead to the rescuer becoming a casualty.

- Reassure the casualty.
- Do not panic and create another casualty.
- Call higher headquarters at the earliest opportunity, and request a specialist engineer and medical help as soon as possible.
- Extract yourself and mark the path as you go.
- Reenter along the marked path.
- Clear a path to the casualty.
- Administer first aid.
- Recover the casualty.

- Mark the area after it is clear, record it on the map, and report to higher headquarters.

Medical personnel and combat lifesavers should anticipate and train for the following injuries:

- Blast injuries with fragments embedded.
- Burns.
- Fractures.
- Traumatic amputations.
- Blunt trauma.
- Psychological anxiety reaction.

LEADER

Leader skills in effective planning, risk management, and mine-data tracking/ dissemination can greatly enhance force protection and reduce casualties in a mined environment. Unit leaders must train themselves and their subordinates on mine-awareness tasks.

RISK MANAGEMENT

Risk management reduces the frequency of mine and UXO strikes and diminishes the physical effects when they do occur. It is used to identify hazards, define risks, identify methods for control, and identify responsibilities for implementation. The risk-management process enables commanders and staffs to define acceptable risk levels and implement controls until risks are commensurate with the mission. Risk management is a simple, five-step process:

- Identify hazards.
- Assess the risk of each hazard.
- Make risk decisions and develop controls.
- Implement controls.
- Supervise.

Identify Hazards

This is often the most difficult part of risk management. A mine hazard is the condition that results from the interactions of a mine, a catalyst (such as activation from a soldier or vehicle), and a common spatial relationship. These hazards are defined in terms of mine types (AT, AP, fuses, metal content, AHD), how a soldier might encounter mines (dismounted or mounted and the type of vehicle), and the locations where encounters would be most likely. Mines are seldom employed in isolation, so complete hazard definition includes the complex obstacle and the covering fires.

Assess the Risk of Each Hazard

This requires determining the probability of a mine strike and the effects of the strike. An effective risk assessment is critical for evaluating the combat effectiveness of a unit in a mined environment. Risk-assessment criteria is developed by using Table F-1. A sample risk assessment is shown in Table F-2, page F-19.

Table F-1. Risk-assessment criteria

			Hazard Probability				
			Frequent (A)	**Probable (B)**	**Occasional (C)**	**Remote (D)**	**Improbable (E)**
	Single Item		Likely to occur frequently	Will occur several times during life of item	Likely to occur sometime during life of item	Unlikely, but possible to occur during life of item	So unlikely, can assume occurrence will not take place
	Fleet/Inventory		Occurs continuously	Will occur frequently	Will occur several times	Unlikely but reasonably expected to occur	Unlikely to occur, but possible
Hazard Severity	CATASTROPHIC (death or system loss)	I	High	High	High	High	Medium
Hazard Severity	CRITICAL (severe injury or occupational illness or major damage)	II	High	High	High	Medium	Low
Hazard Severity	MARGINAL (minor injury or occupational illness or minor damage)	III	High	Medium	Medium	Low	Low
Hazard Severity	NEGLIGIBLE (less than minor injury, illness, or damage)	IV	Medium	Low	Low	Low	Low

Make Risk Decisions and Develop Controls

This step requires decision makers to identify actions that can reduce the probability and/or severity to acceptable levels. This may be accomplished by taking actions to reduce the probability of a mine strike or by providing more protection to the soldier or materiel to reduce severity of a mine strike. Often, it is a combination of the two.

Implement Controls

Leaders must apply the identified controls to reduce the probability and severity of a mine strike.

Table F-2. Sample risk assessment

DIRECTIONS: Circle the appropriate number in each section and total them. Appropriate directions are at the end of the assessment block.

TRAINING (Circle One)
1. Experience working in mined areas
2. No experience working in mined areas, but trained
3. No experience or training working in or around mined areas

TYPE OF AREA (Circle One)
1. Area known by friendly forces to be clear of mines
2. Old confrontation line or suspected mined area
3. Area known to be mined

LIGHT AND WEATHER (Circle One)
1. Daylight with clear weather
2. Daylight with poor weather
3. Darkness

MISSION (Circle One)
1. One that your soldiers have done before
2. One that your subordinate leaders have done before
3. An unfamiliar type of mission

ON OR OFF ROAD (Circle One)
1. Approved division/corps route
2. Paved road
3. Unpaved road
4. Cross-country

SLEEP (Circle One)
1. 6 hours of sleep in last 24
2. 4 hours of sleep in last 24
3. 2 hours of sleep in last 24
4. Awake for more than 24 hours.

TYPE OF VEHICLES IN CONVOY (Circle One)
1. Armored
2. Mixed, armored vehicles leading
3. Wheeled
4. On foot

AVAILABILITY OF MINE INFORMATION (Circle One)
1. Updated mined-area graphics in each vehicle
2. Updated mined-area graphics in the lead vehicle or a reliable, knowledgeable guide is available
3. No upgraded mined-area graphics are available

GROUND COVER (Circle One)
1. Dry, hard ground with short grass
2. Dry, hard ground with long grass or vegetation
3. Wet, muddy ground or snow less than 10 centimeters deep
4. Snow more than 10 centimeters deep

ROAD USE (Circle One)
1. Heavy tracked vehicles or trucks recently used the road
2. Light wheeled vehicles recently used the road
3. No traffic observed on the road; some tire marks
4. No traffic observed on the road; no tire marks

10-16 Continue the mission. Keep following training and common-sense rules that apply to working around mines.
17-24 Continue the mission. Remember to stress safety and mine awareness when you brief soldiers. Ensure that leaders maintain positive control of personnel.
25-35 Consider postponing the mission until better conditions are attained. If you must continue the mission, constantly stress mine awareness and safety. Ask higher headquarters for engineer support to accomplish the mission. Conduct mine-awareness training.

Supervise

This step ensures that controls are implemented and that a measure of quality control exists to ensure a quantified level of clearance.

The key to using risk management successfully is to employ it at each echelon—from the commander, through the tactical planner, to the soldiers executing the mission. Each level identifies hazards, eliminates or reduces hazards as feasible, and communicates the residual hazards to the next lower echelon. As such, each echelon works as a filter to control unacceptable risks.

- Training provides soldiers with an understanding of equipment limitations and plays a critical role in the risk-management process. Capabilities and limitations of Army systems are taken into consideration during the development of doctrine and TTP.
- Risk management at the tactical planning level requires a thorough knowledge and awareness of the hazards and potential controls that can be employed. The planning process requires a methodical and disciplined technique to identify the hazards and develop appropriate controls for operating in a mined environment. The controls for countermine operations, discussed in Chapters 9 and 11 and in FM 3-34.2, provide a framework for risk-managing hazards associated with mines.
- The execution level is the culminating point of risk management. It is where soldiers and leaders employ the systems provided to accomplish the mission. The amount of residual hazards remaining after the filtering process from echelons above may well determine success. The individual soldier is the last element to control any residual hazards.

Optimizing the components of risk management at the tactical planning level is more challenging as emerging technologically dependent systems bring more variables into the mission. While tactical intelligence is the key element in identifying mine-related hazards, technical knowledge is the key element in assessing the risks associated with mine hazards. This knowledge assimilates the tactical intelligence with the capabilities of the unit's equipment, the performance of threat mines, and the protection provided to our soldiers by their vehicles or personal protective equipment. The staff engineer, using his engineer C^2 system to risk-manage each COA, provides the maneuver commander and his staff with information on risks and potential controls early on in the planning process. Each subsequent commander must perform the same analysis and incorporate the mine threat into risk management.

Recording and Mine-Data Tracking

Obtaining and disseminating information are the keys to battlefield management. Units encountering minefields or explosive devices should follow a five-step process—stop, secure, mark, report, and avoid. Units must provide adequate information to their higher headquarters to ensure that follow-on elements are well informed. Information must include known or suspected minefield locations, types of mines (if known), the marking method, the time encountered, and any additional information that may be of use to the clearing unit.

Division and maneuver brigade engineer planning cells must establish a central control cell for mine-clearance information. The central control cell—

- Maintains a current situation map and overlay that depicts friendly and enemy mines and obstacles.
- Maintains and updates information on minefield tracking and route status.
- Receives and maintains minefield recording forms within the unit's AO (includes host-nation minefield data if available).
- Maintains a database of mine information.
- Processes, analyzes, updates, and disseminates the information to subordinate commanders and staff.

MINE-INCIDENT REPORT

A mine incident includes any unplanned activity involving a mine, UXO, or booby trap. It also includes near misses that could have resulted in potential damage or injury. The mine-incident report is a technical report that follows a serious-incident report. The report should be submitted as soon as possible (the local SOP will indicate time requirements). A sample mine-incident report is shown in Figure 11-12, page 11-26.

TRAINING

Modern combat is complex, lethal, and demanding. Soldiers must be capable of performing their missions in any type of battlefield environment. Current doctrine and TTP provide soldiers with guidelines to accomplish their tasks and quality equipment provides the means. The common thread that connects doctrine, tactics, and equipment is quality training. To fight and win, units must train their soldiers to execute all wartime missions successfully. They must use every training opportunity to improve soldier, leader, and unit task performance. Without quality training, no amount of world-class equipment can make the soldier effective or make him survive in a wartime environment. Even the best doctrine in the world is worthless unless soldiers receive effective training. This is especially true with mine-awareness training.

Soldiers must be trained to think mine awareness as well as perform mine-related actions. Decisions, actions, and reactions must become automatic to every soldier. This requires that all soldiers receive mine-awareness training early in their careers. It must begin at early entry training with basic individual tasks and continue through advanced unit training with collective tasks. Soldiers who survive mine threats can survive mine warfare, but it requires continual training.

INDIVIDUAL TRAINING

Basic Mine-Awareness Training

This is the most important phase for preparing soldiers to survive in a mine environment. Individual soldiers must be trained to a level that meets the environment they will face on the modern battlefield. Basic mine-awareness tasks (mine detection, survival rules, minefield indicators, and procedures for self-extraction from a minefield) must begin early in the soldier's career.

With an estimated 80 to 100 million mines deployed worldwide, countermine considerations must become second nature to mission planners. From CSS operations in the rear area to close combat operations in the main battle area, there is no place on the battlefield that is safe from mines. This is especially true with the threat of SCATMINEs that are deployed from aircraft and indirect-fire sources. The need for initial countermine training will get more critical as the modern battlefield becomes more lethal.

Annual Mine-Awareness Training

Units exercising military skills at the National Training Center, Joint Reserve Training Center, or the Combat Maneuver Training Center can reinforce and evaluate their awareness training. Soldiers should be fully trained on the individual and collective tasks that are required to support their unit's mine-related missions before deploying to the combat training centers. An effective method to meet this challenge is a well-planned annual training program that trains soldiers on individual and collective mine-awareness tasks. The plan should address the requirements for initial collective-task training and sustainment training.

Predeployment Mine-Awareness Training

With an ever-changing global environment, the probability of projecting US forces to various geographical locations is very real. It is extremely critical that units be situationally aware. Each flash point in the world contains its own type of threat. Many of these threats can be neutralized or reduced through quality predeployment training. Quality predeployment mine-awareness training enhances the ability of the unit to perform its mission and increases the confidence of soldiers.

Units must train intensively when alerted for a deployment. Predeployment training is the time to polish and refine training and to focus on theater-specific operations. Predeployment training is intended to augment individual and collective training, not replace it.

In-Theater Mine-Awareness Training

Training does not end with predeployment training. Units must train in the country where they deploy. They need to plan and conduct routine sustainment training on individual and collective mine-awareness tasks. This is an essential ingredient of force-protection operations. In-theater mine-awareness training reinforces the soldier's existing skills and places mine survival in the forefront of each soldier's mind.

Leader Training

Officers and NCOs must receive mine-awareness training in basic and career courses. Commissioned and noncommissioned leaders must be trained on the use and the employment of the equipment that they will use in the field. Officer basic courses and officer and NCO advanced courses must include access to modeling/simulation training. Using modeling/simulation enables the leader to practice decision-making, employment, and sustainment operations in a simulated mine environment. This develops the situational awareness required for timely, accurate decisions.

Unit Training

The Army Training and Evaluation Program remains the most effective measure of individual and collective training effectiveness short of actual combat. It provides leaders with the opportunity to verify the effectiveness of virtual and actual training without endangering soldiers with mine effects. Units must include mine-threat scenarios in their home-station training exercises. Basic missions include minefield detection, reduction, marking, proofing, and recording. Commanders are required to assess their unit's state of proficiency in mine-awareness tasks on a routine basis.

Mine awareness is a critical, perishable skill. It must be trained effectively and sustained on a continuous basis. If a unit properly trains its soldiers on mine awareness, it will maintain its force, boost its soldiers' confidence in their abilities, and accomplish its mission more effectively. Mine awareness is not a liability; it is an investment in a unit's future.

Appendix G

Countermine Data

This appendix provides a compilation of countermine information to assist commanders in the decision-making process for countermine operations.

BREACHING ASSETS VERSUS THREAT OBSTACLES

Tables G-1 and G-2, pages G-2 through G-8, provide general guidance on the effect of breaching assets versus threat obstacles.

FOREIGN MINE DATA

Tables G-3 through G-7, pages G-9 through G-12, provide foreign mine characteristics.

FOREIGN MINEFIELD EMPLACEMENT DATA

Table G-8, page G-13, provides foreign minefield emplacement data. Information is presented in the probability of occurrence based on historical data.

FOREIGN MINE DELIVERY SYSTEMS

Table G-9, pages G-14 and G-15, contains characteristics of foreign mine delivery systems.

NOTE: Blank areas in the above tables indicate that the information is not applicable or that it was unavailable when this manual was published.

Table G-1. Mounted breaching assets versus threat obstacles

Asset	Track-Width AT Mine	Full-Width AT Mine	Side-Atk AT Mine	Cmd-Det Blast Mine	AT Ditch and Other Gaps	Wire Obstacles	Remarks
MCB with dog-bone assembly	Is more than 90 percent effective to 12 inches deep at speeds less than 10 kph (soil dependent)	Do not use against magnetic-influence mines; is more than 99 percent effective against tilt-rod mines	Do not use	Do not use	May be effective at filling in some shallow gaps or trenches	Do not use	The MCB provides almost 100 percent clearance of simple pressure-fused AT mines when used with the MICLIC. The MCB has minimal impact on the M1's maneuverability and speed when not plowing. The main gun should be traversed to the side to prevent damage in the event of a mine detonation when plowing. The reduction site should be relatively flat and free of large rocks and other obstructions. Plowing operations should begin about 100 m short of the suspected minefield and continue 100 m past its suspected end (typically 2 to 2 1/2 minutes to cover 300 to 400 m overall). Plowing operations should be performed in a straight line to keep the vehicle out of uncleared ground. Fifteen percent of pressure-fused AT-mine encounters across the MCB's width can be expected to detonate under the skid shoe, probably disabling the blade. The MCB is vulnerable to coupled mines and may be stopped by log-hurdle obstacles laid within a minefield as part of a complex obstacle. A second plow should not be used to clear the center of the lane because this will simply push mines into the area cleared by the first plow.

Table G-1. Mounted breaching assets versus threat obstacles (continued)

Asset	Track-Width AT Mine	Full-Width AT Mine	Side-Atk AT Mine	Cmd-Det Blast Mine	AT Ditch and Other Gaps	Wire Obstacles	Remarks
MCB with IDA	See above	Is effective against magnetic-influence mines; is more than 99 percent effective against tilt-rod mines	Do not use	Do not use	See above	Do not use	These AT mines are fitted with magnetic-influence fuses and AHDs may disable the blade. Dismounts should expect to encounter some pressure-fused AP mines in the breached lane that may be missed by the blade. The MCB clears a 180 inch-wide lane but leaves a 26 inch-wide area uncleared of simple pressure-fused mines and a 54 inch-wide area uncleared of double-impulse mines (under the skid shoes) in the middle of the breached lane. The berms and furrows left by the MCB aid in immediately marking the breached lane. Live mines remain in both berms and may roll back into the cleared lane. Over time, these may vibrate free and roll back into the breached lane.
MCR (including Panther) with dog-bone assembly	Is more than 90 percent effective to 12 inches deep at speeds less than 5 to 15 kph (soil dependent) Has reduced effectiveness against some (mostly Italian) blast-hardened mines Do not use against double-impulse mines	Do not use against magnetic-influence mines; is more than 99 percent effective against tilt-rod mines	Do not use	Do not use	Do not use	Do not use	Normally used to detect the leading edge of a conventionally laid tactical minefield and to proof lanes/routes cleared by other means. Use caution when crossing on an AVLB. May become stuck or ineffective in muddy conditions. The MCR is vulnerable to coupled mines. The MCR is designed to withstand up to two detonations per side. The MCR clears a 160 inch-wide lane but leaves a 72 inch-wide area uncleared in the middle of the breached lane. **NOTE: The M113 also has 72 inches between tracks and must use extreme caution when following the MCR.**

Table G-1. Mounted breaching assets versus threat obstacles (continued)

Asset	Track-Width AT Mine	Full-Width AT Mine	Side-Atk AT Mine	Cmd-Det Blast Mine	AT Ditch and Other Gaps	Wire Obstacles	Remarks
MCR with IDA	See above	Is effective against magnetic-influence mines; is more than 99 percent effective against tilt-rod mines	Do not use	Do not use	Do not use	Do not use	Multiple rollers are required to provide the overlap necessary for proofing the route's entire width when the MCR is used for proofing during route-clearance operations. Vehicle speed is limited to 5 to 15 kph. The main gun should be traversed to the side to prevent damage in the event of a mine detonation when rolling a suspected mined area. The IDA may be damaged by detonated mines.

Table G-2. Dismounted breaching assets versus threat obstacles

Asset	Pressure-Fused AP Mine	Trip-Wire/Break-Wire-Fused AP Mine	Influence-Fused AP Mine	Cmd-Det Blast Mine	Wire Obstacles	Remarks
APOBS	Effectively clears 95 percent, including blast-resistant types	Effectively cuts 99 percent of trip wires in its path	Is ineffective	Is ineffective	Clears a footpath	Provides a 0.6- x 45-m footpath from a 25-m standoff Can be deployed by two men in 2 minutes Consists of two manpack units weighing approximately 55 lbs each
Bangalore torpedo	Effectively clears 85 percent of simple pressure-fused mines; is less effective against blast-hardened mines	Effectively cuts 99 percent of trip wires in its path	Is ineffective	Is ineffective	Clears a footpath	Provides a 0.6- x 15-m footpath Can be deployed by a squad in 5 minutes Consists of ten 1.5-m-long sections weighing 13 lbs each Is ineffective against AT mines
Grapnel hook and WLGH	Is ineffective	Effectively clears 99 percent of trip wires/break wires in a single pass	Is ineffective; however, it may snag the wires between the control unit and its mines	Is ineffective; however, it may snag the wires between the observer and his mines	Is somewhat effective at clearing trip-wire-fused mines integrated with wire obstacles	Has a range of 80 to 100 m when launched Can be thrown 15 to 20 m when hand thrown Must consider the lethal range/radius (up to 90 m in some cases) of fragmenting AP mines that may be encountered, and select a covered position or move back an adequate distance before pulling the grapnel Helps increase confidence by grappling an area several times Is hampered by thick vegetation/overhead obstructions which may reduce the range/ effectiveness of the WLGH

Table G-2. Dismounted breaching assets versus threat obstacles (continued)

Asset	Pressure-Fused AP Mine	Trip-Wire/Break-Wire-Fused AP Mine	Influence-Fused AP Mine	Cmd-Det Blast Mine	Wire Obstacles	Remarks
Probe	Is effective at locating all buried mines; however, extreme caution should be used when probing for small mines (less than 3-inch dia) and sensitive mines (more than 10 lbs actuating force)	Is ineffective; the soldier must rely on clearance by other means and visual detection	Do not use	Ensure that observer is neutralized before approaching	Is ineffective	
Wire cutter	Is ineffective	Do not cut a slack wire until both ends have been checked if break-wire-fused mines may be present; do not cut taught trip wires until both ends have been checked for tension-release FDs	Do not use	Ensure that observer is neutralized before approaching	Is effective; sound can be minimized by wrapping the wire with a cloth before cutting it	
Gunfire	Is effective against surface-laid munitions	Is effective against surface-laid munitions	Do not use	Is effective against surface-laid munitions	Is ineffective	Is used as a last resort Is a relatively slow method for clearing visually detected ordnance Ensures protection of weapons operator Works better with larger caliber weapons (up to 25 mm), incendiary ammo works best May be the only option open to artillery, MP, and transportation units for dealing with SCATMINEs in the absence of engineers and reduction assets

Table G-2. Dismounted breaching assets versus threat obstacles (continued)

Asset	Pressure-Fused AP Mine	Trip-Wire/Break-Wire-Fused AP Mine	Influence-Fused AP Mine	Cmd-Det Blast Mine	Wire Obstacles	Remarks
MiniFlail	Is effective against simple pressure-fused AP mines Is less effective against blast-hardened AP mines	Is effective	Is ineffective	Is armored against fragmenting threats Is vulnerable to command-detonated blast mines	Do not use	Do not use if the presence of AT mines is suspected, because AT mines can destroy the MiniFlail. Can clear at a rate of about 1,500 square meters per hour
AN/PSS-12	Is very effective at finding metallic mines Is less effective against low-metal mines	Cannot detect monofilament fishing line used for trip wires; it is undetectable by the AN/PSS-12. Visual detection and other neutralization means should be used first in any area suspected of having trip wires.	Do not use	Ensure that observer is neutralized before approaching	Is ineffective	Is less effective against low-metal mines and in areas contaminated with metal fragments (artillery, mortar, small arms) or areas that have heavily mineralized soils

Table G-2. Dismounted breaching assets versus threat obstacles (continued)

Asset	Pressure-Fused AP Mine	Trip-Wire/Break Wire-Fused AP Mine	Influence-Fused AP Mine	Cmd-Det Blast Mine	Wire Obstacles	Remarks
Demolitions	Place at least a 1-pound charge within 6 inches of simple pressure-fused mines Ensure that the charge is placed within an inch of blast-hardened mines	Place at least a 1-lb charge within 6 inches of the mine after the mine at the end of a trip wire has been located Can use elevated charges if necessary against the claymore and stake-type mines	Do not use	Ensure that observer is neutralized before approaching Can use elevated charges if necessary against claymore-type mines	Is effective when used to make an improvised bangalore torpedo	
Blast overshoes	May be effective against very small (up to 1 oz of explosive) mines Is ineffective against mines with more than 1-oz explosive weight	Can reduce casualty severity against fragmenting mines	Can reduce casualty severity against fragmenting mines	Ensure that observer is neutralized before approaching Can reduce casualty severity	Is ineffective	
IBASIC	Will protect personnel near the individual who detonates a pressure-fused AP mine	Can reduce casualty severity against fragmenting mines	Can reduce casualty severity against fragmenting mines	Ensure that observer is neutralized before approaching Can reduce casualty severity	Is ineffective	

Table G-3. Foreign track-width AT mines

Origin	Mechanically Planted	Scatterable	Metallic	Nonmetallic	Simple Pressure	Double Impulse	Blast Hardened	AHD
Russia	TM-46/57/62	PTM-1	TM-46/57/62M	TM-62B/D/P	TM-46/57/62		TM-46/57	TMN-46
China	Type 72	Type 84B	Type 72	Type 72	Type 84B	Type 72	Type 72	
Yugo	TMM-1		TMM-1	TMA-1/2/3/4/5	TMA-1/2/5		TMA-3/4	TMM-1, TMA-1/2/3/4/5
Czech	PT-MI-K, PT-MI-BA III		PT-MI-K, PT-MI-D	PT-MI-BA II/III	PT-MI-BA III, PT-MI-D		PT-MI-K, PT-MI-BA II	PT-MI-K, PT-MI-D
Italy	VS-1.6/2.2, TC-.6/6, SB-81	VS-1.6, SB-81		VS-1.6/2.2, TC-.6/6, SB-81			VS-1.6/2.2, TC-3.6/6, SB-81	VS-1.6/2.2
France	M-51, M-52			M-51, M-52	M-51, M-52			M-51
UK	L9 Barmine, MK-7		MK-7	L9 Barmine	L9 Barmine, MK-7	L9 Barmine, MK-7		L9 Barmine
Ger	PM-60, DM-11		DM-21	PM-60, DM-11	PM-60, DM-11, DM-21			PM-60, DM-11, DM-21

Table G-4. Foreign full-width AT mines

Origin	Mechanically Planted	Scatterable	Tilt Rod	Magnetic Influence	Seismic Influence	Coupled	AHD
Russia	TM-62/72/89	PTM-3	TMK-2	TM-62/72/89	TM-62/72/89		TM-62/72/89
China			Type 84C	Type 84A			Type 84A
Yugo	TMRP-6	KB-2	TMRP-6	TMRP-6, KB-2			TMRP-6, KB-2
Czech			PT-MI-P				
France	HPD series	Lance	M-51, M-52, Model 1954	HPD series		Model 1953	HPD series, M-51, Lance
Hungary	UKA-63		UKA-63				UKA-63
Italy	VS-HCT/HCT-2, SB-MV	VS-SATM1, VS-HCT		VS-HCT/HCT-2, SB-MV, VS-SATM1		Fuse option	VS-HCT/HCT-2, SB-MV, VS-SATM1
Sweden	FFV-028			FFV-028		Fuse 15	FFV-028
UK	L9 Barmine	HB-876	L9 Barmine, MK-7	L9 Barmine, HB-876			L9 Barmine, MK-7, HB-876
Germany	DM-31	AT-2, MIFF		DM-31, AT-2, MIFF			DM-31, AT-2, MIFF
Romania						MC-71	

Table G-5. Foreign side-attack AT mines

Origin	Command	Break Wire	Influence	AHD
Russia	MON-100/200/500, TM-83	LMG	TM-73/83	
Yugo		TMRP-6, COBRA		
France	MIACAH F1	MIACAH F1	MIACAH F1	
UK	L14	L14	AJAX	AJAX
Germany			DM-12	
Sweden		FFV-016, AT-4		

Table G-6. Foreign pressure-fused AP mines

Origin	Scatterable	Low Metal	Simple Pressure	Small Diameter	Low-Operating Threshold	Pronged Fuse	Blast Hardened	AHD
Russia	PFM-1		PMN, PMD-6		PMN, PMD-6, PFM-1	OZM series	PMN-2	
China	GLD-111	Type 72A, GLD-111	Type 72A, Type 58, GLD-111	GLD-111	Type-72, Type 58, GLD-111			Type 72A/B
Yugo		PMA-1/2/3	PMA-1, PMD-1		PMA-1/3, PMD-1	PROM-1/2, PMR-3	PMA-2	
Czech			PP-MI-D, PP-MI-BA		PP-MI-D	PP-MI-SR, PP-MI-SR II		
Italy	TS-50, VS-50, SB-33, VS-MK2	TS-50, SB-33, VS-MK2				Valmara 59/69	TS-50, VS-50, SB-33, VS-MK2	VS-50, SB-33, VS-MK2
France		M-59		M-59	M-59		M-59	
UK	Ranger		Ranger	Ranger				
Germany		DM-11	PPM-2		DM-11		DM-11	

Table G-7. Foreign trip-wire/break-wire-fused AP mines

Origin	Scatterable	Trip Wire	Break Wire
Russia	POM-2	POMZ-2, OZM series, MON series	OZM series, MON series
China		Type 58, Type 59, Type 66, Type 69	
Yugo		PROM-1/2/KD, PMR-2/3/4, MRUD	PROM-1/2/KD, PMR-2/3/4, MRUD
Czech		PP-MI-SR, PP-MI-SR II, PP-MI-SK	
Italy	VS-SAPFM3	Valmara 59/69, BM-85, P-25, P-40, VS-DAFM1	
France		M1951/1955, MAPED F1	M1951/1955
UK		PAD	
Germany		DM-31, SM-70	

Table G-8. Foreign emplaced minefields

	Russia	Iraq	North Korea	Bosnia
Manually				
Antitrack AT	1	1	1	1
Antihull AT	2	3	3	2
Side-attack AT	3	4	3	4
Blast AP	1	1	1	1
Directional AP	1	4		1
Bounding AP	1	1	2	1
Stake AP	1	4	1	1
Mechanically				
Antitrack AT	1	1	1	4
Antihull AT	2	3		1
Vehicle Scattered				
Antitrack AT	1			
Antihull AT	1			
Blast AP				
Directional AP				
Bounding AP				
Simple Frag	1			
Artillery/MLRS				
Antitrack AT	1	4		
Antihull AT		1		1
Blast AP				
Directional AP				
Bounding AP				
Stake AP	1			
Helicopter/Fixed-Wing Aircraft				
Antitrack AT	1	1		
Antihull AT	1			
Blast AP	1	1		
Directional AP				
Bounding AP				
Stake AP	1			
NOTE: Scored from 1 (most common) to 4 (very rare).				

Table G-9. Foreign mine delivery systems

Origin	System	Type[1]	Range	Mine Payload	Density[2]	Remarks
Russia	BM-21	R	20 km	PTM-1/3, POM-2	1.5	1,000 x 500 m per battery
	BM -22	R	35 km	PTM 1/3, POM-2	1.5	
	PKPI	F, H	NA	PTM 1/3, PFM-1, POM-2	1.5	Lays relatively narrow strips
	KSMU	F	NA	PTM 1/3, PFM-1, POM-2	1.5	
	UMZ	V	NA	PTM 1/3, POM-2	1.5	
	GMZ/PMZ-4	MP	NA	Varies	0.4 to 1.0	3 systems can emplace 1,200 m with 624 mines in 26 minutes
	VMR-1/2	H	NA	Varies	0.4 to 1.0	3 Mi-8s can lay 1,200 to 3,000 m with 1,200 mines
China	Type 83	R	15 km	Type 84	0.5	900 x 800 m per battery
Yugo	M87	R	50 km	KB-2	0.5	900 x 500 m per battery
Italy	FIROS 25	R	22 km	VS-HCL, VS-Mk, VS-SAPFM 3	0.5	1,000 x 500 m per battery
	FIROS 30	R	35 km	VS-HCL, VS-Mk, VS-SAPFM 3	0.5	
	DAT	H	NA	Varies	1.2	Lays relatively narrow strips of up to 320 AT mines
	Istrice	V	NA	SATM1, VS-Mk 2 EL, VS-SAPFM 3	Max 1.6	Typically 360 x 140 m
	ST-AT/V	MP	NA	Varies	0.3 to 1.5	240 mines per hour
	Grillo 90	MD	NA	SATM1, VS-Mk 2 EL	0.5	5 AT or 15 AP mines per dispenser
Sweden	FFV-5821	MP	NA	FFV-028	0.4 to 1.4	300 buried mines per hour or 500 surface-laid mines per hour
France	Minotaur	V	NA	Lance	0.7	1,200 x 600 m with 800 mines
	EBG	V	NA	Lance	1.5	60 x 600 m with 40 mines
	155-mm How	A	18 km	Lance		8 mines per round
	Matenin	MP	NA	HPD series	0.4 to 1.4	500 mines per hour
UK	JP-233	F	NA	HB 876		430 mines per Tornado aircraft
	Barmine layer	MP	NA	L9 Barmine	0.4 to 0.9	Up to 700 mines per hour
	Ranger	V	NA	L10		1,296 mines per dispenser

Table G-9. Foreign mine delivery systems (continued)

Origin	System	Type[1]	Range	Mine Payload	Density[2]	Remarks
Germany	Skorpion	V, H	NA	AT-2	0.4	1,500 x 50 m with 600 mines
	MW-1	F	NA	MIFF, MUSPA	0.5	55 to 500 m wide and 200 to 2,500 m long with 892 AT mines
	LARS	R	14 km	AT-2	0.5	1,836 AT mines per launcher, 1,000 x 500 m per battery
	MARS	R	30 km	AT-2	0.3	1,000 x 500 m with 336 AT mines per launcher
	FFV-5821	MP	NA	DM-31	0.4 to 1.4	300 buried mines per hour or 500 surface-laid mines per hour

[1] A = artillery, H = helicopter, F = fixed-wing aircraft, R = rocket, V = vehicle dispensed, MD = man-portable dispenser, MP = mechanical planter

[2] AT mines per meter of front

Appendix H

Metric Conversion Chart

This appendix complies with current Army directives which state that the metric system will be incorporated into all new publications. Table H-1 is a conversion chart.

Table H-1. Metric conversion chart

US Units	Multiplied By	Metric Units
Feet	0.3048	Meters
Inches	2.54	Centimeters
Inches	0.0254	Meters
Inches	25.4001	Millimeters
Miles	1.6093	Kilometers
Pounds	453.59	Grams
Pounds	0.4536	Kilograms
Metric Units	**Multiplied By**	**US Units**
Centimeters	0.3937	Inches
Grams	0.03527	Ounces
Kilograms	2.2046	Pounds
Kilometers	0.62137	Miles
Meters	3.2808	Feet
Meters	39.37	Inches
Meters	1.0936	Yards
Millimeters	0.03937	Inches

Glossary

1LT	first lieutenant
1SG	first sergeant
2LT	second lieutenant
AA	avenue of approach
ACE	armored combat earthmover
ACR	armored cavalry regiment
AD	armor division
ADA	air-defense artillery
ADAM	area-denial artillery munition
ADE	assistant division engineer
AHD	antihandling device
AI	area of interest
ALO	air liaison officer
ammo	ammunition
AO	area of operations
AP	antipersonnel
APB	antipersonnel blast
APC	armored personnel carrier
APDS	armor-piercing discarding sabot
APF	antipersonnel fragmentation
APL	antipersonnel land mine
APOBS	Antipersonnel Obstacle Breaching System
AR	Army regulation
ASP	ammunition supply point

ASTAMIDS	Airborne Standoff Minefield Detection System
AT	antitank
atk	attack
ATP	ammunition transfer point
attn	attention
AVLB	armored vehicle-launched bridge
AVLM	armored vehicle-launched MICLIC
BAI	battlefield air interdiction
bde	brigade
BICC	battlefield information control center
BLPS	ballistic and laser protective spectacles
bn	battalion
BOS	Battlefield Operating System
BP	battle position
BRDEC	US Army Belvoir Research, Development, and Engineering Center
C	Celsius
C^2	command and control
CA	civil affairs
cal	caliber
CAS	close air support
cdr	commander
CEV	combat engineer vehicle
CG	commanding general
CHS	combat health support
CL	centerline
cm	centimeter(s)
cmd	command
CMMC	corps material-management center

co	company
COA	course of action
COB	civilian on the battlefield
comp	composition
CS	combat support
CSR	controlled supply rate
CSS	combat service support
CTCP	combat trains command post
cu	cubic
Czech	Czechoslovakia
d	day
DA	Department of the Army
DAO	division ammunition officer
DCU	dispenser control unit
det	detonated
det	detonating
dia	diameter
div	division
DMMC	division material-management center
DOD	Department of Defense
DODIC	Department of Defense identification code
DST	decision support template
DTG	date-time group
DZ	drop zone
EA	engagement area
ea	each
EBA	engineer battlefield assessment
EDD	explosive detector dog

EFP	explosive-formed penetrating
EID	end item code
EM	enlisted member
engr	engineer
ENS	Explosive Neutralization System
EOD	explosive ordnance disposal
ESMB	explosive standoff minefield breacher
F	Fahrenheit
FCP	forward command post
FCS	fire control system
FD	firing device
FECS	Field-Expedient Countermine System
FIST	fire-support team
FLOT	forward line of own troops
FM	field manual
FM	frequency-modulated
FMFM	Fleet Marine Force manual
FPF	final protective fires
frag	fragmentation
FRAGO	fragmentary order
FSB	forward support battalion
FSC	federal stock classification
FSCL	fire-support coordination line
FSCOORD	fire-support coordinator
FSE	fire-support element
ft	foot, feet
ft	fort
FWF	former warring faction

g	gram(s)
G2	Assistant Chief of Staff, G2 (Intelligence)
G3	Assistant Chief of Staff, G3 (Operations and Plans)
G4	Assistant Chief of Staff, G4 (Logistics)
GDP	general defense plan
Ger	Germany
GPS	global-positioning system
H	hour of execution
HE	high-explosive
HEI-T	high-explosive incendiary tracer
HEMMS	hand-emplaced minefield marking set
HEMTT	heavy expanded mobility tactical truck
HEXJAM	hand-emplaced, expandable jammer
HMMWV	high-mobility, multipurpose, wheeled vehicle
HQ	headquarters
hr	hour(s)
HTF	how to fight
hvy	heavy
IBASIC	Improved Body Armor System, Individual Countermine
ID	infantry division
IDA	improved dog-bone assembly
IFV	infantry fighting vehicle
ILLUM	illumination
IMU	inertial measurement unit
in	inch(es)
inf	infantry
IOE	irregular outer edge
IP	improved plow

IPB	intelligence preparation of the battlefield
IR	infrared
ITV	improved TOW vehicle
IVMMD	interim vehicle-mounted mine detector
JAAT	joint air-attack team
Jan	January
Jul	July
K-Kill	catastrophic kill
kg	kilogram(s)
km	kilometer(s)
kph	kilometers per hour
KW	Kiowa Warrior
lb	pound(s)
LBE	load-bearing equipment
LE	low explosive
LED	light-emitting diode
LIN	line item number
LM	landmark
LOC	lines of communication
LOGPAC	logistics package
LRA	local reproduction authorized
LRP	logistics release point
LRSD	long-range surveillance device
LZ	landing zone
m	month
m	meter(s)
M-Kill	mobility kill
M-S	Miznay-Schardin

mag	magnetic
Mar	March
MBA	main battle area
MC	mobility corridor
MCAP	mine-clearing/armor-protection kit
MCB	mine-clearing blade
MCD	magnetic coupling device
MCO	Marine Corps order
MCOO	modified combined obstacle overlay
MCR	mine-clearing roller
MCRP	Marine Corps reference publication
MDI	modernized demolition initiation
MDV	mine-detection vehicle
Met+VE	meteorological data/velocity error
METL	mission-essential task list
METT-TC	mission, enemy, terrain, troops, time available, and civilian considerations
MHE	material handling equipment
MICLIC	mine-clearing line charge
MIDAP	minefield-detection algorithm and processor
MILSTD	military standard
min	minute(s)
min	minimum
MLRS	Multiple-Launched Rocket System
mm	millimeter(s)
MOBA	military operations in built-up areas
MOD	mobile obstacle detachment
MOPMS	Modular Pack Mine System
MOPP	mission-oriented protective posture

MOOTW	military operations other than war
MOUT	military operations on urbanized terrain
MP	military police
MRB	motorized rifle battalion
MRC	motorized rifle company
MRR	motorized rifle regiment
MSD	movement support detachment
MSR	main supply route
NA	not applicable
NAI	named area of interest
NATO	North Atlantic Treaty Organization
NAVSEA	Naval Sea Systems Command
NBC	nuclear, biological, chemical
NCO	noncommissioned officer
NCOIC	noncommissioned officer in charge
NE	northeast
NGO	nongovernment organization
NLT	no later than
nm	nautical mile
no	number
NSN	national stock number
OBSTINTEL	obstacle intelligence
Oct	October
OIC	officer in charge
OP	observation post
OPCON	operational control
OPLAN	operation plan
OPORD	operation order

P	pace(s)
pam	pamphlet
PIR	priority intelligence requirement
PIRS	passive infrared sensor
PL	platoon leader
PL	phase line
PLS	palletized load system
plt	platoon
PSG	platoon sergeant
psi	pounds per square inch
PSYOP	psychological operations
pt	point
PVO	private volunteer organization
QASAS	quality assurance ammunition specialist
R&S	reconnaissance and surveillance
RAAM	remote antiarmor mine
RAC	risk-assessment code
RCU	remote-control unit
RDX	cyclonite
recon	reconnaissance
rep	representative
RF	radio frequency
ROKUS	Republic of Korea, United States
RP	reference point
RSO	range safety officer
RTO	radio-telephone operator
S&A	safing and arming

S&H	safety and handling
S&T	supply and transport
S2	Intelligence Officer (US Army)
S3	Operations and Training Officer (US Army)
S4	Supply Officer (US Army)
SAW	squad automatic weapon
SCATMINE	scatterable mine
SCATMINWARN	scatterable minefield warning
sct	scout
SD	self-destruct
SE	southeast
SEAD	suppression of enemy air defense
sec	second(s)
SEE	small emplacement excavator
SFC	sergeant first class
SFC	sergeant first class
SFF	self-forging fragmentation
SIR	serious incident report
SITEMP	situation template
SLAM	selectable lightweight attack munition
SOEO	scheme of engineer operations
SOF	special-operations forces
SOP	standard operating procedure
SOSR	suppress, obscure, secure, and reduce
SPECS	Special Protective Eyewear Cylindrical System
SPOTREP	spot report
SSG	staff sergeant
SSN	social security number

STANAG	International Standardization Agreement
STP	soldier training publication
T/MDV	towing/mine-detection vehicle
TACC	tactical air control center
TAHQ	theater army headquarters
TAI	target area of interest
TC	track commander
TC	training circular
TCP	traffic-control post
TCP	tactical command post
TF	task force
tm	team
TM	technical manual
TMD	tactical-munition dispenser
TNT	trinitrotoluene
TO	theater of operations
TO	Air Force technical order
TOC	tactical operations center
TOW	tube-launched, optically tracked, wire-guided missile
TRADOC	United States Army Training and Doctrine Command
TRP	target reference point
TTP	tactics, techniques, and procedures
UAV	unmanned aerial vehicle
UK	United Kingdom
UL	Underwriters Laboratories, Incorporated
US	United States
USAF	United States Air Force
USN	United States Navy

UTM	Universal Transverse Mercator
UXO	unexploded ordnance
vs	versus
W	width
w/	with
WLGH	weapon-launched grapnel hook
WO	warning order
WP	white phosphorus
XO	executive officer
y	year
Yugo	Yugoslavia

References

SOURCES USED

These are the sources quoted or paraphrased in this publication.

Joint and Multiservice Publications

AR 385-63. *Policies and Procedures for Firing Ammunition for Training, Target Practice and Combat {MCO P3570.1A}*. 15 October 1983.

FM 20-11. *Military Diving {NAVSEA 0910-LP-708-8000}*. 20 January 1999.

FM 90-3. *Desert Operations {FMFM 7-27}*. 24 August 1993.

FM 90-13. *River-Crossing Operations {MCWP 3-17.1}*. 26 January 1998.

FM 101-5-1. *Operational Terms and Graphics {MCRP 5-2A}*. 30 September 1997.

STANAG 2036. *Land Mine Laying, Marking, Recording and Reporting Procedures*. Edition 5. 12 February 1987.

STANAG 2889. *Marking of Hazardous Areas and Routes Through Them*. Edition 3. 26 March 1984.

STANAG 2990. *Principles and Procedures for the Employment in Land Warfare of Scatterable Mines With a Limited Laid Life - ATP-50*. Edition 1. 15 September 1988.

TM 5-6665-298-10. *Operator's Manual AN/PSS-12 Mine Detecting Set {TM 5-6665-298-10}*. 28 February 1995.

TM 9-1345-203-12&P. *Operator's and Organizational Maintenance Manual (Including Repair Parts and Special Tools List) for Land Mines {TO 11A8-3-1}*. 14 January 1977.

TM 9-2590-509-10. *Operator's Manual Mine Clearing Blade for M1, IPM1, or M1A1 Abrams Tank (NSN 2590-01-230-8862) {TM 2590-10/1}*. 12 April 1990.

Army Publications

AR 40-10. *Health Hazard Assessment Program in Support of the Army Materiel Acquisition Decision Process*. 1 October 1991.

AR 385-10. *The Army Safety Program*. 23 May 1988.

AR 385-16. *System Safety Engineering and Management*. 3 May 1990.

DA Pam 350-38. *Standards in Weapons Training*. 3 July 1997.

FM 1-113. *Utility and Cargo Helicopter Operations*. 12 September 1997.

FM 3-34.2 (formerly FM 90-13-1). *Combined-Arms Breaching Operations*. 31 August 2000.

FM 5-7-30. *Brigade Engineer and Engineer Company Combat Operations (Airborne, Air Assault, Light)*. 28 December 1994.

FM 5-10. *Combat Engineer Platoon*. 3 October 1995.

FM 5-34. *Engineer Field Data*. 30 August 1999.

FM 5-71-2. *Armored Task-Force Engineer Combat Operations*. 28 June 1996.

FM 5-71-3. *Brigade Engineer Combat Operations (Armored)*. 3 October 1995.

FM 5-100. *Engineer Operations*. 27 February 1996.

FM 5-170. *Engineer Reconnaissance*. 5 May 1998.

FM 5-250. *Explosives and Demolitions*. 30 July 1998.

FM 90-5 (HTF). *Jungle Operations (How to Fight)*. 16 August 1982.

FM 90-7. *Combined Arms Obstacle Integration*. 29 September 1994.

FM 90-10 (HTF). *Military Operations on Urbanized Terrain (MOUT) (How to Fight)*. 15 August 1979.

FM 90-10-1. *An Infantryman's Guide to Combat in Built-up Areas*. 12 May 1993.

FM 101-5. *Staff Organization and Operations*. 31 May 1997.

MILSTD 882C (Revision). *System Safety Program Requirements.* 19 January 1993.

STP 5-12B1-SM. *Soldier's Manual: MOS 12B, Combat Engineer, Skill Level 1*. 14 February 2001.

TC 25-8. *Training Ranges*. 25 February 1992.

TC 31-34. *Demining Operations*. 24 September 1997.

TM 1-1520-237-10. *Operator's Manual for UH-60A Helicopters, UH-60L Helicopters, and EH-60A Helicopters*. 31 October 1996.

TM 9-1095-208-10-1. *Operator's Manual for Dispenser, Mine: M139 (NSN 1095-01-235-3139) (EIC: 3V8) with Mounting Kits for 5-Ton Vehicle (1095-01-252-2818) (EIC: 3V9) and M548A1 Vehicle (1095-01-331-6755) (Ground Volcano)*. 10 July 1992.

TM 9-1300-206. *Ammunition and Explosives Standards*. 30 August 1973.

TM 9-1345-203-12. *Operator's and Unit Maintenance Manual for Land Mines*. 30 October 1995.

TM 9-1345-209-10. *Operator's Manual for Modular Pack Mine System (MOPMS) Consisting of Dispenser and Mine, Ground: M131 (NSN 1345-01-160-8909) Control, Remote, Land Mine System: M71 (1290-01-161-3662) and Dispenser and Mine, Ground Training: M136 (6920-01-162-9380)*. 31 March 1992.

TM 9-1375-213-12. *Operator's and Unit Maintenance Manual (Including Repair Parts and Special Tools List): Demolition Materials*. 30 March 1973.

TM 9-1375-215-14&P. *Operator's, Unit, Direct Support and General Support Maintenance Manual (Including Repair Parts and Special Tools List) for Demolition Kit, Mine Clearing Line Charge (MICLIC)*. 31 January 1992.

TM 43-0001-36. *Army Ammunition Data Sheets for Land Mines (FSC 1345)*. 1 September 1994.

DOCUMENTS NEEDED

These documents must be available to the intended users of this publication.

DA Form 1355. *Minefield Record*. March 1987.

DA Form 1355-1-R. *Hasty Protective Row Minefield Record (LRA)*. September 2001.

DA Form 2028. *Recommended Changes to Publications and Blank Forms*. February 1974.

READINGS RECOMMENDED

These readings contain relevant supplemental information.

FM 5-490. *Engineer Diving Operations.* 31 March 1992.

FM 6-20-40. *Tactics, Techniques, and Procedures for Fire Support for Brigade Operations (Heavy).* 5 January 1990.

FM 6-20-50. *Tactics, Techniques, and Procedures for Fire Support for Brigade Operations (Light).* 5 January 1990.

TM 5-2590-214-10. *Operator's Manual for Roller Kit, Mine (NSN 2590-01-134-3724) (EIC: AKZ) and Mounting Kit, Mine Clearing Roller (2590-01-235-5458) (EIC: AKZ).* 30 March 1993.

TM 9-1095-209-10. *Operator's Manual for Mine Dispenser: M138 (Flipper).* 29 March 1990.

Index

I

L

M

N

O

P

T

U

V

W

Field Manual
No. 3-36

Headquarters
Department of the Army
Washington, DC, 25 February 2009

Electronic Warfare in Operations

Contents

Distribution Restriction: Approved for public release; distribution is unlimited.

Figures

Preface

PURPOSE

FM 3-36 provides Army doctrine for electronic warfare (EW) planning, preparation, execution, and assessment in support of full spectrum operations. Users of FM 3-36 must be familiar with full spectrum operations established in FM 3-0; the military decisionmaking process established in FM 5-0; the operations process established in FMI 5-0.1; commander's visualization described in FM 6-0; and electronic warfare described in JP 3-13.1.

SCOPE

FM 3-36 is organized into seven chapters and six appendixes. Each chapter addresses a major aspect of Army EW operations. The appendixes address aspects of EW operations that complement the operational doctrine. A glossary contains selected terms.

- Chapter 1 discusses the nature and scope of electronic warfare and the impact of the electromagnetic environment on Army operations.
- Chapter 2 offers a discussion of EW support to full spectrum operations, combat power, the warfighting functions, and information tasks.
- Chapter 3 introduces the organizational framework for command and control of EW operations.
- Chapter 4 describes how commanders integrate EW operations throughout the operations process.
- Chapter 5 discusses the coordination required to synchronize and deconflict EW operations effectively.
- Chapter 6 provides the baseline for integrating EW operations into joint and multinational operations.
- Chapter 7 discusses the enabling activities that support EW operations, such as command and control, intelligence, logistics, technical support and EW training.
- Appendix A discusses the electromagnetic environment.
- Appendix B illustrates an EW appendix to an operation order.
- Appendix C illustrates an EW running estimate.
- Appendix D discusses EW related reports and messages.
- Appendix E offers a reference guide to Army and joint EW capabilities.
- Appendix F discusses EW-related tools and resources.

APPLICABILITY

FM 3-36 provides guidance on EW operations for commanders and staffs at all echelons. This FM serves as an authoritative reference for personnel who—

- Develop doctrine (fundamental principles and tactics, techniques, and procedures), materiel, and force structure.
- Develop institutional and unit training.
- Develop standing operating procedures for unit operations.
- Conduct planning, preparation, execution and assessment of electronic warfare.

FM 3-36 applies to the Active Army, Army National Guard/Army National Guard of the United States, and U.S. Army Reserve, unless otherwise stated.

Chapter 1

Electronic Warfare Overview

This chapter provides an overview of electronic warfare and the conceptual foundation that leaders require to understand the electromagnetic environment and its impact on Army operations.

OPERATIONAL ENVIRONMENTS

1-1. An *operational environment* is a composite of the conditions, circumstances, and influences that affect the employment of capabilities and bear on the decisions of the commander (JP 3-0). An operational environment includes physical areas—the air, land, maritime, and space domains. It also includes the information that shapes the operational environment as well as enemy, adversary, friendly, and neutral systems relevant to a joint operation. Joint planners analyze operational environments in terms of six interrelated operational variables: political, military, economic, social, information, and infrastructure. To these variables Army doctrine adds two more: physical environment and time. (See FM 3-0 for additional information on the operational variables). Army leaders use operational variables to understand and analyze the broad environment in which they are conducting operations.

1-2. Army leaders use mission variables to synthesize operational variables and tactical-level information with local knowledge about conditions relevant to their mission. They use mission variables to focus analysis on specific elements that directly affect their mission. Upon receipt of a warning order or mission, Army tactical leaders narrow their focus to six mission variables known as METT-TC. They are mission, enemy, terrain and weather, troops and support available, time available and civil considerations. The mission variables outline the situation as it applies to a specific Army unit.

1-3. Commanders employ and integrate their unit's capabilities and actions within their operational environment to achieve a desired end state. Through analyzing their operational environment, commanders understand how the results of friendly, adversary, and neutral actions may impact that end state. During military operations, both friendly and enemy commanders depend on the flow of information to make informed decisions. This flow of information depends on the electronic systems and devices used to communicate, navigate, sense, store, and process information.

INFORMATION AND THE ELECTROMAGNETIC SPECTRUM

1-4. Commanders plan for and operate electronic systems and the weapon systems that depend on them in an intensive and nonpermissive electromagnetic environment. They ensure the flow of information required for their decisionmaking. (Appendix A further discusses the electromagnetic environment.) Within the electromagnetic environment, electronic systems and devices operate in the electromagnetic spectrum. (See figure 1-1, page 1-2.)

1-5. The electromagnetic spectrum has been used for commercial and military applications for over a century. However, the full potential for its use as the primary enabler of military operations is not yet fully appreciated. New technologies are expanding beyond the traditional radio frequency spectrum. They include high-power microwaves and directed-energy weapons. These new technologies are part of an electronic warfare (EW) revolution by military forces. Just as friendly forces leverage the electromagnetic spectrum to their advantage, so do capable enemies use the electromagnetic spectrum to threaten friendly force operations. The threat is compounded by the growth of a wireless world and the increasingly sophisticated use of commercial off-the-shelf technologies.

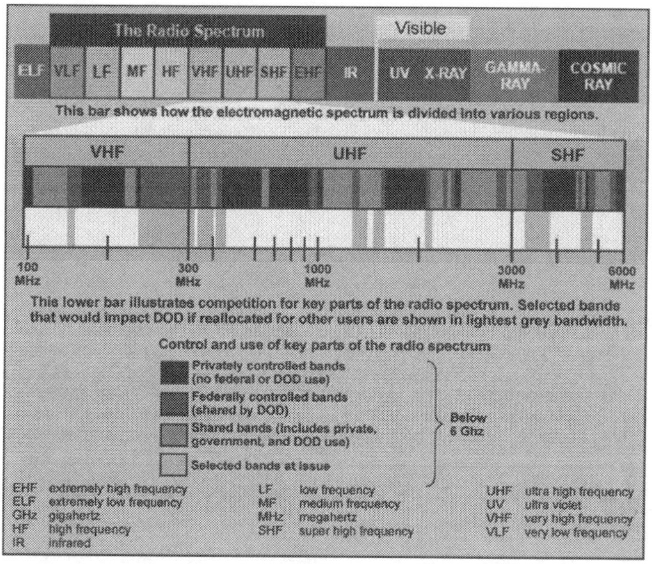

Figure 1-1. The electromagnetic spectrum

1-6. Adversaries and enemies, from small and single actors to large state, multinational, and nonstate actors, use the most modern technology. Such technology is moving into the cellular and satellite communications area. Most military and commercial operations rely on electromagnetic technologies and are susceptible to the inherent vulnerabilities associated with their use. This reliance requires Army forces to dominate the electromagnetic spectrum (within their operational environment) with the same authority that they dominate traditional land warfare operations. Emerging electromagnetic technologies offer expanded EW capabilities. They dynamically affect the electromagnetic spectrum through delivery and integration with other types of emerging weapons and capabilities. Examples are directed-energy weapons, high-powered microwaves, lasers, infrared, and electro-optical and wireless networks and devices.

1-7. In any conflict, commanders attempt to dominate the electromagnetic spectrum. They do this by locating, targeting, exploiting, disrupting, degrading, deceiving, denying, or destroying the enemy's electronic systems that support military operations or deny the spectrum's use by friendly forces. The increasing portability and affordability of sophisticated electronic equipment guarantees that the electromagnetic environment in which forces operate will become even more complex. To ensure unimpeded access to and use of the electromagnetic spectrum, commanders plan, prepare, execute, and assess EW operations against a broad set of targets within the electromagnetic spectrum. (See figure 1-2.)

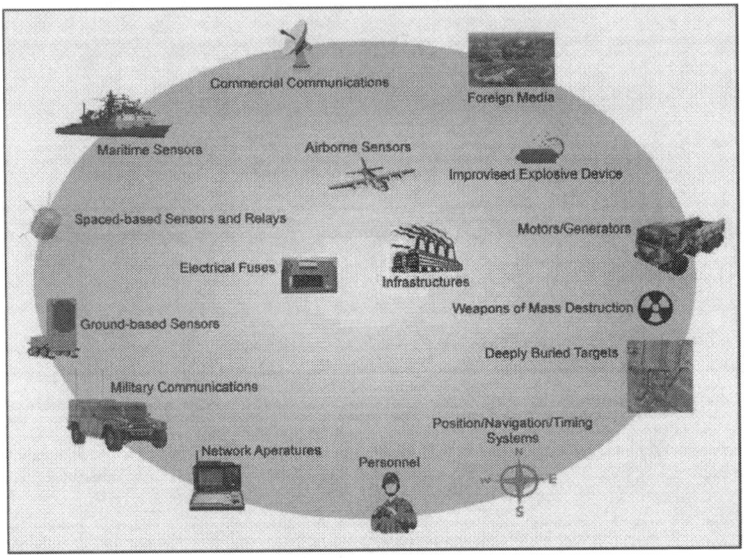

Figure 1-2. Electromagnetic spectrum targets

DIVISIONS OF ELECTRONIC WARFARE

1-8. *Electronic warfare* is defined as military action involving the use of electromagnetic and directed energy to control the electromagnetic spectrum or to attack the enemy. Electronic warfare consists of three divisions: electronic attack, electronic protection, and electronic warfare support (JP 3-13.1). (See figure 1-3.)

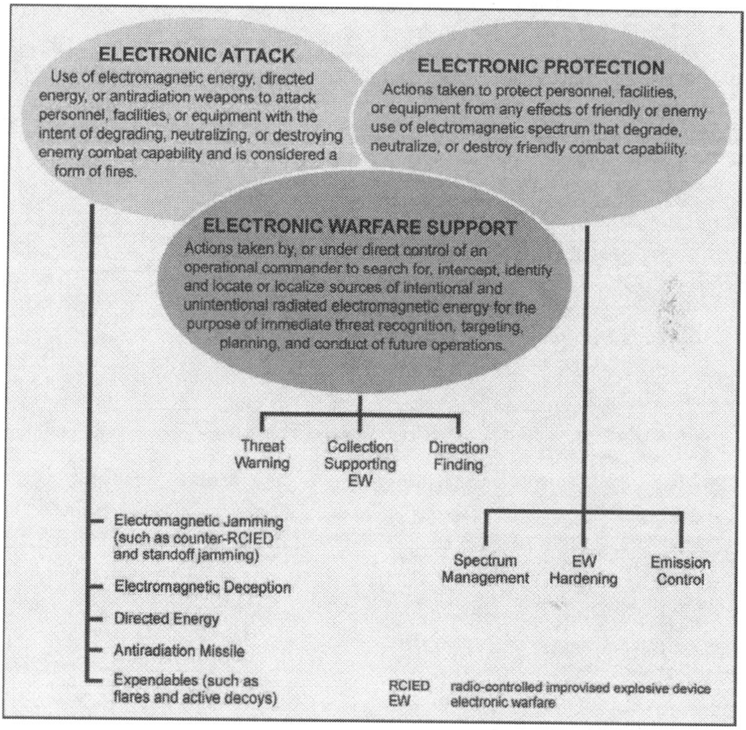

Figure 1-3. The three subdivisions of electronic warfare

ELECTRONIC ATTACK

1-9. *Electronic attack* is a division of electronic warfare involving the use of electromagnetic energy, directed energy, or antiradiation weapons to attack personnel, facilities, or equipment with the intent of degrading, neutralizing, or destroying enemy combat capability and is considered a form of fires (JP 3-13.1). Electronic attack includes—

- Actions taken to prevent or reduce an enemy's effective use of the electromagnetic spectrum, such as jamming and electromagnetic deception.
- Employment of weapons that use either electromagnetic or directed energy as their primary destructive mechanism (lasers, radio frequency weapons, particle beams).
- Offensive and defensive activities including countermeasures.

1-10. Common types of electronic attack include spot, barrage, and sweep electromagnetic jamming. Electronic attack actions also include various electromagnetic deception techniques such as false target or duplicate target generation. (See paragraphs 1-23 to 1-31 for further discussion of electronic attack activities.)

1-11. *Directed energy* is an umbrella term covering technologies that relate to the production of a beam of concentrated electromagnetic energy or atomic or subatomic particles (JP 1-02). A directed-energy weapon uses directed energy primarily as a direct means to damage or destroy an enemy's equipment, facilities, and personnel. In addition to destructive effects, directed-energy weapon systems support area denial and crowd control. (See appendix A for more information on directed energy.)

1-12. Examples of offensive electronic attack include—

- Jamming enemy radar or electronic command and control systems.
- Using antiradiation missiles to suppress enemy air defenses (antiradiation weapons use radiated energy emitted from the target as their mechanism for guidance onto targeted emitters).
- Using electronic deception techniques to confuse enemy intelligence, surveillance, and reconnaissance systems.
- Using directed-energy weapons to disable an enemy's equipment or capability.

1-13. Defensive electronic attack uses the electromagnetic spectrum to protect personnel, facilities, capabilities, and equipment. Examples include self-protection and other protection measures such as use of expendables (flares and active decoys), jammers, towed decoys, directed-energy infrared countermeasure systems, and counter-radio-controlled improvised-explosive-device systems. (See JP 3-13.1 for more discussion of electronic attack.)

ELECTRONIC PROTECTION

1-14. *Electronic protection* is a division of electronic warfare involving actions taken to protect personnel, facilities, and equipment from any effects of friendly or enemy use of the electromagnetic spectrum that degrade, neutralize, or destroy friendly combat capability (JP 3-13.1). For example, electronic protection includes actions taken to ensure friendly use of the electromagnetic spectrum, such as frequency agility in a radio, or variable pulse repetition frequency in radar. Electronic protection should not be confused with self-protection. Both defensive electronic attack and electronic protection protect personnel, facilities, capabilities, and equipment. However, electronic protection protects from the effects of electronic attack (friendly and enemy), while defensive electronic attack primarily protects against lethal attacks by denying enemy use of the electromagnetic spectrum to guide or trigger weapons.

1-15. During operations, electronic protection includes, but is not limited to, the application of training and procedures for countering enemy electronic attack. Army commanders and forces understand the threat and vulnerability of friendly electronic equipment to enemy electronic attack and take appropriate actions to safeguard friendly combat capability from exploitation and attack. Electronic protection measures minimize the enemy's ability to conduct electronic warfare support (electronic warfare support is discussed in paragraphs 1-18 to 1-20) and electronic attack operations successfully against friendly forces. To protect friendly combat capabilities, units—

- Regularly brief force personnel on the EW threat.
- Ensure that electronic system capabilities are safeguarded during exercises, workups, and predeployment training.
- Coordinate and deconflict electromagnetic spectrum usage.
- Provide training during routine home station planning and training activities on appropriate electronic protection active and passive measures.
- Take appropriate actions to minimize the vulnerability of friendly receivers to enemy jamming (such as reduced power, brevity of transmissions, and directional antennas).

1-16. Electronic protection also includes spectrum management. The spectrum manager works for the G-6 or S-6 and plays a key role in the coordination and deconfliction of spectrum resources allocated to the force. Spectrum managers or their direct representatives participate in the planning for EW operations.

1-17. The development and acquisition of communications and electronic systems includes electronic protection requirements to clarify performance parameters. Army forces design their equipment to limit inherent vulnerabilities. If electronic attack vulnerabilities are detected, then units must review these programs. (See DODI 4650.01 for information on the spectrum certification process and electromagnetic compatibility.)

ELECTRONIC WARFARE SUPPORT

1-18. *Electronic warfare support* is a division of electronic warfare involving actions tasked by, or under the direct control of, an operational commander to search for, intercept, identify, and locate or localize sources of intentional and unintentional radiated electromagnetic energy for the purpose of immediate threat recognition, targeting, planning, and conduct of future operations (JP 3-13.1).

1-19. Electronic warfare support systems are a source of information for immediate decisions involving electronic attack, electronic protection, avoidance, targeting, and other tactical employments of forces. Electronic warfare support systems collect data and produce information or intelligence to—

- Corroborate other sources of information or intelligence.
- Conduct or direct electronic attack operations.
- Initiate self-protection measures.
- Task weapon systems.
- Support electronic protection efforts.
- Create or update EW databases.
- Support information tasks.

1-20. Electronic warfare support and signals intelligence missions use the same resources. The two differ in the detected information's intended use, the degree of analytical effort expended, the detail of information provided, and the time lines required. Like tactical signals intelligence, electronic warfare support missions respond to the immediate requirements of a tactical commander. Signals intelligence above the tactical level is under the operational control of the National Security Agency and directly supports the overarching national security mission. Resources that collect tactical-level electronic warfare support data can simultaneously collect national-level signals intelligence. See FM 2-0 for more information on signals intelligence.

ACTIVITIES AND TERMINOLOGY

1-21. Although new equipment and tactics, techniques, and procedures continue to be developed, the physics of electromagnetic energy remains constant. Hence, effective EW activities remain the same despite changes in hardware and tactics. Principal EW activities are discussed in the following paragraphs.

PRINCIPAL ACTIVITIES

1-22. Principal EW activities support full spectrum operations by exploiting the opportunities and vulnerabilities inherent in the use of the electromagnetic spectrum. The numerous EW activities are categorized by the EW subdivisions with which they are most closely associated: electronic attack, electronic warfare support, and electronic protection. JP 3-13.1 discusses these principal activities in detail.

Electronic Attack Activities

1-23. Activities related to electronic attack are either offensive or defensive and include—

- Countermeasures.
- Electromagnetic deception.
- Electromagnetic intrusion.
- Electromagnetic jamming.
- Electromagnetic pulse.
- Electronic probing.

Countermeasures

1-24. *Countermeasures* are that form of military science that, by the employment of devices and/or techniques, has as its objective the impairment of the operational effectiveness of enemy activity (JP 1-02). They can be deployed preemptively or reactively. Devices and techniques used for EW countermeasures include electro-optical-infrared countermeasures and radio frequency countermeasures.

1-25. *Electro-optical-infrared countermeasures* consist of any device or technique employing electro-optical-infrared materials or technology that is intended to impair or counter the effectiveness of enemy activity, particularly with respect to precision guided weapons and sensor systems. Electro-optical-infrared is the part of the electromagnetic spectrum between the high end of the far infrared and the low end of ultraviolet. Electro-optical-infrared countermeasures may use laser and broadband jammers, smokes/aerosols, signature suppressants, decoys, pyrotechnics/pyrophorics, high-energy lasers, or directed infrared energy countermeasures (JP 3-13.1).

1-26. *Radio frequency countermeasures* consist of any device or technique employing radio frequency materials or technology that is intended to impair the effectiveness of or counter enemy activity, particularly with respect to precision guided weapons and sensor systems (JP 3-13.1).

Electromagnetic Deception

1-27. *Electromagnetic deception* is the deliberate radiation, reradiation, alteration, suppression, absorption, denial, enhancement, or reflection of electromagnetic energy in a manner intended to convey misleading information to an enemy or to enemy electromagnetic-dependent weapons, thereby degrading or neutralizing the enemy's combat capability (JP 3-13.4). Among the types of electromagnetic deception are the following:

- Manipulative electromagnetic deception involves actions to eliminate revealing, or convey misleading, electromagnetic telltale indicators that may be used by hostile forces.
- Simulative electromagnetic deception involves actions to simulate friendly, notional, or actual capabilities to mislead hostile forces.
- Imitative electromagnetic deception introduces electromagnetic energy into enemy systems that imitates enemy emissions.

Electromagnetic Intrusion

1-28. *Electromagnetic intrusion* is the intentional insertion of electromagnetic energy into transmission paths in any manner, with the objective of deceiving operators or of causing confusion (JP 1-02).

Electromagnetic Jamming

1-29. *Electromagnetic jamming* is the deliberate radiation, re-radiation, or reflection of electromagnetic energy for the purpose of preventing or reducing an enemy's effective use of the electromagnetic spectrum, with the intent of degrading or neutralizing the enemy's combat capability (JP 1-02).

Electromagnetic Pulse

1-30. *Electromagnetic pulse* is the electromagnetic radiation from a strong electronic pulse, most commonly caused by a nuclear explosion that may couple with electrical or electronic systems to produce damaging current and voltage surges (JP 1-02).

Electronic Probing

1-31. *Electronic probing* is the intentional radiation designed to be introduced into the devices or systems of potential enemies for the purpose of learning the functions and operational capabilities of the devices (JP 1-02). This activity is coordinated through joint or interagency channels and supported by Army forces.

Electronic Warfare Support Activities

1-32. Activities related to electronic warfare support include—

- Electronic reconnaissance.
- Electronic intelligence.
- Electronics security.

Electronic Reconnaisance

1-33. *Electronic reconnaissance* is the detection, location, identification, and evaluation of foreign electromagnetic radiations (JP 1-02).

Electronic Intelligence

1-34. *Electronic intelligence* is technical and geolocation intelligence derived from foreign noncommunications electromagnetic radiations emanating from other than nuclear detonations or radioactive sources (JP 1-02).

Electronics Security

1-35. *Electronics security* is the protection resulting from all measures designed to deny unauthorized persons information of value that might be derived from their interception and study of noncommunications electromagnetic radiations, e.g., radar (JP 1-02).

Electronic Protection Activities

1-36. Activities related to electronic protection include—

- Electromagnetic hardening.
- Electromagnetic interference.
- Electronic masking.
- Electronic warfare reprogramming.
- Emission control.
- Spectrum management.
- Wartime reserve modes.
- Electromagnetic compatibility.

Electromagnetic Hardening

1-37. *Electromagnetic hardening* consists of action taken to protect personnel, facilities, and/or equipment by filtering, attenuating, grounding, bonding, and/or shielding against undesirable effects of electromagnetic energy (JP 1-02).

Electromagnetic Interference

1-38. *Electromagnetic interference* is any electromagnetic disturbance that interrupts, obstructs, or otherwise degrades or limits the effective performance of electronics and electrical equipment. It can be induced intentionally, as in some forms of electronic warfare, or unintentionally, as a result of spurious emissions and responses, intermodulation products and the like (JP 1-02).

Electronic Masking

1-39. *Electronic masking* is the controlled radiation of electromagnetic energy on friendly frequencies in a manner to protect the emissions of friendly communications and electronic systems against enemy electronic warfare support measures/signals intelligence, without significantly degrading the operation of friendly systems (JP 1-02).

Electronic Warfare Reprogramming

1-40. *Electronic warfare reprogramming* is the deliberate alteration or modification of electronic warfare or target sensing systems, or the tactics and procedures that employ them, in response to validated changes in equipment, tactics, or the electromagnetic environment. These changes may be the result of deliberate actions on the part of friendly, adversary, or third parties; or may be brought about by electromagnetic interference or other inadvertent phenomena. The purpose of electronic warfare reprogramming is to maintain or enhance the effectiveness of electronic warfare and target sensing system equipment. Electronic warfare reprogramming includes changes to self-defense systems, offensive weapons systems, and intelligence collection systems (JP 3-13.1).

Emission Control

1-41. *Emission control* is the selective and controlled use of electromagnetic, acoustic, or other emitters to optimize command and control capabilities while minimizing transmissions for operations security: a. detection by enemy sensors; b. mutual interference among friendly systems; and/or c. enemy interference with the ability to execute a military deception plan (JP 1-02).

Electromagnetic Spectrum Management

1-42. *Electromagnetic spectrum management* is planning, coordinating, and managing joint use of the electromagnetic spectrum through operational, engineering, and administrative procedures. The objective of spectrum management is to enable electronic systems to perform their functions in the intended environment without causing or suffering unacceptable interference (JP 6-0).

Wartime Reserve Modes

1-43. *Wartime reserve modes* are characteristics and operating procedures of sensors, communications, navigation aids, threat recognition, weapons, and countermeasures systems that will contribute to military effectiveness if unknown to or misunderstood by opposing commanders before they are used, but could be exploited or neutralized if known in advance. Wartime reserve modes are deliberately held in reserve for wartime or emergency use and seldom, if ever, applied or intercepted prior to such use (JP 1-02).

Electromagnetic Compatibility

1-44. *Electromagnetic compatibility* is the ability of systems, equipment, and devices that utilize the electromagnetic spectrum to operate in their intended operational environments without suffering unacceptable degradation or causing unintentional degradation because of electromagnetic radiation or response. It involves the application of sound electromagnetic spectrum management; system, equipment, and device design configuration that ensures interference-free operation; and clear concepts and doctrines that maximize operational effectiveness (JP 1-02).

Application Terminology

1-45. EW capabilities are applied from the air, land, sea, and space by manned, unmanned, attended, or unattended systems. Units employ EW capabilities to achieve the desired lethal or nonlethal effect on a given target. Units maintain freedom of action in the electromagnetic spectrum while controlling the use of it by the enemy. Regardless of the application, units employing EW capabilities must use appropriate levels of control and protection of the electromagnetic spectrum. In this way, they avoid adversely affecting friendly forces. (Improper EW actions must be avoided because they may cause fratricide or eliminate high-value intelligence targets.)

1-46. In the context of EW application, units use several terms to facilitate control and protection of the electromagnetic spectrum. Terms used in EW application include control, detection, denial, deception, disruption and degradation, protection, and destruction. The three subdivisions of EW—electronic attack, electronic protection, and electronic warfare support—are specified within the following descriptions.

Control

1-47. In the context of EW, control of the electromagnetic spectrum is achieved by effectively coordinating friendly systems while countering enemy systems. Electronic attack limits enemy use of the electromagnetic spectrum. Electronic protection secures use of the electromagnetic spectrum for friendly forces, and electronic warfare support enables the commander's accurate assessment of the situation. All three are integrated for effectiveness. Commanders ensure maximum integration of communications; intelligence, surveillance, and reconnaissance; and information tasks.

Detection

1-48. In the context of EW, detection is the active and passive monitoring of the operational environment for radio frequency, electro-optic, laser, infrared, and ultraviolet electromagnetic threats. Detection is the first step in EW for exploitation, targeting, and defensive planning. Friendly forces maintain the capability to detect and characterize interference as hostile jamming or unintentional electromagnetic interference.

Denial

1-49. In the context of EW, denial is controlling the information an enemy receives via the electromagnetic spectrum and preventing the acquisition of accurate information about friendly forces. Degradation uses traditional jamming techniques, expendable countermeasures, destructive measures, or network applications. These range from limited effects up to complete denial of usage.

Deception

1-50. In the context of EW, deception is confusing or misleading an enemy by using some combination of human-produced, mechanical, or electronic means. Through use of the electromagnetic spectrum, EW deception manipulates the enemy's decision loop, making it difficult to establish accurate situational awareness.

Disruption and Degradation

1-51. In the context of EW, disruption and degradation techniques interfere with the enemy's use of the electromagnetic spectrum to limit enemy combat capabilities. This is achieved with electronic jamming, electronic deception, and electronic intrusion. These enhance attacks on hostile forces and act as force multipliers by increasing enemy uncertainty, while reducing uncertainty for friendly forces. Advanced electronic attack techniques offer the opportunity to nondestructively disrupt or degrade enemy infrastructure.

Protection

1-52. In the context of EW, protection is the use of physical properties; operational tactics, techniques, and procedures; and planning and employment processes to ensure friendly use of the electromagnetic spectrum. This includes ensuring that offensive EW activities do not electronically destroy or degrade friendly intelligence sensors or communications systems. Protection is achieved by component hardening, emission control, and frequency management and deconfliction. Frequency management and deconfliction include the capability to detect, characterize, geolocate, and mitigate electromagnetic interference that affects operations. Protection includes other means to counterattack and defeat enemy attempts to control the electromagnetic spectrum. Additionally, organizations such as a joint force commander's EW staff or a joint EW coordination cell enhance electronic protection by deconflicting EW efforts.

Destruction

1-53. Destruction, in the context of EW, is the elimination of targeted enemy systems. Sensors and command and control nodes are lucrative targets because their destruction strongly influences the enemy's perceptions and ability to coordinate actions. Various weapons and techniques ranging from conventional munitions and directed energy weapons to network attacks can destroy enemy systems that use the electromagnetic spectrum. Electronic warfare support provides target location and related information. While destroying enemy equipment can effectively deny the enemy use of the electromagnetic spectrum, the duration of denial will depend on the enemy's ability to reconstitute. (See JP 3-13.1.)

MEANS VERSUS EFFECTS

1-54. EW means are applied against targets to create a full range of lethal and nonlethal effects. (See figure 1-4.) Choosing a specific EW capability depends on the desired effect on the target and other considerations, such as time sensitivity or limiting collateral damage. EW capabilities provide commanders with additional options for achieving their objectives. During major combat operations there may be circumstances where commanders want to limit the physical damage on a given target. Under such circumstances, the EW staff articulates clearly to the commander the lethal and nonlethal effects EW capabilities can achieve. For example, a target might be enemy radar mounted on a fixed tower. Two EW options to defeat the radar could be to jam the radar or destroy it with antiradiation missiles. If the commander desired to limit damage to the tower, an electronic attack jamming platform would be preferred. In circumstances where commanders cannot sufficiently limit undesired effects such as collateral damage, they may be constrained from applying physical force. The EW staff articulates succinctly how EW capabilities can support actions to achieve desired effects and provide lethal and nonlethal options for commanders.

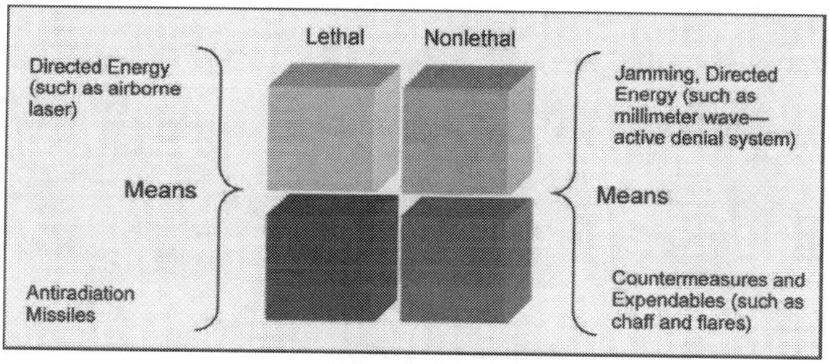

Figure 1-4. Means versus effects

SUMMARY

1-55. As the modern battlefield becomes more technologically sophisticated, military operations continue to be executed in an increasingly complex electromagnetic environment. Therefore, commanders and staffs need to thoroughly understand and articulate how the electromagnetic environment impacts their operations and how friendly EW operations can be used to gain an advantage. Commanders and staffs use the terminology presented in this chapter to describe the application of EW. This ensures a common understanding and consistency within plans, orders, standing operating procedures, and directives.

Chapter 2

Electronic Warfare in Full Spectrum Operations

Information technology is becoming universally available. Most enemies rely on communications and computer networks to make and implement decisions. Radios remain the backbone of tactical military command and control architectures. However, most communications relayed over radio networks are becoming digital as more computers link networks through transmitted frequencies. Therefore, the ability to dominate the electromagnetic spectrum is central to full spectrum operations. This chapter describes how commanders apply electronic warfare capabilities to support full spectrum operations.

THE ROLE OF ELECTRONIC WARFARE

2-1. Army electronic warfare (EW) operations seek to provide the land force commander with capabilities to support full spectrum operations. Full spectrum operations consist of the purposeful, simultaneous combination of offense, defense, and stability or civil support. The goal of full spectrum operations is to change the operational environment so that peaceful processes are dominant. Nonetheless, operational environments are complex; commanders must conduct operations across the entire spectrum of conflict. The Army maintains flexible forces with balanced capabilities and capacities. These flexible and balanced forces remain able to conduct major operations while executing other day-to-day smaller-scale operations. (See FM 3-0.)

2-2. Figure 2-1 (page 2-2) shows the weight of effort for using EW during operations. This figure adapts the elements of full spectrum operations (offense, defense, and stability or civil support) as described in FM 3-0. Overseas, Army forces conduct full spectrum operations (offensive, defensive, and stability) simultaneously as part of a joint force. Within the United States, Army forces conduct homeland defense and civil support operations as part of homeland security. Army electronic warfare (EW) operations seek to provide the land force commander with capabilities to support full spectrum operations. As noted in figure 2-1, statutory law limits the use of EW capabilities in support of civil support operations.

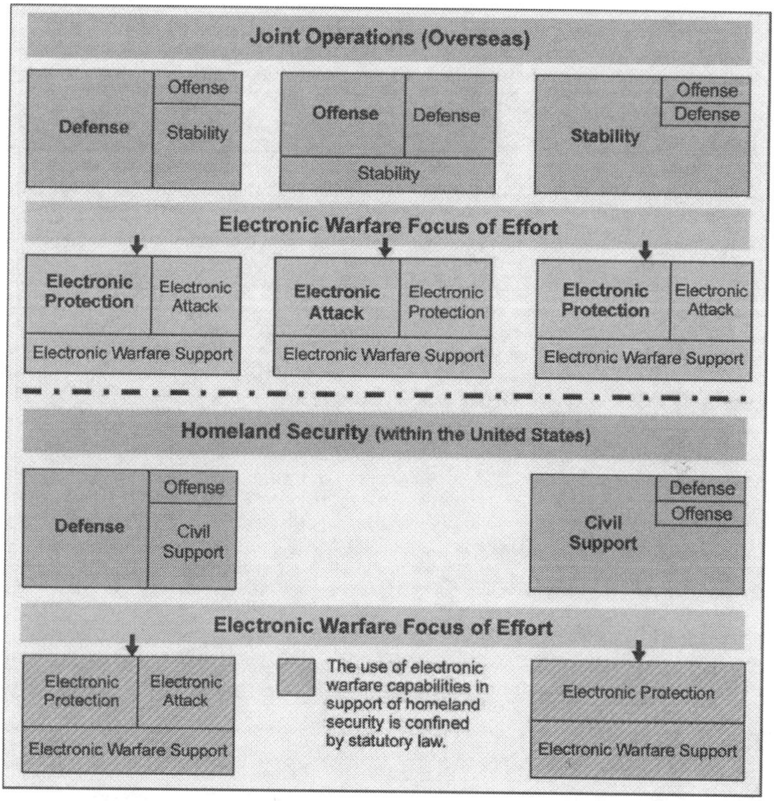

Figure 2-1. Electronic warfare weight of effort during operations

2-3. Full spectrum operations involve more than executing all elements of operations simultaneously. They require that commanders and staffs consider their unit's capabilities and capacities relative to each of the elements of full spectrum operations. Commanders consider how much can be accomplished simultaneously, how much can be phased, and what nonorganic resources may be available to solve problems. The same applies to EW in support of full spectrum operations. Commanders and staffs determine which resident and joint force EW capabilities to leverage in support of each element of full spectrum operations. Weighting the EW focus of effort within each of the elements assists commanders and their staffs in visualizing how EW capabilities can support their operations. Commanders combine offensive, defensive, and stability or civil support operations to seize, retain, and exploit the initiative. As they apply the appropriate level of EW effort to support these elements, commanders can seize, retain, and exploit the initiative within the electromagnetic environment.

THE APPLICATION OF ELECTRONIC WARFARE

2-4. To support full spectrum operations and achieve the goal of electromagnetic spectrum dominance, commanders fully integrate EW capabilities and apply them across the elements of combat power. Leadership and information are applied through, and multiply the effects of, the other six elements of combat power. Paragraphs 2-5 through 2-16 discuss the elements of combat power and how EW capabilities can support them.

IN SUPPORT OF LEADERSHIP

2-5. Leadership initiates the conditions for success. Commanders balance the ability to mass the effects of lethal and nonlethal systems with the requirements to deploy and sustain the units that employ those systems. Generating and maintaining combat power throughout an operation is essential. Today's operational environments require leaders who are competent, confident, and informed in using and protecting combat capabilities that operate within the electromagnetic spectrum. Commanders plan, prepare, execute, and assess EW operations to dominate the electromagnetic spectrum within their operational environment. To accomplish this domination, commanders effectively apply and integrate EW operations across the warfighting functions.

IN SUPPORT OF INFORMATION TASKS AND CAPABILITIES

2-6. Information is the element of combat power consisting of meaningful facts, data, and impressions used to develop a common situational understanding, to enable battle command, and to affect the operational environment. (See FM 3-0 for a discussion of combat power.) In modern conflict, gaining information superiority has become as important as lethal action in determining the outcome of operations. *Information superiority* is the operational advantage derived from the ability to collect, process, and disseminate an uninterrupted flow of information while exploiting or denying an adversary's ability to do the same (JP 3-13). To achieve this operational advantage, Army commanders direct efforts that contribute to information superiority. These efforts fall into four primary areas: Army information tasks; intelligence, surveillance, and reconnaissance; knowledge management; and information management. (See FM 3-0 for a discussion of information superiority.)

2-7. The Army information tasks are used to shape a commander's operational environment. These tasks are information engagement, command and control warfare, information protection, operations security, and military deception. Information capabilities can be used to produce both destructive and constructive effects. For example, destructive actions use information capabilities against the enemy's command and control system and other assets to reduce their combat capability. Constructive actions use information capabilities to inform or influence a particular audience or as a means to affect enemy morale. Although applicable to all elements of full spectrum operations, EW capabilities play a major role in enabling and supporting the execution of the command and control warfare and information protection tasks.

2-8. *Command and control warfare* is the integrated use of physical attack, electronic warfare, and computer network operations, supported by intelligence, to degrade, destroy, and exploit an enemy's or adversary's command and control system or to deny information to it (FM 3-0). It includes operations intended to degrade, destroy, and exploit an enemy's or adversary's ability to use the electromagnetic spectrum and computer and telecommunications networks. *Information protection* is active or passive measures that protect and defend friendly information and information systems to ensure timely, accurate, and relevant friendly information. Information protection denies enemies, adversaries, and others the opportunity to exploit friendly information and information systems for their own purposes (FM 3-0). Table 2-1 shows capabilities, intended effects, staff responsibilities, and functional cells for the command and control warfare and information protection tasks. (For further information on the information tasks, refer to FM 3-0.)

Table 2-1. Two Army information tasks: command and control warfare and information protection

Army Information Tasks	Capabilities	Staff Responsibility	Functional Coordinating Cell	Intended Effects	Integrating Process
Command and Control Warfare	Physical Attack Electronic Attack Computer Network Attack Electronic Warfare Support Computer Network Exploitation	G-3/G-2	Fires	Degrade, disrupt, destroy and exploit enemy command and control	Operations Process
Information Protection	Information Assurance Computer Network Defense Electronic Protection	G-6	Network Operations	Protect friendly computer networks and communication means	
G-2 assistant chief of staff, intelligence G-3 assistant chief of staff, operations				G-6 assistant chief of staff, signal	

2-9. To support these information tasks, commanders ensure EW is coordinated, integrated, and synchronized with all other tasks. This occurs within the operations process through the various functional and integrating cells. Table 2-2 illustrates EW capabilities, actions, and objectives that support the command and control warfare and information protection tasks.

Table 2-2. Electronic warfare support to two Army information tasks

Information Tasks	Command and Control Warfare	Information Protection
Electronic warfare supports by	Locating and identifying threat command and control systems. Denying, disrupting, degrading, and/or destroying the enemy's command and control system. Supporting and complementing computer network attack and computer network exploitation operations.	Deconflicting spectrum usage with the spectrum manager. Hardening equipment against electromagnetic interference. Emissions control.
Action	Electronic attack (jamming, antiradiation missiles). Directed energy and electromagnetic spectrum area denial systems. Expendables (chaff, decoys, and flares). Electronic warfare support/signals intelligence.	Frequency agility in radios. Electronic shielding for systems. Electronic masking. Processes to counter intrusion. Implementing emissions control procedures to safeguard friendly systems and facilities from the effects of friendly and enemy electronic attack.
Objective or Effect	Detect, deny, disrupt or degrade, and destroy.	Control and protection.

In Support of the Warfighting Functions

2-10. EW capabilities support each of the six warfighting functions. Examples of specific supporting capabilities are given in the following paragraphs.

Movement and Maneuver

2-11. The *movement and maneuver warfighting function* is the related tasks and systems that move forces to achieve a position of advantage in relation to the enemy. Direct fire is inherent in maneuver, as is close combat (FM 3-0). EW capabilities that enable the movement and maneuver of Army forces include—

- Suppression and destruction of enemy integrated air defenses.
- Denial of enemy information systems and intelligence, surveillance, and reconnaissance sensors.
- Target designation and range finding.
- Protection from effects of friendly and enemy EW.
- Lethal and nonlethal effects against enemy combat capability (personnel, facilities, and equipment).
- Threat warning and direction finding.
- Use of the electromagnetic spectrum to counter improvised explosive device operations.
- Electromagnetic spectrum obscuration, low observability, and multispectral stealth.

Intelligence

2-12. The *intelligence warfighting function* is the related tasks and systems that facilitate understanding of the operational environment, enemy, terrain, and civil considerations (FM 3-0). It includes tasks associated with intelligence, surveillance, and reconnaissance. EW capabilities that enable the intelligence warfighting function include—

- Increased access for intelligence collection assets (systems and personnel) by reducing antiaccess, antipersonnel, and antisystems threats.
- Increased capability to search for, intercept, identify, and locate sources of radiated electromagnetic energy in support of targeting, information tasks, and future operations.
- Increased capability in providing threat recognition and threat warning to the force.
- Indications and warning of threat emitters and radar.
- Denial and destruction of counter-intelligence, -surveillance, and -reconnaissance systems.

Fires

2-13. The *fires warfighting function* is the related tasks and systems that provide collective and coordinated use of Army indirect fires, joint fires, and command and control warfare, including nonlethal fires, through the targeting process (FM 3-0). It includes tasks associated with integrating command and control warfare. EW capabilities that enable the fires warfighting function include—

- Detection and location of targets radiating electromagnetic energy.
- Disruption, degradation, and destruction options for servicing targets. This includes information systems, targets requiring precision strike (such as minimal collateral damage and minimal weapons signature), hard and deeply buried targets, weapons of mass destruction, and power generation and infrastructure targets.
- Control, dispersion, or neutralization of combatant and noncombatant personnel with nonpersistent effects and minimum collateral damage (scalable and nonlethal).
- Area denial capabilities against vehicles, vessels, and aircraft.

Sustainment

2-14. The *sustainment warfighting function* is the related tasks and systems that provide support and services to ensure freedom of action, extend operational reach, and prolong endurance (FM 3-0). EW capabilities that enable the sustainment warfighting function include—

- Protection of sustainment forces from friendly and adversary use of EW in static or mobile environments.
- Enhanced electromagnetic environment situational awareness through the interception, detection, identification, and location of adversary electromagnetic emissions and by providing indications and warnings. (This information can assist in convoy planning, asset tracking, and targeting of potential threats to sustainment operations.)
- Countering improvised explosive devices to support ground lines of communication (includes counter-radio-controlled improvised-explosive-device systems and countering other threats triggered through the electromagnetic spectrum, such as lasers).
- Spectrum deconfliction and emissions control procedures in support of sustainment command and control.
- Electromagnetic spectrum obscuration, low-observability, and multispectral stealth (These capabilities provide protection during sustainment operations).

Command and Control

2-15. The *command and control warfighting function* is the related tasks and systems that support commanders in exercising authority and direction (FM 3-0). EW capabilities that enable the command and control warfighting function include—

- Protection of friendly critical information systems and command and control nodes, personnel, and facilities from the effects of friendly and adversary EW operations.
- Control of friendly EW systems through—
  - Frequency deconfliction.
  - Asset tracking.
  - Employment execution.
  - Reprogramming of EW systems.
  - Registration of all electromagnetic spectrum emitting devices with the spectrum manager (both prior to deployment and when new systems or devices are added to the deployed force).
- The development of EW command and control tools to enhance required coordination between Army and joint EW operations.
- EW operations integration, coordination, deconfliction, and synchronization through the EW working group (see chapter 3).
- Increased commander situational understanding through improved common operational picture input of electromagnetic spectrum- and EW-related information.
- EW operations monitoring and assessment.

Protection

2-16. The *protection warfighting function* is the related tasks and systems that preserve the force so the commander can apply maximum combat power (FM 3-0). EW capabilities and actions that enable the protection warfighting function include—

- Enhanced electromagnetic spectrum situational awareness through the interception, detection, identification, and location of adversary electromagnetic emissions used to providing indications and warnings of threat emitters and radars.
- Denial, disruption, or destruction of electromagnetic-spectrum-triggered improvised explosive devices and enemy air defense systems.
- Deception of enemy forces.
- Electromagnetic spectrum obscuration, low-observability, and multispectral stealth.
- EW countermeasures for platform survivability (air and ground).
- Area denial capabilities (lethal and nonlethal) against personnel, vehicles, and aircraft.
- Protection of friendly personnel, equipment, and facilities from friendly and enemy electronic attack, including friendly information systems and information. (This includes the coordination and use of both airborne and ground-based electronic attack with higher and adjacent units.)

SUMMARY

2-17. Army EW operations provide the land force commander capabilities to support full spectrum operations (offensive, defensive, and stability or civil support operations). EW supports full spectrum operations by applying EW capabilities to detect, deny, deceive, disrupt, or degrade and destroy enemy combat capability and by controlling and protecting friendly use of the electromagnetic spectrum. These capabilities—when applied across the warfighting functions—enable commanders to address a broad set of electromagnetic-spectrum-related targets to gain and maintain an advantage within the electromagnetic spectrum.

Chapter 3
Electronic Warfare Organization

A flexible organizational framework and capable, proficient electronic warfare personnel enable the commander's electronic warfare capability on the battlefield. This chapter discusses a framework that ensures coordination, synchronization, and integration of electronic warfare into full spectrum operations. This electronic warfare organizational framework supports current operations and is adaptable for future operations.

ORGANIZING ELECTRONIC WARFARE OPERATIONS

3-1. Operational challenges across the electromagnetic spectrum are expanding rapidly. As Army electronic warfare (EW) capabilities expand to meet these challenges, the organizational design required to coordinate, synchronize, integrate, and deconflict these capabilities must transform as rapidly. To meet current and future requirements, command and control of EW operations is built around the concept of EW working groups. Figure 3-1, page 3-2, illustrates the EW coordination organizational framework.

ARMY SERVICE COMPONENT COMMAND, CORPS, AND DIVISION LEVELS

3-2. A *working group* is a temporary grouping of predetermined staff representatives who meet to coordinate and provide recommendations for a particular purpose or function (FMI 5-0.1). The EW working group, when established, is responsible to the G-3 through the fires cell. An EW working group usually includes representation from the G-2, G-3, G-5, G-6, and G-7. (Joint doctrine calls this organization the EW coordination cell.) The EW working groups depicted in figure 3-1 (page 3-2) facilitate the internal (Army) and external (joint) integration, synchronization, and deconfliction of EW actions with fires, command and control, movement and maneuver, intelligence, sustainment and protection warfighting functions. Normally, EW working groups do not add additional structure to an existing organization. As depicted in figure 3-1, working groups vary in size and composition based on echelon.

3-3. Normally, the senior EW officer heads the EW working group and is accountable to the G-3 for integrating EW requirements. Working within the fires cell, the EW officer coordinates directly with the fire support coordinator for the integration of EW into the targeting process. This ensures EW capabilities are fully integrated with all other effects. Additional staff representation within EW working groups may include a fire support coordinator, a spectrum manager, a space operations officer, and liaison officers as required. Depending on the echelon, liaisons could include joint, interagency, and multinational representatives. When an Army headquarters serves as the headquarters of a joint task force or joint force land component command, the Army headquarters' working group becomes the joint force EW coordination cell.

3-4. When Army forces are employed as part of a joint or multinational force, they normally have EW representatives supporting higher headquarters' EW coordination organizations. These organizations may include the joint force commander's EW staff or the information operations cell within a joint task force. Sometimes a component EW organization may be designated as the joint EW coordination cell. (Chapter 6 discusses joint electronic warfare operations in more detail.) The overall structure of the combatant force and the level of EW to be conducted determine the structure of the joint EW coordination cell. The organization to accomplish the required EW coordination and functions varies by echelon.

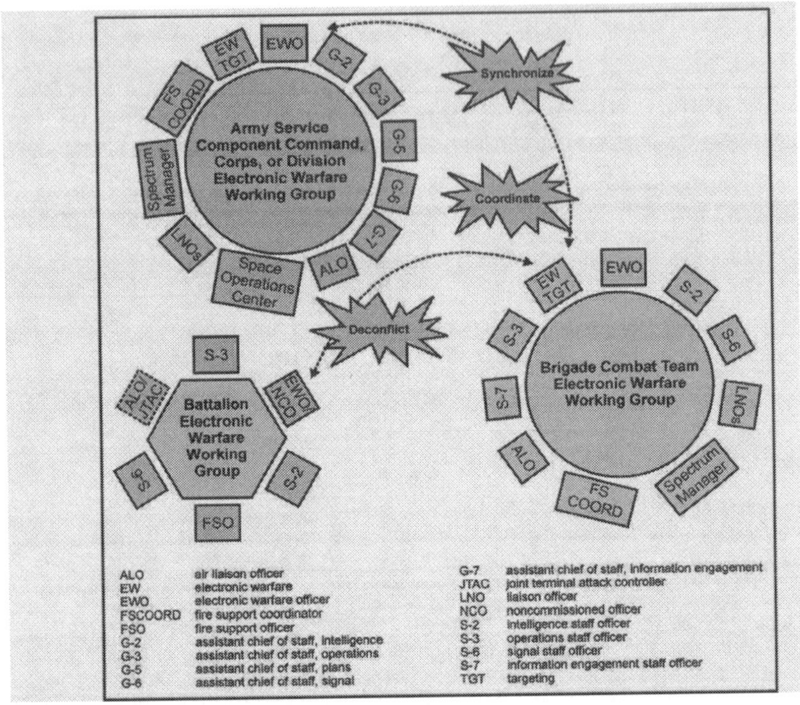

Figure 3-1. Electronic warfare coordination organizational framework

3-5. Regardless of the organizational framework employed, EW working groups perform specific tasks. Table 3-1 (page 3-3) details the functions of the EW working groups by echelon from battalion to Army Service component command. There is no formal organizational framework for EW at the company level (see paragraph 3-9).

Table 3-1. Functions of electronic warfare working groups

EW Working Group	Functions
Division and Above ALO EWO EW targeting G-2 G-3 G-5 G-6 G-7 FSCOORD LNOs Spectrum manager Space support officer	**Peacetime: Division—ASCC** • Conduct long-range electronic warfare planning in support of theater or combatant command requirements. • Integrate electronic warfare into operation plans and concept plans. • Develop electronic warfare supporting plans to operation plans and contingency plans. • Coordinate joint electronic warfare training and exercises. • Develop information and knowledge necessary to support contingency planning (for example, joint restricted frequency list development, spectrum management, and deconfliction). **Wartime: Division—ASCC** • Serve as the joint force land component or joint task force EW working group. • When directed, serve as the jamming control authority. • Develop and promulgate electronic warfare policies and support higher level policies. • Identify and coordinate intelligence support requirements for electronic warfare. • Plan, coordinate, and assess offensive and defensive electronic warfare requirements. • Plan, coordinate, synchronize, deconflict, and assess electronic warfare operations. • Maintain current assessment of electronic warfare resources available to the commander. • Prioritize electronic warfare effects and targets. • Predict effects of friendly and enemy electronic warfare. • Coordinate spectrum management and radio frequency deconfliction with G-6 and J-6. • Plan, assess, and implement friendly electronic security measures. • Plan, coordinate, integrate, and deconflict electronic warfare effects within the operations process.
Brigade S-3 EWO EW targeting FSCOORD S-2 S-6 ALO LNOs S-7 Spectrum manager	**Peacetime:** • Develop electronic warfare supporting requirements to operations plans and exercises. **Wartime:** • Support electronic warfare policies. • Plan, prepare, execute, and assess electronic warfare operations. • Integrate electronic warfare intelligence preparation of the battlefield into the operations process. • Identify and coordinate intelligence support requirements for BCT and subordinate units' electronic warfare operations. • Assess offensive and defensive electronic warfare requirements. • Maintain current assessment of electronic warfare resources available to unit. • Prioritize BCT and subordinate units' electronic warfare targets. • Plan, coordinate, and assess friendly electronic warfare operations. • Implement friendly electronic security measures (for example, electromagnetic spectrum mitigation and network protection). • When directed, serve as the jamming control authority.
Battalion EWO/NCO FSO S-2 S-3 S-6 JTAC	**Peacetime:** • Support BCT electronic warfare requirements to operations and exercises. **Wartime:** • Evaluate electronic warfare offensive, defensive, and support requirements. • Coordinate electronic warfare operations with higher headquarters. • Identify and coordinate intelligence support requirements with higher headquarters. • Execute electronic warfare in support of current operations. • Assess electronic warfare operations.

ALO	air liaison officer	G-7	assistant chief of staff, information engagement
ASCC	Army service component command	J-6	communications system directorate of a joint staff
BCT	brigade combat team	JTF	joint task force
EW	electronic warfare	JTAC	joint terminal attack controller
EWO	electronic warfare officer	LNO	liaison officer
FSCOORD	fire support coordinator	NCO	noncommissioned officer
G-2	assistant chief of staff, intelligence	S-2	intelligence staff officer
G-3	assistant chief of staff, operations	S-3	operations staff officer
G-4	assistant chief of staff, logistics	S-6	signal staff officer
G-5	assistant chief of staff, plans	S-7	information engagement staff officer
G-6	assistant chief of staff, signal		

Brigade Level

3-6. At the brigade level, the EW officer heads the EW working group and is accountable to the S-3 for integrating EW requirements. Additional staff representation within EW working groups at the brigade combat team level may include the fire support coordinator, EW targeting technician, S-2, S-6, spectrum manager, S-7, and liaison officers as required.

3-7. The EW working group at the brigade combat team coordinates with the higher echelon EW working groups. The brigade working group plays an important role in requesting and integrating joint air and ground EW support. It also manages the brigade's organic EW "fight" within the fires cell. The EW officer works as part of the brigade combat team staff. In this position, the EW officer synchronizes, integrates, and deconflicts brigade combat team EW actions with the EW working group at division level. Although EW falls under the control of the S-3, EW officers are fully immersed in fires targeting and planning to ensure proper use and coordination of EW. See table 3-1, page 3-3, for an outline of the functions of the brigade combat team EW working group.

Battalion Level

3-8. At the battalion level, the EW officer or noncommissioned officer leads the EW working group and is accountable to the S-3 for integrating EW requirements. Additional staff representation within EW working groups at the battalion level may include the S-2, S-6, fire support officer, and a joint terminal attack controller when assigned. The battalion EW working group coordinates battalion EW operations with the brigade combat team EW working group. See table 3-1, page 3-3, for an outline of the functions of the battalion EW working group.

Company Level

3-9. At the company level, trained EW personnel holding an additional skill identifier of 1K (tactical EW operations) or 1J (operational EW operations) perform several tasks. They advise the commander on the employment of EW equipment, track EW equipment status, assist operators in the use and maintenance of EW equipment, and coordinate with higher headquarters EW working groups.

PLANNING AND COORDINATING ELECTRONIC WARFARE ACTIVITIES

3-10. Key personnel involved in the planning and coordination of EW activities are—

- G-3 and S-3 staff.
- EW officer.
- Fire support coordinator.
- G-2 and S-2 staff.
- G-6 and S-6 staff.
- Electromagnetic spectrum manager.
- Liaisons.

G-3 or S-3 Staff

3-11. The G-3 or S-3 staff is responsible for the overall planning, coordination, and supervision of EW activities, except for intelligence. The EW officer is part of the G-3 or S-3 staff. The G-3 or S-3 staff—

- Plans for and incorporates EW into operation plans and orders, in particular within the fire support plan and the information operations plan (in joint operations).
- Tasks EW actions to assigned and attached units.
- Exercises control over electronic attack, including integration of electromagnetic deception plans.

- Directs electronic protection measures the unit will take based on recommendations from the G-6 or S-6, the EW officer, and the EW working group.
- Coordinates and synchronizes EW training with other unit training requirements.
- Coordinates and synchronizes EW training with other unit training requirements.
- Issues EW support tasks within the unit intelligence, surveillance, and reconnaissance plan. These tasks are according to the collection plan and the intelligence synchronization matrices developed by the G-2 or S-2 and the collection manager.
- Coordinates with the EW working group to ensure planned EW operations support the overall tactical plan.
- Integrates electronic attack as a form of fires within the fires cell.

ELECTRONIC WARFARE OFFICER

3-12. As a member of the G-3 or S-3 staff, the EW officer plans, coordinates, and supports the execution of EW. The EW officer—

- Leads the EW working group.
- Plans, coordinates, and assesses EW offensive, defensive, and support requirements.
- Supports the G-2 or S-2 during intelligence preparation of the battlefield.
- Supports the fire support coordinator to ensure electronic attack fires are integrated with all other effects.
- Plans, assesses, and implements friendly electronics security measures.
- Prioritizes EW effects and targets with the fire support coordinator.
- Plans and coordinates EW operations across functional and integrating cells.
- Deconflicts EW operations with the spectrum manager.
- Maintains a current assessment of available EW resources.
- Participates in other cells and working groups (as required) to ensure EW integration.
- Serves as EW subject matter expert on existing EW rules of engagement.
- When designated, serves as the jamming control authority.
- Prepares, submits for approval, and supervises the issuing and implementation of fragmentary orders for EW operations.

G-2 OR S-2 STAFF

3-13. The G-2 or S-2 staff advises the commander and staff on the intelligence aspects of EW. The G-2 or S-2 staff—

- Provides threat data to support programming of unit EW systems and deconfliction of their use by the EW working group.
- Ensures that electronic order of battle requirements are included in the intelligence collection plan.
- Determines enemy EW organizations, disposition, capabilities, and intentions via collection and analysis.
- Determines enemy EW vulnerabilities and high-value targets.
- Assesses effects of friendly EW operations on the enemy.
- Helps prepare the intelligence-related portion of the EW running estimate.
- Provides input to the restricted frequency list by recommending guarded frequencies.
- Provides updates on the rapid electronic order of battle.
- Maintains appropriate threat EW databases.
- Works with the EW working group to ensure that intelligence collection is synchronized with EW requirements and deconflicted with planned EW actions. Ensures that EW threat data is deconflicted with friendly electromagnetic spectrum needs.

Network Operations Officer

3-14. The network operations officer (in the G-6 or S-6 staff) coordinates the communications network for the following services:

- Preparing the electronic protection policy on behalf of the commander.
- Assisting in preparing EW plans and orders.
- Reporting all enemy electronic attack activity detected by friendly communications and electronics elements to the EW working group for counteraction.
- Assisting the unit EW officer with resolving EW systems maintenance and communications fratricide problems.

Spectrum Manager

3-15. The spectrum manager coordinates electromagnetic spectrum use for a wide variety of communications and electronic resources. The spectrum manager—

- Issues the signal operating instructions.
- Provides all spectrum resources to the task force.
- Coordinates for spectrum usage with higher echelon G-6 or S-6, and applicable host-nation and international agencies as necessary.
- Coordinates the preparation of the restricted frequency list and issuance of emissions control guidance.
- Coordinates frequency allotment, assignment, and use.
- Coordinates electromagnetic deception plans and operations in which assigned communications resources participate.
- Coordinates measures to reduce electromagnetic interference.
- Coordinates with higher echelon spectrum managers for electromagnetic interference resolution that cannot be resolved internally.
- Assists the EW officer in issuing guidance in the unit (including subordinate elements) regarding deconfliction and resolution of interference problems between EW systems and other friendly systems.
- Participates in the EW working group to deconflict friendly electromagnetic spectrum requirements with planned EW operations and intelligence collection.

SUMMARY

3-16. The organizational framework for EW coordination and functions varies by echelon. The necessity to form an EW working group is largely based on the overall structure of the combatant force and the level of EW to be conducted. During unified actions, other Service EW officers, signals intelligence officers, and EW asset representatives are invaluable to Army EW working groups in the planning, preparation, execution, and assessment of EW operations. As Army EW capabilities and concepts for employment continue to evolve, so do the organizational designs that ensure their effective command and control and execution in support of operations.

Chapter 4

Electronic Warfare and the Operations Process

The *operations process* consists of the major command and control activities performed during operations: planning, preparing, executing, and continuously assessing the operation. The commander drives the operations process (FM 3-0). These activities occur continuously throughout an operation, overlapping and recurring as required (see figure 4-1). The staff electronic warfare officer is actively involved in the operations process. Electronic warfare planning, preparation, execution, and assessment require collective expertise from operations, intelligence, signal, and battle command. The electronic warfare officer—through the unit's electronic warfare working group—integrates efforts across the warfighting functions. This ensures that electronic warfare operations support the commander's objectives.

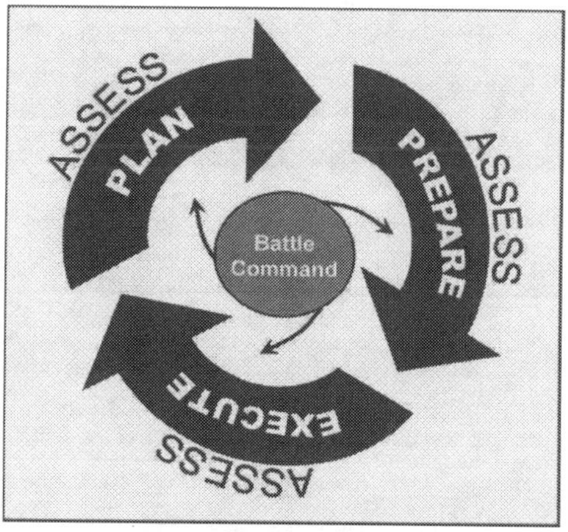

Figure 4-1. The operations process

SECTION I — ELECTRONIC WARFARE PLANNING

4-1. Electronic warfare (EW) planning is based on three main considerations. The first is applying the military decisionmaking process (MDMP). EW planners understand and follow its seven steps. In a time-constrained environment they still follow all seven steps, abbreviating the MDMP process appropriately. Additionally, EW planners apply EW integrating processes. They understand how EW actions contribute to operations. They integrate and synchronize EW activities starting with planning and continuing throughout operations. Finally, EW planners apply EW employment considerations according to the characteristics of EW capabilities.

THE MILITARY DECISIONMAKING PROCESS

4-2. EW planning minimizes fratricide and optimizes operational effectiveness during execution. Therefore, EW planning occurs concurrently with other operational planning during the MDMP. The MDMP synchronizes several processes, including intelligence preparation of the battlefield (IBP) (see FM 34-130), the targeting process (see FM 6-20-10), and risk management (see FM 5-19). These processes occur continuously during operations.

4-3. Depending on the organizational echelon, the staff EW officer leads EW planning through the EW working group. (The EW working group at echelons above brigade is sometimes referred to as an EW coordination cell.) An EW working group is normally supported by representatives from the G-2 or S-2, G-3 or S-3, G-6 or S-6, and other staff as required. Other staff representatives can include the fire support coordinator or fire support officer, spectrum manager, air liaison officer, space officer, and liaison officers. Paragraphs 4-5 through 4-33 outline key EW contributions to the processes and planning actions that occur during the seven steps of the MDMP. (FM 5-0 discusses the MDMP.)

RECEIPT OF MISSION

4-4. Commanders begin the MDMP upon receiving or anticipating a new mission. During this first step, commanders issue their initial guidance and initial information requirements or commander's critical information requirements.

4-5. Upon receipt of a mission, the staff EW officer alerts the staff members supporting the EW working group. The EW officer and support staff begin to gather the resources required for mission analysis. Resources might include a higher headquarters operation order or plan, maps of the area of operations, electronic databases, required field manuals and standing operating procedures, current running estimates, and reachback resources (see appendix F). The EW officer also provides input to the staff's initial assessment and updates the EW running estimate. As part of this update, the EW officer identifies all friendly EW assets and resources and their status. The EW officer also provides this information throughout the operations process. This includes monitoring, tracking, and seeking out information relating to EW operations to assist the commander and staff.

MISSION ANALYSIS

4-6. Planning includes a thorough mission analysis. Both the process and products of mission analysis help commanders refine their situational understanding and determine their restated mission. (See FM 5-0 for more details.) The EW officer and supporting members of the EW working group contribute to the overall mission analysis by participating in IPB and through the planning actions discussed in paragraphs 4-7 through 4-14. (Paragraphs 4-35 to 4-40 discuss EW input to IPB during operations.)

4-7. The EW officer and EW working group members—

- Convene the appropriate EW working group.
- Determine known facts, status, or conditions of forces capable of EW operations as defined in the commander's planning documents, such as a warning order or operation order.
- Identify EW planning support requirements and develop support requests as needed.

4-8. The EW officer and EW working group members support the G-2 and S-2 in IPB by—

- Determining the threat's dependence on the electromagnetic spectrum.
- Determining the threat's EW capability.

- Determining the threat's intelligence system collection capability.
- Determining which threat vulnerabilities relate to the electromagnetic spectrum.
- Determining how the operational environment affects EW operations using the operational variables and mission variables as appropriate.
- Initiating, refining, and validating information requirements and requests for information.

4-9. The EW officer and EW working group members—

- Determine facts and develop necessary assumptions relevant to EW such as the status of EW capability at probable execution and time available.
- Analyze the commander's mission and intent from an EW perspective.
- Identify constraints relevant to EW—
  - Actions EW operations must perform.
  - Actions EW operations cannot perform.
  - Other constraints.
- Analyze mission, enemy, terrain and weather, troops and support available, time available and civil considerations from the EW perspective.

4-10. The EW officer and EW working group members determine enemy and friendly centers of gravity and list their critical capabilities, requirements, and vulnerabilities from an EW perspective. (They determine how EW capabilities can best attack an enemy's command and control system.) The center of gravity analysis process outlined in figure 4-2 helps identify and list the critical vulnerabilities of enemy centers of gravity. The EW officer and EW working group members also list the critical requirements associated with the identified command and control critical capability (or command and control nodes) and then identify the critical vulnerabilities associated with the critical requirements. Through this process, the EW officer and EW working group members help determine which vulnerabilities can be engaged by EW capabilities to produce a decisive outcome.

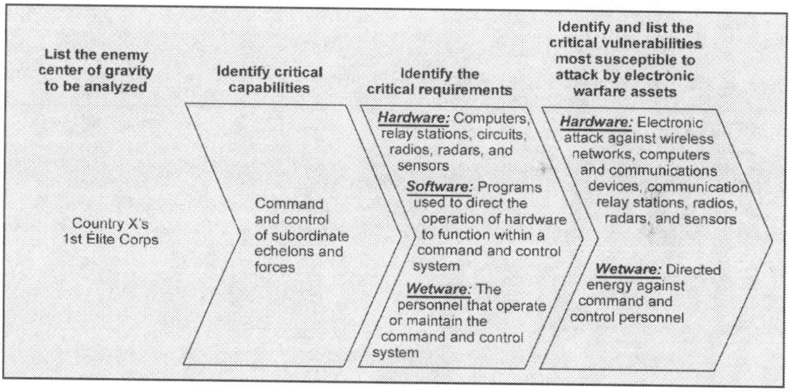

Figure 4-2. Example of analysis for an enemy center of gravity

4-11. Additionally, the EW officer and EW working group members determine how EW can help protect friendly centers of gravity. The center of gravity analysis process as outlined in figure 4-2 can also be used help identify critical vulnerabilities of friendly centers of gravity. The EW officer and EW working group members list the critical requirements associated with the identified friendly command and control critical capability. Then, the EW officer and EW working group members identify the critical vulnerabilities associated with the critical requirements. These vulnerabilities can help determine how to best use EW capabilities to defend or protect friendly centers of gravity from enemy attack. Key to this portion of the analysis is to assess the potential impact of EW operations on friendly information systems such as electromagnetic interference.

4-12. The EW officer and EW working group members identify and list—

- High-value targets that can be engaged by EW capabilities.
- Tasks that EW forces perform according to EW subdivision (electronic attack, electronic warfare support, and electronic protection) in support of the warfighting functions. These include—
  - Determining specified EW tasks.
  - Determining implied EW tasks.

4-13. The EW officer and EW working group members—

- Conduct initial EW force structure analysis to determine if sufficient assets are available to perform the identified EW tasks. (If organic assets are insufficient, they draft requests for support and augmentation.)
- Conduct an initial EW risk assessment and review the risk assessment done by the entire working group.
- Provide EW perspective in the development of the commander's restated mission.
- Assist in development of the mission analysis briefing for the commander.

4-14. By the conclusion of mission analysis, the EW officer and EW working group members generate or gather the following products and information:

- The initial information requirements for EW operations.
- A rudimentary command and control nodal analysis of the enemy.
- The list of EW tasks required to support the mission.
- A list of assumptions and constraints related to EW operations.
- The planning guidance for EW operations.
- EW personnel augmentation or support requirements.
- An update of the EW running estimate.
- EW portion or input to the commander's restated mission.

COURSE OF ACTION DEVELOPMENT

4-15. After receiving the restated mission, commander's intent, and commander's planning guidance, the staff develops courses of action (COAs) for the commander's approval. Figure 4-3 depicts the required input to COA development and identifies the key contributions made by the EW officer and EW working group members during the process and output stages (center and right of figure 4-3). The actions the EW officer and EW working group members perform to support COA development are discussed in more detail in paragraphs 4-16 through 4-20.

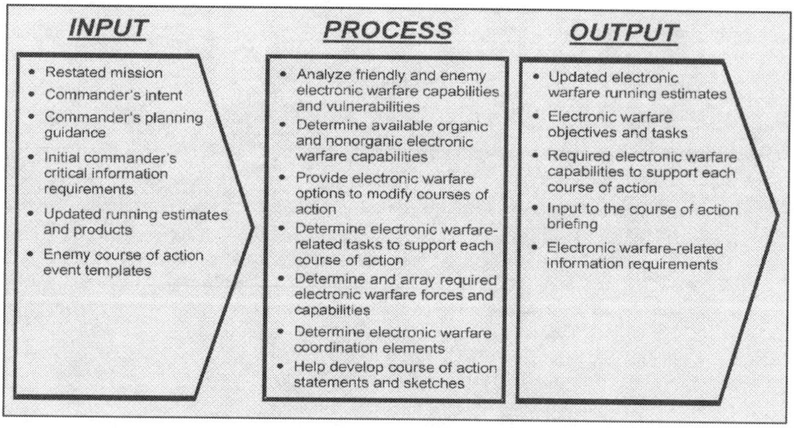

Figure 4-3. Course of action development

4-16. The EW officer and EW working group members contribute to COA development through the following planning actions—

- Determining which friendly EW capabilities are available to support the operation, including organic and nonorganic capabilities for planning.
- Determining possible friendly and enemy EW operations, including identifying friendly and enemy vulnerabilities.

4-17. Additionally, the EW officer and EW working group members help develop initial COA options by—

- Identifying COA options that may be feasible based on their functional expertise (while brainstorming of COAs).
- Providing options to modify a COA to enable accomplishing a requirement within the EW area of expertise.
- Identifying information (relating to EW options) that may impact other functional areas and sharing that information immediately.
- Identifying the EW-related tasks required to support the COA options.

4-18. The EW officer and EW working group members determine the forces required for mission accomplishment by—

- Determining the EW tasks that support each COA and how to perform those tasks based on available forces and capabilities. (Available special technical operations capabilities are considered in this analysis.)
- Providing input and support to proposed deception options.
- Ensuring the EW options provided in support of all possible COAs meet the established screening criteria.

4-19. The EW officer and EW working group members identify EW supporting tasks and their purpose in supporting any decisive, shaping, and sustaining operations as each COA is developed. These EW tasks include those—

- Focused on defeating the enemy.
- Required to protect friendly force operations.

4-20. The EW officer and EW working group members assist in developing the COA briefing as required. By the conclusion of COA development, the EW officer and EW working group members generate or gather the following products and information:

- A list of EW objectives and desired effects related to the EW tasks.
- A list of EW capabilities required to perform the stated EW tasks for each COA.
- The information and intelligence requirements for performing the EW tasks in support of each COA.
- An update to the EW running estimate.

COURSE OF ACTION ANALYSIS (WAR-GAMING)

4-21. The COA analysis allows the staff to synchronize the elements of combat power for each COA and to identify the COA that best accomplishes the mission. It helps the commander and staff to—

- Determine how to maximize the effects of combat power while protecting friendly forces and minimizing collateral damage.
- Further develop a visualization of the battle.
- Anticipate battlefield events.
- Determine conditions and resources required for success.
- Determine when and where to apply force capabilities.
- Focus IPB on enemy strengths and weaknesses as well as the desired end state.
- Identify coordination needed to produce synchronized results.
- Determine the most flexible COA.

Paragraphs 4-22 to 4-23 discuss actions the EW officer and EW working group members perform to support COA analysis. (See FM 5-0 for more information on war-gaming.)

4-22. During COA analysis, the EW officer and EW working group members synchronize EW actions and assist the staff in integrating EW capabilities into each COA. The EW officer and EW working group members address how each EW capability supports each COA. They apply these capabilities to associated time lines, critical events, and decision points in the synchronization matrix (see table 4-1). During this planning phase, the EW officer and EW working group members aim to—

- Analyze each COA from an EW functional perspective.
- Recommend any EW task organization adjustments.
- Identify key EW decision points.
- Provide EW data for synchronization matrix.
- Recommend EW priority intelligence requirements.
- Identify EW supporting tasks to any branches and sequels.
- Identify potential EW high-value targets.
- Assess EW risks created by telegraphing intentions, allowing time for enemy to mitigate effects, unintended effects of electronic attack, and the impact of asset or capability shortfalls.

4-23. By the conclusion of COA analysis (war-gaming), the EW officer and EW working group members generate or gather the following products and information:

- The EW data for the synchronization matrix.
- The EW portion of the branches and sequels.
- A list of high-value targets related to EW.
- A list of commander's critical information requirements related to EW.
- The risk assessment for EW operations in support of each COA.
- An update to the EW running estimate.

Table 4-1. Sample input to synchronization matrix

TIME/EVENT		H - 8	H - hour	H + 8
Enemy Actions		Enemy monitors movements	Defends from Security Zone	Commits reserve
Decision Points		Launch deep attack		
MANEUVER	1st Brigade	Move on route Paula	Cross line of departure	Seize objective Nick
MANEUVER	2d Brigade	Move on route Mike	Cross line of departure	Seize objective Dave
MANEUVER	3rd Brigade	Move on route Sean		Forward passage of lines with 1st Brigade
MANEUVER	Aviation Brigade	Deep attack on objective Rose at H - 1		
MANEUVER	Division Cavalry		Screen northern flank	
FIRES	Fires Brigade	Preparation fires initiated at H - 5		
FIRES	Air Defense	Weapons hold	Weapons tight	Weapons tight
FIRES	C2W - EA - CNA - Physical Attack - CNE - ES	- Locate enemy ISR on maneuver routes - Deny and disrupt enemy ISR of maneuver routes at H - .5 to H - hour - Disrupt and destroy known enemy C2 nodes and IADS	- Activate CREW systems - Jamming (to disrupt/deny enemy C2 nodes) - Electronic deception - Provide indications and warnings to maneuver brigades	Disrupt and destroy enemy C2 system
C2			Tactical CP with lead brigade	

C2	command and control	EA	electronic attack
C2W	command and control warfare	ES	electronic warfare support
CNA	computer network attack	EW	electronic warfare
CNE	computer network exploitation	H-hour	specific time an operation or exercise begins
CP	command post	IADS	integrated air defense system
CREW	counter radio-controlled improvised explosive device electronic warfare	ISR	intelligence, surveillance, and reconnaissance

Note: This is not complete. Its intent is to show how EW can be integrated into a synchronization matrix.

Course of Action Comparison

4-24. COA comparison starts with all staff members analyzing and evaluating the advantages and disadvantages of each COA from their perspectives. Staff members present their findings for the others' consideration. Using the evaluation criteria developed during COA analysis, the staff outlines each COA, highlighting its advantages and disadvantages. Comparing the strengths and weaknesses of the COAs identifies their advantages and disadvantages with respect to each other. (See FM 5-0 for further discussion of COA comparison).

4-25. During COA comparison, the EW officer and EW working group members compare COAs based on the EW-related advantages and disadvantages (see center of figure 4-4). Typically, planners use a matrix to assist in the COA comparisons. The EW officer may develop an EW functional matrix to compare the COAs or to use the decision matrix developed by the staff. Regardless of the matrix used, the evaluation criteria developed prior to war-gaming are used to compare the COAs. Normally, the chief of staff or executive officer weights each criterion used for the evaluation based on its relative importance and the commander's guidance. (See FM 5-0 for more information on COA comparison and a sample decision matrix.)

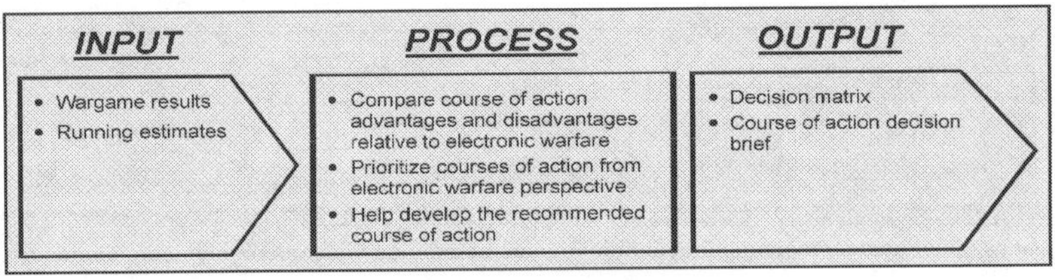

Figure 4-4. Course of action comparison

4-26. By the conclusion of COA comparison, the EW officer and EW working group members generate or gather the following products and information:

- A list of the pros and cons for each COA relative to EW.
- A prioritized list of the COAs from an EW perspective.
- An update to the EW running estimate if required.

Course of Action Approval

4-27. The COA approval process has three components. First, the staff recommends a COA, usually in a decision briefing. Second, the commander decides which COA to approve. Lastly, the commander issues the final planning guidance.

4-28. During COA approval, the EW officer supports the development of the COA decision briefing and the development of the warning order as required. If possible, the EW officer attends the COA decision briefing to receive the commander's final planning guidance. If unable to attend the briefing, the EW officer receives the final planning guidance from the G-3 or S-3. The final planning guidance is critical in that it normally provides—

- Refined commander's intent.
- New commander's critical information requirements to support the execution of the chosen COA.
- Risk acceptance.
- Guidance on priorities for the elements of combat power, orders preparation, rehearsal, and preparation.

4-29. After the COA decision has been made, the EW officer and EW working group members generate or gather the following products and information:

- An updated command and control nodal analysis of the enemy relevant to the selected COA.
- Required requests for information to refine the enemy command and control nodal architecture.
- Latest electronic order of battle tailored to the selected COA.
- Any new direction provided in the refined commander's intent.
- A list of any new commander's critical information requirements that can be used in support of EW operations.
- The warning order to assist developing EW operations required to support the operation order or plan.
- Refined input to the initial intelligence, surveillance, and reconnaissance (ISR) plan, including—
  - Any additional specific EW information requirements.
  - Updated potential collection assets for the unit's ISR plan.

ORDERS PRODUCTION

4-30. Orders production consists of the staff preparing the operation order or plan by converting the selected COA into a clear, concise concept of operations. The staff also provides supporting information that enables subordinates to execute and implement risk controls. They do this by coordinating and integrating risk controls into the appropriate paragraphs and graphics of the order.

4-31. During orders production, the EW officer provides the EW operations input for several sections of the operation order or plan. See appendix B for the primary areas for EW operations input within an Army order or plan. The primary areas for EW input in a joint order, if required, also are shown in appendix B. (See CJCSM 3122.03C for the Joint Operation Planning and Execution System format).

DECISIONMAKING IN A TIME-CONSTRAINED ENVIRONMENT

4-32. In a time-constrained environment, the staff might not be able to conduct a detailed MDMP. The staff may choose to abbreviate the process as described in FM 5-0. The abbreviated process uses all seven steps of the MDMP in a shortened and less detailed manner.

4-33. The EW officer and core members of the EW working group meet as a regular part of the unit battle rhythm. However, the EW officer calls unscheduled meetings if situations arise that require time-sensitive planning. Regardless of how much they abbreviate the planning process, the EW officer and supporting members of the EW working group always—

- Update the EW running estimate in terms of assets and capabilities available.
- Update essential EW tasks with the requirements of the commander's intent.
- Coordinate support requests and intelligence requirements with appropriate staff elements and outside agencies.
- Provide EW input to fragmentary orders through the G-3 or S-3 as necessary to drive timely and effective EW operations.
- Deconflict planned EW actions with other uses of the spectrum, such as communications.
- Synchronize electronic attack and EW support actions.
- Synchronize other intelligence collection in support of EW requirements.
- Deconflict EW activities specifically with aviation operations.
- Synchronize EW support to the command and control warfare and information protection information tasks.

THE INTEGRATING PROCESSES AND CONTINUING ACTIVITIES

4-34. Commanders use several integrating processes and continuing activities to synchronize operations throughout the operations process. (See figure 4-5.) The EW officer ensures EW operations are fully synchronized and integrated within these processes and continuing activities. Other staff members supporting the EW working group assist the EW officer. Paragraphs 4-35 through 4-52 outline some key integrating processes and continuing activities. These processes and activities require EW officer involvement throughout the operations process.

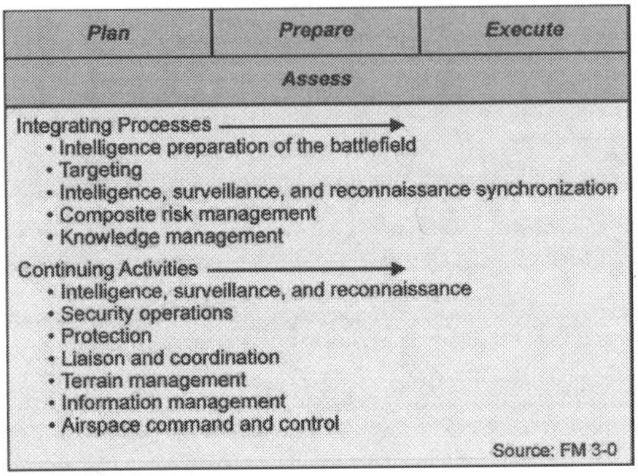

Figure 4-5. Integrating processes and continuing activities

INTELLIGENCE PREPARATION OF THE BATTLEFIELD

4-35. *Intelligence preparation of the battlefield* is the systematic, continuous process of analyzing the threat and environment in a specific geographic area. Intelligence preparation of the battlefield is designed to support the staff estimate and military decisionmaking process. Most intelligence requirements are generated as a result of the intelligence preparation of the battlefield process and its interrelation with the decisionmaking process (FM 34-130). The G-2 or S-2 leads IPB planning with participation by the entire staff. This planning activity is used to define and understand the operational environment and the options it presents to friendly and adversary forces. Only one IPB planning activity exists within each headquarters; all affected staff cells participate. (FM 2-0 provides more information on IPB.) Paragraphs 4-36 through 4-40 discuss how the EW officer and the EW working group support IPB during operations.

4-36. In addition to the input provided to the initial IPB (during step 2 of mission analysis), the EW officer supports IPB throughout the operations process by providing input related to EW operations. (See figure 4-6.) This input includes (but is not limited to) the following EW considerations:

- Evaluating the operational environment from an EW perspective.
- Describing how the effects of the operational environment may impact EW operations.
- Evaluating the threat's capabilities; doctrinal principles; and tactics, techniques, and procedures from an EW perspective.
- Determining threat COAs.

4-37. When evaluating the operational environment from an EW perspective, the EW officer—

- Determines the electromagnetic environment within the defined physical environment:
  - Area of operations.
  - Area of influence.
  - Area of interest.
- Uses electronic databases to identify gaps.
- Identifies adversary fixed EW sites such as EW support and electronic attack sites.
- Identifies airfields and installations that support, operate, or house adversary EW capabilities.
- In coordination with the G-2 or S-2 and G-6 or S-6, helps identify enemy electromagnetic spectrum usage and requirements within the area of operations and area of interest.

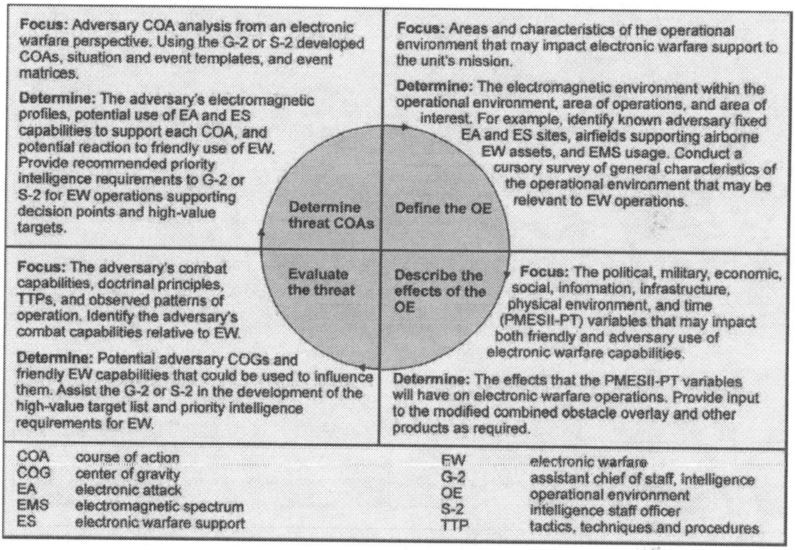

Figure 4-6. Electronic warfare support to intelligence preparation of the battlefield

4-38. When describing how the variables of the operational environment may impact EW operations, the EW officer—

- Focuses on characteristics of both the land and air domains using the factors of observation and fields of fire, avenues of approach, key and decisive terrain, obstacles, and cover and concealment.
- Identifies key terrain that may provide protection for communications and target acquisition systems from exploitation or disruption.
- Identifies how terrain affects line of sight, including effects on both communications and non-communications emitters.
- Evaluates how vegetation affects radio wave absorption and antenna height requirements.
- Locates power lines and their potential to interfere with radio waves.
- Assesses most likely and most dangerous avenues of approach (air, ground) and where EW operations would likely be positioned to support these approaches.

- If operating within urban terrain, considers how the infrastructure—power plants, power grids, structural heights, and communications and media nodes—may restrict or limit EW capabilities.
- Assists the G-2 or S-2 with the development of a modified combined obstacle overlay.
- Determines how weather—visibility, cloud cover, rain, and wind—may affect ground-based and airborne EW operations and capabilities (for example, no-go weather conditions at an airborne EW launch and recovery base).
- Considers all other relevant aspects of the operational environment that affect EW operations, using the operational variables (PMESII-PT—political, military, economic, social, information, infrastructure, physical environment, and time) and mission variables (METT-TC—mission, enemy, terrain and weather, troops and support available, time available, and civil considerations).

4-39. When evaluating enemy capabilities, the EW officer and supporting staff examine doctrinal principles; tactics, techniques and procedures; and observed patterns of operation from an EW perspective. The EW officer—

- Uses the operational variables (PMESII-PT) and mission variables (METT-TC) to help determine the adversary's critical nodes.
- Collects the required data—operational net assessments, electronic order of battle, and electronic databases—to template the command and control critical nodes and the systems required to support and maintain them.
- Assists the G-2 in determining the adversary's EW-related threat characteristics (order of battle) by identifying—
  - Types of communications equipment available.
  - Types of noncommunications emitters.
  - Surveillance and target acquisition assets.
  - Technological sophistication of the threat.
  - Communications network structure.
  - Frequency allocation techniques.
  - Operation schedules.
  - Station identification methods.
  - Measurable characteristics of communications and noncommunications equipment.
  - Command, control, and communications structure of the threat.
  - Tactics from a communication perspective. Examples are how the enemy deploys command, control, and communications assets; whether or not communications systems are remote; and the level of discipline in procedures, communications security, and operations security.
  - Electronic deception capabilities.
  - Reliance on active or passive surveillance systems
  - Electromagnetic profiles of each node.
  - Unique electromagnetic spectrum signatures.
- Assists the G-2 or S-2 in center of gravity analysis. Helps identify the critical system nodes of the center of gravity and determines what aspects of the system should be engaged, exploited, or attacked to modify the system's behavior or to achieve a desired effect.
- Identifies organic and nonorganic EW capabilities available to achieve desired effects on identified high-value targets.
- Submits initial EW-related requests for information that describe the intelligence support required to support EW operations.
- Obtains the high-value target list, threat templates, and initial priority intelligence requirements list to assist in follow-on EW planning.

4-40. When determining adversary COAs, the EW officer—

- Assists the G-2 or S-2 in development of adversary COAs.
- Provides EW input to the situation templates.
- Ensures event templates include EW named areas of interests.
- Assists in providing EW options for target areas of interest.
- Assists in providing EW options to support decision points.
- Provides EW input to the event template and event matrix.

TARGETING

4-41. *Targeting* is the process of selecting and prioritizing targets and matching the appropriate response to them, considering operational requirements and capabilities (JP 3-0). A decide, detect, deliver, and assess methodology is used to direct friendly forces to attack the right target with the right asset at the right time. (See figure 4-7.) Targeting provides an effective method to match the friendly force capabilities against targets. Commander's intent plays a critical role in the targeting process. The targeting working group strives to thoroughly understand the commander's intent to ensure the commander's intended effects on targets are achieved.

4-42. An important part of targeting is identifying potential fratricide situations and performing the coordination measures to manage and control the targeting effort positively. The targeting working group and staff incorporate these measures into the coordinating instructions and appropriate annexes of the operation plans and orders. (FM 6-20-10 has more information on targeting.)

DECIDE		DETECT	
Determine	**Based on**	**Determine**	**Based on**
<u>What (task)</u>: Enemy focused. Determine what EW tasks are essential to the success of the operation (enemy formation or function to influence, and desired targeting effect). <u>Why (purpose)</u>: Friendly focused. Determine the purpose for the use of EA fires (for example, to clear transit routes for maneuvering forces).	• Commanders initial planning guidance • Mission analysis - Specified and implied tasks - Intelligence preparation of the battlefield + target value analysis = enemy courses of action and high-value targets • Commander's intent	<u>Who/Where</u>: Focused on detection. Assets are deployed to detect high-payoff targets. ES collection assets identify and locate targets that can be influenced by EA. Once targets are identified, EA fires can be used to influence the targets based on the identified weaknesses by the target assessment and SIGINT teams.	• COA development • Concept of fires • War-gaming • COA decision • Scheme of fires • High-payoff target list and attack guidance matrix • Reconnaissance and surveillance plan
ASSESS		**DELIVER**	
Determine	**Based on**	**Determine**	**Based on**
<u>Effect</u>: Identifies whether the intended effect achieved by the EA fires was successful or not.	• Operations plans or orders • EW task execution • Effects of EA fires • BDA • MOE • Target assessment teams • SIGINT team assessment	<u>Who/When</u>: Focused on delivery. Addresses the who and when portion of task (such as the jamming of a designated target and the duration desired).	• COA development • Concept of fires • War-gaming • COA decision • Scheme of fires • High-payoff target list and attack guidance matrix • Reconnaissance and surveillance plan

BDA battle damage assessment
COA course of action
EA electronic attack
ES electronic warfare support
EW electronic warfare
MOE measures of effectiveness
SIGINT signals intelligence

Figure 4-7. Electronic warfare in the targeting process

4-43. The EW officer thoroughly integrates electronic attack in the targeting process and integrates electronic attack fires into all appropriate portions of the operation plan, operation order, and other planning products. In support of EW targeting, the EW officer—

- Helps the targeting working group determine electronic attack requirements against specific high-payoff targets and high-value targets.
- Ensures electronic attack can meet the desired effect (in terms of the targeting objective).
- Coordinates with the signals intelligence staff element through the collection manager to satisfy EW support and electronic attack information requirements.
- Prepares the EW tab and the EW portion of the command and control warfare tab to the fires appendix.
- Provides electronic attack mission management through the tactical operations center or joint operations center and the tactical air control party (for airborne electronic attack).
- Provides electronic attack mission management as the jamming control authority for ground or airborne electronic attack when designated.
- Prepares and coordinates the EW annex for operation plans and operation orders.
- Determines and requests theater Army electronic attack support.
- Recommends to the G-3 or S-3 and the fire support coordinator or fire support officer whether to engage a target with electronic attack.
- Expedites electromagnetic interference reports to the targeting working group. (See appendix D for information on electromagnetic interference reporting.)

Decide

4-44. Decide is the first step in the targeting process. This step provides the overall focus for fires, a targeting plan, and some of the priorities for intelligence collection. As part of the staff in the fires cell, the EW officer assists the targeting working group in planning the target priorities for each phase and critical events of the operation. Initially, the targeting working group does not develop electronic attack targets using any special technique or separately from targets for physical destruction. However, as the process continues, these targets are passed through intelligence organizations and further planned using ISR procedures. The planned use of electronic attack is integrated into the standard targeting products (graphic or text-based). Products that involve electronic attack planning may include—

- High-payoff target list.
- Attack guidance matrix.
- Appendix 4 (Electronic Warfare) to Annex P (Information Operations) of the operation order. (At the time this manual was written, this was the current doctrine for operation orders. This appendix will be revised upon publication of the revised FM 5-0.)

Detect

4-45. Based on what the targeting working group identified as high-payoff targets during the decide step, collection assets are then deployed to detect them. The intelligence enterprise pairs assets to targets based on the collection plan and the current threat situation. When conducting electronic attack operations in support of command and control warfare, ISR units perform EW support tasks linked to and working closely with the electronic attack missions. Electronic warfare support units (with support from the target assessment and signals intelligence staff elements) provide the data—location, signal strength, and frequency of the target—to focus electronic attack assets on the intended target. These assets also identify the command and control system vulnerabilities open to attack by electronic attack assets.

Deliver

4-46. Once friendly force capabilities identify, locate, and track the high-payoff targets, the next step in the process is to deliver fires against those targets. Electronic attack assets must satisfy the attack guidance developed during the decide step. Close coordination between those conducting EW support and electronic attack is critical during the engagement. The EW officer facilitates this coordination and ensures electronic attack fires are fully synchronized and deconflicted with other fires. The EW officer remains aware of the potential for unintended effects between adjacent units when conducting electronic attack. The EW officer continually coordinates with adjacent unit EW officers to mitigate and deconflict these effects during cross-boundary operations. Normally, the G-3, S-3, or fire support coordinator provides requirements and guidance for this coordination and synchronization in the attack guidance matrix, intelligence synchronization matrix, spectrum management plan, and the EW input to the operation plan or operation order annexes and appendixes.

Assess

4-47. Once the target as been engaged, the next step is to assess the engagement's effectiveness. This is done through combat assessment, which involves determining the effectiveness of force employment during military operations. It consists of three elements:

- Munitions effects assessment.
- Battle damage assessment.
- Re-attack recommendations.

4-48. The first two elements, munitions effects assessment and battle damage assessment, are used to inform the commander on the effects achieved against targets and target sets. From this information, the G-2 or S-2 continues to analyze the threat's ability to further conduct and sustain combat operations (sometimes articulated in terms of the effects achieved against the threat's centers of gravity). The last element involves the assessment and recommendation whether or not to re-attack the targets.

4-49. The assessment of a jamming mission used against an enemy's command and control system is unlike fires that can be observed visually. The signals intelligence staff element and units executing the electronic attack mission coordinate continuously to assess mission effectiveness. Close coordination between sensor and shooter allows instant feedback on the success or failure of the intended jamming effects. It also can quickly provide the necessary adjustments to produce desired effects.

Intelligence, Surveillance, and Reconnaissance Synchronization

4-50. *Intelligence, surveillance, and reconnaissance synchronization* is the task that accomplishes the following: analyzes information requirements and intelligence gaps; evaluates available assets internal and external to the organization; determines gaps in the use of those assets; recommends intelligence, surveillance, and reconnaissance assets controlled by the organization to collect on the commander's critical information requirements; and submits requests for information for adjacent and higher collection support (FM 3-0). ISR synchronization considers all assets—both internal and external to the organization. It identifies information gaps and the most appropriate assets for collecting information to fill them.

4-51. Planning for ISR operations begins during mission analysis. Although led by the G-3 or S-3, it is supported by the entire staff, subordinate units, and external partners. ISR operations collect, process, store, display, and disseminate information from a multitude of collection sources. The staff thoroughly understands, integrates, and synchronizes the ISR plan across all echelons.

4-52. The EW officer ensures the ISR plan supports the EW-related information requirements determined during the planning process. The EW officer coordinates these requirements with the signals intelligence staff element through the G-2 or S-2.

EMPLOYMENT CONSIDERATIONS

4-53. EW has specific ground-based, airborne, and functional (electronic attack, electronic warfare support, or electronic protection) employment considerations. The EW officer ensures EW-related employment considerations are properly articulated early in the operations process. Each capability employed has certain advantages and disadvantages. The staff plans for all of these before executing EW operations.

Ground-Based Electronic Warfare Considerations

4-54. Ground-based EW capabilities support the commander's scheme of maneuver. Ground-based EW equipment can be employed by a dismounted Soldier or on highly mobile platforms. Due to the short-range nature of tactical signals direction finding, electronic attack assets are normally located in the forward areas of the battlefield, with or near forward units.

4-55. Ground-based EW capabilities have certain advantages. They provide direct support to maneuver units (for example, through counter-radio-controlled improvised-explosive-device EW and communications or sensor jamming). Ground-based EW capabilities support continuous operations and respond quickly to EW requirements of the ground commander. However, to maximize the effectiveness of ground-based EW capabilities, maneuver units must protect EW assets from enemy ground and aviation threats. EW equipment should be as survivable and mobile as the force it supports. Maneuver units must logistically support the EW assets, and supported commanders must clearly identify EW requirements.

4-56. Ground-based EW capabilities have certain limitations. They are vulnerable to enemy attack and can be masked by terrain. They are vulnerable to enemy electromagnetic deceptive measures and electronic protection actions. In addition, they have distance or propagation limitations against enemy electronic systems.

Airborne Electronic Warfare Considerations

4-57. While ground-based and airborne EW planning and execution are similar, they significantly differ in their EW employment time. Airborne EW operations are conducted at much higher speeds and generally have a shorter duration than ground-based operations. Therefore, the timing of airborne EW support requires detailed planning.

4-58. Airborne EW requires the following:

- A clear understanding of the supported commander's EW objectives.
- Detailed planning and integration.
- Ground support facilities.
- Liaisons between the aircrews of the aircraft providing the EW support and the aircrews or ground forces being supported.
- Protection from enemy aircraft and air defense systems.

4-59. Airborne EW capabilities have certain advantages. They can provide direct support to other tactical aviation missions such as suppression of enemy air defenses, destruction of enemy air defenses, and employment of high-speed antiradiation missiles. They can provide extended range over ground-based assets. Airborne EW capabilities can provide greater mobility and flexibility than ground-based assets. In addition, they can support ground-based units in beyond line-of-sight operations.

4-60. The limitations associated with airborne EW capabilities are time-on-station considerations, vulnerability to enemy electronic protection actions, electromagnetic deception techniques, and limited assets (support from nonorganic EW platforms need to be requested).

Electronic Attack Considerations

4-61. Electronic attack includes both offensive and defensive activities. (Chapter 1 provides a full definition of electronic attack). These activities differ in their purpose. Defensive electronic attack protects friendly personnel and equipment or platforms. Offensive electronic attack denies, disrupts, or destroys enemy capability. In either case, certain considerations are involved in planning for employing electronic attack:

- Friendly communications.
- Intelligence collection.
- Other effects.
- Nonhostile local electromagnetic spectrum use.

- Hostile intelligence collection.
- Persistency of effect.

4-62. The EW officer, the G-2 or S-2, the G-3 or S-3, the G-6 or S-6, the spectrum manager, and the G-7 or S-7 coordinate closely to avoid friendly communications interference that can occur when using EW systems on the battlefield. Coordination ensures that electronic attack systems frequencies are properly deconflicted with friendly communications and intelligence systems or that ground maneuver and friendly information tasks are modified accordingly.

4-63. The number of information systems, EW systems, and sensors operating simultaneously on the battlefield makes deconfliction with communications systems a challenge. The EW officer, the G-2 or S-2, the G-6 or S-6, and the spectrum manager plan and rehearse deconfliction procedures to quickly adjust their use of EW or communications systems.

4-64. Electronic attack operations depend on EW support and signals intelligence to provide targeting information and battle damage assessment. However, EW officers must keep in mind that not all intelligence collection is focused on supporting EW. If not properly coordinated with the G-2 or S-2 staff, electronic attack operations may impact intelligence collection by jamming or inadvertently interfering with a particular frequency being used to collect data on the threat, or by jamming a given enemy frequency or system that deprives friendly forces of that means of collecting data. Either can significantly deter intelligence collection efforts and their ability to answer critical information requirements. Coordination between the EW officer, the fire support coordinator, and the G-2 or S-2 is prevents this interference. In situations where a known conflict between the intelligence collection effort and the use of electronic attack exists, the EW working group brings the problem to the G-3 or S-3 for resolution.

4-65. Other forms of effects rely on electromagnetic spectrum. For example, psychological operations may plan to use a given set of frequencies to broadcast messages, or a military deception plan may include the broadcast of friendly force communications. In both examples, the use of electronic attack could unintentionally interfere or disrupt such broadcasts if not properly coordinated. To ensure electronic attack does not negatively impact planned operations, the EW officer coordinates between fires, network operations, and other functional or integrating cells as required.

4-66. Like any other form of electromagnetic radiation, electronic attack can adversely affect local media and communications systems and infrastructure. EW planners consider unintended consequences of EW operations and deconflict these operations with the various functional or integrating cells. For example, friendly jamming could potentially deny the functioning of essential services such as ambulance or fire fighters to a local population. EW officers routinely synchronize electronic attack with the other functional or integrating cells responsible for the information tasks. In this way, they ensure that electronic attack efforts do not cause fratricide or unacceptable collateral damage to their intended effects.

4-67. The potential for hostile intelligence collection also affects electronic attack. A well-equipped enemy can detect friendly EW capabilities and thus gain intelligence on friendly force intentions. For example, the frequencies Army forces jam could indicate where they believe the enemy's capabilities lie. The EW officer and the G-2 or S-2 develop an understanding of the enemy's collection capability. Along with the red team (if available), they determine what the enemy might gain from friendly force use of electronic attack. (A *red team* is an organizational element comprised of trained and educated members that provide an independent capability to fully explore alternatives in plans and operations in the context of the operational environment and from the perspective of adversaries and others. [JP 2-0])

4-68. The effects of jamming only persist as long as the jammer itself is emitting and is in range to affect the target. Normally this time frame is a matter of seconds or minutes, which makes the timing of such missions critical. This is particularly true when jamming is used in direct support of aviation platforms. For example, in a mission that supports suppression of enemy air defense, the time on target and duration of the jamming must account for the speed of attack of the aviation platform. They must also account for the potential reaction time of enemy air defensive countermeasures. The development of directed-energy weapons may change this dynamic in the future. However, at present (aside from antiradiation missiles), the effects of jamming are less persistent than effects achieved by other means.

Electronic Protection Considerations

4-69. Electronic protection is achieved through physical security, communications security measures, system technical capabilities (such as frequency hopping and shielding of electronics), spectrum management, and emission control procedures. The EW officer and EW working group members must consider the following key functions when planning for electronic protection operations:

- Vulnerability analysis and assessment.
- Monitoring and feedback.
- Electronic protection measures and how they affect friendly capabilities.

Vulnerability Analysis and Assessment

4-70. Vulnerability analysis and assessment forms the basis for formulating electronic protection plans. The Defense Information Systems Agency operates the Vulnerability Analysis and Assessment Program, which specifically focuses on automated information systems and can be very useful in this effort.

Monitoring and Feedback

4-71. The National Security Agency monitors communications security. Their programs focus on telecommunications systems using wire and electronic communications. Their programs can support and remediate the command's communications security procedures when required.

Electronic Protection Measures and Their Effect on Friendly Capabilities

4-72. Electronic protection measures include any measure taken to protect the force from hostile electronic attack actions. However, these measures can also limit friendly capabilities or operations. For example, denying frequency usage to counter-radio-controlled improvised-explosive-device EW systems on a given frequency to preserve it for a critical friendly information system could leave friendly forces vulnerable to certain radio-controlled improvised explosive devices. The EW officer and the G-6 or S-6 carefully consider these second-order effects when advising the G-3 or S-3 regarding electronic protection measures.

Electronic Warfare Support Considerations

4-73. The distinction between whether a given asset is performing a signals intelligence or EW support mission is determined by who tasks and controls the assets, what they are tasked to provide, and the purpose for which they are tasked. Operational commanders task assets to conduct EW support for the purpose of immediate threat recognition, targeting, planning the conduct of future operations, and other tactical actions (such as threat avoidance and homing). The EW officer coordinates with the G-2 or S-2 to ensure all EW support needed for planned EW operations is identified and submitted to the G-3 or S-3 for approval by the commander. This ensures that the required collection assets are properly tasked to provide the EW support. In cases where planned electronic attack actions may conflict with the G-2 or S-2 intelligence collection efforts, the G-3, S-3, or commander decides which has priority. The EW officer and the G-2 or S-2 develop a structured process within each echelon for conducting this intelligence gain-loss calculus during mission rehearsal exercises and predeployment work-ups.

Electronic Warfare Reprogramming Considerations

4-74. Electronic warfare reprogramming refers to modifying friendly EW or target sensing systems in response to validated changes in enemy equipment and tactics or the electromagnetic environment. (See paragraph 1-40 for the complete definition.) Reprogramming EW and target sensing system equipment falls under the responsibility of each Service or organization through its respective EW reprogramming support programs. It includes changes to self-defense systems, offensive weapons systems, and intelligence collection systems. During joint operations, swift identification and reprogramming efforts are critical in a rapidly evolving hostile situation. The key consideration for EW reprogramming is joint coordination. Joint coordination of Service reprogramming efforts ensures reprogramming requirements are identified, processed, and implemented consistently by all friendly forces. During joint operations, EW reprogramming coordination and monitoring is the responsibility of the joint force commander's EW staff. (For more information on EW reprogramming, see FM 3-13.10).

SECTION II — ELECTRONIC WARFARE PREPARATION

4-75. *Preparation* consists of activities performed by units to improve their ability to execute an operation. Preparation includes, but is not limited to, plan refinement; rehearsals; intelligence, surveillance, and reconnaissance; coordination; inspections; and movement (FM 3-0). Preparation creates conditions that improve friendly forces' opportunities for success. It facilitates and sustains transitions, including those to branches and sequels.

4-76. During preparation, the EW officer and members of the EW working group focus their actions on the following activities:

- Revising and refining the EW estimate, EW tasks supporting command and control warfare, and EW support to the overall plan.
- Rehearsing the synchronization of EW support to the plan (including integration into the targeting process, request procedures for joint assets, deconfliction procedures, and asset determination and refinement).
- Synchronizing the collection plan and intelligence synchronization matrix with the attack guidance matrix and EW input to the operation plan or order annexes and appendixes.
- Assessing the planned task organization developed to support EW operations, including liaison officers and organic and nonorganic capabilities required by echelon.
- Coordinating procedures with ISR operational elements (such as signals intelligence staff elements).
- Training the supporting staff members of the EW working group during mission rehearsal exercises.
- Completing precombat checks and inspections of EW assets.
- Completing sustainment preparations for EW assets.
- Coordinate with the G-4 or S-4 to develop EW equipment reporting formats.
- Completing briefbacks by subordinate EW working groups on planned EW operations.
- Refining content and format for the EW officer's portion of the battle update assessment and brief.

SECTION III — ELECTRONIC WARFARE EXECUTION

4-77. *Execution* is putting the plan into action by applying combat power to accomplish the mission and using situational understanding to assess progress and make execution and adjustment decisions (FM 3-0). Commanders focus their subordinates on executing the concept of operations by issuing their intent and mission orders.

4-78. During execution, the EW officer and EW working group members—

- Serve as the EW expert for the commander.
- Maintain the running estimate for EW operations.
- Monitor EW operations and recommend adjustments during execution.
- Recommend adjustments to the commander's critical information requirements based on the situation.
- Recommend adjustments to EW-related control measures and procedures.
- Maintain direct liaison with the fires and network operations cells and the command and control warfare working group (if formed) to ensure integration and deconfliction of EW operations.
- Coordinate and manage EW taskings to subordinate units or assets.
- Coordinate requests for nonorganic EW support.

- Continue to assist the targeting working group in target development and recommend targets for attack by electronic attack assets.
- Receive, process, and coordinate subordinate requests for EW support during operations.
- Receive and process immediate support requests for suppression of enemy air defense or EW from joint or multinational forces; coordinate through fire support officer and fire support coordinator with the battlefield coordination detachment and joint or multinational liaisons for support request.
- Coordinate with airspace control section on all suppression of enemy air defense or EW missions.
- Provide input to the overall assessment regarding effectiveness of electronic attack missions.
- Maintain, update, and distribute the status of EW assets.
- Validate and disseminate cease-jamming requests.
- Coordinate and expedite electromagnetic interference reports with the analysis and control element for targeting and the spectrum manager for potential deconfliction.
- Perform jamming control authority function for ground-based EW within the assigned area of operations (when designated by the jamming control authority).

SECTION IV — ELECTRONIC WARFARE ASSESSMENT

4-79. *Assessment* is the continuous monitoring and evaluation of the current situation, particularly the enemy, and progress of an operation (FM 3-0). Commanders, assisted by their staffs, continuously assess the current situation and progress of the operation and compare it with the concept of operations, mission, and commander's intent. Based on their assessment, commanders direct adjustments, ensuring that the operation remains focused on the mission and commander's intent.

4-80. As depicted in figure 4-5 (page 4-10), assessment occurs throughout every operations process activity and includes three major tasks:

- Continuously assessing the enemy's reactions and vulnerabilities.
- Continuously monitoring the situation and progress of the operation towards the commander's desired end state.
- Evaluating the operation against measures of effectiveness and measures of performance.

4-81. The EW officer and supporting members of the EW working group make assessments throughout the operations process. During planning and preparation activities, assessments of EW are made during the MDMP, IPB, targeting, ISR synchronization, and composite risk management integration.

4-82. The EW officer, in conjunction with the G-5 or S-5, helps develop the measures of performance and measures of effectiveness for evaluating EW operations during execution. A *measure of performance* is a criterion used to assess friendly actions that is tied to measuring task accomplishment (JP 3-0). A *measure of effectiveness* is a criterion used to assess changes in system behavior, capability, or operational environment that is tied to measuring the attainment of an end state, achievement of an objective, or creation of an effect (JP 3-0). In the context of EW, an example of a measure of performance is the percentage of known enemy command and control nodes targeted and attacked by electronic attack means (action) versus the number of enemy command and control nodes that were actually destroyed or rendered inoperable for the desired duration (task accomplishment). Measures of effectiveness are used to determine the degree to which an EW action achieved the desired result. This is normally measured through analysis of data collected by both active and passive means. For example, effectiveness is measured by using radar or visual systems to detect changes in enemy weapons flight and trajectory profiles.

4-83. During execution, the EW officer and members of the EW working group participate in combat assessments within the fires cell to determine the effectiveness of electronic attack employment in support of operations. Combat assessment consists of three elements: munitions effects assessment, battle damage assessment, and reattack recommendations. (Paragraphs 4-47 to 4-49 discuss combat assessment.)

SUMMARY

4-84. The EW officer and staff members supporting the EW working group ensure the successful integration of EW capabilities into operations. The EW officer leads the EW integration effort throughout the operations process. The EW officer must be familiar with and participate in the applicable integrating processes and continuing activities discussed within this chapter.

Chapter 5

Coordination, Deconfliction, and Synchronization

Once the commander approves an operation plan or order and preparations are complete, the electronic warfare officer and supporting staff turn to coordinating, deconflicting, and synchronizing the electronic warfare efforts. They ensure electronic warfare actions are carried out as planned or are modified in response to current operations. This chapter discusses major areas and activities that require continuous coordination, deconfliction, and synchronization by the electronic warfare officer and supporting staff of the electronic warfare working groups.

COORDINATION AND DECONFLICTION

5-1. A certain amount of coordination is part of the planning process. However, once a plan is approved and an operation begins, the electronic warfare (EW) staff effort shifts to the coordination and deconfliction necessary to ensure units carry out EW actions as planned or modify actions to respond to the dynamics of the operation.

5-2. The EW officer and members of the EW working group continuously monitor several key areas. These include EW coordination across organizations (higher, lower, and adjacent units), support request coordination, electromagnetic spectrum management, EW asset management, functional coordination between EW subdivisions, EW reprogramming, and EW deconfliction. Normally, EW personnel on watch in the operations center monitor and coordinate activities of these key areas. They alert the EW officer or other EW support personnel to address the required actions.

COORDINATION ACROSS ORGANIZATIONS

5-3. At the joint level, the information operations division of the J-3 performs EW coordination. The EW section of the information operations staff engages in all EW functions. This section performs peacetime contingency planning, completes day-to-day planning and monitoring of routine theater EW activities, and crisis action planning for contingencies as part of emergent joint operations. The EW section coordinates closely with other appropriate staff sections and other larger joint planning groups as required. (JP 3-13.1 discusses joint EW coordination.)

5-4. In the early stages of contingencies, the joint force commander's EW staff assesses the staffing requirements for planning and execution. This staff also coordinates EW planning and course of action development with the joint force commander's components. Services begin component EW planning and activate their EW working groups per combatant command or Service guidelines. When the scope of a contingency becomes clearer, the command EW officer may request that the joint force commander establish a joint EW coordination cell. If a joint EW coordination cell is formed, it normally requires additional augmentation from the Service or functional components. Depending on the size of the force, EW personnel from the division, corps, or theater are expected to augment the joint EW coordination cell to form a representative EW planning and execution organization. The senior Army organization's staff EW officer anticipates this requirement and prepares to support the augmentation if requested.

5-5. Coordination occurs through established EW working groups from theater level to battalion level. Within Army organizations, the coordination of EW activities occurs both horizontally and vertically. At every level, the staff EW officer ensures the necessary coordination. Normally, coordination of EW activities between the Army and joint force air component commander flows through the battlefield coordination detachment at the joint air operations center. EW staffs at higher echelons monitor EW-related activities and resolve conflicts when necessary.

5-6. Normally the senior Army headquarters (ARFOR) G-3 or S-3 coordinates with external EW organizations, unless direct liaison is authorized at lower echelons. Other components requesting Army EW support coordinate their support requirements with the EW officer located at the ARFOR headquarters or tactical operations center. Often, a liaison from the requesting organization completes these requests. If other Service or functional components have an immediate need for Army EW support, they send the request to the operational fires directorate or fires cell and the senior headquarters EW working group (sometimes referred to as an EW coordination cell) via the Global Command and Control System or Global Command and Control System-Army. In support of external EW coordination, the staff EW officer within the J-3, G-3, or S-3—

- Provides an assessment of EW capabilities to other component operation centers.
- Coordinates preplanned EW operations with other Service components (within prescribed time lines).
- Updates preplanned EW operations in coordination with other components as required.

SUPPORT REQUEST COORDINATION

5-7. Units requesting electronic attack support forward requests to the appropriate EW working group. (See appendix D for the electronic attack request format.) Each EW working group prioritizes the requests and forwards them to the higher headquarters. The commander who owns the capability when the requested support is needed approves the requests. The technical data required to support the execution of the request is passed through EW channels at the appropriate level of classification.

5-8. Electronic warfare support requests are prioritized and passed from the EW working groups through G-2 or S-2 channels and are approved by the commander who owns the capability. New EW support requests are integrated into the intelligence synchronization process. If they are approved, they appear in the intelligence synchronization plan and the unit intelligence, surveillance, and reconnaissance plan. See FMI 2-01 for details on the intelligence synchronization process. The technical data required to support EW support requests passes via signals intelligence channels within the G-2 or S-2 by classified means.

ELECTROMAGNETIC SPECTRUM MANAGEMENT

5-9. The electromagnetic spectrum is a finite resource. Once apportioned, this resource must be managed efficiently to maximize the limited spectrum allocated to support military operations. Electromagnetic spectrum operations aim to enable electronic systems to perform their functions in the intended environment without causing or experiencing unacceptable interference. Electromagnetic spectrum operations deconflict all military, national, and host-nation systems being used in the area of operations, including electronic protection systems, communications systems, sensors, and weapon systems.

5-10. Spectrum management involves planning, coordinating, and managing use of the electromagnetic spectrum through operational, engineering, and administrative procedures. Primarily, it involves determining what specific activities will occur in each part of the available spectrum. For example, some frequencies are assigned to the counter radio-controlled improvised-explosive-device EW systems operating in the area of operations. These frequencies then are deconflicted with ground tactical communications. The spectrum manager ensures all necessary functions that require use of the electromagnetic spectrum have sufficient allocation of that spectrum to accomplish their purpose. Where a conflict (two or more functions require the same portion of the spectrum) exists, the spectrum manager resolves the conflict through direct coordination. Figure 5-1 shows the basic procedures the spectrum manager follows to deconflict spectrum use.

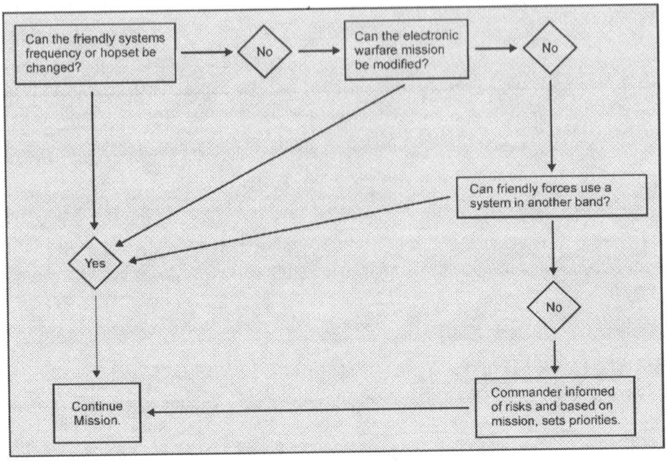

Figure 5-1. Spectrum deconfliction procedures

5-11. The spectrum manager is a member of the G-6 or S-6 section that has staff responsibility for spectrum management in the unit. The spectrum manager is a member of the unit's EW working group. Conflicts regarding spectrum use and allocation that cannot be resolved through direct coordination by the spectrum manager are referred to the G-3 or S-3 for resolution.

JAMMING CONTROL AUTHORITY

5-12. Depending on the operational situation, an Army headquarters may be designated as the jamming control authority. This authority serves as the senior jamming control authority in the area of operations. It establishes guidance for jamming on behalf of the joint force commander. If designated as the jamming control authority, the senior staff EW officer normally is tasked with the following responsibilities:

- Participating in development of and ensuring compliance with the joint restricted frequency list.
- Validating and approving or denying cease-jamming requests.
- Maintaining situational awareness of all jamming-capable systems in the area of operations.
- Acting as the joint force commander's executive agent for developing EW intelligence gain-or-loss recommendations when electronic attack or electronic warfare support conflicts occur.
- Coordinating jamming requirements with joint force components.
- Investigating unauthorized jamming events and implementing corrective measures.

See JP 3-13.1 for further information on jamming control authority.

ASSET MANAGEMENT

5-13. Regardless of echelon, the EW officer monitors and tracks the organization's EW assets and their status. The EW officer makes recommendations to the G-3 or S-3 concerning EW asset allocation and reallocation when required. The EW officer monitors and tracks EW asset status within the EW working group and reports this information to higher echelons via the Army battle command system.

OTHER COORDINATING ACTIONS

5-14. In addition to the functional considerations listed in chapter 4, several coordinating actions must also take place between the EW working groups (at all echelons) and the other planning and execution cells within the headquarters. These actions include—

- Detailed coordination between the EW activities and the intelligence activities supporting an operation.
- Coordination of EW systems reprogramming.
- Coordination with the working groups or cells coordinating the command and control warfare and information protection tasks.

Coordination Between EW Activites and Intelligence Activities

5-15. Most of the intelligence effort, before and during an operation, relies on collection activities targeted against various parts of the electromagnetic spectrum. Electronic warfare support depends on the timely collection, processing, and reporting of intelligence and combat information to alert EW operators and other military activities about intelligence collected in the electromagnetic spectrum. The EW officer and G-2 or S-2 ensure EW collection priorities and EW support collection assets are integrated into a complete intelligence collection plan. This plan ensures that units maximize the use of scarce intelligence and collection assets to support the commander's objectives.

Coordination of EW Systems Reprogramming

5-16. The EW officer and G-2, at division and corps levels, track and coordinate EW systems reprogramming input submitted by lower echelons. This input is then forwarded to the Army Service component command headquarters for submission to the Army Reprogramming Analysis Team. EW officers ensure this input is promptly submitted to ensure urgent reprogramming actions are completed for assigned systems. See FM 3-13.10 for detailed procedures for reprogramming EW and target sensing systems.

Coordination Between EW, Command and Control Warfare, and Information Tasks

5-17. EW working groups coordinate their supporting actions with the elements responsible for the Army information tasks—information engagement, command and control warfare, information protection, operations security, and military deception. Although EW plays a major role in supporting command and control warfare and information protection, it also enhances or provides direct support to other information tasks. For example, enemy radio and television broadcasts can be disrupted or replaced with friendly radio and television messages as part of larger psychological operations in support of information engagement. Electronic deception capabilities can support and enhance an overall military deception operation.

DECONFLICTION

5-18. Friendly forces depend on electromagnetic energy and the electromagnetic spectrum to sense, process, store, measure, analyze, and communicate information. This dependency creates the potential for significant interference between various friendly systems. Without proper deconfliction, interference could damage friendly capabilities or lead to operational failure. This is especially true with regard to EW systems. EW deconfliction includes—

- Friendly electromagnetic spectrum use for communications and other purposes (such as navigation systems and sensors) with electronic attack activities (such as counter-radio-controlled improvised-explosive-device EW systems).
- Electronic attack activities with electronic warfare support activities (potential electromagnetic interference of collection assets).

- Electronic attack and electronic warfare support activities with information tasks involving electromagnetic emissions (such as counter-radio-controlled improvised-explosive-device EW systems interfering with a psychological-operations radio broadcast).
- Electronic attack activities with host-nation electromagnetic spectrum users (such as commercial broadcasters, emergency first responders, and law enforcement).

5-19. The forum for deconfliction is the unit's EW working group. As such, the specific composition of the working group may expand to include more than the standard staff representation described in chapter 3. Regardless of echelon, to perform its critical deconfliction function, the EW working group retains knowledgeable representation from and ready access to decisionmakers. The EW working group also retains knowledge of and access to higher headquarters assistance and reachback capabilities available (See appendix F for more information).

SYNCHRONIZATION

5-20. EW, particularly in electronic attack, can produce both intended and unintended effects. Therefore, units thoroughly synchronize its use with other forms of fires and with friendly systems operating in the electromagnetic spectrum. Through synchronization, units avoid negative effects such as communications fratricide by jammers. The EW officer ensures all EW activities are integrated into the appropriate sections of plans—fires, information protection, command and control warfare, and military deception plans. This officer also synchronizes EW activities for maximum contribution to the commander's desired effects while preventing EW from inhibiting friendly force capabilities. The primary forum for this synchronization is the unit's EW working group. The EW officer attends the regular targeting meetings in the fires cell and may also participate (perhaps as a standing member) in other functional or integrating cells and working groups. These may include fires, information engagement, network operations, or future operations. The EW officer's participation in these other cells and working groups helps to synchronize EW operations.

SUMMARY

5-21. EW capabilities yield many advantages for the commander. The EW working group's sole purpose is to facilitate the integration, coordination, deconfliction, and synchronization of EW operations to ensure advantages are achieved. This effort requires constant coordination with the unit's other functional cells and working groups. As conflicts are identified during the planning and execution of operations, the EW officer and supporting staff members coordinate solutions to those conflicts within the EW working group.

Chapter 6
Integration with Joint and Multinational Operations

Joint warfare is team warfare. It requires the integrated and synchronized application of all appropriate capabilities. During joint operations, Services work together to accomplish a mission. In multinational operations, forces of two or more nations work together to accomplish a mission. During both joint and multinational operations, forces operate under established organizational frameworks and coordination guidelines. This chapter describes the joint and multinational operational frameworks and guidelines for integrating electronic warfare capabilities.

JOINT ELECTRONIC WARFARE OPERATIONS

6-1. One strength of operating as a joint force is the ability to maximize combat capabilities through unified action. However, the ability to maximize the capabilities of a joint force requires guidelines and an organizational framework that can be used to integrate them effectively. JP 3-13.1 establishes the guidelines and organizational framework for joint electronic warfare (EW) operations.

6-2. Joint task forces are task-organized. Therefore, their composition varies based on the mission. Normally the EW organization within a joint force centers on the—

- Component commands.
- Supporting joint centers.
- Joint force staff.
- Joint force commander's EW staff, joint electronic warfare coordination cell, or information operations (IO) cell.

The supporting centers for EW operations may include the joint operations center, joint intelligence center, Joint Frequency Management Office (JFMO), and joint targeting coordination board.

JOINT FORCE PRINCIPAL STAFF FOR ELECTRONIC WARFARE

6-3. In EW, the principal staff consists of the J-2, J-3, and J-6. The J-2 collects, processes, tailors, and disseminates all-source intelligence for EW. The J-3 has primary staff responsibility for EW activity. This director also plans, coordinates, and integrates joint EW operations with other combat disciplines in the joint task force. Normally, the joint force commander's EW staff or a joint EW coordination cell and an IO cell assist the J-3. The joint force staff network operations director (in the J-6) coordinates electromagnetic spectrum use for information systems with electromagnetic-dependent weapons systems used by the joint force. The IO officer is the principal IO advisor to the J-3. This officer is the lead planner for integrating, coordinating, and executing IO. The command EW officer is the principal EW planner on the J-3 staff. This officer coordinates with the IO cell to integrate EW operations fully with other IO core, supporting, and related capabilities (see JP 3-13.1 for further information)

JOINT FORCE COMMANDER'S ELECTRONIC WARFARE STAFF

6-4. A joint force commander's EW staff supports the joint force commander in planning, coordinating, synchronizing, and integrating joint force EW operations. The joint force commander's EW staff ensures that joint EW capabilities support the joint force commander's objectives. The joint force commander's EW staff is an element within the J-3. It consists of representatives from each component of the joint force.

An EW officer appointed by the J-3 leads this element. The joint force commander's EW staff includes representatives from the J-2 and J-6 to facilitate intelligence support and EW frequency deconfliction.

6-5. On many joint staffs, the intra-staff coordination previously accomplished through a joint force commander's EW staff is now performed by an IO cell or similar organization. An IO cell, if established, coordinates EW activities with other IO activities to maximize effectiveness and prevent mutual interference. If both a joint force commander's EW staff and an IO cell exist, a joint force commander's EW staff representative may be assigned to the IO cell to facilitate coordination. For more information about the organization and procedures of the joint IO cell, see JP 3-13.

Joint Electronic Warfare Coordination Cell

6-6. The decision to form a joint EW coordination cell depends on the anticipated role of EW in an operation. When EW is expected to play a significant role in the joint force commander's mission, a component command's EW coordination organization may be designated as the joint EW coordination cell to handle the EW aspects of the operation. The joint EW coordination cell may be part of the joint force commander's staff, be assigned to the J-3 directorate, or remain within the designated component commander's structure. The joint EW coordination cell plans operational-level EW for the joint force commander. (JP 3-13.1 discusses the joint EW coordination cell in more detail.)

Joint Task Force Component Commands

6-7. Joint task force component commanders exercise operational control of their EW assets. Each component is organized and equipped to perform EW tasks in support of its basic mission and to provide support to the joint force commander's overall objectives. If a component command (Service or functional) is designated to stand up a joint EW coordination cell, it executes the responsibilities and functions outlined in JP 3-13.1.

6-8. A major consideration for standing up a joint EW coordination cell at the component command level is access to a special compartmented information facility to accomplish the cell's required coordination functions. Optimal joint EW coordination cell staffing dictates including special technical operations personnel cleared to coordinate and deconflict special technical operations issues. Special technical operations are associated with the planning and coordination of advanced special programs and the integration of new capabilities into operational units.

6-9. Under current force structure, the special technical operations requirement limits the activation of a joint EW coordination cell to organizations at corps and above levels. Organizations below corps level require significant joint augmentation to meet the special technical operations requirement.

Joint Frequency Management Office

6-10. Joint policy tasks each geographic combatant commander to establish a structure to manage spectrum use and establish procedures that support ongoing operations. This structure must include a JFMO. The JFMO may be assigned from the supported combatant commander's J-6 staff, from a component's staff, or from an external command such as the Joint Spectrum Center. The JFMO coordinates the information systems use of the electromagnetic spectrum, frequency management, and frequency deconfliction. The JFMO develops the frequency management plan and makes recommendations to alleviate mutual interference.

6-11. The G-6 or S-6 coordinates the Army's use of the electromagnetic spectrum, frequency management, and frequency deconfliction with the JFMO through the network operations cell. If established, coordination with the joint spectrum management element is required. (See figure 6-1.)

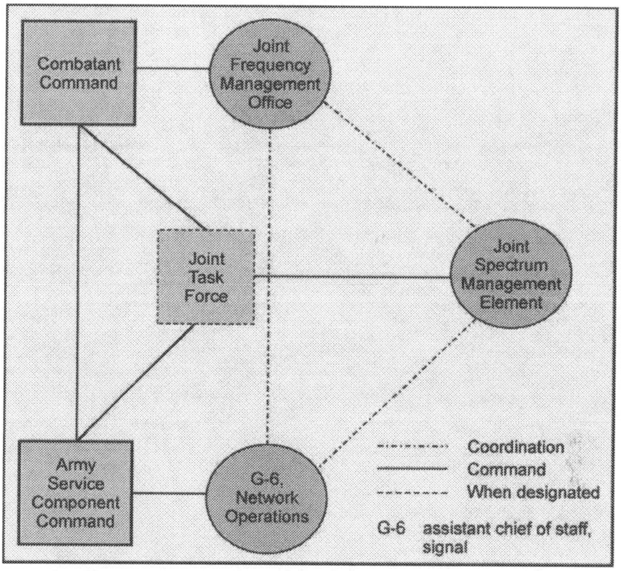

Figure 6-1. Joint frequency management coordination

JOINT INTELLIGENCE CENTER

6-12. The joint intelligence center is the focal point for the intelligence structure supporting the J-2. Directed by the J-2, the joint intelligence center communicates directly with component intelligence agencies and monitors intelligence support to EW operations. This center can adjust intelligence gathering to support EW missions. Within the G-2, EW support requests are coordinated through the requirement cell and then forwarded to the requirements division within the joint intelligence center. (See figure 6-2, page 6-4.)

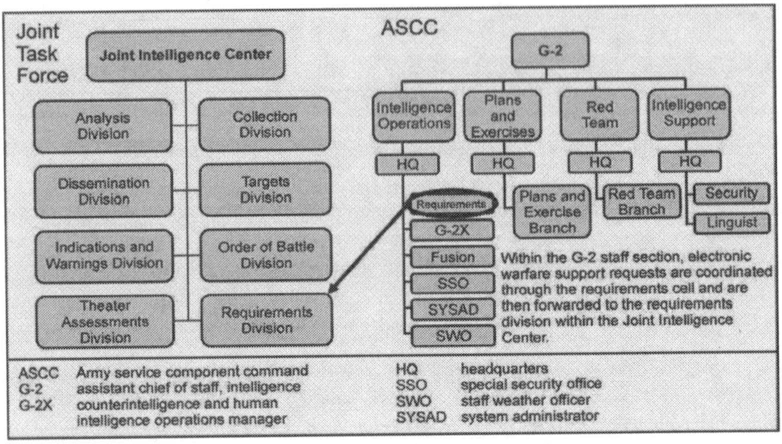

Figure 6-2. Electronic warfare support request coordination

6-13. The composition and focus of each joint intelligence center varies by theater. However, each can perform indications and warning as well as collect, manage, and disseminate current intelligence. Through the joint intelligence center, the ARFOR (Army Service component) headquarters coordinates support from the Air Force, Navy, and Marine Corps and national, interagency, and multinational sources. In addition to its other functions, the joint intelligence center coordinates the acquisition of national intelligence for the joint task force and the combatant command's staff.

JOINT TARGETING COORDINATION BOARD

6-14. The joint targeting coordination board focuses on developing broad targeting priorities and other targeting guidance in accordance with the joint force commander's objectives as they relate operationally. The joint targeting coordination board remains flexible enough to address targeting issues without becoming overly involved in tactical-level decisionmaking. Briefings conducted at the joint targeting coordination board focus on ensuring that intelligence, operations (by all components and applicable staff elements), fires, and maneuver are on track, coordinated, and synchronized. For further information on the joint targeting coordination board, see JP 3-60.

MULTINATIONAL ELECTRONIC WARFARE OPERATIONS

6-15. EW is an integral part of multinational operations (sometimes referred to as combined operations). U.S. planners integrate U.S. and multinational EW capabilities into a single, integrated EW plan. U.S. planners provide multinational forces with information concerning U.S. EW capabilities and provide them EW planning and operational support. However, the planning of multinational force EW is difficult due to security issues, differences in levels of training, language barriers, and terminology and procedural issues. U.S. and North Atlantic Treaty Organization (NATO) EW doctrine provide commonality and a framework for using EW in NATO operations. (See Allied Joint Publication 3.6 for specific information.)

MULTINATIONAL FORCE COMMANDER

6-16. The multinational force commander provides guidance for planning and conducting EW operations to the multinational force through the C-3 and the EW coordination cell. The EW coordination cell is located at multinational force headquarters. An IO cell may also be established to coordinate all IO-related activities, including related EW operations.

Joint Operations Staff Section

6-17. Within the multinational staff, the joint operations section has primary responsibility for planning and integrating EW activities. A staff EW officer is designated with specific responsibilities. These include integrating multinational augmentees, interpreting or translating EW plans and procedures, coordinating appropriate communications connectivity, and integrating multinational force communications into a joint restricted frequency list.

Multinational Electronic Warfare Coordination Cell

6-18. In multinational operations, the multinational force commander uses an EW coordination cell as the mechanism for coordinating EW resources within the area of operations. This cell is an integral part of the multinational joint force headquarters J-3 staff, at whatever level is appropriate. It provides an effective means of coordinating all EW activities by the multinational force. The multinational force EW coordination cell plans and coordinates all in-theater EW activities in close liaison with the J-2, J-5, and J-6.

Electronic Warfare Mutual Support

6-19. Electronic warfare mutual support is the timely exchange of EW information to make the best use of the available resources. It is facilitated by the use of an agreed reference database called the NATO emitter database. Electronic warfare mutual support procedures developed during EW planning include—

- A review of friendly and enemy information data elements that may be exchanged.
- Mechanisms leading to the exchange of data during peace, crisis, and war.
- Development of peacetime exercises to practice the exchange of data.
- Establishment of EW points of contact with adjacent formations and higher and subordinate headquarters for planning purposes, regardless of whether EW resources exist or not.
- Initial acquisition and maintenance of multinational force EW capabilities.
- Exchange of EW liaison teams equipped with appropriate communications.
- Establishment and rehearsal of contingency plans for the exchange of information on friendly and enemy forces.
- Development of communications protocols in accordance with NATO Standardization Agreement (STANAG) 5048.
- Provision of secure, dedicated, and survivable communications.

Other Considerations

6-20. EW in multinational operations addresses other considerations. Soldiers must consider—

- Exchange of EW information.
- Exchange of signals intelligence information.
- Exchange of the electronic order of battle.
- Electronic warfare reprogramming.

6-21. Army forces participating in multinational EW operations must exchange EW information with other forces. They must help develop joint information exchange protocols and use those protocols for conducting operations.

6-22. Exchanging signals intelligence information requires care to avoid violating signals intelligence security rules. The policy and relationship between EW and signals intelligence within NATO are set out in NATO Military Committee (MC) 64.

6-23. In peacetime, before forming a multinational force, the exchange of electronic order of battle information is normally achieved under bilateral agreement. During multinational operations, a representative of the joint EW coordination cell, through the theater joint analysis center or the joint intelligence center, ensures the maintenance of an up-to-date electronic order of battle. The inclusion of multinational forces is based on security and information exchange guidelines agreed upon by the participating nations.

6-24. Electronic warfare reprogramming is a national responsibility. However, the joint EW coordination cell remains aware of reprogramming efforts being conducted within the multinational force. FM 3-13.10 guides the Army's reprogramming effort.

SUMMARY

6-25. Every joint or multinational operation is uniquely organized to accomplish the mission. Army EW officers integrate EW forces and capabilities with the organizations and agencies outlined in this chapter. To coordinate Army EW operations with joint and multinational forces, Army EW officers must understand fully the organizational frameworks, policies, and guidelines established for joint and multinational EW operations.

Chapter 7
Electronic Warfare Capabilities

Electronic warfare capabilities consist of high-demand, low-density assets across the Services. Hence, the conduct of electronic warfare operations requires joint interdependence. This complex interdependence extends beyond the traditional Service capabilities. It includes national agencies—such as the Central Intelligence Agency, National Security Agency, and Defense Intelligence Agency—that constantly seek to identify, catalog, and update the electronic order of battle of enemies and adversaries. To support the joint force commander, the subject matter expertise and unique capabilities provided by each Service, agency, and branch or proponent are integrated with all available electronic warfare capabilities.

SERVICE ELECTRONIC WARFARE CAPABILITIES

7-1. Each Service maintains electronic warfare (EW) capabilities to support operational requirements. During operations, the Army is dependent on organic and nonorganic EW capabilities from higher echelons, joint forces, and national agencies. Army EW planners leverage all available EW capabilities to support Army operations. Although not all-inclusive, appendix E provides a listing of current Army, Marine Corps, Navy, and Air Force EW capabilities and references.

EXTERNAL SUPPORT AGENCIES AND ACTIVITIES

7-2. Army EW planners routinely use and receive support from external organizations to assist in planning and integrating EW operations. Support from these organizations may include personnel augmentation, functional area expertise, technical support, and planning support.

BIG CROW PROGRAM OFFICE

7-3. The Big Crow Program Office was established in 1971 to provide testing environments for U.S. military radio frequency sensor, communication, and navigation systems. Today, the Big Crow Program Office provides customers with joint, multifunctional support for testing communications, sensors, information operations, and related weapon systems in support of Department of Defense (DOD), the individual Services, the National Aeronautics and Space Administration, the National Reconnaissance Office, and others. This support includes replicating information operations and EW threat environments as well as providing telemetry recording, technology prototyping, proof-of-concept demonstrations, and information operations and EW training. Big Crow's mission and capabilities now span the electromagnetic spectrum, encompassing EW, telemetry, radar, and electro-optical systems. Mobile and worldwide deployable, the Big Crow Program Office offers a variety of capabilities.

DEFENSE INFORMATION SYSTEMS AGENCY

7-4. The Defense Information Systems Agency is a combat support agency. It plans, develops, fields, operates, and supports command, control, communications, and information systems. These systems serve the President, the Secretary of Defense, the Joint Chiefs of Staff, the combatant commanders, and other DOD components. The Defense Information Systems Agency also operates the Vulnerability Analysis and Assessment Program. This program specifically focuses on automated information systems.

Joint Communications Security Monitor Activity

7-5. The Joint Communications Security Monitor Activity was created in 1993 by a memorandum of agreement between the Services' operations deputies, Directors of the Joint Staff, and the National Security Agency. The Joint Communications Security Monitor Activity monitors (collects, analyzes, and reports) communications security of DOD telecommunications and automated information systems as well as related noncommunications signals. Its purpose is to identify potentially exploitable vulnerabilities and to recommend countermeasures and corrective actions. The Joint Communications Security Monitor Activity supports real world operations, joint exercises, and DOD systems monitoring.

Joint Information Operations Warfare Command

7-6. The Joint Information Operations Warfare Command (JIOWC) was activated in 2006 as a functional component to the United States Strategic Command (USSTRATCOM). JIOWC integrates joint information operations into military plans, exercises, and operations across the spectrum of conflict. It is a valuable resource for commanders during the planning and execution of joint information operations. JIOWC deploys information operations planning teams when the commander of USSTRATCOM approves a request for support. This center delivers tailored, highly skilled support and sophisticated models and simulations to joint commanders and provides information operations expertise in joint exercises and contingency operations.

7-7. JIOWC also fields the Joint Electronic Warfare Center. This center provides specialized expertise in EW. It is an innovation center for existing and emerging EW capabilities and tactics, techniques, and procedures via a network of units, labs, test ranges, and academia. The Joint Electronic Warfare Center also has EW reprogramming oversight responsibilities for the Joint Staff. This oversight includes organizing, managing, and exercising joint aspects of EW reprogramming and facilitating the exchange of joint EW reprogramming data. The actual reprogramming of equipment, however, is a Service responsibility.

Joint Spectrum Center

7-8. The Joint Spectrum Center was activated in 1994 under the direction of the joint staff's J-6. The Joint Spectrum Center assumed all the missions and responsibilities previously performed by the Electromagnetic Compatibility Center plus additional responsibilities. Personnel in the Joint Spectrum Center are experts in spectrum planning, electromagnetic compatibility and vulnerability, electromagnetic environmental effects, information systems, modeling and simulation, operations support, and system acquisition. The Joint Spectrum Center provides complete, spectrum-related services to combatant commanders, Services, and other government agencies. The Joint Spectrum Center deploys teams in support of the combatant commanders and serves as the DOD focal point for supporting spectrum supremacy aspects of information operations. It assists Soldiers in developing and managing the joint restricted frequency list and helps to resolve operational interference and jamming incidents. The Joint Spectrum Center can also provide databases of friendly force command and control systems for use in planning electronic protection. The Joint Spectrum Center is a field office within the Defense Spectrum Organization under the Defense Information Systems Agency.

Joint Warfare Analysis Center

7-9. The Joint Warfare Analysis Center is a Navy-sponsored joint command under the J-3 established in 1994. The Joint Warfare Analysis Center assists the Chairman of the Joint Chiefs of Staff and combatant commanders in preparing and analyzing joint operational plans. It provides analysis of engineering and scientific data and integrates operational analysis with intelligence.

Marine Corps Information Technology and Network Operations Center

7-10. The Marine Corps Information Technology and Network Operations Center is the Marine Corps' enterprise network operations center. The Marine Corps Information Technology and Network Operations Center is the nerve center for the central operational direction and configuration management of the Marine

Corps enterprise network. It is co-located with the Marine Corps forces computer network defense, the component to the joint task force for computer network operations, and the Marine Corps computer incident response team. This relationship provides a strong framework for integrated network management and defense.

NATIONAL SECURITY AGENCY

7-11. The National Security Agency/Central Security Service is America's cryptologic organization. This organization protects U.S. government information systems and produces foreign signals intelligence information. Executive Order 12333, 4 December 1981, describes the responsibility of the National Security Agency/Central Security Service in more detail. The resources of National Security Agency/Central Security Service are organized for two national missions:

- The Information Assurance Mission provides the solutions, products, and services, and conducts defensive information operations, to achieve information assurance for information infrastructures critical to U.S. national security interests.
- The Signals Intelligence Mission allows for an effective, unified organization and control of all the foreign signals collection and processing activities of the United States. The National Security Agency is authorized to produce signals intelligence in accordance with objectives, requirements, and priorities established by the Director of National Intelligence in consultation with the President's Foreign Intelligence Advisory Board.

7-12. The Director, National Security Agency is the principal signals intelligence and information security advisor to the Secretary of Defense, Director of National Intelligence, and the Chairman of the Joint Chiefs of Staff. The Director, National Security Agency provides signals intelligence support to combatant commanders and others in accordance with their expressed formal requirements.

SUMMARY

7-13. This chapter and appendix E provide a sampling of available joint and Service EW capabilities, activities, and agencies that support ground force commanders in full spectrum operations. To leverage these capabilities for EW support, Army EW officers acquire a working knowledge of the capabilities available and the procedures for requesting support. Additionally, appendix F provides information on available EW related tools and other resources.

Appendix A
The Electromagnetic Environment

Electromagnetic energy is both a natural and manmade occurrence. This energy, in the form of electromagnetic radiation, consists of oscillating electric and magnetic fields and is propagated at the speed of light. Electromagnetic radiation is measured by the frequency of its wave pattern's repetition within a set unit of time. The standard term for the measurement of electromagnetic radiation is the hertz (Hz), the number of repetitions (cycles) per second. The electromagnetic spectrum refers to the range of frequencies of electromagnetic radiation.

OVERVIEW OF THE ELECTROMAGNETIC ENVIRONMENT

A-1. The electromagnetic environment is the resulting product of the power and time distribution, in various frequency ranges, of radiated or conducted electromagnetic emission levels. Within their intended operational environment, a military force, system, or platform may encounter these emissions while performing tasks during operations. The electromagnetic environment is the sum of—

- Electromagnetic interference.
- Electromagnetic pulse.
- Hazards of electromagnetic radiation to personnel, ordnance, and volatile materials.
- Natural phenomena effects of lightning and precipitation static. (*Precipitation static* is charged precipitation particles that strike antennas and gradually charge the antenna, which ultimately discharges across the insulator, causing a burst of static [JP 3-13.1]).

THE ELECTROMAGNETIC SPECTRUM

A-2. The *electromagnetic spectrum* is the range of frequencies of electromagnetic radiation from zero to infinity. It is divided into 26 alphabetically designated bands (JP 1-02). The spectrum is a continuum of all electromagnetic waves arranged according to frequency and wavelength. The electromagnetic spectrum extends from below the frequencies used for modern radio (at the long-wavelength end) through gamma radiation (at the short-wavelength end). It covers wavelengths from thousands of kilometers to a fraction of the size of an atom. Figure A-1 shows the spectrum regions and wavelength segments associated with the electromagnetic spectrum.

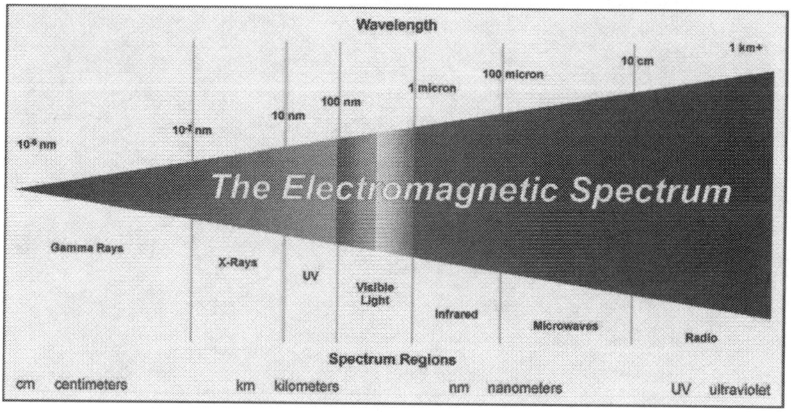

Figure A-1. The electromagnetic spectrum

A-3. Included within the radio and microwave regions of the electromagnetic spectrum are the radio frequency and radar bands. These bands are routinely referred to by their band designators. For example, high frequency radios are HF radios and K-band radars are radars that operate between 18 and 27 gigahertz. Civilian agencies and military forces throughout the world use several different designator systems, which can result in confusion. Table A-1 shows the radio frequency band designators and their associated frequency ranges. It also shows radar band designators, associated frequency ranges, and typical usage. These are standard designations used by the United States.

Table A-1. Radio and radar designators and frequency bands

Radio Frequency Band Designator	Radio Frequency Range	Radar Band Designator*	Frequency Range	Typical Usage
ULF	lower than 3 Hz	VHF	50-330 MHZ	Very long-range surveillance
ELF	3 Hz - 3 kHz	UHF	300-1,000 MHz	Very long-range surveillance
VLF	3 - 30 kHz	L	1-2 Ghz	Long-range surveillance, enroute traffic control
LF	30 - 300 kHz	S	2-4 Ghz	Moderate-range surveillance, terminal traffic control, long-range weather
MF	300 kHz - 3 MHz	C	4-8 Ghz	Long-range tracking, airborne weather
HF	3 - 30 MHz	X	8-12 Ghz	Short-range tracking, missile guidance, mapping, marine radar, airborne intercept
VHF	30 - 300 MHz	K_u	12-18 Ghz	High resolution mapping, satellite altimetry
UHF	300 MHZ - 3 GHz	K	18-27 Ghz	Little use
SHF	3 - 30 GHz	K_a	27-40 Ghz	Very high resolution mapping, airport surveillance
EHF	30 - 300 GHz			
Sub-millimeter	300 Ghz - 1 THz			

EHF extremely high frequency
ELF extremely low frequency
GHz Gigahertz
HF high frequency
Hz hertz
kHz kilohertz
LF low frequency
MF medium frequency
MHz megahertz
SHF super high frequency
THz terahertz
UHF ultra high frequency
ULF ultra low frequency
VHF very high frequency

* Radar band designators relate back to the early development of radar in World War II when the letter designators were used for purposes of secrecy. After the requirement for secrecy was no longer needed, these letter band designators remained.

MILITARY OPERATIONS AND THE ELECTROMAGNETIC ENVIRONMENT

A-4. The impact of the electromagnetic environment upon the operational capability of military forces, equipment, systems, and platforms is referred to as electromagnetic environmental effects. Electromagnetic environmental effects encompass all electromagnetic disciplines, including—

- Electromagnetic compatibility and electromagnetic interference.
- Electromagnetic vulnerability.
- Electromagnetic pulse.
- Electronic protection.
- Hazards of electromagnetic radiation to personnel, ordnance, and volatile materials (such as fuels).
- Natural phenomena effects of lightning and precipitation static.

A-5. *Electromagnetic vulnerability* consists of the characteristics of a system that cause it to suffer a definite degradation (incapability to perform the designated mission) as a result of having been subjected to a certain level of electromagnetic environmental effects (JP 3-13.1). Electronic warfare support plays a key role in identifying the electromagnetic vulnerability of an adversary's electronic equipment and systems. Friendly forces take advantage of these vulnerabilities through electronic warfare operations.

DIRECTED ENERGY

A-6. Directed energy refers to technologies that produce of a beam of concentrated electromagnetic energy or atomic or subatomic particles (see chapter 1). *Directed-energy warfare* is military action involving the use of directed-energy weapons, devices, and countermeasures to either cause direct damage or destruction of enemy equipment, facilities, and personnel, or to determine, exploit, reduce, or prevent hostile use of the electromagnetic spectrum through damage, destruction, and disruption. It also includes actions taken to protect friendly equipment, facilities, and personnel and retain friendly use of the electromagnetic spectrum (JP 1-02). A *directed-energy weapon* is a system using directed energy primarily as a direct means to damage or destroy enemy equipment, facilities, and personnel (JP 1-02). In addition to destructive effects, directed-energy weapons can also support area denial, crowd control, and obscuration.

A-7. The application of directed energy includes lasers, radio-frequency weapons, and particle-beam weapons. As directed-energy weapons evolve, the tactics, techniques, and procedures for their use also evolve to ensure their safe, effective employment. In electronic warfare, most directed-energy applications fit into the category of electronic attack. However, other applications can be categorized as electronic protection or even electronic warfare support. Examples include the following:

- Applications used for electronic attack, which may include—
  - A laser designed to blind or disrupt optical sensors.
  - A millimeter wave directed-energy weapon used for crowd control.
  - A laser-warning receiver designed to initiate a laser countermeasure to defeat a laser weapon.
  - A millimeter wave obscuration system used to disrupt or defeat a millimeter wave system.
  - A device used to counter radio-controlled improvised explosive devices.
- A laser-warning receiver designed solely to detect and analyze a laser signal is used for electronic warfare support.
- A visor or goggle designed to filter out the harmful wavelength of laser light is used for electronic protection.

A-8. As the use of destructive directed-energy weapons grows, Army forces require the capability to collect information on them. Additionally, Army forces require tactics, techniques, and procedures to mitigate directed-energy weapon effects. Currently, the definitions and terms relating to directed energy are articulated within electronic warfare doctrine. As the technologies related to directed energy expand, joint and Army doctrine may discuss employing directed energy under other doctrinal subjects.

Appendix B

Electronic Warfare Input to Operation Plans and Orders

This appendix discusses electronic warfare input to Army and joint plans and orders.

ARMY PLANS AND ORDERS

B-1. This paragraph lists the electronic warfare (EW) information required for Army operation plans and orders. (See figure B-1 on page B-2 for the EW appendix format.) This discussion is based on current doctrine from FM 5-0. When it is republished, FM 5-0 will state where to place EW-related information in the revised plans and orders format. In addition to the appendix 4 (Electronic Warfare) to Annex P (Information Operations), the following components of operation plans and orders may require EW input:

- **Base order or plan**:
  - Sub-subparagraph (2) (Fires) to subparagraph a (Concept of Operations) to paragraph 3 (Execution).
  - Sub-subparagraph (7) (Information Operations) to subparagraph a (Concept of Operations) to paragraph 3 (Execution).
- **Annex D (Fire Support)**:
  - Sub-subparagraph (4) (Electronic Warfare) to subparagraph b (Air Support) to paragraph 3 (Execution)
  - Appendix 1 (Air Support).
- **Annex L (Intelligence, Surveillance, and Reconnaissance)**:
  - Sub-subparagraph (2) (Fires) to subparagraph a (Concept of Operations) to paragraph 3 (Execution).
  - Sub-subparagraph (7) (Information Operations) to subparagraph a (Concept of Operations) to paragraph 3 (Execution).
- **Annex N (Space)**: Sub-subparagraph (10) (Electronic Warfare) to subparagraph b (Space Activities) to paragraph 3 (Execution).
- **Annex P (Information Operations)**:
  - Sub-sub-subparagraph (d) (Electronic Warfare) to sub-subparagraph (8) to subparagraph a (Concept of Support) to paragraph 3 (Execution).
  - Sub-subparagraph (3) (List of Tasks to Electronic Warfare Units) to subparagraph b (Tasks to Subordinate Units) to paragraph 3 (Execution).

[Classification]

Appendix 4 (Electronic Warfare) to Annex P (Information Operations) to OPORD No____

1. SITUATION.

   a. Enemy.
   - Identify the vulnerabilities of enemy information systems and electronic warfare systems.
   - Identify the enemy capability to interfere with accomplishment of the electronic warfare mission.

   b. Friendly.
   - Identify friendly electronic warfare assets and resources that affect electronic warfare planning by subordinate commanders.
   - Identify friendly foreign forces with which subordinate commanders may operate.
   - Identify potential conflicts within the friendly electromagnetic spectrum, especially if conducting joint or multinational operations. Identify and de-conflict methods and priority of spectrum distribution.

   c. Attachments and detachments.
   - List the electronic warfare assets that are attached or detached.
   - List the electronic warfare resources available from higher headquarters.

2. MISSION. State how electronic warfare will support the commander's objectives.

3. EXECUTION.

   a. Scheme of support. State the electronic warfare tasks.

   b. Tasks to subordinate units. Identify the electronic warfare tasks for each unit.

   c. Coordinating instructions.
   - Identify electronic warfare instructions applicable to two or more units.
   - Identify the requirements for the coordination of electronic warfare actions between units.
   - Identify the emission control guidance.

4. SERVICE SUPPORT. Identify service support for electronic warfare operations.

5. COMMAND AND SIGNAL.

   a. Command.

   b. Signal. Identify if any, the special or unusual electronic warfare-related communications requirements.

[Classification]

Figure B-1. Appendix 4 (Electronic Warfare) to annex P (Information Operations) instructions

JOINT PLANS AND ORDERS

B-2. If required to provide EW input to portions of a joint order, the primary areas for input are the following:

- Paragraph 3 (Execution) to appendix 3 (Information Operations) to Annex C (Operations).
- Tab B (Electronic Warfare) to appendix 3 (Information Operations) to Annex C (Operations).

B-3. See CJCSM 3122.03C for the Joint Operations Planning and Execution System format.

Appendix C

Electronic Warfare Running Estimate

This appendix discusses the electronic warfare running estimate. A *running estimate* is a staff section's continuous assessment of current and future operations to determine if the current operation is proceeding according to the commander's intent and if future operations are supportable (FM 3-0).

C-1. The electronic warfare (EW) running estimate is used to support the military decisionmaking process during planning and execution. During planning, the EW running estimate provides an assessment of the supportability of each proposed course of action from an EW perspective. The format of the EW running estimate closely parallels the steps of the military decisionmaking process. It serves as the primary tool for recording the EW officer's assessments, analyses, and recommendations for EW operations. The EW officer and staff in the EW working group are responsible for conducting the analysis and providing recommendations based on the EW running estimate.

C-2. A complete EW running estimate should contain the information necessary to answer any question the commander may pose. If there are gaps in the EW running estimate, the staff identifies the gaps as information requirements and submits them to the intelligence cell. The EW running estimate can form the basis for EW input required in other applicable appendixes and annexes within operation plans and orders. Figure C-1 on page C-2 provides a sample EW running estimate for use during planning.

1. **MISSION.** Show the restated mission resulting from mission analysis.

2. **SITUATION AND CONSIDERATIONS.**

   a. Characteristics of the area of operations.
      - Weather. State how the weather may impact EW operations.
      - Terrain. State how aspects of the terrain may impact EW operations.
      - Civil Considerations. State how rules of engagement and civil emergency responder frequency restrictions may impact EW operations.

   b. Enemy forces. Discuss enemy dispositions, composition, strength, capabilities, and courses of action (COAs) as they affect EW operations. Identify enemy EW vulnerabilities.

   c. Friendly forces.
      - List the current status of the forces EW resources.
      - List the current status of additional EW support resources.
      - Provide a comparison of EW support requirements with available capabilities and recommend solutions for any discrepancies.
      - Identify friendly forces EW vulnerability and recommend solutions.

   d. Assumptions. List any assumptions used that may affect the employment of EW capabilities.

3. **COURSES OF ACTION.**

   a. List the friendly COAs that were wargamed.
   b. List the evaluation criteria identified during the COA analysis.

4. **ANALYSIS.** Analyze each COA using the evaluation criteria identified during COA analysis.

5. **COMPARISION.** Compare each COA. Rank order the COAs for each EW key consideration identified.

6. **RECOMMENDATION AND CONCLUSIONS.** This paragraph translates the "best" course of action (as determined in paragraph 5) into a complete recommendation. It should outline who, what, where, when, how, and why from the EW point of view. It states which course of action can best be supported by friendly EW, and is less vulnerable to enemy EW force capabilities.

   a. Recommend the most supportable COA from an EW perspective.
   b. List any EW related issues, deficiencies and risks and provide recommendations to reduce their impact.

ANNEXES: Include annexes as required. Annexes with pertinent details should be used to the extent practical to support the contents of the estimate. These annexes may be in considerable detail with only the high points included in the body of the estimate. Annexes should add depth to the contents of the estimate, but should not be used as a substitute for key points that should be included in the bodv of the estimate.

Figure C-1. Example of an electronic warfare running estimate

C-3. Once the commander approves the order, the EW running estimate is used to inform current and future operations. During execution the EW running estimate is used to help determine if current EW operations are proceeding according to plan and if future EW operations are supportable. Figure C-2, page C-3, shows a sample of the information that might be used to update the EW running estimate during execution. The EW officer and supporting staff members within the EW working group produce and update the running estimate.

Current operation order and fragmentary orders

- Define the battlefield environment. Focus on the aspects of the terrain and weather that could assist or enable electronic warfare operations from both a friendly and threat viewpoint.
  - Maintain updated weather and terrain data.
  - Locate terrain for communications and non-communications sites, line of sight.
  - Identify aspects of terrain and weather that may have an impact on the electromagnetic spectrum.
- Define the threat.
  - Communications systems, including threat radio nets and network nodes.
  - Noncommunications emitters.
  - Electronic support systems.
  - Electronic attack systems.
- Identify host-nation use of the electromagnetic spectrum (restricted frequencies such as government, industry, and emergency responders).
- Identify friendly capabilities, shortfalls and readiness.
  - Electronic attack capabilities and status (joint and Army).
  - Location and availability of organic friendly electronic warfare capabilities (such as Prophets and counter-radio-controlled IED EW systems.
  - Electronic warfare vulnerabilities.
  - Equipment updates, both hardware and software.
- Identify enemy capabilities, shortfalls and readiness.
  - Electronic warfare capabilities and status (if known).
  - Electronic warfare vulnerabilities.
  - Electronic order of battle.

Electronic Warfare Target Folder

- Describe the targeted capability, its associated vulnerabilities, and the friendly capabilities used to engage them.
- Maintain updated high-value target and high-payoff target lists.
- Develop a prioritized target list based on high-value targets and high-payoff targets.
- EW target folders are split between traditional and asymmetric targets.
  - Traditional targets might include integrated air defense systems, communications nodes, and radar facilities. Traditional targets are normally fixed or less mobile than asymmetric targets and are easier to develop.
  - Asymmetric targets might include individual cell phones, radio-controlled improvised explosive devices, global positioning systems and wireless networks. Asymmetric targets can be either stationary or mobile and are typically harder to develop than traditional targets during the targeting phase.

Figure C-2. Sample update information to the electronic warfare running estimate

Appendix D

Electronic Warfare-Related Reports and Messages

This appendix provides information and references for electronic warfare and electronic warfare-related reports and message formats.

MESSAGES AND SUMMARIES

D-1. The following messages and summaries are associated with the planning, synchronization, deconfliction, and assessment of EW operations.

ELECTRONIC ATTACK DATA MESSAGE

D-2. An electronic attack data message reports an electronic attack strobe from an affected or detecting unit's position to an aircraft emitting an electronic attack. It is used to determine the location of a hostile or unknown aircraft emitting an electronic attack. The detecting unit reports its detection to all units using a given network when the data link is degraded or not operational.

D-3. Upon receipt of several messages, the source of enemy electronic attack can be determined by comparing lines of bearing from the different origins (triangulation).

D-4. See FM 6-99.2, page 83, for the format.

ELECTRONIC ATTACK REQUEST FORMAT

D-5. Electronic fires fall within three categories: preplanned, preplanned on-call, and immediate. Requesting airborne electronic attack support for ground operations is similar to requesting close air support. Requests for an electronic attack are sent via the normal joint air request process. Requesters use either a joint tactical air strike request or joint tactical air support request. (See FM 3-09.32 for a sample.) A theater-specific electronic attack request format may complement a joint tactical air strike request.

D-6. When submitting the request, the following information must be provided in the remarks section (section 8):

- Target location.
- Prioritized target description and jam frequencies.
- Time on target (window).
- Joint terminal attack controller.
- Jamming control authority call sign and frequency.
- Friendly force disposition (for example, troop movement route).
- Friendly frequency restrictions.
- Remarks.

ELECTRONIC WARFARE FREQUENCY DECONFLICTION MESSAGE

D-7. An EW frequency deconfliction message promulgates a list of protected, guarded, and taboo frequencies. This list allows friendly forces to use the frequency spectrum without adverse impact from friendly electronic attack. (See FM 6-99.2, page 86, for the format.)

Electronic Warfare Mission Summary

D-8. The EW mission summary summarizes significant EW missions and reports the status of offensive EW assets. EW and electronic-attack-capable surface and air units use it to provide information on EW operations. Service components use it to report significant events for subsequent analysis. (See FM 6-99.2, page 87, for the format.)

Electronic Warfare Requesting Tasking Message

D-9. Joint task force commanders use the electronic warfare requesting tasking message to task component commanders to perform EW operations in support of the joint EW plan and to support component EW operations. Component commanders use this message to request EW support from sources outside their command.

Joint Tactical Air Strike Request or Joint Tactical Air Support Request

D-10. Use a joint tactical air strike request or joint tactical air support request to request electronic attack. These requests require the information listed in paragraph D-6. Organizations without an automated capability submit these requests using DD Form 1972 (Joint Tactical Air Strike Request). See JP 3-09.3 and FM 3-09.32 for more information.

Joint Spectrum Interference Resolution

D-11. The joint spectrum interference resolution program replaced the DOD meaconing, intrusion, jamming, and interference program in June, 1992. Follow guidance in CJCSI 3320.02C to report incidents of spectrum interference.

Joint Restricted Frequency List

D-12. Operational, intelligence, and support elements use the joint restricted frequency list to identify the level of protection desired for various networks and frequencies. The list should be limited to the minimum number of frequencies necessary for friendly forces to accomplish objectives.

D-13. See Annex A to appendix B to JP 3-13.1 for the joint restricted frequency list format. The format is used by the joint automated communications-electronics operations instruction system. The format is unclassified but should show the proper classification of each paragraph when filled in. (See CJCSI 3320.01B and JP 3-13.1 for additional information.)

COUNTER-IMPROVISED-EXPLOSIVE-DEVICE ACTIVITIES

D-14. Certain reports and references are associated with counter-improvised-explosive-device activities. Most of these reports include information pertinent to counter-radio-controlled improvised-explosive-device EW activities. EW working groups have the responsibility to monitor these reports to assess planned counter-radio-controlled improvised-explosive-device EW operations and to support future operations. These reports typically use formats established in FM 6-99.2 modified to include improvised explosive device considerations and current operations. See GTA 90-10-046 for examples of reports and references applicable to counter-radio-controlled improvised-explosive-device EW operations.

Appendix E

Army and Joint Electronic Warfare Capabilities

This appendix provides information on Army and other Service electronic warfare capabilities. It is not an all-inclusive list. Due to the evolving nature of electronic warfare equipment and systems, this information is perishable and should be augmented, updated, and maintained by the unit electronic warfare officer.

ARMY

E-1. The Army is currently expanding its electronic warfare (EW) capability. It maintains several EW systems in its inventory. Currently, all units whose sole purpose is to conduct EW operations are assigned to 1st Information Operations Command. When requested, these capabilities are provided to combatant commands for employment at corps and lower echelons.

Counter-Radio-Controlled Improvised-Explosive-Device EW Systems

E-2. Counter-radio-controlled improvised-explosive-device EW systems form a family of electronic attack systems. Army forces use these systems to prevent improvised explosive device detonation by radio frequency energy. The Army maintains both a mounted and dismounted counter-radio-controlled improvised-explosive-device EW capability to protect personnel and equipment. For a detailed description of these systems, see appendix F.

Aircraft Survivability Equipment

E-3. Aircraft survivability equipment aims to reduce aircraft vulnerability, thus allowing aircrews to accomplish their immediate mission and survive. Army aviation maintains a suite of aircraft survivability equipment that provides protection against electronic attack. This protection can include radio frequency warning and countermeasures systems, a common missile warning system, information requirement countermeasures systems, and laser detection and countermeasure systems. For a detailed description of aircraft survivability equipment EW-related systems, see appendix F.

Intelligence Systems

E-4. The intelligence community maintains many systems that provide data for use in EW operations. Signals intelligence systems provide most of this required data. These assets are dual use. Usually the data collected is categorized as signals intelligence. It is maintained within sensitive compartmented information channels and governed by the National Security Agency/Central Security Service. The data sometimes support EW or, more specifically, electronic warfare support. Paragraphs E-5 through E-7 illustrate some intelligence systems that (when tasked) can provide electronic warfare support data to support electronic attack and electronic protection actions. For a detailed description of other intelligence and EW-support-related systems, see appendix F.

Guardrail Common Sensor

E-5. The Guardrail common sensor is a corps-level airborne signals intelligence collection and location system. (See figure E-1.) It provides tactical commanders with near real-time targeting information. Key features include the following: integrated communications intelligence and electronic intelligence reporting, enhanced signal classification and recognition, near real-time direction finding, precision emitter location, and an advanced integrated aircraft cockpit. Preplanned product improvements include frequency extension, computer-assisted online sensor management, upgraded data links, and the capability to exploit a wider range of signals. The Guardrail common sensor shares technology with the ground-based common sensor, airborne reconnaissance-low, and other joint systems.

Figure E-1. Guardrail common sensor

Aerial Common Sensor

E-6. The aerial common sensor is the Army's programmed airborne intelligence, surveillance, and reconnaissance system. (See figure E-2.) It will replace the current RC-7 airborne reconnaissance-low and Guardrail common sensor programs. The aerial common sensor uses the operational and technical legacies of the airborne reconnaissance-low and Guardrail common sensor systems as well as some technological improvements. This sensor will then provide a single, effective, and supportable multiple-intelligence system for the Army. The aerial common sensor will include a full multiple-intelligence capability, including carrying signals intelligence payloads, electro-optic and infrared sensors, radar payloads, and hyperspectral sensors.

Figure E-2. Aerial common sensor (concept)

Prophet

E-7. The Prophet system is the division, brigade combat team, and armored cavalry regiment principal ground tactical signals intelligence and EW system. (See figure E-3.) Prophet systems will also be assigned to the technical collection battalion of battlefield surveillance brigades. Prophet detects, identifies, and locates enemy electronic emitters. It provides enhanced situational awareness and actionable 24-hour information within the unit's area of operations. Prophet consists of a vehicular signals intelligence receiver mounted on a high mobility multipurpose wheeled vehicle, plus a dismounted-Soldier-portable version. The dismounted Soldier portable version is used for airborne insertion or early entry to support rapid reaction contingency and antiterrorist operations. Future Prophet systems are planned to include an electronic attack capability.

Figure E-3. Prophet (vehicle-mounted)

MARINE CORPS

E-8. The Marine Corps has two types of EW units: radio battalions (often called RADBNs), and Marine tactical EW squadrons (referred to as VMAQs). Paragraphs E-9 through E-24 discuss the units' missions, their primary tasks, and capabilities currently being employed. (For further information on the Marine Corps EW units and systems, see MCWP 2-22.)

RADIO BATTALION

E-9. Radio battalions are the Marine Corps' tactical level ground-based EW units. During operations, teams from radio battalions are most often attached to the command element (or senior headquarters) of Marine expeditionary units. Each radio battalion has the following mission, tasks, and equipment.

Mission and Tasks

E-10. The mission of the radio battalion is to provide communications security monitoring, tactical signals intelligence, EW, and special intelligence communication support to the Marine air-ground task force (MAGTF). The radio battalion's tasks include—

- Executing interception; radio direction finding; recording and analysis of communications and noncommunications signals; and signals intelligence processing, analysis, production, and reporting.
- Conducting EW against enemy or adversary communications.
- Helping protect MAGTF communications from enemy exploitation by conducting communications security monitoring, analysis, and reporting on friendly force communications.
- Providing special intelligence communications support and cryptographic guard (personnel and terminal equipment) in support of the MAGTF command element. Normally, the communications unit supporting the MAGTF command element provides communications connectivity for special intelligence communications.
- Providing task-organized detachments to MAGTFs with designated signals intelligence, EW, special intelligence communication, and other required capabilities.
- Exercising technical control and direction over MAGTF signals intelligence and EW operations.
- Providing radio reconnaissance teams with specialized insertion and extraction capabilities (such as combat rubber raiding craft, fast rope, rappel, helocast, and static-line parachute) for specified signals intelligence and limited electronic attack support during advance force, preassault, or deep postassault operations.
- Coordinating technical signals intelligence requirements and exchanging technical information and material with national, combatant command, joint, and other signals intelligence units.
- Providing intermediate, third, and fourth echelon maintenance of the radio battalion's signals intelligence and EW equipment.

Equipment

E-11. The following illustrate EW capabilities a radio battalion uses to accomplish the mission and perform the tasks in support of the MAGTF:

AN/ULQ-19(V)2 Electronic Attack Set

E-12. The AN/ULQ-19(V)2 electronic attack set allows operators to conduct spot or sweep jamming of single-channel voice or data signals. To provide the required jamming, the system must be employed and operated from a location with an unobstructed signal line of sight to the target enemy's communications transceiver.

AN/MLQ-36 Mobile Electronic Warfare Support System

E-13. The AN/MLQ-36 mobile electronic warfare support system provides a multifunctional capability that gives signals intelligence and EW operators limited armor protection. This equipment can provide signals intelligence and EW support to highly mobile mechanized and military operations in urban terrain where maneuver or armor protection is critical. This system is installed in a logistic variant of the Marine Corps's light armored vehicle. It consists of the following:

- Signals intercept system.
- Radio direction finding system.
- Electronic attack system.
- Secure communication system.
- Intercom system.

AN/MLQ-36A Mobile Electronic Warfare Support System (Product Improved)

E-14. The product-improved AN/MLQ-36A mobile electronic warfare support system (sometimes called the AN/MLQ-36A MEWSS PIP) is an advanced signals intelligence and EW system integrated into the Marine Corps's light armored vehicle. (See figure E-4.) This system replaces the equipment in the AN/MLQ-36.

E-15. The AN/AMLQ-36A has the following capabilities:

- Detect and evaluate enemy communications emissions.
- Detect and categorize enemy noncommunications emissions (such as battlefield radars).
- Determine lines of bearing.
- Degrade enemy tactical radio communications.

When mission-configured and working cooperatively with other AN/MLQ-36As, the system can provide precision location of battlefield emitters.

E-16. This system and its future enhancements will provide the capability to exploit new and sophisticated enemy electronic emissions and conduct electronic attack in support of existing and planned national, combatant command, fleet, and MAGTF signals intelligence and EW operations.

Figure E-4. AN/MLQ-36A mobile electronic warfare support system

Marine Tactical Electronic Warfare Squadron

E-17. Marine tactical electronic warfare squadrons are the Marine Corps's airborne tactical EW units. Each squadron has the following mission, tasks, and capabilities.

Mission and Tasks

E-18. The mission of the electronic warfare squadron is to provide EW support to the MAGTF and other designated forces. The squadron conducts tactical jamming to prevent, delay, or disrupt the enemy's ability to use the following kinds of radars: early warning, acquisition, fire or missile control, counterfire, and battlefield surveillance. Tactical jamming also denies and degrades enemy communication capabilities. The squadron conducts electronic surveillance operations to maintain electronic orders of battle. These include both selected emitter parameters and nonfriendly emitter locations. The squadron also provides threat warnings for friendly aircraft, ships, and ground units. Squadron tasks include—

- Providing airborne electronic attack and EW support to the aviation combat element and other designated operations by intercepting, recording, and jamming threat communications and noncommunications emitters.
- Processing, analyzing, and producing routine and time-sensitive electronic intelligence reports for updating and maintaining enemy electronic order of battle.

- Providing liaison personnel to higher staffs to assist in squadron employment planning.
- Providing an air EW liaison officer to the MAGTF EW coordination cell.
- Conducting electronic attack operations for electronic protection training of MAGTF units.

E-19. The squadron's EW division supports EA-6B Prowler tactical missions with intelligence, the tactical electronic reconnaissance processing and evaluation system (TERPES), and the joint mission planning system. All systems support premission planning and postmission processing of collected data, and production of pertinent intelligence reports. Working with squadron intelligence, these systems provide required electronic intelligence and electronic order of battle intelligence products to the aviation combat element, MAGTF, and other requesting agencies.

Equipment

E-20. Marine tactical electronic warfare squadrons maintain the following equipment:

- EA-6B Prowler.
- Joint mission planning system.
- Tactical electronic reconnaissance processing and evaluation system.

EA-6B Prowler

E-21. The EA-6B Prowler is a subsonic, all-weather, carrier-capable aircraft. (See figure E-5.) The crew consists of one pilot and three electronic countermeasure officers. The EA-6B has two primary missions. One is collecting and processing designated threat signals of interest for jamming and subsequent processing, analysis, and intelligence reporting. The other is employing the AGM-88 high-speed antiradiation missile against designated targets. The EA-6B's AN/ALQ-99 tactical jamming system incorporates receivers for the reception of emitted signals and external jamming pods for the transmission of energy to jam targeted radars (principally those associated with enemy air defense radars and associated command and control). In addition to the AN/ALQ-99, the EA-6B also employs the USQ-113 communications jammer to collect, record, and disrupt threat communications.

Figure E-5. EA-6B Prowler

Joint Mission Planning System

E-22. The joint mission planning system helps the EA-6B aircrew plan and optimize receivers, jammers, and high-speed antiradiation missiles. This system allows an operator to—

- Maintain area of operations emitter listings.
- Edit emitter parameters.
- Develop mission-specific geographic data and electronic order of battle to—
  - Tailor or create high-speed antiradiation missile direct attack libraries, or manually modify entries or new threat cards.
  - Plan target selection.
- Perform postflight mission analysis to—
  - Identify electronic emitters using various electronic parameter databases and electronic intelligence analytical techniques.
  - Localize emitters by coordinates with a certain circular error of probability for each site.
  - Correlate new information with existing data.
  - Gather postflight high-speed antiradiation missile information. This information includes aircraft launch parameters, predicted seeker footprint, and the onboard system detection of a targeted signal at impact.

AN/TSQ-90 Tactical Electronic Reconnaissance Processing and Evaluation System

E-23. The TERPES (AN/TSQ-90) is an air and land transportable, single-shelter electronic intelligence processing and correlation system. Each of the four Marine tactical electronic warfare squadrons includes a TERPES section.

E-24. A TERPES section consists of Marines, equipment, and software. The section identifies and locates enemy radar emitters from data collected by EA-6B aircraft and those received from other intelligence sources. It processes and disseminates EW data rapidly to MAGTF and other intelligence centers and provides mission planning and briefing support. Section support areas include operational support, intelligence analysis support, data fusion, fusion processing, and intelligence reporting. The section provides the following operational support:

- Translates machine-readable, airborne-collected, digital data into human- and machine-readable reports (such as paper, magnetic tape, secure voice, plots, and overlays).
- Receives and processes EA-6B mission tapes.
- Accepts, correlates, and identifies electronic emitter data from semiautomatic or automatic collection systems using various electronic parameter databases and various analysis techniques.
- Provides tactical jamming analysis.

AIR FORCE

E-25. The Air Force has two primary platforms that provide EW capability: the EC-130H Compass Call and RC-135V/W Rivet Joint. (For further information on Air Force EW equipment, see AFDD 2-5.1.)

EC-130H COMPASS CALL

E-26. The EC-130H Compass Call is an airborne tactical weapon system. (See figure E-6.) Paragraphs E-27 through E-31 discuss the EC-130H missions, primary tasks, and capabilities.

Mission and Tasks

E-27. The EC-130H's mission is to disrupt enemy command and control information systems and limit the coordination essential for force management. The EC-130H's primary task is to employ offensive counterinformation and electronic attack capabilities in support of U.S. and multinational tactical air, surface, and special operations forces.

Figure E-6. EC-130H Compass Call

Capabilities

E-28. The EC-130H is designed to deny, degrade, and disrupt adversary command and control information systems. This includes denial and disruption of enemy surveillance radars; denial and disruption of hostile communications being used in support of enemy ground, air, or maritime operations; and denial and disruption of many modern commercial communication signals that an adversary might employ.

Compass Call During Operation Iraqi Freedom

During Operation Iraqi Freedom, much speculation appeared in the press about why Iraqi forces failed to ignite the oil facilities they had wired for destruction. During the coalition's seizure of Al Faw, Compass Call disrupted the Iraqi regime's control of its troops by jamming its communications. Instead of receiving orders to detonate the oil terminals, Iraqi troops heard only the ratcheting static of Compass Call jamming until coalition ground troops had secured the area. In addition to the conquest of the Al Faw Peninsula, successful military operations supported by Compass Call in Operation Iraqi Freedom included the seizure of four airfields; two successful prisoner of war rescues; and the ground offensive from Basrah to Nasariyah, Najaf, Baghdad, and Tikrit. In all these instances, Compass Call jamming prevented a trained, experienced enemy from coordinating actions against coalition forces.

"EC-130H Compass Call: A textbook example of Joint Force integration at its best", Electronic Warfare Working Group, U.S. House of Representatives, Issue Brief #17, 11 Mar 2004. (Available at http://www.house.gov/pitts/initiatives/ew/Library/Briefs/brief17.htm)

RC-135V/W Rivet Joint

E-29. Paragraphs E-30 through E-31 discuss the missions, primary tasks, and capabilities of the RC-135V platforms.

Mission and Tasks

E-30. The RC-135V/W Rivet Joint is a combatant-command-level surveillance asset that responds to national-level taskings. (See figure E-7.) Its mission is to support national consumers, combatant commanders, and combat forces with direct, near real-time reconnaissance information and electronic warfare support. It collects, analyzes, reports, and exploits information from enemy command and control information systems. During most contingencies, it deploys to the theater of operations with the airborne elements of the theater air control system.

RC-135V/W Rivet Joint, U.S. Air Force Fact Sheet, September, 2007.

Figure E-7. RC-135V/W Rivet Joint

Capabilities

E-31. The RC-135V/W is equipped with an extensive array of sophisticated intelligence gathering equipment that enables monitoring of enemy electronic activity. The aircraft is integrated into the theater air control system via data links and voice (as required). Refined intelligence data can be transferred from Rivet Joint to an Airborne Warning and Control System platform through the tactical digital information link. Alternatively, this data can be placed into intelligence channels via satellite and the tactical information broadcast service (a near real-time combatant command information broadcast). The aircraft has secure ultrahigh frequency, very high frequency, and high frequency (commonly known as UHF, VHF, and HF respectively) as well as satellite communications. It can be refueled in the air.

NAVY

E-32. The Navy's primary airborne EW platforms are the EA-6B Prowler and its planned replacement, the E/A-18G Growler. E/A-18G fielding is scheduled to begin in 2009 and is scheduled to replace the Navy's carrierborne EA-6B aircraft. The Navy also maintains both surface and subsurface EW shipboard systems for offensive and defensive missions in support of the fleet. (For further information on Navy missions and equipment, see NWP 3-13.)

EA-6B PROWLER

E-33. Paragraphs E-34 through E-39 discuss the missions, primary tasks, and capabilities of the Navy's EA-6B Prowler platforms. (See figure E-8.)

Figure E-8. Navy EA-6B Prowler

Mission and Tasks

E-34. The mission of the Navy's EA-6B Prowler is to ensure survivability of U.S. and multinational forces through suppression of enemy air defenses (using the radar-jamming AN/ALQ-99 tactical jamming system), lethal suppression (using the AGM-88 high-speed antiradiation missile), and communications jamming (using the USQ-113 radio countermeasures set). Prowlers have supported U.S. and multinational forces operating from various expeditionary sites throughout the world while maintaining full presence on all Navy aircraft carriers.

Capabilities

E-35. The Navy's EA-6B Prowlers are outfitted with either the improved capability II or improved capability III systems. The following lists the major capability upgrades these systems provide.

Improved Capability II

E-36. The improved capability II program was initiated in the 1980s. It was completed across the fleet of EA-6B aircraft (including U.S. Marine Corps aircraft) in the 1990s. The program incorporated incremental capability improvements that include communications, navigation, and computer interface upgrades; a high-speed antiradiation missile capability; and improved jamming pods. Several system interfaces were also upgraded in preparation for the improved capability III improvements.

Improved Capability III

E-37. The improved capability III program incorporates a highly evolved receiver system and provides upgraded EA-6B aircraft with increased signal detection, geolocation capability, a new selective reactive-jamming capability, and better reliability. High-speed antiradiation missile employment is also improved due to the speed of the receiver and its geolocation accuracy. Increased battlefield situational awareness of joint forces is also provided through Link-16. The improved capability III program provides a new ALQ-218 receiver system, integration of the USQ-113 and the multifunctional information distribution system (often called MIDS). This system incorporates Link-16 and various connectivity avionics into the Prowler. The major EW-related subsystems are the AN/ALQ-99 (V) tactical jamming countermeasures set and AN/USQ-113 (V) radio countermeasures set.

E-38. The AN/ALQ-99 (V) tactical jamming countermeasures set has upgraded receivers and processors to provide the following:

- Improved frequency coverage.
- Direction-of-arrival determination capability.
- Narrower frequency discrimination to support narrowband jamming.
- Enhanced interface with onboard systems.

E-39. The AN/USQ-113 (V) radio countermeasures set will enhance the aircraft's jamming capability through its integration with the tactical display system. This will enable the crew to display AN/USQ-113 communications jamming data as well as control AN/USQ-113 operations through the tactical display system.

E/A-18G Growler

E-40. The E/A-18G Growler is the Navy's replacement aircraft for the EA-6B Prowler. Paragraphs E-41 and E-42 discuss the missions, primary tasks, and capabilities of the Navy's E/A-18G Growler. (See figure E-9.) E/A-18G fielding began in 2008. The first operational E/A-18G deployment will occur in 2009, as the Navy begins to replace its carrierborne EA-6B aircraft.

Figure E-9. EA-18 Growler

Mission and Tasks

E-41. The EA-18G can detect, identify, locate, and suppress hostile emitters. It will provide enhanced connectivity to national, combatant command, and strike assets. Additionally, the EA-18G will provide organic accurate emitter targeting using on-board suppression weapons, such as the high-speed antiradiation missile.

Capabilities

E-42. The following is a list of the E/A-18G's general capabilities:

- Suppression of enemy air defenses. The EA-18G will counter enemy air defenses using both reactive and preemptive jamming techniques.
- Stand-off and escort jamming. The EA-18G will be highly effective in the traditional stand-off jamming mission, but with the speed and agility of a Super Hornet, it will also be effective in the escort role.
- Integrated air and ground airborne electronic attack. Enhanced situational awareness and uninterrupted communications will enable the EA-18G to achieve a higher degree of integration with ground operations than previously.
- Self-protect and time-critical strike support. With its active electronically scanned array radar, digital data links, and air-to-air missiles, the EA-18G will be able to protect itself and effectively identify and prosecute targets.
- Growth. High commonality with the F/A-18E and F/A-18F, nine available weapon stations, and modern avionics enable cost-effective synergistic growth, setting the stage for continuous capability enhancement.

E-43. The following is a list of the E/A-18G's airborne electronic attack capabilities:

- Entire spectrum. The EA-18G's ALQ-218 wideband receiver combined with the ALQ-99 tactical jamming system will be effective against any surface-to-air threat.
- Precision airborne electronic attack. Selective-reactive technology enables the EA-18G to rapidly sense and locate threats much more accurately than before. This improved accuracy enables greater concentration of energy against threats.
- Advanced communication countermeasures. Its modular communication countermeasure set enables the EA-18G to counter a wide range of communication systems and is readily adaptable to an ever changing threat spectrum.
- Interference cancellation system. This system dramatically enhances aircrew situational awareness by enabling uninterrupted communications during jamming operations.

CAPABILITIES SUMMARY

E-44. Table E-1 lists Army and joint EW capabilities. (Bold text indicates capabilities not described in the preceding paragraphs.) EW officers, noncommissioned officers, and supporting staff members should be familiar with these capabilities and how they can support Army operations. Additional information on the EW capabilities listed in table E-1 is found in the Web sites listed in table E-2, page E-12.

Table E-1. Army and joint electronic warfare capabilities

	Army	Air Force	Navy	Marine Corps
Airborne	RC-12 Guardrail	**EC-130J Commando Solo**	EA-6B Prowler	EA-6B Prowler
	airborne common sensor	EC-130H Compass Call	EA-18G Growler	
		RC-135V/W Rivet Joint	**EP-3E Aries II**	
		F-16CJ		
		E-8 JSTARS		
Unmanned aircraft system*	**RQ-5A/MQ-5B Hunter (Corps)**	**RQ-4A (Joint) Global Hawk**	**RQ-2 Pioneer**	
	RQ-7A/B Shadow (brigade)	**RQ-1L (Joint) Predator**	**MQ-8B Fire Scout Vertical Take-off**	
	MQ-1C/Sky Warrior (replacement for Hunter)	**RQ-11 Raven**	**Silver Fox**	
	MQ-8 Fire Scout			**RQ-11 Raven**
	RQ-11 Raven (battalion) Hand Launched			**Scan Eagle**
Ground	AN/MLQ-40 Prophet		**CREW Systems (Joint)**	**AN/MLQ-36 MEWSS**

Note: *Other Services may refer to unmanned aircraft systems as unmanned aerial systems or vehicles.
CREW counter radio-controlled improvised explosive device electronic warfare
MEWSS mobile electronic warfare support system
SOF special operations forces

Table E-2. Electronic warfare systems and platforms resources

Army platforms and systems http://www.sed.monmouth.army.mil/avionics/ http://www.sec.army.mil/secweb/fact_sheets/fact_sheets.php
Air Force platforms and systems http://www.af.mil/factsheets/factsheet.asp?fsID=182 http://www.airforce-technology.com/projects/#Unmanned_Aerial_Vehicles_(UAV_/_UCAV)
Navy systems platforms and systems http://acquisition.navy.mil/programs http://www.naval-technology.com/projects/ http://www.navy.mil/navydata/fact.asp
Marine Corps platforms and systems http://www.marcorsyscom.usmc.mil/sites/cins/INTEL/USMC%20CREW/index.html
Joint programs https://www.jieddo.dod.mil

Appendix F

Tools and Resources Related to Electronic Warfare

This appendix provides information on tools and reachback resources related to electronic warfare. Electronic warfare officers, noncommissioned officers, and supporting staff members should be familiar with these tools and resources and how to use them to support electronic warfare operations. Some tools and resources require an approved user account prior to being granted access.

ARMY REPROGRAMMING ANALYSIS TEAM

F-1. The Army Reprogramming Analysis Team (ARAT) supports tactical commanders. It provides timely reprogramming of any Army-supported software used for target acquisition, target engagement, measurement and signature intelligence, and vehicle and aircraft survivability (including that operated by other Services). The team provides software changes not readily possible by operator input to respond to rapid deployments or changes in the operational environment. See their Web site at https://ako.sec.army.mil/arat/index.html (Army Knowledge Online login required).

F-2. ARAT provides reprogramming support to counter-radio-controlled improvised-explosive-device (IED) electronic warfare (EW) (sometimes referred to as CREW), and other electronic systems.

F-3. The team is accessible via the Army Reprogramming Analysis Team's Warfighter Survivability Software Support Portal. A secure Internet protocol router network (SIPRNET) account is required to access the portal.

NATIONAL GROUND INTELLIGENCE CENTER

F-4. The National Ground Intelligence Center provides all-source analysis of the threat posed by IEDs produced and used by foreign terrorist and insurgent groups. The center supports U.S. forces during training, operational planning, deployment, and redeployment.

F-5. The center maintains a counter-IED targeting program (often called CITP) portal on its SIPRNET site. This portal provides information concerning IED activities and incidents as well as IED assessments.

ELECTRONIC ORDER OF BATTLE

F-6. An electronic order of battle details all known combinations of emitters and platforms in a particular area of responsibility. It consists of several reachback resources:

- National Security Agency-Electronic Intelligence Parameter Query.
- U.S. electromagnetic systems database.
- National Ground Intelligence System parametric information relational intelligence tool database.
- Military equipment parametrics and engineering database.

E-SPACE

F-7. E-Space is a Department of Defense (DOD) entity housed in the National Security Agency. It provides intelligence assistance (primarily signals intelligence) to deployed EW officers. E-Space is a reachback capability available to EW officers and spectrum managers that can be leveraged to provide all-source intelligence products and answers to requests for information and spectrum interference questions.

JOINT ELECTRONIC WARFARE CENTER

F-8. The Joint Electronic Warfare Center is DOD's only joint EW center of expertise. It provides EW subject matter expertise from a range of backgrounds, including people with current multi-Service operational experience. The center has a limited capability to perform modeling and simulation studies and EW red team support. It can deploy in a support role if approved by the U.S. Strategic Command.

JOINT IMPROVISED EXPLOSIVE DEVICE DEFEAT ORGANIZATION

F-9. The Joint Improvised Explosive Device Defeat Organization (known as JIEDDO) leads, advocates, and coordinates all DOD actions in support of efforts by combatant commanders and their joint task forces to defeat IEDs as weapon of strategic influence.

JOINT SPECTRUM CENTER

F-10. The Joint Spectrum Center ensures DOD effectively uses the electromagnetic spectrum in support of national security and military objectives. The center serves as DOD's center of excellence for electromagnetic spectrum management matters in support of the combatant commands, military departments, and DOD agencies in planning, acquisition, training, and operations.

F-11. The center maintains databases and provides data about friendly force command and control information system locational and technical characteristics. This information is used to plan electronic protection measures. These databases provide EW planners with information covering communication, radar, navigation, broadcast, identification, and EW systems operated by the DOD, other government agencies, and private businesses and organizations.

F-12. The center provides information on a quick-reaction basis in various formats and media to support EW planners and spectrum managers.

KNOWLEDGE AND INFORMATION FUSION EXCHANGE

F-13. The Knowledge and Information Fusion Exchange (sometimes called KnIFE) is a program sponsored by U.S. Joint Forces Command. It provides Soldiers with observations, insights, and lessons from operations around the world.

ADDITIONAL INFORMATION

F-14. Further information on the above tools and resources can be accessed through Army Knowledge Online. The links to these Web sites can be viewed by first accessing the "Army Operational Electronic Warfare Course" on Army Knowledge Online at http://www.us.army.mil/suite/page/400055 and then clicking on Folders >Links>EW links.

Glossary

SECTION I – ACRONYMS AND ABBREVIATIONS

ARAT	Army Reprogramming Analysis Team
C-3	operations directorate of a multinational (combined) staff
CJCSI	Chairman of the Joint Chiefs of Staff instruction
CJCSM	Chairman of the Joint Chiefs of Staff manual
COA	course of action
DD	Department of Defense (official forms only)
DOD	Department of Defense
DODI	Department of Defense Instruction
EW	electronic warfare
FM	field manual
FMI	field manual, interim
G-2	assistant chief of staff, intelligence
G-3	assistant chief of staff, operations
G-5	assistant chief of staff, plans
G-6	assistant chief of staff, signal
G-7	assistant chief of staff, information engagement
GTA	graphic training aid
HF	high frequency
Hz	hertz
IED	improvised explosive device
IO	information operations
IPB	intelligence preparation of the battlefield
ISR	intelligence, surveillance, and reconnaissance
J-2	intelligence directorate of a joint staff
J-3	operations directorate of a joint staff
J-5	plans directorate of a joint staff
J-6	communications system directorate of a joint staff
JFMO	Joint Frequency Management Office
JIOWC	Joint Information Operations Warfare Center
JP	joint publication
MAGTF	Marine air-ground task force
MC	Military Committee (NATO)
MCWP	Marine Corps warfighting publication
MDMP	military decisionmaking process

NATO	North Atlantic Treaty Organization
S-2	intelligence staff officer
S-3	operations staff officer
S-6	signal staff officer
S-7	information engagement staff officer
SIPRNET	SECRET Internet Protocol Router Network
STANAG	standardization agreement (NATO)
TERPES	tactical electronic reconnaissance processing and evaluation system
U.S.	United States

SECTION II – TERMS

communications security

(joint) The protection resulting from all measures designed to deny unauthorized persons information of value that might be derived from the possession and study of telecommunications, or to mislead unauthorized persons in their interpretation of the results of such possession and study. (JP 6-0)

computer network operations

(joint) Comprised of computer network attack, computer network defense, and related computer network exploitation enabling operations. (JP 3-13)

directed energy

(joint) An umbrella term covering technologies that relate to the production of a beam of concentrated electromagnetic energy or atomic or subatomic particles. (JP 3-13.1)

electromagnetic environment

(joint) The resulting product of the power and time distribution, in various frequency ranges, of the radiated or conducted electromagnetic emission levels that may be encountered by a military force, system, or platform when performing its assigned mission in its intended operational environment. It is the sum of the electromagnetic interference; electromagnetic pulse; hazards of electromagnetic radiation to personnel, ordnance, and volatile materials; and natural phenomena effects of lightning and precipitation static. (JP 3-13.1)

electromagnetic environmental effects

The impact of the electromagnetic environment upon the operational capability of military forces, equipment, systems, and platforms. It encompasses all electromagnetic disciplines, including electromagnetic compatibility and electromagnetic interference; electromagnetic vulnerability; electromagnetic pulse; electronic protection, hazards of electromagnetic radiation to personnel, ordnance, and volatile materials; and natural phenomena effects of lightning and precipitation static. (JP 3-13.1)

electromagnetic spectrum

(joint) The range of frequencies of electromagnetic radiation from zero to infinity. It is divided into 26 alphabetically designated bands. (JP 1-02)

electromagnetic vulnerability

(joint) The characteristics of a system that cause it to suffer a definite degradation (incapability to perform the designated mission) as a result of having been subjected to a certain level of electromagnetic environmental effects. (JP 1-02)

electronic attack

(joint) Division of electronic warfare involving the use of electromagnetic energy, directed energy, or antiradiation weapons to attack personnel, facilities, or equipment with the intent of degrading, neutralizing, or destroying enemy combat capability and is considered a form of fires. (JP 3-13.1)

electronic protection

(joint) Division of electronic warfare involving actions taken to protect personnel, facilities, and equipment from any effects of friendly or enemy use of the electromagnetic spectrum that degrade, neutralize or destroy friendly combat capability. (JP 3-13.1)

electronic warfare

(joint) Military action involving the use of electromagnetic and directed energy to control the electromagnetic spectrum or to attack the enemy. Electronic warfare consists of three divisions: electronic attack, electronic protection, and electronic warfare support. (JP 3-13.1)

electronic warfare support

(joint) Division of electronic warfare involving actions tasked by, or under direct control of, an operational commander to search for, intercept, identify, and locate or localize sources of intentional and unintentional radiated electromagnetic energy for the purpose of immediate threat recognition, targeting, planning, and conduct of future operations. (JP 3-13.1)

emission control

(joint) The selective and controlled use of electromagnetic, acoustic, or other emitters to optimize command and control capabilities while minimizing, for operations security: a. detection by enemy sensors; b. mutual interference among friendly systems; and/or c. enemy interference with the ability to execute a military deception plan. (JP 1-02)

joint restricted frequency list

(joint) A time a geographically-oriented listing of TABOO, PROTECTED, and GUARDED functions, nets, and frequencies. It should be limited to the minimum number of frequencies necessary for friendly forces to accomplish objectives. (JP 3-13.1)

working group

(Army) A temporary grouping of predetermined staff representatives who meet to coordinate and provide recommendations for a particular purpose or function. (FMI 5-0.1)

References

REQUIRED PUBLICATIONS

These documents must be available to intended users of this publication.

FM 1-02 (101-5-1). *Operational Terms and Graphics.* 21 September 2004.

JP 1-02. *Department of Defense Dictionary of Military and Associated Terms.* 12 April 2001. (As amended through 4 March 2008.)

JP 3-13.1. *Electronic Warfare.* 25 January 2007.

FM 3-0. *Operations.* 27 February 2008.

FM 5-0 (101-5). *Army Planning and Orders Production.* 20 January 2005.

FM 6-0. *Mission Command: Command and Control of Army Forces.* 11 August 2003.

FMI 5-0.1. *The Operations Process.* 31 March 2006.

RELATED PUBLICATIONS

These documents contain relevant supplemental information.

JOINT AND DEPARTMENT OF DEFENSE PUBLICATIONS

Most joint publications are available online: <http://www.dtic.mil/doctrine/jpcapstonepubs.htm.>

CJCSI 3320.01B *Electromagnetic Spectrum Use in Joint Military Operations.* 01 May 2005

CJCSI 3320.02C. Joint Spectrum Interference Resolution (JSIR). 27 January 2006 (with change 1 as of 25 February 2008).

CJCSI 3320.03A *Joint Communications Electronics Operation Instructions.* 11 June 2005.

CJCSM 3122.03C. *Joint Operation Planning and Execution System Volume II, Planning Formats and Guidance.* 17 August 2007.

CJCSM 3320.01B *Joint Operations in the Electromagnetic Battlespace.* 25 March 2006.

CJCSM 3320.02A *Joint Spectrum Interference Resolution (JSIR) Procedures.* 16 February 2006.

DODI 4650.01. Policy and Procedures for Management and Use of the Electromagnetic Spectrum. 09 January 2009.

JP 2-0. *Joint Intelligence.* 22 June 2007.

JP 2-01. *Joint and National Intelligence Support to Military Operations.* 07 October 2004.

JP 3-0. *Joint Operations.* 17 September 2006.

JP 3-09. *Joint Fire Support.* 13 November 2006.

JP 3-09.3. *Joint Tactics, Techniques, and Procedures for Close Air Support (CAS).* 03 September 2003.

JP 3-13. *Information Operations.* 13 February 2006.

JP 3-13.3. *Operations Security.* 29 June 2006.

JP 3-13.4 (JP 3-58). *Military Deception.* 13 July 2006.

JP 3-60. *Joint Targeting.* 13 April 2007.

JP 6-0. *Joint Communications System.* 20 March 2006.

ARMY PUBLICATIONS

Most Army doctrinal publications are available online: <http://www.army.mil/usapa/doctrine/Active_FM.html>.

FM 2-0 (34-1). *Intelligence.* 17 May 2004.

FM 3-09.32. JFIRE: *Multi-Service Tactics, Techniques, and Procedures for the Joint Application of Firepower.* 20 December 2007.

FM 3-13 (100-6). *Information Operations: Doctrine, Tactics, Techniques, and Procedures.* 28 November 2003.

FM 3-13.10 (3-51.1). *Multi-Service Tactics, Techniques, and Procedures for the Reprogramming of Electronic Warfare and Target Sensing Systems.* 22 January 2007.

FM 5-19 (100-14). *Composite Risk Management.* 21 August 2006.

FM 6-20-10. *Tactics, Techniques, and Procedures for the Targeting Process.* 8 May 1996.

FM 6-99.2 (101-5-2). *U.S. Army Report and Message Formats.* 30 April 2007.

FM 34-130. *Intelligence Preparation of the Battlefield.* 8 July 1994.

FMI 2-01. *Intelligence, Surveillance, and Reconnaissance (ISR) Synchronization.* 11 November 2008.

GTA 90-10-046. *MNC-I Counter IED Smart Book.* September 2008.

NATO PUBLICATIONS

Allied Joint Publication 3.6. *Allied Joint Electronic Warfare Doctrine.* December 2003.

MC 64. *NATO Electronic Warfare (EW) Policy.* 26 April 2004.

STANAG 5048 C3 (Edition 5). *The Minimum Scale of Convectivity for Communications and Information Systems for NATO Land Forces.* 16 February 2000.

OTHER PUBLICATIONS

AFDD 2-1.9. *Targeting.* 8 June 2006.

AFDD 2-5.1. *Electronic Warfare.* 5 November 2002.

Executive Order 12333. *United States Intelligence Activities.* 4 December 1981.

MCWP 2-22 (2-15.2). *Signals Intelligence.* 13 July 2004.

NWP 3-13. *Navy Information Operations.* June 2003.

SOURCES USED

Electronic Warfare Working Group, U.S. House of Representatives, Issue Brief #17. "Compass Call During Operation Iraqi Freedom." 11 March 2004. Available online at http://www.house.gov/pitts/initiatives/ew/Library/Briefs/brief17.htm.

PRESCRIBED FORMS

None

REFERENCED FORMS

DA Forms are available on the APD website (www.apd.army.mil. DD forms are available on the OSD website (www.dtic.mil/whs/directives/infomgt/forms/formsprogram.htm).

DA Form 2028. *Recommended Changes to Publications and Blank Forms.*

DD Form 1972. *Joint Tactical Air Strike Request.*

Index

Entries are by paragraph number unless specified otherwise.

Entries are by paragraph number unless specified otherwise.

Entries are by paragraph number unless specified otherwise.

Entries are by paragraph number unless specified otherwise.

Entries are by paragraph number unless specified otherwise.

Entries are by paragraph number unless specified otherwise.

Made in the USA
Monee, IL
05 November 2026

67858273R00328